ALLE ZEIT WACH
1842

# Atlas of Deformational and Metamorphic Rock Fabrics

Edited by
Graham J. Borradaile M. Brian Bayly
Chris McA. Powell

With 649 Figures

Springer-Verlag
Berlin Heidelberg New York 1982

Graham J. Borradaile, Associate Professor
Department of Geology, Lakehead University
Thunder Bay, Ontario, P7B 5E1, Canada

M. Brian Bayly, Professor
Department of Geology
Rensselaer Polytechnic Institute
Troy, New York 12181, USA

Chris McA. Powell, Associate Professor
Geology Department, School of Earth Sciences
Macquarie University
North Ryde, NSW 2113, Australia

ISBN-13:978-3-642-68434-0 e-ISBN-13:978-3-642-68432-6
DOI:10.1007/978-3-642-68432-6

Library of Congress Cataloging in Publication Data
Main entry under title: Atlas of deformational and metamorphic rock fabrics. Bibliography: p. Includes index. 1. Rocks, Metamorphic–Atlases. 2. Petrofabric analysis–Atlases. I. Borradaile, G. J. II. Powell, C. McA. (Chris McA.), 1943–. III. Bayly, B. M. (Brian M.), 1929–.
QE475.A2A87 552'.4 81-18543 ISBN-13:978-3-642-68434-0 AACR2



Softcover reprint of the hardcover 1st edition 1982



Reproduction of the figures: Gebrüder Czech, München

2132/3130-543210

# Preface

In May 1976 Lucian B. Platt organized a highly successful Penrose Conference on *The Formation of Rock Cleavage* at Bryn Mawr College in Pennsylvania, U.S.A. The meeting drew together about 70 specialists from both sides of the Atlantic and from Australasia, who contributed discussions on various aspects of rock cleavage and its formation. Even early in the meeting it became clear to the participants that they lacked a common terminology, that often the same technical word implied different things to different people and that observables and descriptors were loosely defined. In an attempt to improve communication the present editors contacted about 190 workers after the conference with a view to compiling a set of photographs with captions to illustrate exactly what workers were talking about. As a result the compilation was published as a limited edition by an inexpensive offset process at the University of Tasmania. The success of that provisional edition of the *Atlas of Rock Cleavage* and the responses of the readers prompted us to make a more extensive collection of material, contact a wider range of workers and, with the support of Dr. Konrad Springer, to publish the present higher-quality reproduction of the contributors' plates.
In the present volume we have tried to be very broad while remaining narrow: a narrow intention has been retained, to produce a volume that will be useful to people studying anisotropic materials, but broad ranges have been encompassed in scale, material, and texture. Work concerning ice, sediments and igneous rocks is represented; scale runs from outcrops to electron microscopy; and any texture that might throw light on the development of textural planar anisotropy has been included. We hope the volume will be useful not only to metamorphic petrologists but also to workers on other materials, by shedding light on their textures and problems from a perhaps unfamiliar point of view.
On the other hand, while emphasizing breadth both of content and of applicability, we have tried to stay within our own limited realm of competence. In particular, the introductory text is narrower, and discusses specifically *cleavage* as conceived in metamorphic petrology. When it comes to analyzing usage and separating helpful terms from those that are cloudy or misleading, any commentary is risky, and commentary lacking a tightly defined topic is almost bound to do harm. So the text focusses on cleavage; for people working on non-rock material, we have tried to make clear what we consider reasonable use of our own terms, but to avoid straying abroad – the risk of foot-in-mouth becomes too large. Thus the range of features illustrated is broad, but the range of terms on whose use we have commented is rather narrow.

**Acknowledgements**

Production of this Atlas was made easier through the technical support offered by Wendy Bons, John de Lathower and Sam Spivak at Lakehead

University, Thunder Bay, Ontario. Costs involved by the editors were defrayed by a grant from Canadian Pacific Railways administered by the Senate Research Committee of Lakehead University and by grant A6861 from the Natural Sciences and Engineering Research Council of Canada to Graham Borradaile. Beside the people involved in this volume being indebted to each other, we all acknowledge help from those who contributed to the earlier *Atlas of Rock Cleavage* and thus helped the present volume to come into being.

Spring 1982

G. J. Borradaile
M. B. Bayly
C. McA. Powell

# Contents

# Contributors' Addresses


| | |
|---|---|
| AHMAD, RADI | School of Earth Sciences, University of Melbourne, Parkville, Victoria 3052, Australia |
| AINSWORTH, J. | Department of Geology and Geophysics, University of California, Berkeley, California 96720, U.S.A. |
| ALVAREZ, WALTER | Department of Geology, and Geophysics, University of California, Berkeley, California 94720, U.S.A. |
| BATES, D. E. B. | Geology Department, University College of Wales, Aberystwyth, United Kingdom |
| BAYLY, M. B. | Department of Geology, Rensselaer Polytechnic Institute, Troy, New York 12181, U.S.A. |
| BEACH, A. | Department of Geology, The University, P.O. Box 147, Liverpool L6P 3BX, United Kingdom |
| BELL, T. H. | Department of Geology, P.O. James Cook University, Queensland 4811, Australia |
| BISHOP, D. G. | N. Z. Geological Survey, Box 5342, Dunedin, New Zealand |
| BOND, W. D. | Ontario Geological Survey, Ministry of Natural Resources, 77 Grenville Street, Toronto, Ontario M5S 1B3, Canada |
| BORRADAILE, GRAHAM JOHN | Department of Geology, Lakehead University, Thunder Bay, Ontario P7B 5E1, Canada |
| BOUCHEZ, JEAN-LUC | Laboratoire de Tectonophysique, Université de Nantes, 44072 Nantes Cedex, France |
| BOUDIER, FRANÇOISE | Laboratoire de Tectonophysique, 38 Boulevard Michelet, 44072 Nantes, France |
| BOULTER, C. A. | Department of Geology, University of Western Australia, Nedlands, Western Australia 6009, Australia |
| BRADBURY, H. J. | Department of Geology, Yale University Box 6666, Newhaven, Connecticut 06511, U.S.A. |
| BREAKS, F. W. | Ontario Geological Survey, Ministry of Natural Resources, 77 Grenville Street, Toronto, Ontario M5S 1B3, Canada |
| BURCHFIEL, B. C. | Department of Earth and Planetary Sciences, Massachusetts Institute of Technology, Cambridge, Massachusetts 02139, U.S.A. |
| BURGER, H. ROBERT | Department of Geology, Smith College, Northampton, Massachusetts 01063, U.S.A. |
| CONAGHAN, P. J. | School of Earth Sciences, Macquarie University, North Ryde, New South Wales 2113, Australia |

Cook, D. Institute of Sedimentary and Petroleum Geology, Geological Survey of Canada, 3303 33rd Street NW, Calgary, Canada

Cosgrove, John William Geology Department, Imperial College of Science and Technology, Prince Consort Road, London, SW7 2AZ, United Kingdom

Cox, S. F. Department of Earth Sciences, Monash University, Clayton, Victoria 3168, Australia

Crook, Keith A. W. Department of Geology, Australian National University, GPO Box 4, Canberra ACT 2600, Australia

Davies, W. Geology Department, University College of Wales, Aberystwyth, United Kingdom

De Boer, R. B. Koninklijke/Shell Exploratie en Produktie Laboratorium, Volmerlaan 6, Rijswijk, The Netherlands

Dennis, John G. Department of Geological Sciences, California State University, Long Beach, California 90840, U.S.A.

Durney, D. W. School of Earth Sciences, Macquarie University, North Ryde, New South Wales 2113, Australia

Engelder, Terry Lamont-Doherty Geological Observatory, Palisades, New York 10964, U.S.A.

Etheridge, M. A. Department of Earth Sciences, Monash University, Clayton, Victoria 3168, Australia

Finley, Sharon Department of Geology, Rensselaer Polytechnic Institute, Troy, New York 12181, U.S.A.

Friedman, M. Department of Geology and Center for Tectonophysics, Texas A & M University, College Station, Texas 77843, U.S.A.

Gardner, D. A. C. Gulf Canada Resources, Inc., P.O. Box 130, Calgary, Alberta T2P 2H7, Canada

Ghent, E. D. Department of Geology, University of Calgary, Calgary, Alberta T2N 1N4, Canada

Gill, Greta Department of Geology and Geophysics, Boston College, Chestnut Hill, Massachusetts 02167, U.S.A.

Glen, R. A. Geological Survey of New South Wales, State Office Block, Phillip Street, Sydney, New South Wales 2000, Australia

Granath, James W. Department of Earth and Space Sciences, State University of New York, Stony Brook, New York 11974, U.S.A.

Gray, D. R. Department of Geological Sciences, Virginia Polytechnic Institute and State University, Blacksburg, Virginia 24061, U.S.A.

Gregg, William J. Department of Geology and Geological Engineering, Michigan Technological University, Houghton, Michigan 49931, U.S.A.

Hancock, P. L. — Department of Geology, University of Bristol, Queen's Building, University Walk, Bristol BS8 1TR, United Kingdom

Harris, A. L. — Department of Geology, University of Liverpool, Liverpool L69 3BX, United Kingdom

Heard, H. C. — Lawrence Livermore Laboratory, University of California, Livermore, California 94550, U.S.A.

Higgs, N. G. — Department of Geology and Center for Tectonophysics, Texas A & M University, College Station, Texas 77843, U.S.A.

Holcombe, R. J. — Department of Geology and Mineralogy, University of Queensland, St. Lucia, Queensland 4067, Australia

Hudleston, P. J. — Department of Geology and Geophysics, University of Minnesota, Minneapolis, Minnesota 55455, U.S.A.

Hughes, K. C. — Department of Geology, University of Western Australia, Nedlands, Western Australia 6009, Australia

Jackson, P. A. — Department of Geology, Lakehead University, Thunder Bay, Ontario P7B 5E1, Canada

Kehlenbeck, M. M. — Department of Geology, Lakehead University, Thunder Bay, Ontario P7B 5E1, Canada

Kennedy, Myra C. — Department of Geology, Lakehead University, Thunder Bay, Ontario P7B 5E1, Canada

Kissin, S. A. — Department of Geology, Lakehead University, Thunder Bay, Ontario P7B 5E1, Canada

Knipe, R. J. — Department of Earth Sciences, The University, Leeds LS2 9JT, United Kingdom

Lebedeva, N. B. — Department of Geology, Moscow State University, 117234 Moscow, U.S.S.R.

Lincoln, Beth Z. — Department of Earth and Space Sciences, University of California, Los Angeles, Los Angeles, California 90405, U.S.A.

Maltman, Alex J. — Department of Geology, University College of Wales, Aberystwyth, Wales, United Kingdom

Mancktelow, N. S. — Department of Geology and Mineralogy, University of Adelaide, G.P.O. Box 84, Adelaide 5001, Australia

Martin, D. — School of Earth Sciences, Macquarie University, North Ryde, NSW 2113, Australia

Maxwell, J. C. — Department of Geological Sciences, The University of Texas at Austin, Austin, Texas 78712, U.S.A.

McClay, K. R. — Department of Geology, University of London Goldsmiths College, New Cross, London SE14 6NW, United Kingdom

Means, W. D. — Department of Geological Sciences, State University of New York at Albany, 1400 Washington Avenue, Albany, New York 12222, U.S.A.

MIMRAN, Y. — Geological Survey, 30 Malkhei Yisrael Street, Jerusalem, Israel

MOENCH, ROBERT H. — U.S. Geological Survey, MS 930, Box 25046, Federal Center, Denver, Colorado 80225, U.S.A.

MORRITT, R. F. C. — Western Mining Corporation, P.O. Box 71, Kalgoorlie, Western Australia 6430, Australia

NAGTEGAAL, P. J. C. — Koninklijke/Shell Exploratie en Produktie Laboratorium, Volmerlaan 6, Rijswijk, The Netherlands

NICKELSEN, RICHARD P. — Department of Geology and Geography, Bucknell University, Lewisburg, Pennsylvania 17837, U.S.A.

NICOLAS, ADOLPHE — Laboratoire de Tectonophysique, 38 Boulevard Michelet, 44072 Nantes, France

OLDERSHAW, A. E. — Department of Geology, University of Calgary, Calgary, Alberta T2N 1N4, Canada

PFIFFNER, O. ADRIAN — Geological Institute, E.T.H., CH-8092 Zurich, Switzerland

PLATT, LUCIAN B. — Bryn Mawr College, Bryn Mawr, Pennsylvania 19010, U.S.A.

POULSEN, K. H. — Department of Geological Sciences, Queen's University, Kingston, Ontario K7L 3N6, Canada

POWELL, C. McA. — School of Earth Sciences, Macquarie University, North Ryde, New South Wales 2113, Australia

PUPPOLO, DAVID — Department of Geology, Rensselaer Polytechnic Institute, Troy, New York 12181, U.S.A.

ROBERTS, DAVID — Norges Geologiske Undersøkelse, Postboks 3006, 7001 Trondheim, Norway

RUBENACH, M. J. — Department of Geology, P. O. James Cook University, Queensland 4811, Australia

RUSSELL-HEAD, D. S. — Faculty of Engineering, University of Melbourne, Parkville, Victoria 3052, Australia

SANSONE, STANLEY A. — Department of Geology and Geophysics, The University of Connecticut, Storrs, Connecticut 06268, U.S.A.

SEYMOUR, D. B. — Geological Survey, 66 Adelaide Terrace, Perth, Western Australia 6000, Australia

SIMON, R. I. — Department of Geological Sciences, Virginia Polytechnic Institute and State University, Blacksburg, Virginia 24061, U.S.A.

SOLÉ-SUGRAÑES, LUIS — Instituto Jaime Almera, C.S.I.C., Egipciacas 11, Barcelona, Spain

SPANG, J. H. — Department of Geology and Center for Tectonophysics, Texas A & M University, College Station, Texas 77843, U.S.A.

STEUER, MARK R. — Department of Geology, Bryn Mawr College, Bryn Mawr, Pennsylvania 19010, U.S.A.

| | |
|---|---|
| Stout, M. Z. | Department of Geology, University of Calgary, Calgary, Alberta T2N 1N4, Canada |
| Talbot, C. J. | Department of Geology, The University, Dundee, Scotland, United Kingdom |
| Twombly, G. | Department of Geology and Geophysics, University of Calgary, Calgary, Alberta T2N 1N4, Canada |
| Vernon, R. H. | School of Earth Sciences, Macquarie University, North Ryde, New South Wales 2113, Australia |
| Wardlaw, N. C. | Department of Geology and Geophysics, University of Calgary, Calgary, Alberta T2N 1N4, Canada |
| Watkinson, A. J. | Department of Geology, Washington State University, Pullman, Washington 99163, U.S.A. |
| Weber, K. | Geologisch-Paläontologisches Institut, Goldschmidtstraße 3, D-3400 Göttingen, Federal Republic of Germany |
| Wenk, H. R. | Department of Geology and Geophysics, University of California, Berkeley, California 94720, U.S.A. |
| White, S. H. | Department of Geology, Imperial College, London, United Kingdom |
| Williams, Paul F. | Department of Geology, University of New Brunswick, Fredericton, New Brunswick, Canada |
| Williams, P. R. | G.P.O. Box 124B, Hobart, Tasmania 7001, Australia |
| Williams, V. A. | C.R.A. Exploration P/Ltd., 9th Floor, Babcock House, 140 Pacific Highway, North Sydney, New South Wales 2060, Australia |
| Wilson, C. J. L. | School of Earth Sciences, University of Melbourne, Parkville, Victoria 3052, Australia |
| Wright, Ellen K. | Department of Geology, The George Washington University, Washington, D.C. 20052, U.S.A. |
| Wright, Thomas O. | National Science Foundation, Washington, D.C. 20550, U.S.A. |

# Chapter I

## Introduction

Our knowledge of metamorphic and deformational rock fabrics has been acquired largely by studying the products of deformation in ancient fold belts. As such, our understanding of how rock fabrics form has been built up from inferences about, rather than direct observations of, processes that might have operated, and many of those inferences have proved incorrect. Extension of the deep-sea drilling project to actively deforming continental margins is a step towards providing information about rock fabrics that are currently forming, as is the attempt to measure stress, strain and nature of pore fluid in modern fold belts, but these approaches are very much in their infancy. Similarly, direct observations of fabric changes have begun using analogue materials. The problem arising from such an inferential study is that we have no prior knowledge of the natural categories or divisions that exist. The major categories of rock fabric have, by necessity, been established arbitrarily in the hope that they provide a useful subdivision. As novel observations or techniques open new vistas, earlier ideas have had to be modified or discarded.

In the past decade there has been a great upsurge in interest in rock fabrics, and many earlier ideas about deformational processes and their products have changed. Rock fabrics formerly separated into different categories have been united, and other fabrics previously given the same name have been shown to have separate origins. Today, there are still many competing ideas about how particular rock fabrics form.

To bring order into this confusion, we must distinguish *descriptions* of features that can be observed from *inferences* about their likely origin. The present chapter deals only with observables; Chapter II deals with imaginable processes and Chapter III deals with the possibility of inferring processes from observations. In this chapter, we have elaborated and modified the morphological scheme of *rock cleavage* outlined by POWELL (1979) who pointed out that most cleaved rocks have a domainal structure consisting of zones that tend to split or cleave (*cleavage domains*) separating less cleavable zones (*microlithons*). The primary subdivision of this classification depends on whether the tendency to cleave is evenly distributed, or *continuous* throughout the rock at the scale of observation, or whether cleavage is spaced. Further subdivision depends on the shape of the cleavage, and, in spaced cleavage, the fabric of the microlithons and the proportion of the rock occupied by cleavage domains.

### Spaced Versus Continuous Cleavage

A cleavage is said to be spaced if, at the scale of observation, there are domains where the tendency to cleave is greater than in between. Cleavages that result from the statistical alignment of mineral grains evenly distributed throughout a rock, or from cleavage domains too closely spaced to be resolved at the scale of observation, are said to be continuous. The scale of observation is paramount in deciding whether to call a particular cleavage spaced or continuous, because a cleavage that appears continuous in hand specimen or outcrop may be found to be spaced at microscopic or submicroscopic level, consisting of narrow domains of aligned minerals separating wider domains of more randomly oriented minerals. Some cleavages are continuous at all scales, but commonly we find that continuous cleavages are spaced if examined at higher powers of reso-

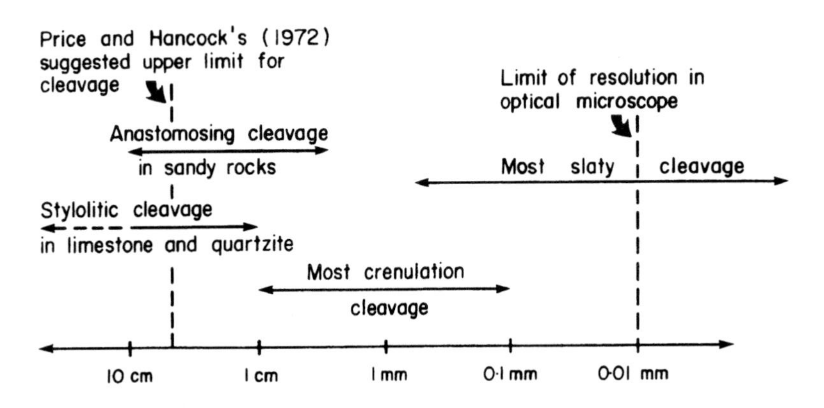


**Fig. 1.1.** Average spacing of cleavage domains. (Modified from POWELL 1979)

lution. Or where a cleavage is due to spaced domains we may find that, within a single domain, cleavage is continuous down to the scale of single mineral grains. The spacing of some common cleavages is shown in Fig. 1.1.

Subdivision of the range of continuous cleavage is simply on the basis of the mean grain size of the minerals forming the cleavage, i.e., whether it is fine, as in some slates, or coarse, as in gneisses and schists. In practice, domainal cleavage spaced more closely than 0.01 mm appears continuous in an optical microscope, and its spaced character can be seen only with an electron microscope. At the other end of the scale spectrum, PRICE and HANCOCK (1972) and HANCOCK (this Vol.) have suggested that an upper limit of 5 cm separation should be used to distinguish cleavage from joints, with a further restriction that the distance between the cleavage surfaces should be less than one-twentieth of the thickness of the bed in which they occur. Structures with larger spacing are called joints. This limit seems to us restrictive, because it arbitrarily separates the more widely spaced members of what may be essentially the same group of structures. Widely spaced rock cleavage has received little attention, and thus we prefer to wait until sufficient measurements have been made to determine the natural limit to spacing of rock cleavage.

A practical problem concerns determining the average spacing of any particular cleavage. Each cleavage has its own peculiarities which may be measured in order to describe the geometry completely. For example, in the three cleavages illustrated in Fig. 1.2, for the parallel cleavage we could measure the average width between the cleavage traces, for the anastomosing or reticulate cleavage we could measure the average width of each of the lensoid microlithons (avoiding the tapered ends), and for the trapezoidal or conjugate cleavage we could average the maximum width of the trapezoidal microlithons. For simplicity and comparison between different cleavages, we recommend that the average spacing of a cleavage be calculated by counting the number of cleavage traces crossed on a traverse normal to the general cleavage plane. The anastomosing or trapezoidal nature of the cleavage can be noted as an additional feature, and quantified if required. For comparative purposes, sections should be cut in common orientations, either parallel or perpendicular to local fold axes or mineral elongations.

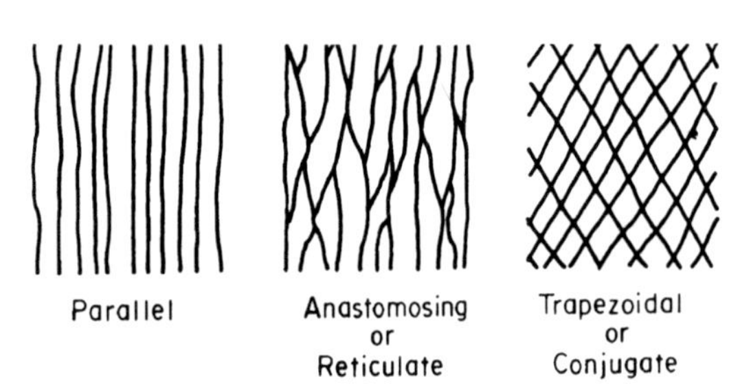


**Fig. 1.2.** Three different cleavage shapes with the same average spacing

## Disjunctive and Crenulation Cleavage

The second subdivision of spaced cleavage depends on the relationship of the cleavage domains to the fabric of the microlithons. Where the microlithons contain a pre-existing grain alignment that is deformed systematically into microfolds, the cleavage is called a *crenulation* cleavage. In rocks where there is no pre-existing preferred orientation, or where there is no systematic relationship between the pre-existing preferred orientation and the new fabric, the cleavage is called *disjunctive*. In most crenulation cleavages, the surfaces of parting lie along the limbs of the microfolds. Crenulation and disjunctive cleavages are subdivided further on the basis of the shape of the cleavage domains, the proportion of the rock occupied by the cleavage domains, and the abrupt or gradual nature of the transition from the microlithons into the cleavage.

### Shape of the Cleavage Domains

The shape of the cleavage domains can be described in terms of two characteristics: (1) the degree of planarity or unevenness of the cleavage surface, and (2) the patterns of cleavage surfaces. The range of planarity or unevenness is reflected in terms such as smooth, rough, wiggly and stylolitic (Fig. 1.3). *Smooth* cleavage

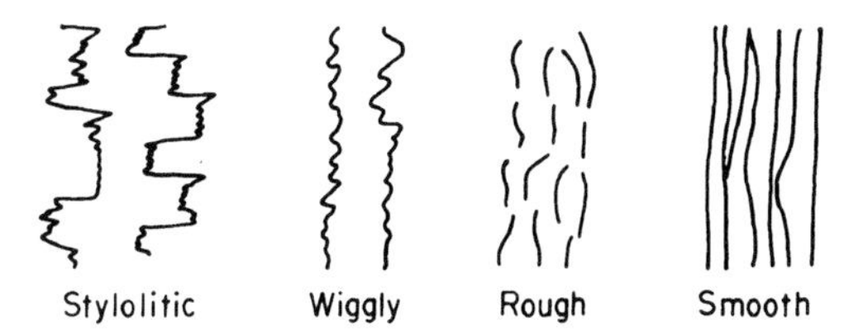


**Fig. 1.3.** Degree of planarity or unevenness of cleavage

shapes are generally more closely spaced than *stylolitic* varieties, and smooth cleavage may be more continuous along its trace than *rough* cleavage. Rough cleavage commonly results from steps as the surface of parting jumps from one cleavage trace to the next. Smooth cleavage is typical of slate, and rough cleavage of deformed sandstones.

The second characteristic describes the patterns of cleavage traces. Three common interconnecting shapes are *anastomosing* or *reticulate, trapezoidal,* and *conjugate* (Fig. 1.4), and to some degree they represent an increasing degree to which two distinct directions can be recognized in a cleavage. Anastomosing or

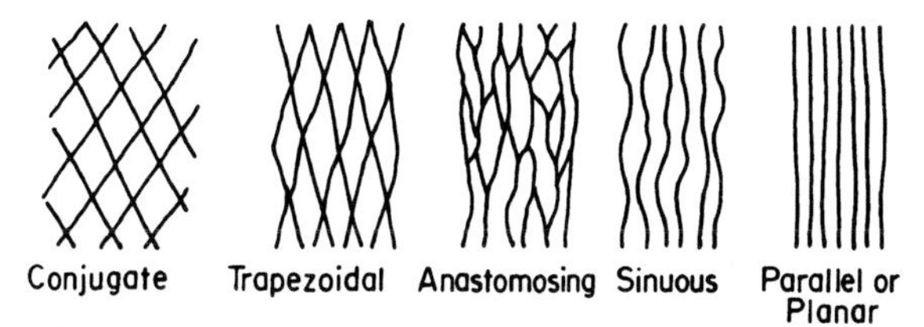


**Fig. 1.4.** Patterns of cleavage surfaces

reticulate morphology is found in many different rocks including deformed mudstone (CROOK 1964), sandstone, gneiss and granite. Trapezoidal cleavage is common in deformed sandstone, and conjugate cleavage, though less widely reported, has been found in deformed sandstone and crenulated slate and schist.

### Relative Width of the Cleavage Domains

The relative width of cleavage domains compared with the microlithons can be measured by counting the proportion of the rock occupied by cleavage domains in traverses normal to the general cleavage trace. This proportion can vary from nearly zero in rocks where the cleavage appears as widely spaced, wispy or hairline cracks to nearly 100% in rocks with wide cleavage domains. Where the cleavage has substantial width relative to the microlithons, the cleavage is called *zonal*, and the cleavage zones can be further qualified as broad or narrow. If the cleavage domains have expanded to occupy the entire fabric, the cleavage is continuous. Transitions from narrow, spaced cleavage domains into continuous cleavage can be observed where schists grade into gneisses, or at lower metamorphic grades where anastomosing disjunctive cleavage in sandstone passes into continuous cleavage in interbedded slate.

### Gradual or Abrupt Change from Microlithon into Cleavage

The change in passing from the microlithons into cleavage domains can be abrupt or *discrete* where there is a sharp boundary between the two domains, or it can be *gradational* as in many crenulation cleavages where the cleavage is merely the aligned limbs of harmonic microfolds. In many gradational crenulation cleavages there is a gradual change in the relative concentration of leucocratic minerals (quartz and feldspar) in sympathy with the gradual change in orientation of the micaceous minerals defining the microfolds such that the hinge areas are enriched in leucocratic minerals compared with the microfold limbs. In some cases, the mineralogical differentiation can produce a *differentiated layering* or *mineral segregation* parallel to the new cleavage, and in extreme cases the differentiation can obliterate the earlier fabric.

## Morphology of Common Spaced Cleavages

The morphology of common varieties of spaced cleavage can be represented on a three-dimensional diagram where the axes correspond to the degree of planarity of the cleavage domains, the fraction of the rock they occupy, and the nature of their margins (Fig. 1.5A). Description of each cleavage involves specifying three parameters that refer to a single point in the cube space. Thus, we can distinguish discrete, stylolitic, crack-like cleavage from smooth, gradational, zonal cleavage, and so on. The fields of common varieties of crenulation and disjunctive cleavage are shown in Figs. 1.5B and 1.5C. Boxes with solid outlines show common varieties, and boxes with broken outlines show cleavage intermediate between end members, or cleavage thought by us to be less common. Some fields are unlikely to be present in nature (e.g., stylolitic gradational cleavage or smooth gradational cleavages occupying a negligible fraction of the rock volume). The diagrams show that there are more varieties of discrete cleavage than gradational cleavage, and that all gradational cleavage occupies a significant proportion of the rock in which it occurs.

We believe that the concepts and usage in Fig. 1.5 are common, and we have tried to reflect present practice rather than be innovative. By way of summary, there follows a glossary of words that have been used in this chapter the use of which is favored for describing cleavage. In addition to these, there are available to observers many words that can be used as occasion warrants, without forming part of a standardized system. Any word free of process connotations can be a useful descriptor of rock cleavage. A *chart of possibilities* or *decision tree* is illustrated to facilitate the use of these terms (Fig. 1.6).

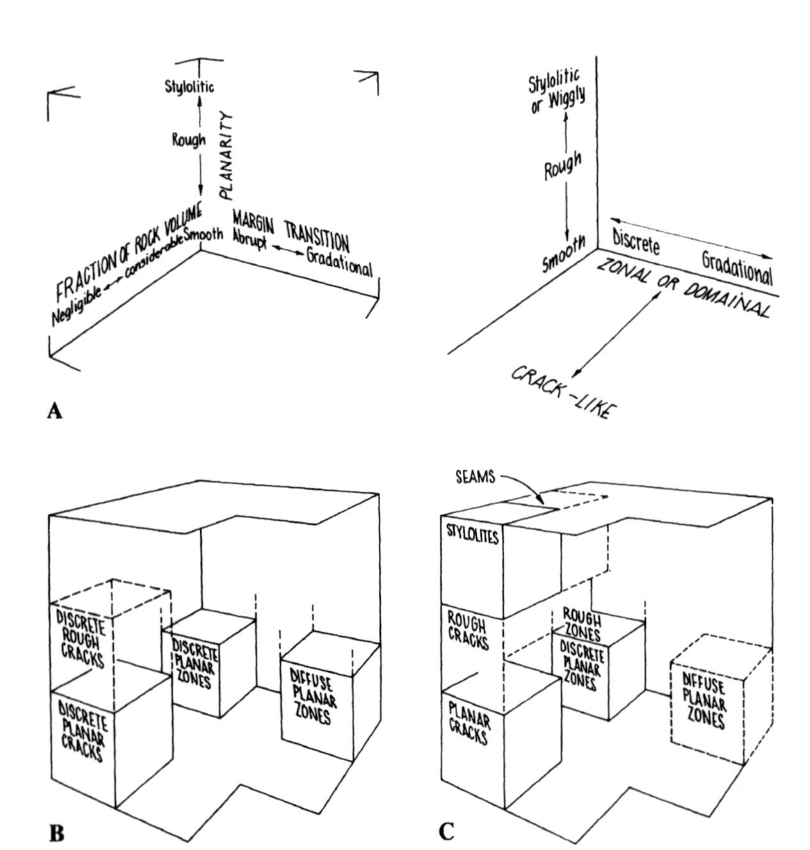


**Fig. 1.5.** Common varieties of spaced cleavage.
**A** Three characteristics used to define a cube space.
**B** Common crenulation cleavages.
**C** Common disjunctive cleavages

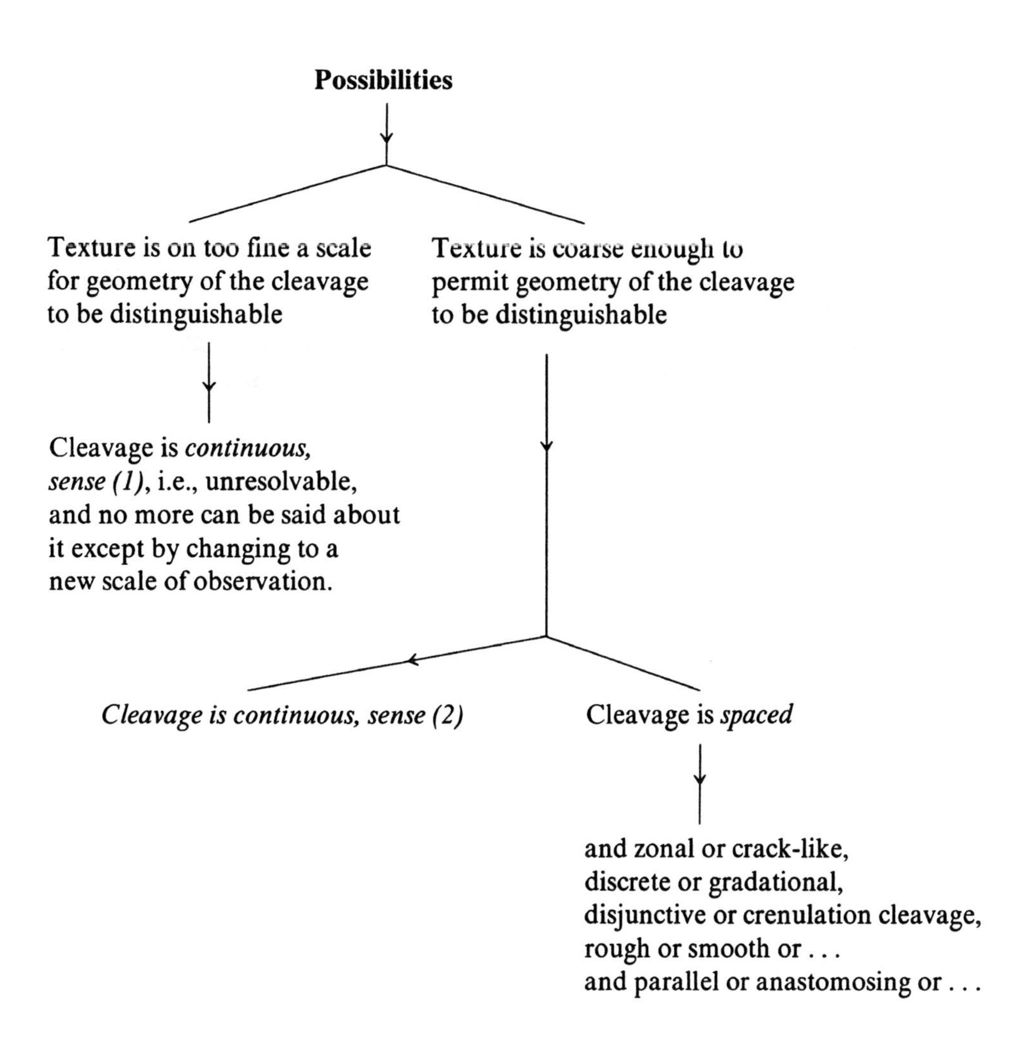


**Fig. 1.6.** Chart illustrating the application of the cleavage terms suggested in the glossary

## Glossary of Cleavage Terms

*Anastomosing cleavage* Quasi-planar cleavage with a reticulating or interconnecting pattern, commonly spaced at the mm to cm scale (CROOK 1964).

*Cleavage domains* Domains along which a rock parts or cleaves easily compared with the rock slices between (=microlithons). Cleavage domains (also called folia, films, or seams) are commonly marked by stronger preferred mineral orientation than the microlithons they enclose.

*Conjugate cleavage* A pair of cleavages formed during a single deformation and having similar form, symmetrically inclined about a plane of symmetry. The plane bisecting the acute angle between conjugate cleavages is commonly the cleavage plane in associated rock types.

*Continuous cleavage (1)* Cleavage that appears continuous at the scale of observation, with no microlithons or cleavage-free domains big enough to be seen.

*Continuous cleavage (2)* Cleavage resulting from parallelism of planar fabric elements that are distributed throughout the rock and not concentrated in any special locations (CHIDESTER 1962, DENNIS 1972 p 179). *Comment* These two meanings are quite distinct and in no way equivalent: continuous cleavage (1) means that domainal fabric cannot be seen but may exist, whereas continuous cleavage (2) means that cleavage domains do not exist in the sample at hand.

*Crenulation cleavage* Cleavage surfaces, whether micaceous layers or sharp breaks, that are separated by thin slices of rock containing a crenulated cross lamination (RICKARD 1961, p 325). Discrete crenulation cleavages are sharply defined discontinuities in the crenulated fabric, and truncate the pre-existing fabric, whereas gradational crenulation cleavages are laminar domains, coincident with fold limbs, the zones having diffuse, gradational boundaries through which the pre-existing fabric is continuous.

*Differentiated layering* A secondary layering formed by mineral segregation in zonal cleavage.

*Discrete* Cleavage domains that have sharply defined edges. Opposite of gradational.

*Disjunctive cleavage* Spaced cleavage independent of any pre-existing mineral orientation.

*Foliation* 1. Planar location fabric usually defined by compositional variation. (This Atlas.) 2. In some cases may refer to a penetrative preferred dimensional orientation in e.g. homogeneous schists. (Not used in this sense in this Atlas.)

*Gradational* Cleavage domains that grade into the microlithons with no sharply defined boundary. "Gradational" overlaps in sense with the term "zonal" as used by GRAY (1977a), but not with zonal as used in this glossary.

*Microlithon* A thin slice of rock enclosed by cleavage domains. Microlithons commonly have differently, and more poorly, oriented minerals compared with cleavage domains.

*Rock cleavage* A set of closely spaced secondary, planar, parallel fabric elements that impart mechanical anisotropy to the rock, without apparent loss of cohesion (DENNIS, in BAYLY et al. 1977).

*Rough cleavage* Cleavage with an irregular or rough surface of parting. GRAY (1977b) has described one type of rough cleavage that consists of short, discontinuous cleavage domains enveloping detrital grains in deformed sandstone: the present usage preserves a sense close to the every-day meaning of the word.

*Schistosity* Planar anisotropy in metamorphic rocks resulting predominantly from grain orientation (DENNIS, in BAYLY et al. 1977).

*Smooth cleavage* Cleavage with an even or smooth surface of parting. Smooth cleavage is commonly planar or sinuous with continuous domains free from irregularity or unevenness at the scale of observation. Many slates contain smooth cleavage.

*Spaced cleavage* Cleavage spaced at finite intervals, at the scale under consideration (DENNIS, in BAYLY et al. 1977). The practical lower limit for distinguishing cleavage domains in field outcrop is between 0.1 mm and 1 mm, and with an optical microscope is 0.01 mm, so that cleavages with domains more closely spaced appear continuous at those scales. The upper spacing limit is not defined, although nearly all cleavages have a mean spacing of less than 10 cm.

*Stylolitic cleavage* Cleavage with irregular, toothed surfaces of stylolitic form.

*Trapezoidal cleavage* Cleavage enlosing irregular quadrilateral-shaped microlithons.

*Zonal cleavage* Cleavage that occupies a considerable fraction of the rock volume. The cleavage zones can be called broad or narrow, depending on their width relative to the width of the microlithons. This use of "zonal" contains part of the meaning of GRAY (1977a), but we have separated the concepts *wide* and *narrow* from the concepts *gradational* or *discrete*, and use *zonal* only as a width descriptor.

# Chapter II

## Processes Contributing to Development of Cleavage

An analysis of possible processes is offered that attempts two properties, (a) it is exhaustive (b) the processes listed are nonoverlapping. The intention is not to *change* present usage, but to emphasize successful terms or practices (that quickly and accurately transfer an idea from writer to reader) and to guard against troublesome terms or practices (that seem likely to cloud the quick, accurate transfer of ideas).

At the outset it is important to separate geometrical processes from material processes. For example, geometrical processes include change of shape and change of size. By contrast, material processes include diffusion of atoms through a crystal lattice and transport of atoms as solutes by a moving solvent. The plan will be to list geometrical processes first and material processes later.

By way of extra attention to this point, we note that a grain boundary has no mass. Grains or atoms are material things, with mass, but the boundary is more of a geometrical abstraction; the grain's long axis is another. Thus by *geometrical process* we mean change in the disposition of things with no mass, such as boundaries or long axes, while by *material process* we mean change in the disposition of things that have mass, such as atoms, groups of atoms or whole grains. When a whole grain suffers a rigid-body movement, the movement of its boundary is so closely related to the movement of the material that it seems unnecessary to try to keep the boundary and the material as separate concepts; but most of the changes to be discussed are more complicated than this, and keeping geometry and material separate proves convenient.

### Possible Geometrical Changes

We begin by considering a finite time and listing ways in which the end state may differ from the starting state. We defer until later any thought about movement paths or history by which the change occurs.

The range of possible changes can be sketched in terms of two end members

i. change in arrangement of grain centers with no change in texture;
ii. change in texture with no change in arrangement of grain centers.

The first could be achieved with close-packed identical spheres; see Fig. 2.1.

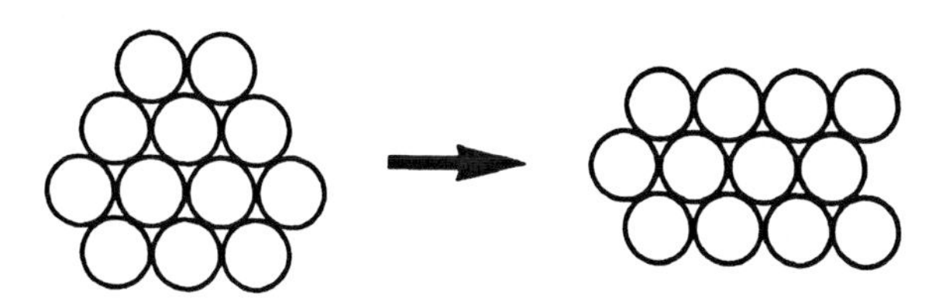

**Fig. 2.1.** A rearrangement of spheres

The second, also highly artificial, is shown by example in Fig. 2.2.

**Fig. 2.2.** A rearrangement of grain boundaries

The purpose of showing these two end members is to introduce the idea that any real change in geology is somewhere between the two; but some changes are more toward type (i) – much deformation without much change of texture, as in the flow of loose granular aggregates and slurries, while some changes are more toward type (ii) – much change of texture without much overall deformation, as in thermal meta-

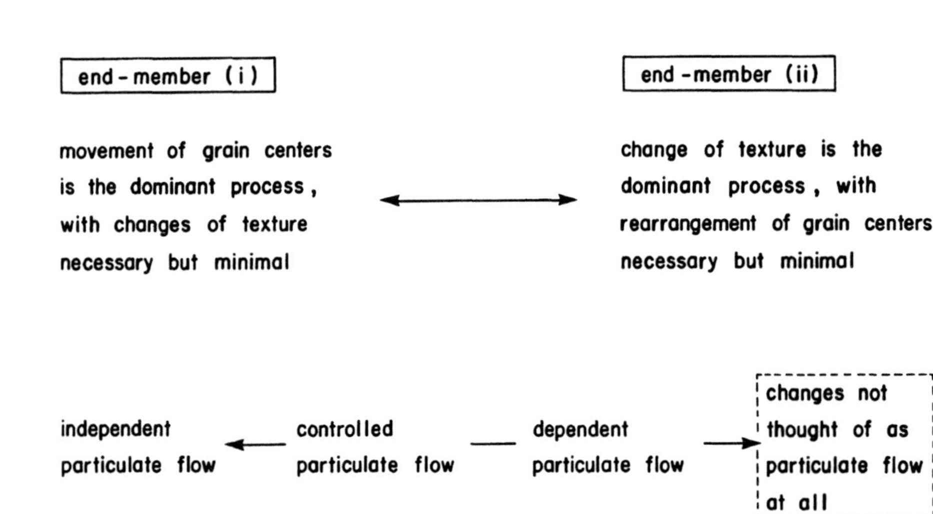


**Fig. 2.3.** A continuum of material processes

morphism. A more careful way of describing the range of possibilities is thus:

*Type (i).* movement of grain centers is the dominant process. A certain amount of grain-boundary adjustment has to occur, or there has to be a small amount of freely deforming interstitial material, but these effects occur as needed, to the minimal extent just sufficient to permit the flow of grains.

*Type (ii).* rearrangement of texture is the dominant process. As the texture changes, there have to be displacements of grain centers but these occur to the minimal extent just sufficient to permit change of texture.

Again, the purpose of describing two endmembers is to lead to the thought that most real processes will be somewhere between the two. The range of possibilities is partly covered by Borradaile (1979a, 1981) who distinguished and described

independent particulate flow,
controlled particulate flow,
dependent particulate flow,
see Fig. 2.3.

## Possible Details in a Geometrical Change

If grain centers are treated as simple points, the way their end state differs from their initial state can be shown simply by an appropriate number of displacement vectors; but when we switch to *texture*, the way the end state differs from the initial state is more complex. Between start and finish, there is likely to be change in any or all of the following:

size of a typical grain,
shape of a typical grain,
length orientation of a typical grain,
lattice orientation of a typical grain,
homogeneity of the aggregate,
anisotropy of the aggregate,
degree of likeness of grains in respect to size,
degree of likeness of grains in respect to shape,
degree of likeness of grains in respect to orientation,
distribution of intergranular discontinuities.

It is to be emphasized that these changes normally occur together. There is no reason to expect that any will ever occur in isolation, in the natural world; but each can be completely isolated as a concept. As examples, the distinctions among shape, size, and orientation are illustrated in Fig. 2.4, changes of homogeneity,

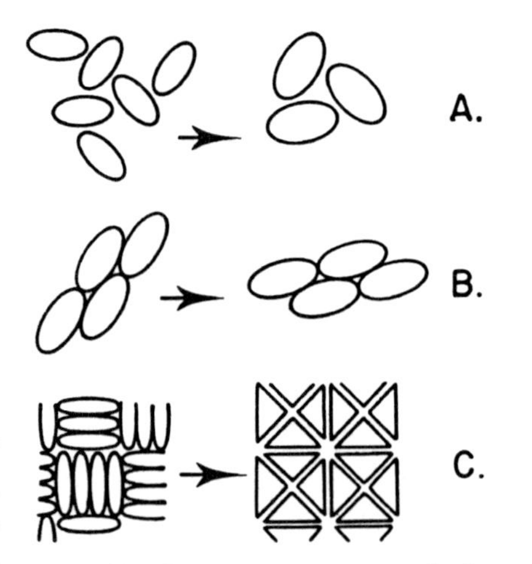


**Fig. 2.4.** Separate aspects of change in texture. **A** Change of size (no change of shape or orientation); **B** Change of orientation (no change of shape or size); **C** Change of shape (no change of size or orientation)

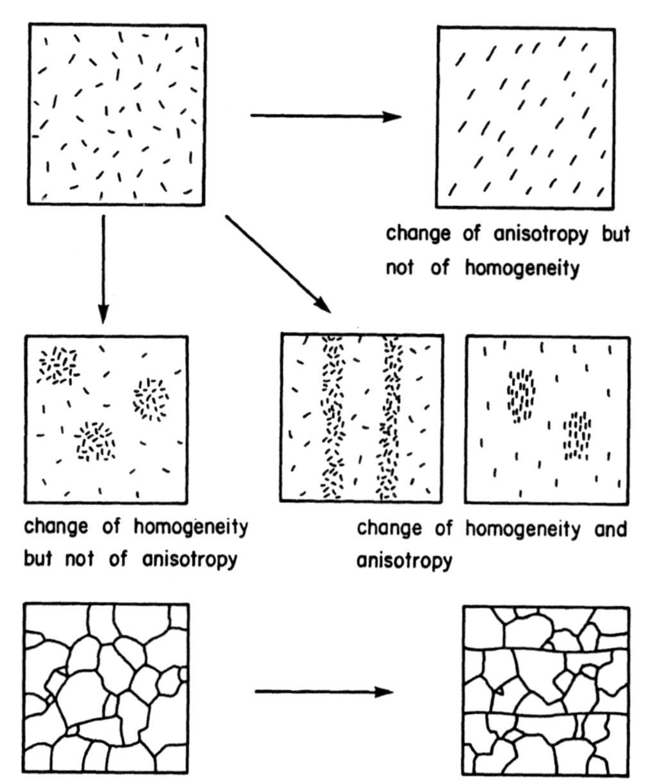


**Fig. 2.5.** Separate aspects of change in texture, continued

anisotropy and patterns of discontinuities are shown in Fig. 2.5, and change of alikeness is shown in Fig. 2.6. Of course, among these, some are probable and some improbable, and some relevant to cleavage development and some less relevant; but as concepts, all are clearly distinguishable and independent of the others.

The distinctions just made are most easily thought of with a monomineralic rock in mind, but apply also to polymineralic rocks. Also if we go on to envisage bulk transfer of material into or out of the volume under study, the foregoing list of possibilities continues to be sufficient. If we go on to envisage metamorphic reactions and the growth of grains of a mineral that was not present initially, we come to some difficulties, but before trying to surmount these, let us consider material processes and complete a preliminary sketch of the overall picture. The matter of changes of geometry during reaction can be picked up later.

## Possible Geometrical Changes: Discussion

An alternative, more usual approach to the foregoing matters is on the lines: movement of grains past each other involves adjustment of grain geometry, and this can be achieved by bending, twinning, granulation, slip, partition into subgrains etc. Why have these common terms not been used to list possible geometrical changes? Why has a less familiar list been used instead? The reason has to do with terms being fundamental or primary. Among colors, red, yellow, and blue are primary while orange and green are not; both orange and green contain a primary ingredient, yellow, but yellow and red have no common part. Similarly among numbers, 2, 3, and 5 are primary, while 6 and 15 are

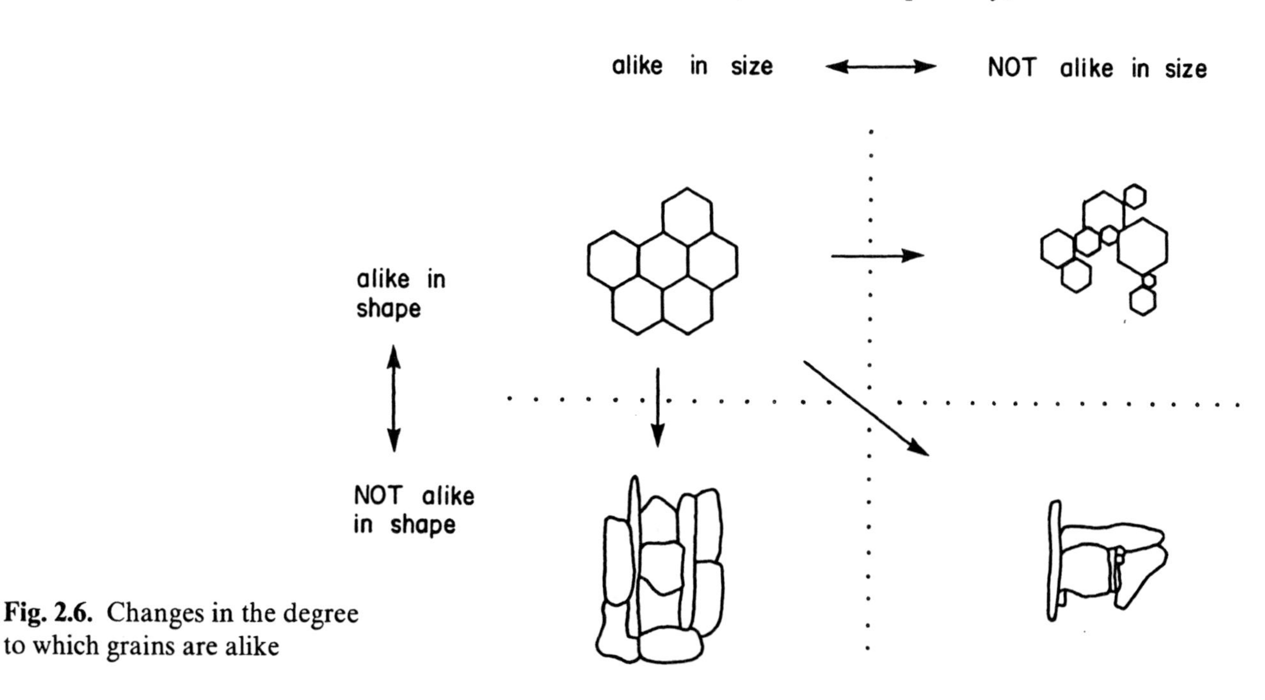


**Fig. 2.6.** Changes in the degree to which grains are alike

not; both 6 and 15 contain a prime factor, 3, but 3 and 5 have no common factor (except 1). Similarly change of size and change of lattice orientation are primary, while bending and twinning are not; both bending and twinning contain some change of lattice orientation, but change of lattice orientation and change of size have no common part.

Why do we need terms that are fundamental or primary in this sense? The answer has to do with being cautious or conservative. An objective is to make assertions as follows: "I can see ... in the rock and conclude that ... has occurred." In such an assertion, an objective is to conclude no more than what the observation permits. If all one's terms carry two or three implications, one is easily led into implying more than one intends; if one favors terms that have just one implication each, it is easier to state just what is so. For example, to state from Fig. 2.7, "From the pseudomorph, I conclude that there has been bending" is risky; "From the pseudomorph, I conclude that there has been change of shape" is on firmer ground. One may, of course, go on to note the *possibility* that there has been bending, if one wishes to do so.

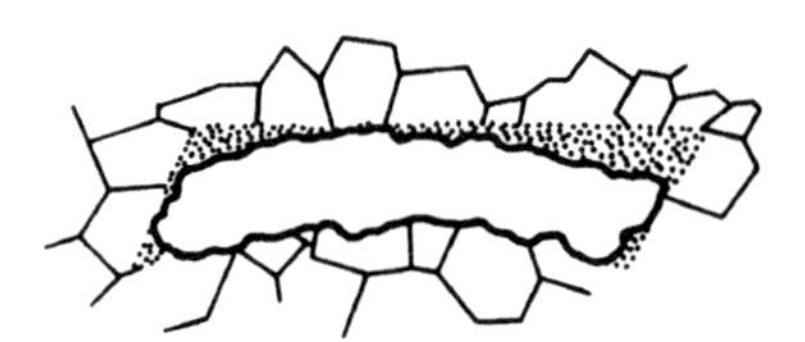

**Fig. 2.7.** A pseudomorph with straight-line boundary and a crystal with curved-line boundary

## Possible Material Processes

We have in mind here such things as

i. diffusion of single atoms inside a grain,
ii. dissolution at a crystal surface,
iii. translation of an entire grain, and so on – processes that are specified in terms of things that have mass.

As in the section on geometrical change, we shall try to form a list of processes in which all items are independent or primary – no process listed should imply the occurrence of another process listed; at the same time we shall try to make the list comprehensive, so that any process that might contribute to development of cleavage is included.

First let us divide a rock into (i) the crystalline grains (ii) the less ordered or non-crystalline material between the grains; see Fig. 2.8. Having defined a boundary (even though the material properties change in a continuous manner) we can now clarify, for example, *dissolution* as specifically the jump of material across this boundary. (Where the composition is the same on either side, the process is called melting, but let us stick with the idea that the disordered phase is richer in volatiles or, in atom terms, richer in hydrogen, carbon etc. – sometimes chlorine.) Then, for example, a finite change of grain geometry might involve diffusion of atoms in the disordered phase, plus dissolution, plus creep of dislocations in the ordered phase.

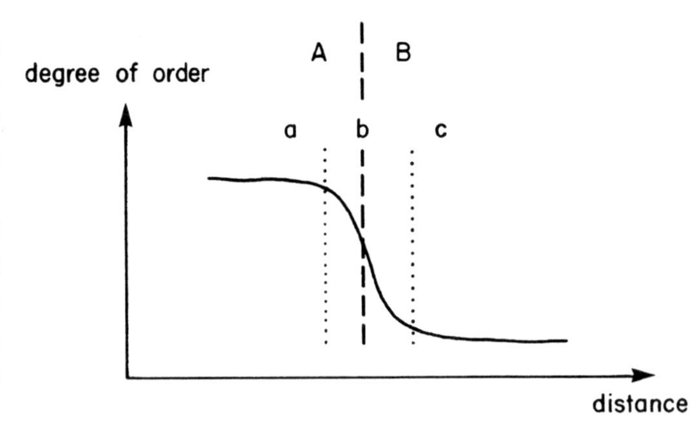


**Fig. 2.8.** Two possible descriptions of a crystal boundary
The *dotted lines* separate regions *a, b* and *c*, which could be called *a* body of crystal, *b* disordered surface phase, *c* intergranular fluid. Alternatively, the *dashed line* separates regions *A* and *B* which could be called *A* body of crystal, *B* less ordered intercrystalline material. Either system is arbitrary, because the boundary is in fact gradational. For the purposes of this discussion, the two-way split will be used in preference to the three-way split

Using the statement above of what we mean by *crystal boundary*, we can offer a list of ten pos-

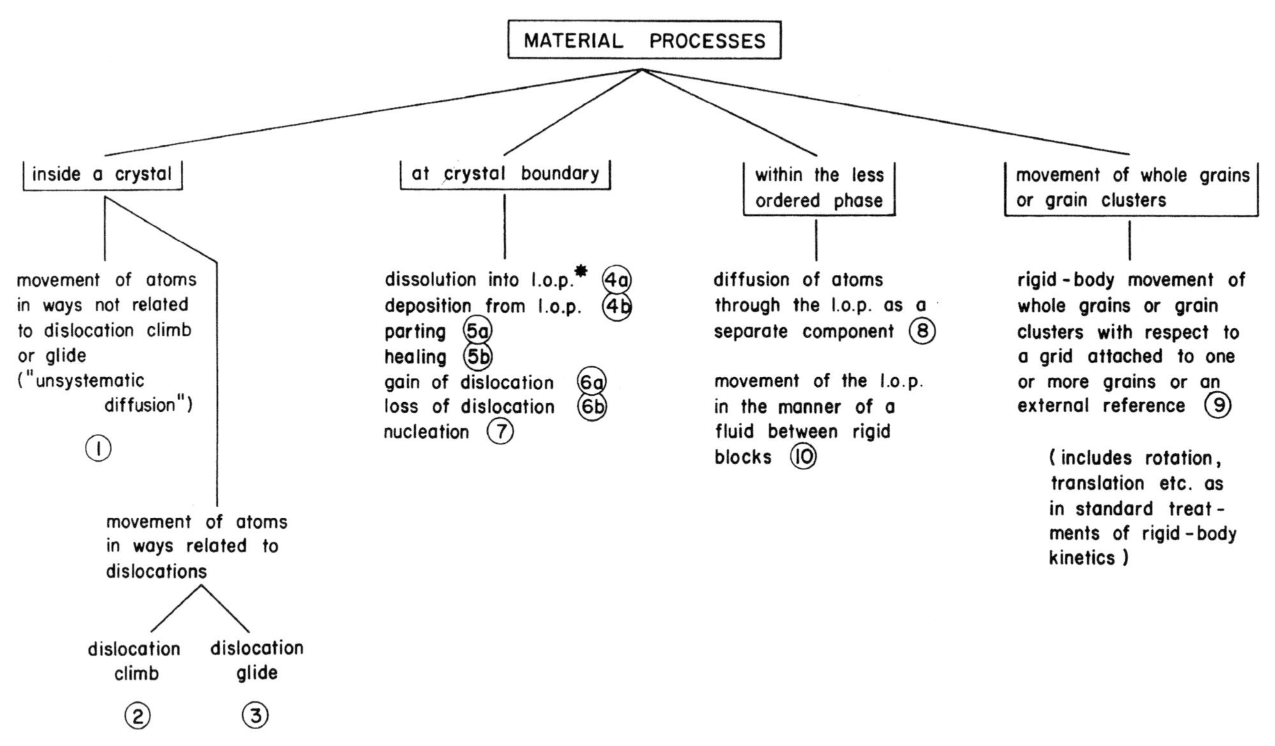


**Fig. 2.9.** Diagram showing possible material processes. * l.o.p. = less ordered phase, i.e. part B in Figure 2.8

sible processes in Fig. 2.9. A first example of the use of these ten is to discuss the geometrical change in Fig. 2.10.

**Fig. 2.10.** A geometrical change involving several grains

This change could occur by (10), rigid-body movement of a grain or cluster of grains, or by (4a) (8) (7) (4b), dissolution-diffusion-nucleation-deposition. A second example is shown in Fig. 2.11.

**Fig. 2.11.** A geometrical change involving a single grain

If the geometrical change is a change of lattice orientation as well as changes of shape and length orientation, we might suggest that the change occurred by dislocation processes (6a) (3) (6b) and sum the whole thing up as twinning; if the change of shape and length orientation has occurred without change of lattice orientation, we might still think of dislocation glide, or perhaps of the group (4a) (8) (4b), dissolution-diffusion-deposition (without any nucleation). Except for (10), the processes are compared and contrasted in diagram form in Fig. 2.12. The claim that other terms in use such as *recrystallization, kinking, polygonization* could if necessary be spelled out and explained using these ten processes and the geometrical terms from pages 7–9 has been tested by attempting a partial glossary, pages 15–16.

## Cataclasis, Recovery, and Recrystallization

There are several ways in which a cluster of, say, 10 grains can convert to a cluster of 15 grains. We try here to identify separate possibilities without getting hung up on the question of what words designate which processes. Two possible consequences of bending a crystal are shown in Fig. 2.13. Sequence A, B, C illustrates an ideal sequence where, under bend-

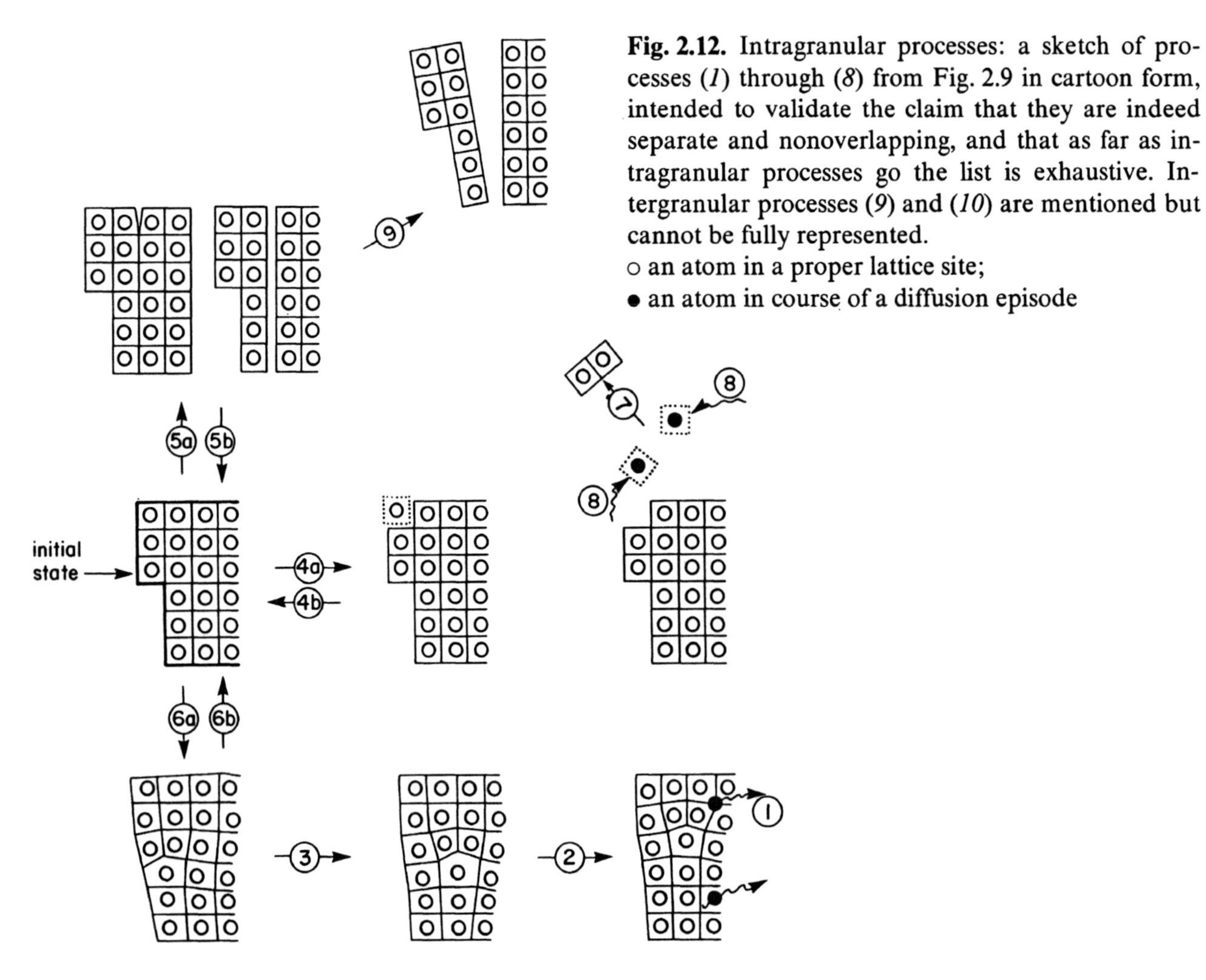


**Fig. 2.12.** Intragranular processes: a sketch of processes (*1*) through (*8*) from Fig. 2.9 in cartoon form, intended to validate the claim that they are indeed separate and nonoverlapping, and that as far as intragranular processes go the list is exhaustive. Intergranular processes (*9*) and (*10*) are mentioned but cannot be fully represented.
○ an atom in a proper lattice site;
● an atom in course of a diffusion episode

ing, bond lengths and angles change but no atoms change neighbors. Ultimately some bonds break (stage B) and new crystal surface is created. Subsequently the two new surfaces can be joined and healed by local atom migration processes so that final porosity is negligible (stage C) but the essential feature is that atoms do not change neighbors in course of conversion from one grain in (A) to two grains in (B). Sequence D, E, F shows by contrast a sequence where, under bending, atoms do change neighbors. By atom movements such as dislocation glide and climb, dislocations increase in number and segregate into clusters. With migration of dislocations into more and more concentrated groups, the crystal develops into two or more volumes where the density of dislocations is low, separated by layers of material so full of dislocations as to be called a less ordered intergranular phase. The end results of ABC and DEF are not very different, yet the processes are in concept completely different as regards change of atoms' neighbors. They can of course occur simultaneously: a rapid rise of stress would favor sequence ABC and a slow rise would favor DEF, but under suitable stress

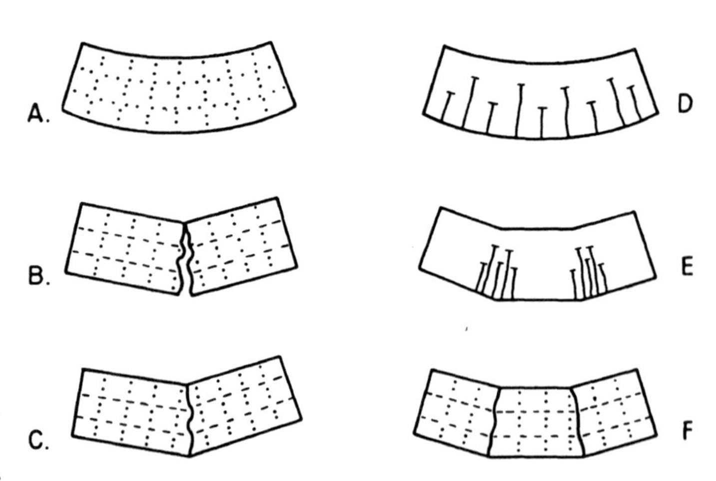


**Fig. 2.13.** Two different ways of forming "subgrains"

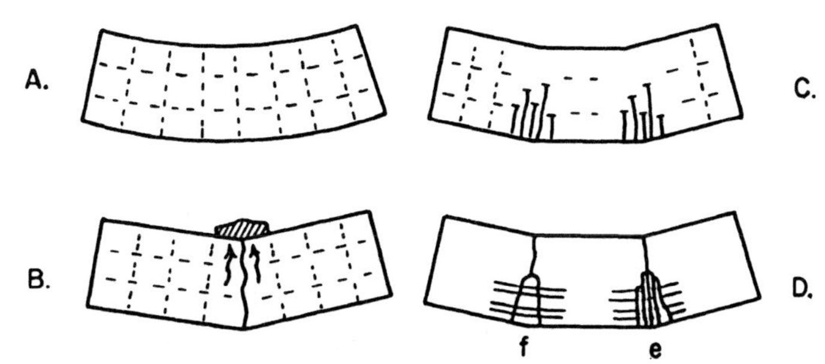


**Fig. 2.14.** Two different possibilities for nucleation

conditions a crystal could separate into two crystals by a mixture of processes.

We turn now to the possibility of nucleation: Fig. 2.14 shows how nucleation may be associated with either of the two possibilities already discussed. The main requirement for nucleation is the right material in a high energy state. The strained bonds in Fig. 2.13 ABC are a source of energy for nucleation and for subsequently driving migration to feed the growth of the nuclei (Fig. 2.14A, B). In Fig. 2.13 D, E, F, the zones with high density of dislocation are, in a similar way, zones of high energy where new nuclei can form and grow (Fig. 2.14C, D). In Fig. 2.14D the right-hand new grain boundary has a mini-grain *e* in it whose orientation could hardly arise except by nucleation. The left-hand new grain boundary has a mini-grain *f* whose origin is less clear-cut: grain *f* could be newly nucleated, its growth having occurred by its boundaries sweeping outward through the host in a cannibalistic way, or grain *f* could simply be a fourth low-strain volume developed as in sequence (Fig. 2.13 D, E, F).

A fourth way in which the number of grains can increase is shown in Fig. 2.15.

**Fig. 2.15.** An increase in number of grains that does not involve nucleation

If again we try to use Fig. 2.9, the four processes for an increase in number of grains can be distinguished as:

| | | |
|---|---|---|
| i. | 5 a (±5 b, 4 a, 8, 4 b) | Fig. 2.13 ABC |
| ii. | 1, 2, 3, 6 b | Fig. 2.13 DEF |
| iii. | 7 (±4, 5, 6, 8) | Fig. 2.14 |
| iv. | 4 a, 8, 4 b, "disconnection" | Fig. 2.15 |

In words, *cataclasis* fairly well designates (i) and *nucleation and growth* can be used to designate (iii); at the time of writing, there seem to be no one-word names for processes (ii) and (iv) that can be relied on to carry a standard, widely recognized meaning. In particular, *recovery* and *recrystallization* serve well enough to indicate the general coverage of the foregoing section, but at present are in the precarious state of meaning different things to different people.

### Extensions and Examples

We recall that the task currently in hand is to list processes that possibly contribute to development of rock fabrics, and cleavage in particular. There remain three aspects not yet considered explicitly, namely, mineral reaction, metasomatism, and change in number of grains present. It will be argued that no new material processes have to be considered in admitting these effects.

#### Mineral Reaction

Processes already envisaged include dissolution of a component into a less ordered phase, nucleation, and deposition of components from the less ordered phase. This group of processes covers instances where the nucleating and growing grain is a different mineral species from the grains that are losing components, as well as instances where it is just a new grain growing at the expense of old grain's of the same species. Thus, at the level of Fig. 2.9 reaction involves just the same processes as recrystallization.

#### Metasomatism

The processes listed in Fig. 2.9 include several by which the contents within a specified envelope can change. In particular, dissolution and deposition can go on at different rates, with consequent change in bulk chemistry of the total mass of crystals, and flow or creep of the less ordered phase can occur into or out of any specified space.

## Change in Number of Grains

In the absence of metasomatism, change in the number of grains is closely tied to mean grain size: as one goes up, the other goes down, except for minor variations that can perhaps be produced by change in the manner in which the population of grain sizes is distributed about a given mean. But where material is being lost, the number of grains can remain fixed while all grains diminish in size, or the number can diminish as well as the size diminishing, and so on. Thus change in number of grains is a geometrical change independent of those listed on p. 8, that for comprehensiveness should be added to that list.

## Examples

In light of the foregoing material, a comprehensive description of a change in texture might be as follows: "The change of texture was due 10% to reaction, 5% to removal of material, and 85% due to change at constant mineral content" (or non-numerical statements of the same type). "The main change involved strong flattening by particulate flow; there was diminution of grain size by cataclasis with no significant nucleation but with extensive change in the texture of pre-existing grains. These improved in length orientation and lattice orientation by twinning and rigid-body rotation; grain-boundary misfits that were not accommodated by twinning were taken care of by grain-boundary slip and minor componental movements at grain margins (dissolution and redeposition). Neither atoms nor dislocations migrated significantly within the bodies of grains except in the twinning process. In course of deformation, the rock became inhomogeneous and anisotropic, the reaction products being localized on discrete surfaces that were probably surfaces with considerable slip."

This example reminds us how complicated a change of texture is, in terms of processes. Sometimes we are lucky and can pursue our ends without recording all aspects, but when we need a full statement, it is by necessity rather long and involved. This, of course, is all concerned with expressing a hypothesis; presenting evidence in favor of the hypothesis, or against it, would be separate again. On a positive note, we suggest that when people take the necessary time to spell out the process they envisage in detail, as here, they find a large measure of agreement with fellow workers and are able to make progress on small significant points of difference. When people abbreviate their hypothesis by a catchword or two, differences may be magnified and one worker may find himself unable to contribute usefully to another's thinking; or again, a difference may be obscured, and this also destroys an opportunity when one person's experience could usefully add to another's.

## Tests of Usefulness

Any system for describing the natural world by a short list of separate possibilities is of course open to objection – no system is free of faults. But the system put forward so far, even though imperfect, may be useful. We may test its usefulness by asking, "Are there terms with cloudy or overlapping meanings that can be clarified by using the system outlined?" An example follows:

Suppose Worker A states that, in a given rock, cleavage is due to segregation of mica, while Worker B states that it is due to porphyroblastesis. We wonder if they are agreeing or disagreeing; or, supposing they agree to some extent, we wonder on what points they agree and on what points they disagree. Suppose Worker A analyzes "segregation" on the lines of the system offered and Worker B analyzes "porphyroblastesis." If thereafter each sees clearly where they agree, where they differ in emphasis but not content, and where they actually entertain ideas that are in conflict, then the analysis and the terms involved have been useful; if they are more confused at the end than at the start, the system offered has failed.

## A Partial Glossary

(The following entries are not intended to rival or supersede entries in the American Geological Institute Glossary or similar works. The purpose here is to test the adequacy of the preceding analyses: can commonly used terms indeed be clarified using the geometrical- and material-process ideas just given? The glossary entries should be considered with this particular purpose in mind; they are not necessarily good explanations for the general enquirer).

*Blastesis* nucleation and growth of a new piece of lattice (in contrast to extension or modification of crystals already present in the rock; may or may not involve neomineralization) (7+4b)

*Cataclasis* cracking into many pieces (5a)

*Cataclastic flow* change in arrangement of grain centers in space by rolling, cracking and sliding, with no change of grain shape except by parting (9+5a)

*Coble creep* deformation of a coherent mass of crystals by processes localized at their boundaries (4–9; 10 and 1–3 minor). Less neighbor switching than in *dependent particulate flow*

*Controlled particulate flow* between *dependent* and *independent* particulate flow, q.v.

*Creep* slow flow

*Deformation lamella* a lamella inside a crystal within which deformation is more intense than in the crystal as a whole.
(Deformation occurs by 1 and/or 2 and/or 3 with 6a and possibly 6b)

*Dependent particulate flow* change in arrangement of grain centers in space by rolling and sliding, with some change of grain shape by parting and some by boundary processes; rolling and sliding made easier by grain-boundary adjustments such as removal of asperities (9 extensive, 4–8 minor, 1–3 negligible)

*Dewatering* loss of water from a mass of grains previously assembled loosely by sedimentary processes, as the grains rearrange into a less porous geometry (9+10) (4a, 5a minor)

*Disaggregation* increase in the amount of less ordered phase amongst a mass of crystals OR reduction in viscosity of l.o.p. OR both, as would follow from incursion of a low-viscosity fluid (10)

*Flow* change in shape without development of a new family of discontinuities
In *continuum flow* there are no discontinuities, e.g., idealized fluids;
In *particulate flow* each particle is surrounded by discontinuities on particle scale, but no mega-discontinuities develop on bulk scale;
In *flow of a crystal* the megascopic shape changes without the crystal parting into two crystals (though the process may again depend on the existence of many discontinuities within the flowing mass – in this instance, slip planes – or on deformation lamellae)

**Folding* a change from planar to nonplanar

*Fracture* (5a)

***Grain flow* a particulate flow (q.v.) where the grain-size distribution does not include a large proportion of fine-grain particles (9)
(contrast with slurry flow)
(includes sand flow, gravel flow etc.)

*Independent particulate flow* change in arrangement of grain centers in space by rolling and sliding with negligible change of grain shape; rolling and sliding made easy by an excess of less ordered phase (l.o.p.) (9+10)

*Kinking of a crystal* special case of flow of a crystal where the flow is localized in planar slabs and controlled in amount by the continuity requirements of the crystal lattice (much 3, minor 1+2)
(not as strictly linked to atom-row angles as twinning is; differs from a deformation lamella by being bigger)

**Kinking of laminar rocks* a change from planar laminae to nonplanar by rotations on narrow hinges that lie in well defined parallel planes

*Melting* dissolution into a less ordered phase of closely related composition (4a)

*Nabarro-Herring creep* deformation of a coherent mass of crystals by processes, especially diffusion, that affect the interior of each crystal as much as its boundaries (as opposed to deformation by processes that are most active along boundaries)
(1, 2 or 3 dominant; 4–10 minor or absent)

*Neighbor switching* conversion from

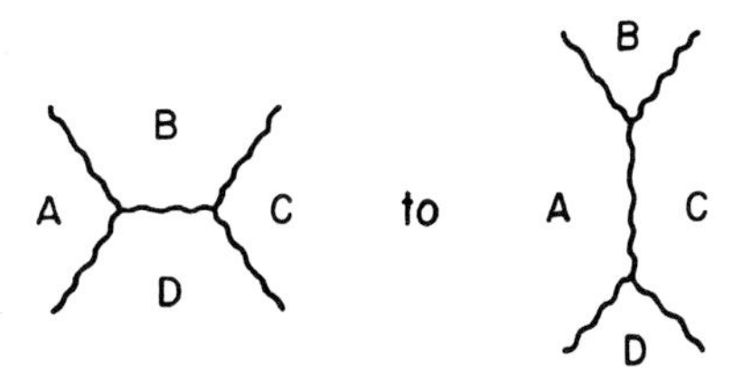


or the analogous change in three dimensions, where A, B, C, D are grains in a group of grains. The term names a geometrical process that can be produced by a variety of material processes

*Neomineralization* the appearance of a new mineral species (7+4b)

*Particulate flow* flow of a sample of material that consists of a mass of particles in a more mobile matrix, where i. the ratio of grains to matrix is large, and ii. strain within individual particles is small in comparison with the strain suffered by the sample as a whole (9)
Item (i) distinguishes particulate flows from suspensions; item (ii) distinguishes particulate flows

from flows, such as those in glacier ice and cold-worked metals, where strain within particles is not so small in comparison with the strain suffered by the sample as a whole. Particulate flows include grain flow, slurry flow etc.

*Pressure solution* dissolution in circumstances where, in the writer's opinion, the effect of pressure is a noteworthy part of the combination of effects leading to local undersaturation (4)

*Rolling* movement of one particle past another without sliding (where two surfaces are in contact, movement of one with respect to the other without sliding at the point of contact) (9)

*Rotation* (of a unit of material such as a grain or grain-cluster) a material process (9)

*Rotation* (of a grain's long axis or other geometrical feature that has no mass) a geometrical process
For either kind of rotation to occur, other processes normally have to go on simultaneously

*Sand flow* a particulate flow where the particles are largely independent and sand-sized; see *Independent particulate flow* (9)

*Sliding* where two bodies are in contact, movement of one surface past the other in a direction parallel to the plane of contact (9)
Because real surfaces are imperfect, sliding is normally accompanied by other processes, thus 9+10 or 9+5 a or etc.

*Slip* = sliding. (By custom, *slip* is used more where the surfaces are smooth, and *sliding* is used more where the surfaces are rough or irregular; boulders slide while playing cards slip) (9)

*Slurry flow* a particulate flow (q.v.) where the grain-size distribution includes a large proportion of fine-grain particles (contrast with grain flow) (9)

*Superplastic flow* a particulate flow where grain-boundary processes are more important than intra-grain deformation, and where grain-boundary sliding is more important than rolling or dissolution/deposition or cushioning by l.o.p. (9+3 etc.)

* *Transposition* (of a set of planes that trend in a direction, say, D) a modification of the set of planes so that, although the trend remains on gross scale still parallel to D, the planes run in detail for the most part parallel to a new direction E

*Twinning* 1. a static geometrical condition "Many instances of twinning exist..."
2. a special kind of shape change in a crystal, where all the atoms in a subvolume change position by amounts that satisfy the symmetry requirements of the crystal

*Work hardening* 1. a change in the interior configuration of a material that is caused by deformation and causes an increase in elastic modulus,
2. (less commonly) a change caused by deformation that causes an increase in viscosity.

* process on scale of a rock unit rather than on grain scale

** Note on *grain flow* in sedimentology
We have to distinguish flows dominated by inertia forces from flows dominated by viscous forces and, separately, to distinguish flow of a mass of particles that are all coarse from flow of a mass of particles containing many in the fine-grain tail; one is a mechanical distinction and the other a textural distinction. Among textures we have

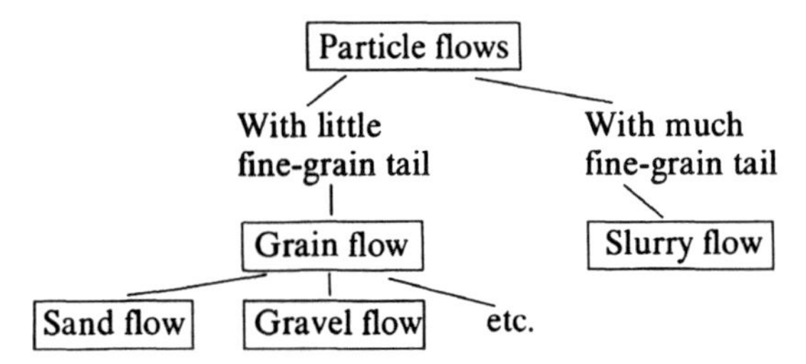


In the present Atlas, almost all grain flows considered are viscous flows, whereas in sedimentology, grain flows are more often inertia flows; but this should not prevent the term *grain flow* from being used with the same meaning in both contexts.

# Chapter III

## Possible Links Between Observables and Processes

The main basis for proceeding at all is the thought that a preliminary effort may serve as a root: we look to other hands to nurse it into a flower.

This chapter is different in a fundamental way from the two preceding chapters on *observables* and *processes*; essentially, it cannot be written. We do not at present know how to infer a group of processes from a group of observables. The purpose of this Atlas is to improve the groundwork from which people attack this task. If the Atlas is effective, it will help in the current trend toward exchange of opinions about how different cleavages form, and perhaps at some future time, the combined knowledge and insight of many people will result in a robust system of patterns of inference; but at present we do not have such a system – it is the absence of such a system that provoked the creation of the Atlas in the first place. (By *robust system of patterns of inference* we mean something like the Facies Classification of metamorphic rocks. For all its defects, this constituted a framework within which people could compare notes, distinguish the commonplace from the exceptional, and assemble a pattern of dominant tendencies. To a reasonable extent, it combined many people's observations into a digested form, and so helped an individual observer to know where he stood. It was the feeling that students of microstructure are currently groping for or working toward such a synthesis, but currently lack one, that provoked this groundwork-level effort.)

A consequence of the state of affairs just outlined is that the present chapter is *highly tentative*. As editors we feel marginally competent to assemble the series of contributors' plates, and to offer the chapters on Observables and Processes (Chaps. I and II). But as regards patterns of inference, we sense quicksands at our feet.

### The Structure of the Problem

At the outset we note scarcity of one-to-one correspondence: one process does not lead to just one possible resulting appearance. One appearance can result from process A or process B, and also one process can lead to appearance X or appearance Y. But our situation has not even the simplicity of a basic many-one/one-many mapping: a single appearance has several possibilities for its history and each history involves several processes. The structure of the situation is shown in Fig. 3.1.

Diagram A shows two possible histories for one appearance, stating "zonal cleavage can arise from rigid rotation plus dissolution or from rigid rotation plus deposition of a solute". Diagram B shows a fuller structure, including two statements of the type in diagram A and also the new type "rigid rotation plus deposition of solute can generate either a zonal cleavage or a continuous cleavage." The basic many-one/one-many structure is shown by just the left-hand part (the triangles, circles, and dotted lines). It is helpful if all concerned agree that our problem is not just of this simple kind.

### Should We Try to be Exhaustive or Submit to Being Fragmentary?

Let us look at two examples: "Half a crinoid ossicle reveals dissolution and/or fracture" and "A quartz pod or lens can result from deforming a quartzose pebble or from boudinage of a quartzose layer". One way to assemble current opinions on cleavage would be to gather a group of such statements. Any such group would be incomplete, and any statement in it

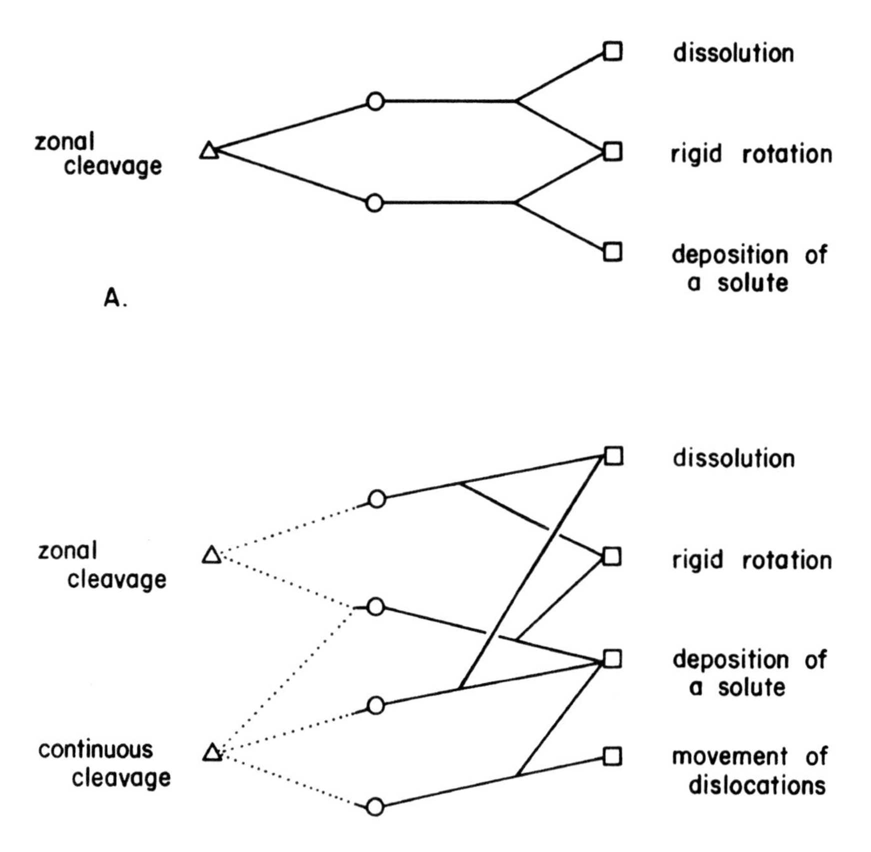


**Fig. 3.1.** The nature of the links between process and result. □ a single process as in Fig. 2.9. △ a single type of cleavage as in Fig. 1.5 or 1.6. ○ a combination of processes. Each *circle* represents a combination such as "dissolution and rigid rotation in equal amounts" or "dissolution of quartz and rigid rotation of mica" or "extensive grain flow with rigid rotation of mica followed by dissolution of quartz from points where mica flakes impinge". How much detail is needed to give a circle maximum usefulness as a building block is not obvious: for discussion, see accompanying text

would be subject to correction; any such group would cry out for rewriting in a more careful form, with proper *if's* and *but's*; any short system has the weakness of being fragmentary.

Another way to assemble current opinions would be more analytical. Any rock sample must be layered or not layered: if layered, the layers must be due to sedimentation or . . .; if not layered, the sample must either be homogeneous or carry inhomogeneities, and these must be either . . . Following this approach, completeness can be assured at each step but the cumbersomeness is monumental.

When our communal knowledge is better, it may be possible to create a statement of an intermediate kind, starting from either end. Working by accretion, we may learn to put in *all* the particular statements worth making and avoid fragmentariness; or working by analysis, we may learn to streamline and prune out, so as to create a manageable system and avoid cumbersomeness. Somewhere in the middle ground, we hope continued effort will discover a pattern of inferences that is reasonably exhaustive and yet manageable – this is the main hope. But in our present state of knowledge, we seem condemned to be either fragmentary or cumbersome. A fragmentary picture has been prepared by accretion and forms the rest of this chapter. Meanwhile, an exhaustive system has been conceived and partly implemented; its nature is indicated in Appendix I.

### Instances Where a Process can be Inferred

To start the process of assembling current opinions on cleavage using the piecemeal or accretionary method, seven instances are put forward:

Close-spaced parallel partings in sandstone or limestone;

Rock with local concentrations of minerals;

Rock with displaced parts that were formerly continuous

1. where the displaced parts come from units that were of limited extent in all directions (lumps rather than layers)

2. offset layering (where extent is unbounded parallel to the layers);

crenulated rocks with compositional zoning, continuous cleavage, rocks rich in platy grains, continuous cleavage, rocks poor in platy grains.

**Close-spaced Parallel Partings in Sandstone or Limestone.** Where a rock is a mass of equant grains, e.g., sandstone or limestone, and simple in texture except for parallel partings, the two main possibilities are (A) purely mechanical processes such as a loading–unloading cycle (B) localized dissolution. The fracturing in hypothesis A may be strongly or not-so-strongly hydraulic, and a given instance may have a history in which both (A) and (B) have operated in sequence. Even where the present geometry is strongly stylolitic, one may, if one chooses, assume that a mechanical process created the partings, and subsequent dissolution simply took advantage of them. However, most of the processes shown in Fig. 2.9 can be rejected as not having contributed to cleavage development. (See Plates 65 B, 66 C, 70 A.)

**Rock with Local Concentrations of Minerals.** Where a rock has local concentrations of a mineral that seems to be present in more dispersed form around the site of greatest concentration, the two main possibilities are (A) selective removal of other minerals so that the concentration seen is a residue (B) selective transport of the concentrated mineral to the concentration site. Where the site of concentration could be a through-going channel, both (A) and (B) are possibilities. Where the site seems to be isolated, the geometry of the concentration and the mineral identity help to distinguish (A) from (B). In Fig. 3.2 there is a strong presumption that the quartz has been selectively brought to its present site, while the magnetite has been selectively left behind as a residue. (See also Plates 78, 80, 121, 200).

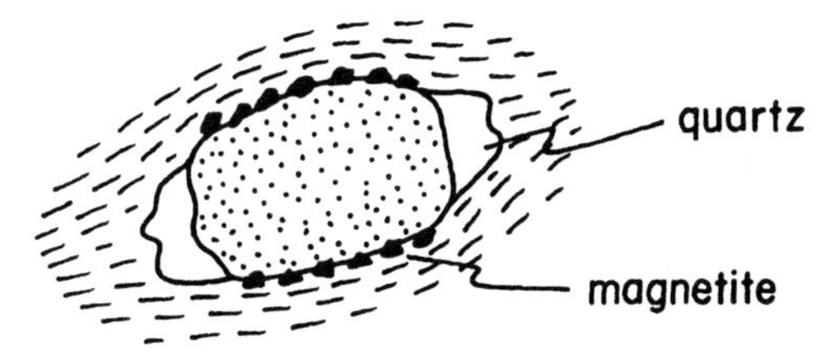


**Fig. 3.2.** A megacryst with quartz and magnetite apparently redistributed

If such a rock has cleavage, the cleavage may be a consequence of such a selective migration of quartz; another possibility is that cleavage developed first and admitted active fluids, the rearrangement of quartz and magnetite occurring thereafter as a consequence. Dissolution and transport can be clearly inferred, but the "cause" of the cleavage cannot be inferred so clearly.

A comparable situation is shown in Plates 115 C and 137 C. Here evidence suggests that the dark material has been concentrated as a residue by dissolution and removal of pale grains; but films of dark concentrated residue extend through pale layers where they probably did not originate: the appearance suggests that a fluid not only created the residue but also transported it to new locations. Once again, the origin of the *cleavage* is somewhat more open to doubt, even if the origin of the dark films is considered straightforward.

**Rock with Displaced Parts that were Formerly Continuous Units of Limited Extent.** Where a rock contains part of an object that was originally equant or at least bounded, two possibilities arise (A) a remainder can be found; in this case we can see both the displacement and the extent to which the parts recombine to form a whole (Plate 123 A) OR (B) no remainder can be found (Plates 93, 94 B, 95 B, 96 B). In (B) there is no limit to the possible displacement or to the possible loss by solution, crushing, etc., whereas in (A) we are able to set upper limits on both displacement and material loss. As has been noted elsewhere, we may find an example where it is very clear that half a fossil has been destroyed and almost certainly by solution; also the rock has cleavage – but we may yet find it hard to decide whether the rock's developing cleavage was a cause or a consequence of the solution process. Of course, situation B, the "unpaired remnant" is less informative and far more common.

**Offset Layering.** A layer differs from a bounded object in that, once parted, its parts can be displaced in the plane of the layer by any amount without the displacement being detectable afterward. Hence, the kinetics of a broken plane cannot be reconstructed as definitely as the kinetics of a broken fossil, etc. (though three intersecting planes can be reconstructed very nicely). Where a layer is offset at a planar discontinuity, there is normally a suite of possible displacement schemes, involving more or less loss of material at the fault by dissolution together with displacement parallel to the fault.

The plane of discontinuity may carry an abnormal concentration of a mineral that is otherwise distributed uniformly through the host-rock (Plates 98A, 117). In such a case, further restrictions exist on the amount of material removed by dissolution and hence on the amount of zone-parallel displacement. If one can show for several such zones that zone-parallel displacement is negligible, one can argue that the *main* effect has been dissolution and shortening perpendicular to the zones; and if the rock has cleavage, one can attribute the cleavage to the dissolution (or, if the cleavage planes were preexisting, they were at least enhanced by dissolution).

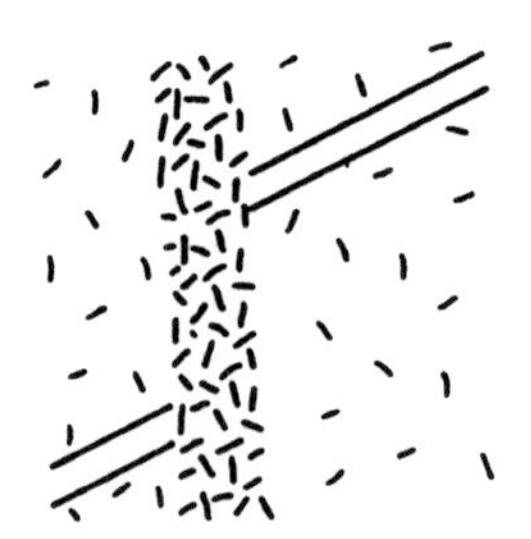

**Fig. 3.3.** A plane of discontinuity with concentration of one mineral species

An angle of intersection seen in the rock today may have had a different magnitude earlier in the rock's history; kinematic reconstructions are therefore usually open to a measure of doubt.

**Crenulated Rocks with Compositional Zoning.** Where a rock is crenulated, compositionally zoned, and confined between non-crenulated layers, the main possibility for the crenulation is internal buckling; if the zoning is parallel to the crenulation axial planes, the main possibility for the zoning is material migration during folding; the most readily acceptable idea is dissolution in the limbs and either total removal of the solute or deposition in an adjacent hinge. Crenulated layering may still be continuously traceable through the attenuated limbs so as to eliminate micro-faulting parallel to crenulation axial planes. If such a rock has a spaced cleavage that coincides with crenulation limbs, we have an instance where the process leading to cleavage can be rather confidently spelled out.

**Continuous Cleavage, in Rocks Rich in Platy Grains.** Where a rock has an ideal continuous cleavage, i.e., the texture is statistically homogeneous down to the scale of a single grain, and the rock contains enough platy grains to be responsible for the cleavage (e.g., Plates 8, 9, 11B,C) processes that may have contributed include:

i. Rigid rotation of platy grains
ii. Selective dissolution of grains in disfavored sites, orientations or conditions of strain
iii. Selective growth of grains in favored sites, orientations or conditions
iv. Nucleation of new grains in favored sites, orientations or conditions
v. Change of orientation by internal strain of grains.

Concurrent particulate flow may have been large or small in amount, and it is also possible to assign to particulate flow a dominant or a passive mechanical role. At the least, a minimal amount of passive particulate flow would necessarily accompany any of the five changes listed; also, a greater amount of particulate flow could have occurred simultaneously with cleavage development, without being necessary for its development – a sort of neutral accompaniment; or one might wish to propose particulate flow as the rock's primary response to some external change, and to view the five changes listed as necessary accommodations provoked by the particulate flow as dominant effect. Plates 132, 134, 135 and 137 show rocks where the idea of extensive particulate flow can be supported from observable features, but the point here emphasized is that in almost any rock, extensive particulate flow is a possibility to be included among one's working hypotheses.

Where a platy homogeneous rock is coarse-grained, as in Plate 11A, the possibility exists that the rock was inhomogeneous and domainal when finer grained. For example, a slate may develop with domainal texture, as in Plate 101B, the original fissility developing along

with the development of domains. Subsequent grain growth may be enough to consume whole domains and generate a rock homogeneous except on grain scale; yet the original process responsible for the cleavage would have been the early heterogeneous one.
Summing up this section, we find that apparently simple textures do not necessarily have simple histories, and perhaps after all, this type of rock is an instance where no particular process can be inferred, where inference is inconclusive.

**Continuous Cleavage, Rocks Poor in Platy Grains.** Where a rock has an ideal continuous cleavage and the rock is too purely quartz or calcite, with too few platy grains to be responsible for the cleavage, the cleavage may be due (1) to preferred lattice orientation with little length orientation or (2) to preferred length orientation with little lattice orientation, or (3) to a combination, Plates 3, 146, 149 ff. For (3), possibilities are (i) strain of grains by twinning or other dislocation movements, (ii) selective dissolution, (iii) selective growth, (iv) nucleation of new grains in the favored orientation. Rigid rotation is less likely to be responsible for the cleavage than in rocks with platy grains, though of course unknown amounts may have occurred during cleavage development; the same goes for particulate flow, in line with the discussion given in the preceding section.

## Commentary

Any discussion of specific instances of cleavage, such as those in the preceding section, brings two thoughts to the fore:
**1. Diverse opinions are abundant.** Any assertion "that x... can clearly be inferred" is likely to provoke rejoinders of two types (a) "x... cannot safely be inferred because..." and (b) "y... can be inferred as well or instead." The purpose of this Atlas is to provoke such rejoinders – they are the essence of the accretion method, despite their seeming to lead us away from a condition where we reach rapid, simple, firm conclusions.

**2. Some processes leave no trace.** If our objective is to determine what process in the past led to a rock's present cleavage, looking at the rock is a trap. Looking at the rock automatically fixes our attention on certain processes that have left visible traces in the present texture; and this takes our attention away from the possibility of other processes that, by their nature, do not normally leave traces. Two such processes are dilation and independent particulate flow. Many hypotheses of cleavage development hinge on more water having been present in the rock at former times than is present now. Perhaps the grains used to be farther apart, perhaps there was a time when they could switch neighbors in a free-floating way. The dilation could be short-lived, and one swarm of cracks might close as another opened – one does not have to believe in longstanding 10% porosity; one does not in fact have to believe the cracks *were* ever wider than as seen now. The point is that they *could have been*, and present texture will never tell us; a fluid-filled crack is like morning dew – it is the very nature of the thing to disappear without trace. (In the absence of fossils, homogeneous strain can also leave little record, or the record can be destroyed by a later effect, such as grain growth.) Fig. 3.4 is a diagram version of the same point. Regardless of the foregoing cautions, we hope the reader enjoys using the plates.

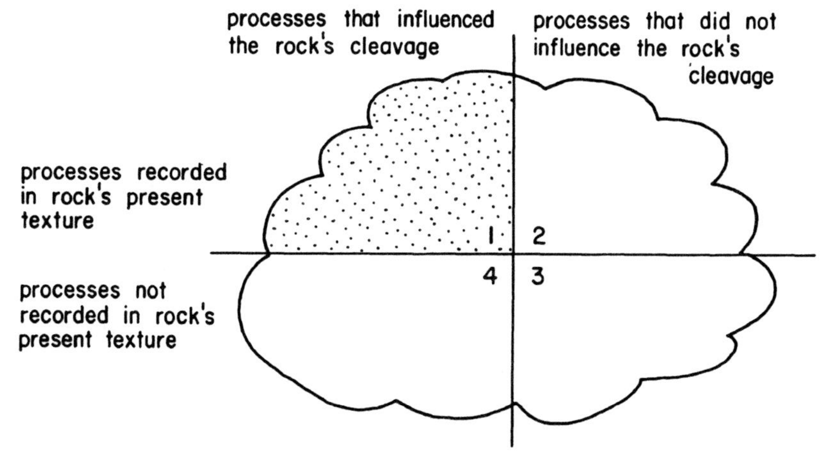


**Fig. 3.4.** The curly envelope embraces all processes that have affected a rock. There are no grounds for supposing that we live perpetually in quadrant 1

# Appendix I:

## A System for Drawing Conclusions from Observables

The following system is not particularly recommended. It was developed as an experiment in trying to be comprehensive. It clearly has a tremendous disadvantage in being cumbersome. Somewhat more of the system has been created than is reproduced here, and people interested to pursue the matter may communicate with the publishers or M.B.B. privately.

The system centers on the worker at the microscope. He has an observable in front of him – a bent fossil or whatever. He knows what processes a bent fossil suggests in *his* experience; the purpose of this system is to pool efforts and list what processes a bent fossil suggests in other people's experience. Thus it is an observable from which the search begins, rather than a process.

Of course *later*, or on other occasions, the man at the microscope will have a single process in mind and will ask, "What should I seek that, if found, would confirm that this process has operated?" Thus there is need also for a presentation organized by process. But we feel that open minds and helpful exchange of experience among colleagues are favored by making the system organized by observables the primary system, with the other in a minor or auxiliary role.

### A System of Observables

The first thing the observer chooses to give attention to may be:

| | | |
|---|---|---|
| I. | planar | e.g. a flake, a rock layer, a fissure, a kink band |
| II. | equant | e.g. a megacryst |
| III. | elongate | e.g. a mullion, a quartz rod, a tremolite needle |
| IV. | modified | e.g. bent, cracked, partly replaced |
| V. | an anisotropy | e.g. a preferred orientation, a layering |
| VI. | a heterogeneity | e.g. a layering, a clustering |
| VII. | ...... | |
| VIII. | ...... | |

This outline and the part of the system which we choose to illustrate here is indicated in Fig. 3.5 (opposite).

According to which one he chooses to focus on first, he should go to the appropriate page in the following set of Roman numeral pages. Since we are only illustrating a concept here only pages I and IV are included as examples. The system here initiated would be very large if worked out exhaustively. A small portion follows, so as to demonstrate the concept and give readers the opportunity to consider the potential benefits and drawbacks. (See Fig. 3.5.)

The system, if properly completed, would allow for different starting approaches. For example, faced with a layered rock, an observer might choose to enter the system at page I or at page V or at page VI; whichever he chose, he would be led to a reasonable sequence of next steps and options. Ideally, with, say, four observables, he would be brought to the same conclusion regardless of the order in which he chose to use them as clues.

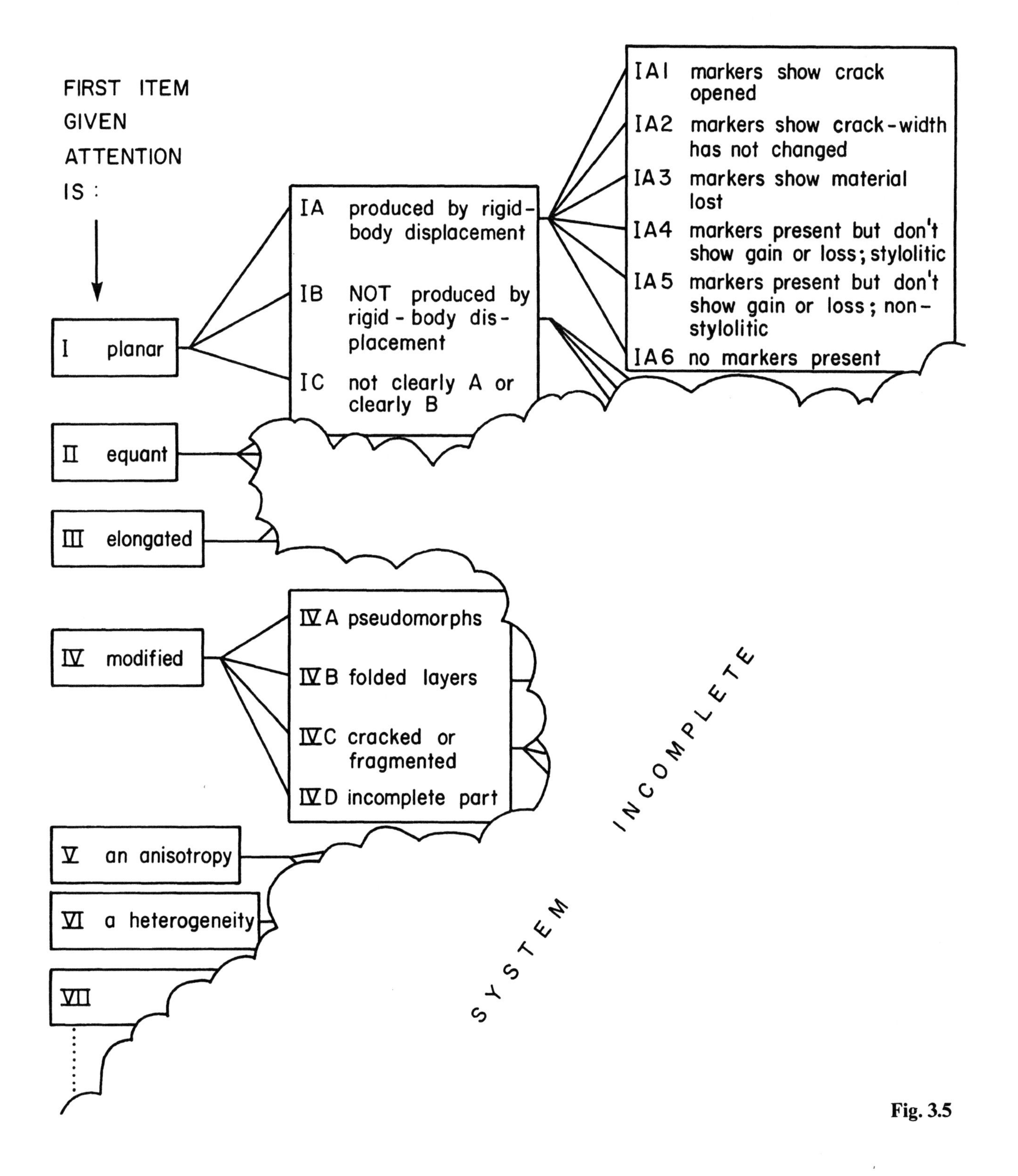
FIRST ITEM GIVEN ATTENTION IS :
I planar
II equant
III elongated
IV modified
V an anisotropy
VI a heterogeneity
VII
IA produced by rigid-body displacement
IB NOT produced by rigid-body displacement
IC not clearly A or clearly B
IA1 markers show crack opened
IA2 markers show crack-width has not changed
IA3 markers show material lost
IA4 markers present but don't show gain or loss; stylolitic
IA5 markers present but don't show gain or loss; non-stylolitic
IA6 no markers present
IVA pseudomorphs
IVB folded layers
IVC cracked or fragmented
IVD incomplete part
SYSTEM INCOMPLETE

Fig. 3.5

# Appendix I

## Possibility I

## The Thing that Catches Attention is Planar

If it is clearly a feature produced by rigid-body displacement, go to I A, p 26
(It may be a fracture, joint or crack, a shear zone in a rock or a kink band in a crystal, a twin lamella, . . .)

If it is clearly not *formed* by rigid-body displacement (though displacement may have affected it) go to I B, p . . . (not included)
(It may be a sedimentary layer, a mica flake, a fish scale, a shale pebble (eroded flat), a quartzite pebble (flat by deformation), . . .)

If it is neither clearly A nor clearly B, go to I C, p . . . (not included)
(It may be a lenticular pocket of quartz grains, a zone in zonal cleavage such as a mica-rich zone or clay-rich film, . . .)

# Appendix I

## Possibility IV

## The Thing that Catches Attention is a Modified Version of Something whose Former State can be Postulated

If it is a pseudomorph, go to IV A, p ...... (not included)

a folded layer, go to IV B, p ...... (not included)

a cracked or fragmented object (crystal, pebble, fossil, a boudinaged layer), go to IV C, p ...... (not included)

*part* of an object, with no sign of the other part or parts, go to IV D, p ...... (not included)

# Appendix I

## Possibility I A

## A Planar Feature Where some Clue is Present Showing it to be Produced by Displacement (Fault, Fracture, Shear Zone, Kink Band, Twin Lamella)

1. If geometry shows that a gap has opened, go to I A.1, p 27 (Either the walls fit 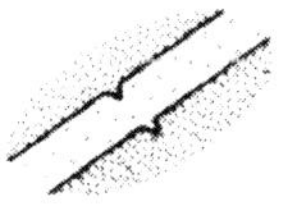 or the rock contains markers 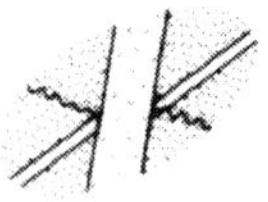 and there is now an air-gap or an infilling material)

2. If geometry shows that gap-width is as first formed, either with slip  or just a joint with no slip, 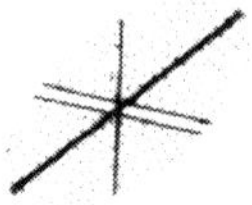 i.e., no material has been added or removed, go to I A.2, p 28

3. If geometry shows, after due consideration of the third dimension, that material has been removed go to I A.3, p 29

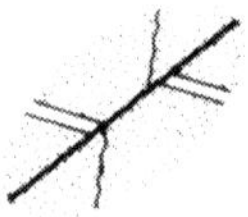

4. If markers show that there has been appreciable displacement but do not tell about addition or removal of material, and the interface is stylolitic go to I A.4, p 30

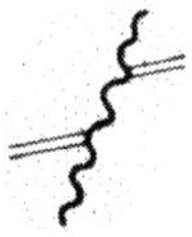

5. If markers show that there has been appreciable displacement but do not tell about addition or removal of material, and the interface is a simple non-stylolitic surface 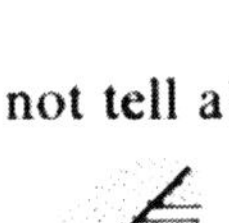 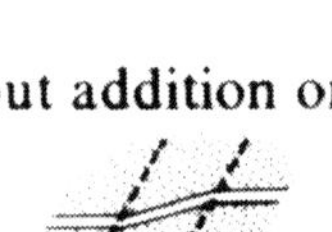 go to IA.5, p . . . . . . (not included here)

6. If the item seen is a simple crack in a rock that has no internal markers to reveal displacement, go to IA.6, p . . . . . . (not included)

Where lenticular material is present but it is not clear whether it is a crack-filling the present section I A should not be used; I C or VI should be used instead, see p. 23.

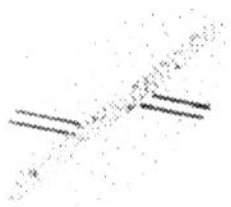

# Appendix I

## Possibility I A.1

## Geometry Shows that a Gap has Opened

| If there is: | Process was: |
|---|---|
| a. An open air-gap | Parting and dilation |
| b. Sandstone filling a crack in siltstone ("sand dike") | Parting, dilation, and infilling |
| c. Quartz, calcite, or other mineral filling a crack in rock, where | |
| (i) The filling is equant grains and not fibrous | Parting and dilation, plus nucleation and deposition from solution |
| (ii) The filling is fibrous | Parting, dilation, and deposition (little or no nucleation) |
| d. Mineral filling a crack in a mineral grain, where | |
| (i) The filling is separate crystals | Same as at (c) |
| (ii) The infilling is an overgrowth and not separate crystals | Parting, dilation, and deposition without nucleation |

# Appendix I

## Possibility I A.2
## Geometry or Other Evidence Suggests that There has been Little Introduction or Removal of Material

For shear zones and kink bands, comparison of bulk composition inside and outside the high-strain zone may serve instead of geometrical clues. For twinning, constancy of total bulk is usually not verified by observation but is a simple acceptable postulate.

| If there is: | Process was: |
|---|---|
| a. Simple joint without displacement | Thermal contraction or desiccation, or change in externally applied stress or change in fluid pressure, or a combination of these |
| b. Simple fault or slip-plane | Rigid-body displacement<br>(i) in a simple crystal: by creep of dislocations<br>(ii) in rock: by separation as in (a), or by destruction of asperities with production of rock flour, gouge, etc., or a combination of these |
| c. Shear zone as in gneiss | "Flow of rock". The many interrelated processes are discussed on p. 7–12 |
| d. Kink band as in phyllite | Same as (c) |
| e. Kink band in single crystal (mica, pyroxene, olivine etc.) | Slip of lattice planes without strict crystallographic control on amount or direction |
| f. Twinning | Slip of lattice planes with strict crystallographic control on amount and direction (by creep of dislocations) |

Note: b (ii), c, and d grade into each other. In each of the following respects, a continuous range of possibilities exists from one end-member ot the other:

| | | |
|---|---|---|
| 100% catalasis in the zone | ← gradational to → | 100% preservation of continuous geometry |
| Material originally isotropic (c) | ← gradational to → | material originally highly anisotropic (d) |
| Aspect ratio of zone is high | ← gradational to → | low |

# Appendix I

## Possibility IA.3

## Geometry Shows that Material has been Removed

Process was: Dissolution of rock material and transport in solution;
or grinding, crushing, etc., and transport of fragments as a slurry, paste or other fluidized form;
or removal of a rigid slice, with preservation of all parts (no change except separation from each other).

If there is a concentration of insoluble material or a stylolitic profile, as in IA.4, there probably has been dissolution.

# Appendix I

## Possibility I A.4

## Stylolitic Surface where Marker Shows that There has been Displacement

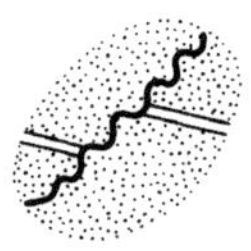

Here we are interested in the possibility that the stylolitic geometry reveals that dissolution has occurred. The alternative is that displacement occurred on a planar surface that later became undulose.

| If there is: | Process was: |
|---|---|
| A large ratio of amplitude to wavelength | Dissolution and removal of material as a solute |
| A concentration of insoluble material at the interface with the character of a residue, not a later introduction | " |
| Mismatch between rock texture and the stylolitic wavelength so that undulations cannot be attributed to recrystallization, overgrowth, etc. | " |

# Appendix I

The system offered seems capable of assembling many observables in a manner that is systematic and yet allows for varying behavior among different observers; and it provides for conclusions about processes, insofar as these can be reached from the observables. A weakness is that it is not directed to a narrow purpose: it is a system incorporating many possible conclusions from many possible observables. Any worker will normally have a narrower purpose in mind, and would benefit if the system could be separated into independent, more compact sub-systems. But such separation is not practicable – workers' purposes are too variable. Also, the broad range has some advantages. For example, a worker who seeks evidence about the development of cleavage surfaces might be inclined to disregard a rock's pseudomorphs; but a pseudomorph may indicate a hydration process, and even if the rock's cleavage is discrete and there seems no geometrical or physical link between the pseudomorph and the cleavage, evidence of hydration is evidence that water has been migrating amongst the grains – a fact that would properly be included in thoughts about how the cleavage developed. Following this example, we argue that almost any observable in a cleavable rock has some bearing on the question of how the cleavage arose. If this is so, then tracing the consequences of all relevant observations must necessarily be a cumbersome task. Though there are few observables that in themselves directly show that a particular process has operated, there are some which exclude certain processes and many that play some part in indicating what is probable or possible. Hence the largeness of the system is inherent in the problem: rock textures cannot be subjected to logical analysis except through a many-branched decision tree.

# Chapter IV

## The Plates

In the Introduction we discussed schemes for classifying tectonic and metamorphic fabrics using features observable at a variety of scales. We have emphasized that descriptors used to describe such observables carry no genetic implications. Furthermore the descriptors should not overlap in meaning with other descriptors.

In understanding the geometric similarities or relationships between different types of cleavage and schistosity it was necessary to resort to three-dimensional diagrams (Fig. 1.5, p 4). However, the relationships are often gradational in more than three directions and in an ideal universe we would like several more dimensions in which to depict the possible transitions from one type of cleavage to another.

In presenting the reader with this collection of plates the first difficulty we had to face was the linear format. The plates must be bound sequentially and this constrains the gradation between textures and cleavage types to one direction. Any given plate can only be linked directly with the plates immediately before it and after it. This is unsatisfacory because the plate in question may show a gradual transition to a texture illustrated perhaps fifty pages away. One can often see reasons for placing a given plate in several positions in the text. Unfortunately we had to choose just one place in the sequence for each plate! Paying due attention to the captions we have arranged the plates so that similarities in texture (at whatever scale is present) draw plates together. Fourteen major groups appear in this system, and the first fourteen sections of plates correspond to these groups. In Fig. 4.1, cartoon sketches summarize the essential feature which has drawn plates together in that particular group. The numbers against each sketch correspond to the number of the section in the sequence of plates. In many cases the first and last plates in a group show features transitional to plates in the preceding and subsequent groups.

A second aspect is that some plates represent features off the main theme that nevertheless show essential features parallel with the main theme. Thus there are four further sections (Sects. 15 to 18) covering blastesis (15), genesis and strain (16), polyphase deformation (17), and cleavage-layering relationships (18).

One further point is worthy of mention. Any of the fourteen morphological categories of "cleavage" shown in Fig. 4.1 may occur on a very fine scale. If it is not resolvable on outcrop scale the label *continuous on outcrop scale* could be applied to any of these textures.

Common abbreviations in the captions include:
PPL: plane polarized light
NX: crossed nicols
SEM: scanning electron microscope
TEM: transmission electron microscope

Fig. 4.1. Morphological types of planar tectonic rock fabrics. Numbers refer to sections in the sequence of plates.

CONTINUOUS

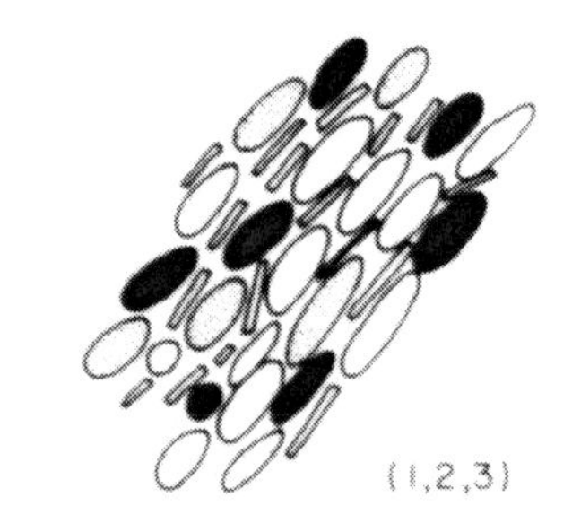


CRENULATION

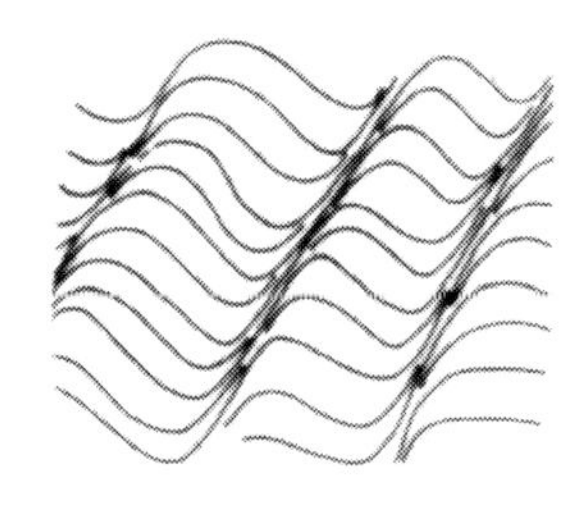


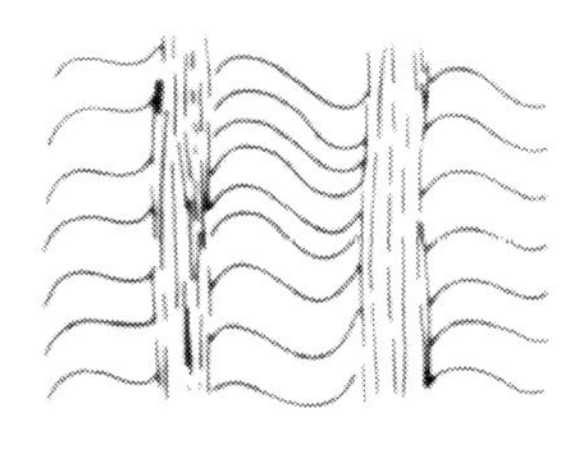


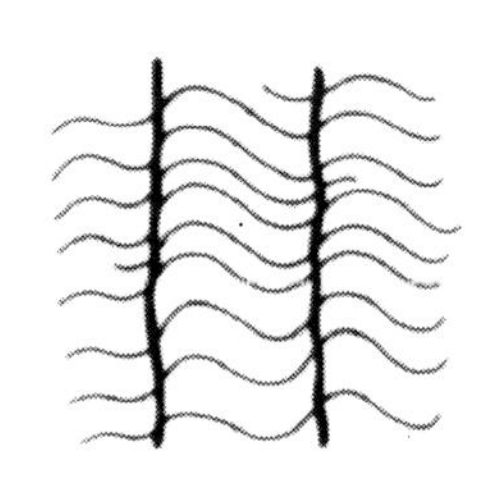


DISJUNCTIVE: CRACK-LIKE

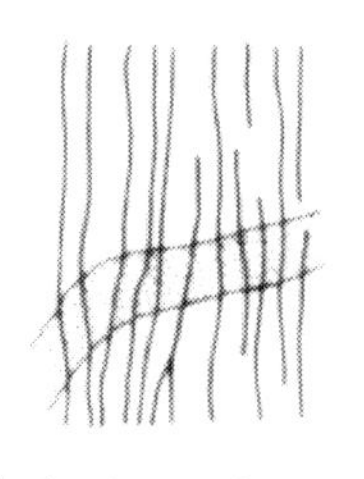


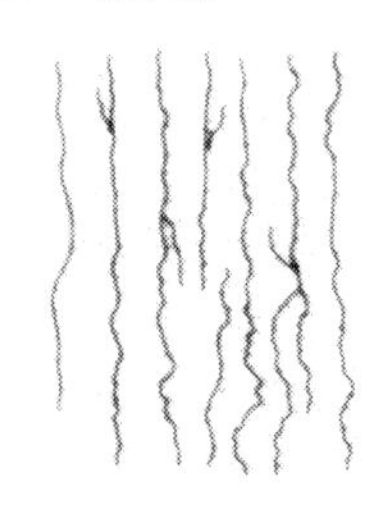


DIFFERENTIATED: ZONAL (9-13: compositional differentiation)

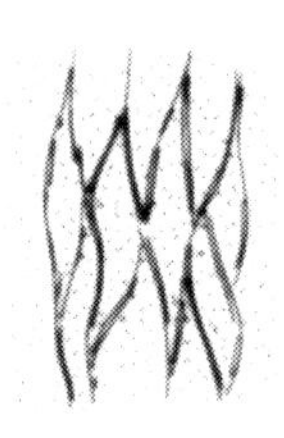

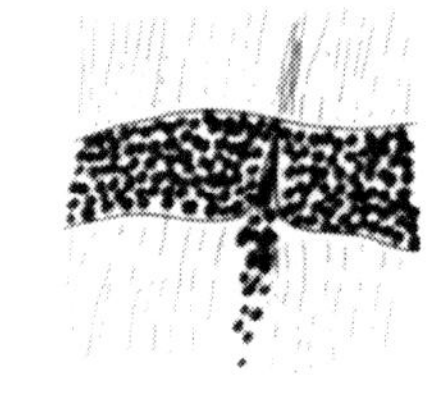


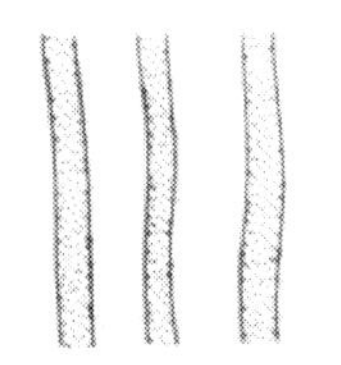

# Continuous Cleavage

## Section 1: Continuous Cleavage Formed by Coarse Aligned Grains

(Plates 1 to 7)

A small group of plates are here chosen to represent *continuous* fabrics in coarse-grained rocks. Other examples occur in other sections of the book, where they are located to give emphasis to other aspects of their texture. The present group of seven plates are selected to illustrate the extreme range of rock types which may have preferred dimensional grain orientations or preferred crystallographic orientations. These include gneisses of regional and local (shear zone) extent, marbles (experimentally deformed) and primary sedimentation fabrics.

---

**Plates**

1 Schistosity and shear zones
2 Continuous planar fabric in granulite facies gneiss
3 Calcite fabrics in experimental shear zones (I)
4 Calcite fabrics in experimental shear zones (II)
5 Calcite fabrics in experimental shear zones (III)
6 Bedding-plane foliation (I)
7 Bedding-plane foliation (II)

# 1. Schistosity and Shear Zones

G. J. Borradaile

Meta-granite of the Navado-Filabride thrust-nappe complex, Lubrín, Provincía di Almeria, Spain.

Heterogeneously deformed and metamorphosed granitic rocks are believed to have been thrust into supracrustal rocks in the Alpine Betic Cordillera. Over about 20 $km^2$ the granitic rocks show, in various stages of development, a gently dipping schistosity which contains a shape-fabric and mineral lineation. Mostly the rocks may be described as tourmaline granite-gneisses although metamorphosed igneous textures may be recognised.

This illustration shows that the transition from granite to granite-gneiss is a continuous tectonic one. It occurs in shear zones where heterogeneous ductile deformation causes the schistosity, defined by preferred dimensional and crystallographic orientation of grains, to increase in intensity toward the center of the zone. The grain size also decreases toward the center of the shear zones. The photograph shows the transition from isotropic granite in the foreground to gneiss at the top of the picture, which is the center of a shear zone parallel to the scale ruler. The other half of this shear zone is out of the field of view of the camera beyond the top of the photograph.

The strain variation on the margins of shear zones five to ten times larger than the one illustrated was investigated using an original technique (Borradaile 1976b). There the degree of fabric development was comparable to the foreground and center of the present illustration and the following range in local, total strain was found:

| | $X$ | $Y$ | $Z$ |
|---|---|---|---|
| near center of shear zone | 2.29 | 1.89 | 0.23 |
| away from shear zone | 2.27 | 0.76 | 0.58 |

Within the limits of detection, the schistosity was found to be parallel to the $XY$ plane of the *local* strain ellipsoid.

Scale: 15 cm

1

## 2. Continuous Planar Fabric in Granulite Facies Gneiss

M. M. KEHLENBECK

Lac Rouvray, Quebec, Canada.

This outcrop of a granulite facies, orthopyroxene-bearing quartzofeldspathic gneiss is from the basement gneiss complex associated with the Lac Rouvray anorthosite mass (KEHLENBECK 1972). The Lac Rouvray anorthosite is a part of the Lac St. Jean anorthosite massif which is located centrally in the Grenville Province of Quebec.

The homogeneous anisotropic fabric, subhorizontal in the photograph, is defined by preferred dimensional orientation of grains. A later, weak crenulation fabric with axial surfaces nearly vertical in the photograph affects this continuous fabric.

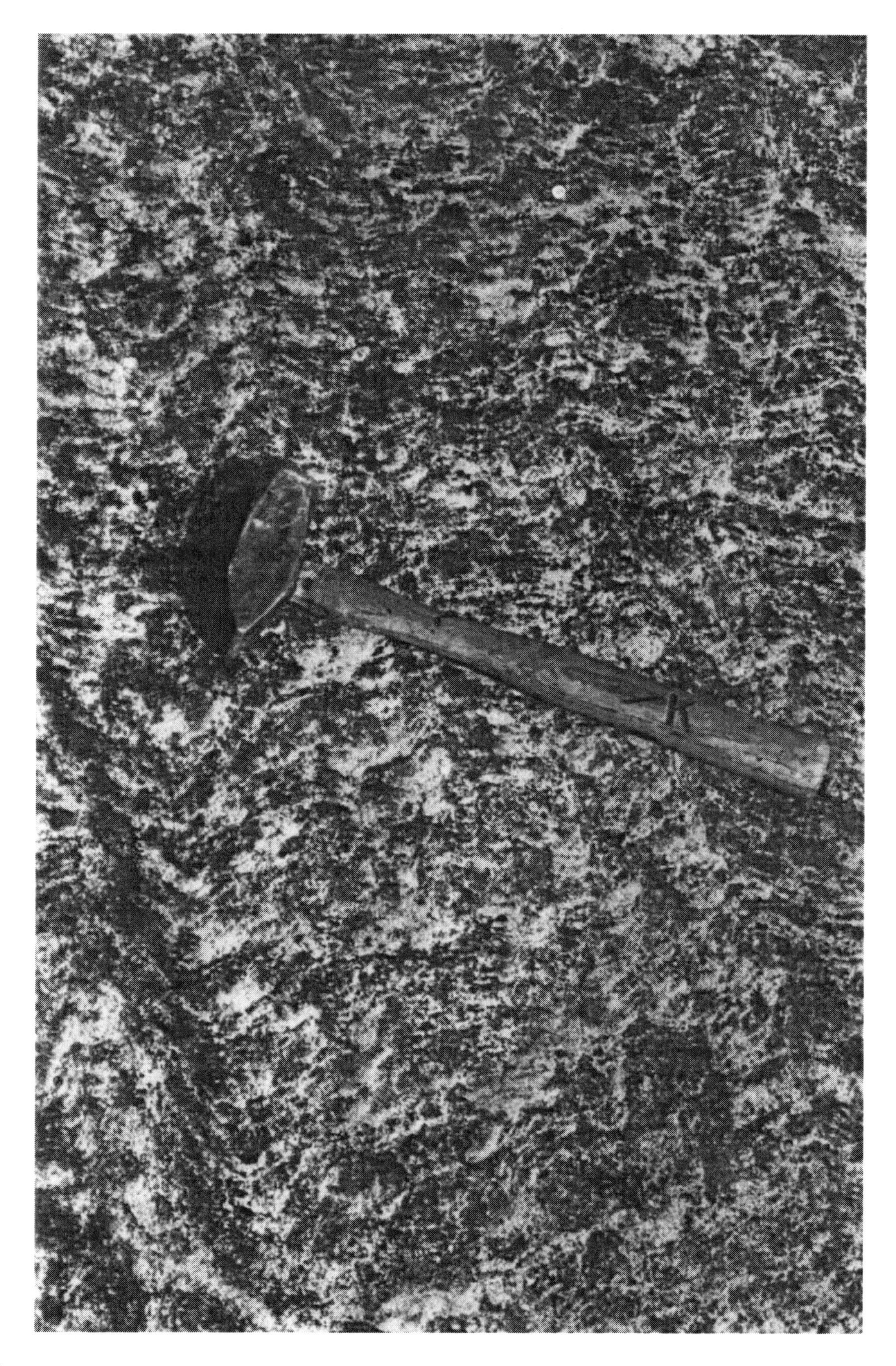

2

# 3. Evolution of Calcite Fabrics in Experimental Shear Zones (I)

M. FRIEDMAN and N. G. HIGGS

Cylindrical specimens of Tennessee sandstone with dry crushed calcite along 35°-precut surfaces were deformed at 200 MPa confining pressure, from 25° to 910 °C, and a shear strain rate of $10^{-2}$ $s^{-1}$ (FRIEDMAN and HIGGS 1981). Inelastic deformation is confined to the calcite layer. Shear displacements (dextral and indicated by arrows in the following photographs) were 1.5 to 3.0 mm with shear strain ($\gamma$) $\leqq$5.7. The compacted gouge (about 0.5 mm thick) deformed in simple shear with noncoaxial incremental stress, strain and finite strain axes. Sliding mode changed from stick-slip to stable sliding between 250° and 400 °C. Below 250 °C the calcite deformed primarily by cataclasis with Riedel shears and conspicuous microfractures and twin lamellae in the porphyroclasts.

Between 250° and 650 °C slip on *r* $\{10\bar{1}1\}$ and twin gliding on *e* $\{01\bar{1}2\}$ predominated and produced highly elongated porphyroclasts with strong preferred dimensional and crystallographic orientations. Average axial ratios change from about 2 at $\leqq$250 °C to as much as 16 at 650 °C ($\gamma$=4.96); average apparent long axes increase two to threefold, but average nominal grain-areas remain constant. The grain shape fabrics fit predicted fabrics based upon homogeneous simple shear. C-axes of the porphyroclasts are oriented within 20° of the axis of finite maximum shortening, and track the axis of shortening with increasing shear strain.

At 550°–600 °C, recrystallization occurs, after shearing by slip. Mosaics of very fine-grained neoblasts are produced with axial ratios <2.0. At 805° to 910° two populations of neoblasts occur in each of four specimens. The first consists of larger, conspicuously twinned neoblasts with a ghost structure that tracks the finite strains and a strong crystallographic fabric that tracks neither the incremental stress nor the finite strain. These neoblasts are syntectonic and appear to be continually re-formed during shearing. The second group of neoblasts are smaller, strain-free and more equant and exhibit a very scattered crystallographic orientation. They are probably due to late annealing.

---

**A. Cataclasis.** Cataclasis in shear zone. High- and low-angle Riedel shears are indicated ($R_2$ and $R_1$). Specimen 55, 25 °C. Scale: 0.1 mm PPL

**B. Riedel shears.** Low-angle Riedel shears (sense of shear given by arrows) and high-angle Riedel shears. Note calcite porphyroclasts aligned at about 25° to margins of shear zone. Specimen 100, 250 °C. Scale: 0.1 mm PPL

**C. Porphyroclasts.** Low-angle Riedel shear offsets elongated porphyroclast. Specimen 104, 400 °C, shear strain $\gamma$= 1.65. Scale: 0.1 mm PPL

**D. Twinned porphyroclasts.** Elongate porphyroclasts at about 20° to shear zone boundary. Note planar *e*-twin lamellae in largest porphyroclast are subparallel to shear zone boundary. Specimen 64, 525 °C, shear strain $\gamma$=2.46. Scale: 0.1 mm PPL

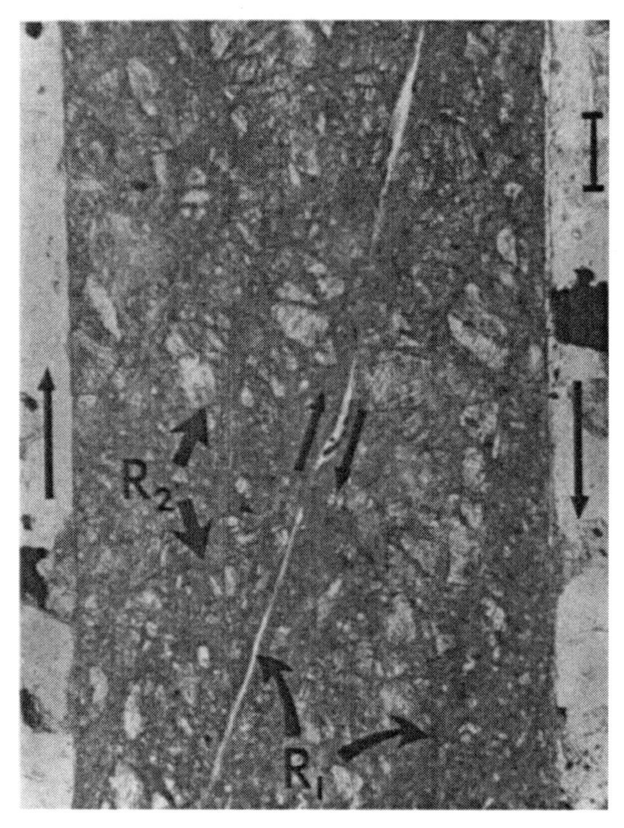


3A

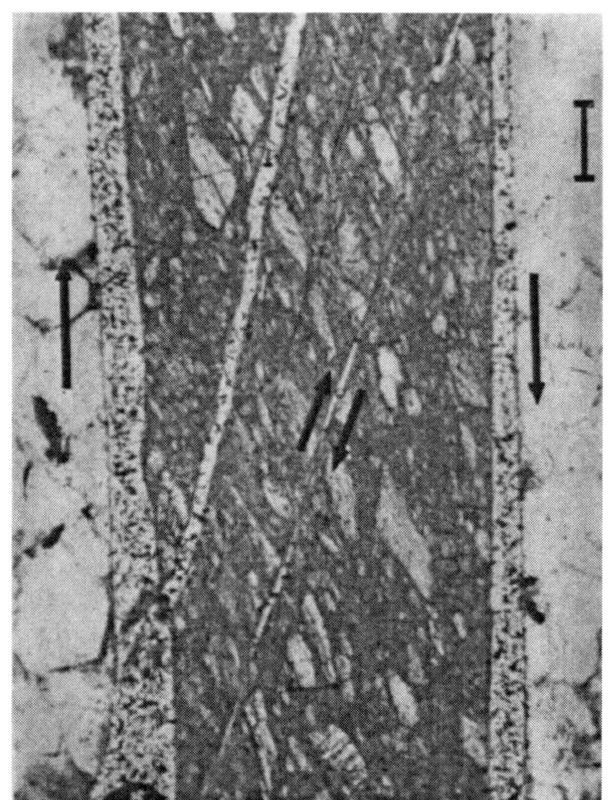

3B

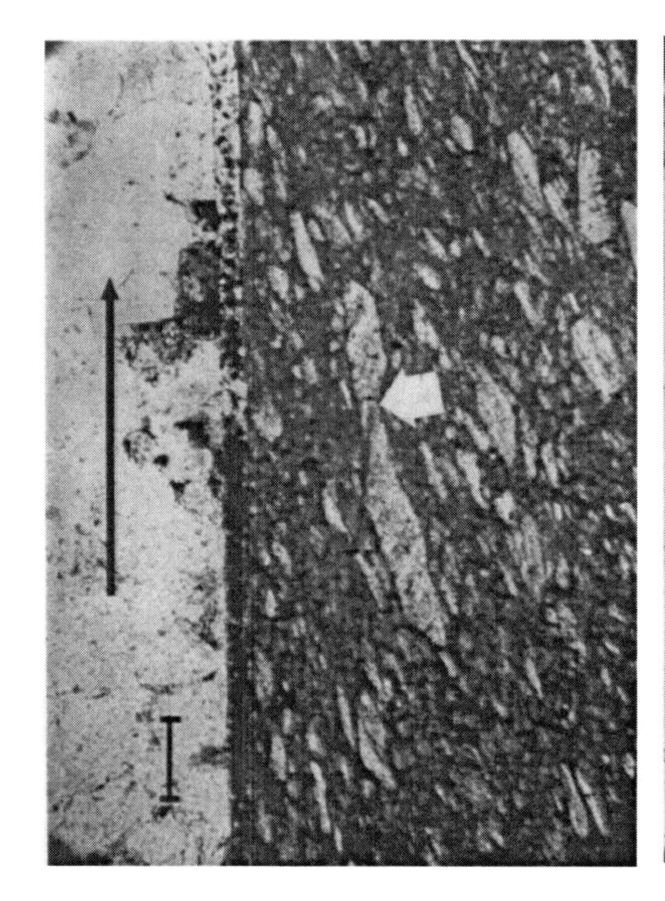

3C

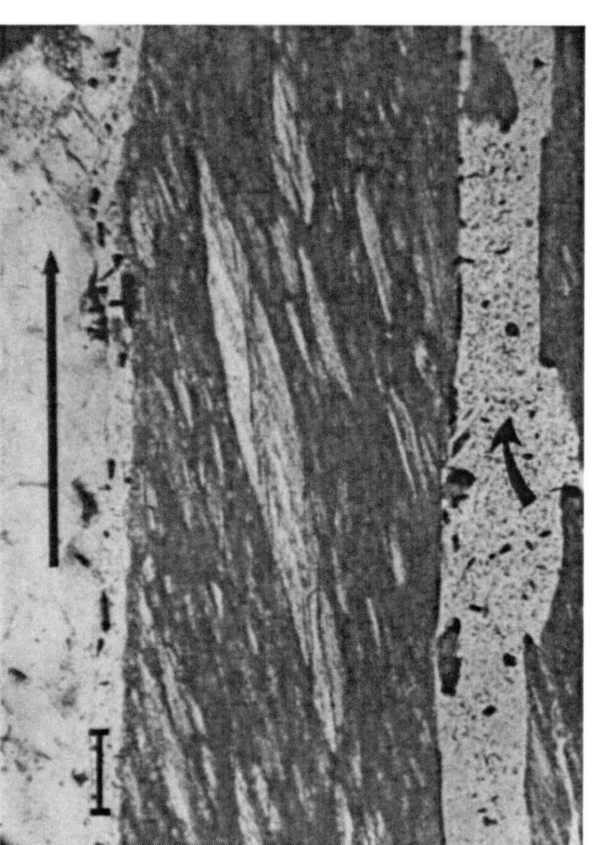

3D

# 4. Evolution of Calcite Fabrics in Experimental Shear Zones (II)

M. FRIEDMAN and N. G. HIGGS

(Dextral shear zones oriented horizontally in these photographs)

**A. Twinned porphyroclasts.** Highly elongated, completely twinned porphyroclasts with irregular *e*-twin lamellae (*curved arrow*) and late planar twin lamellae (*straight arrow*). Specimen 207, 650 °C, shear strain $\gamma = 4.96$. Scale: 0.1 mm PPL

**B. Incipient recrystallization.** Indicated by *white arrow*. Specimen 107, 550 °C, shear strain $\gamma = 2.98$. Scale: 0.1 mm PPL

**C. Pervasive recrystallization.** Fine-grained neoblasts occur with residual elongate porphyroclasts and high-angle Riedel shears. Specimen 108, 600 °C, shear strain $\gamma = 3.63$. Scale: 0.1 mm PPL

4 A

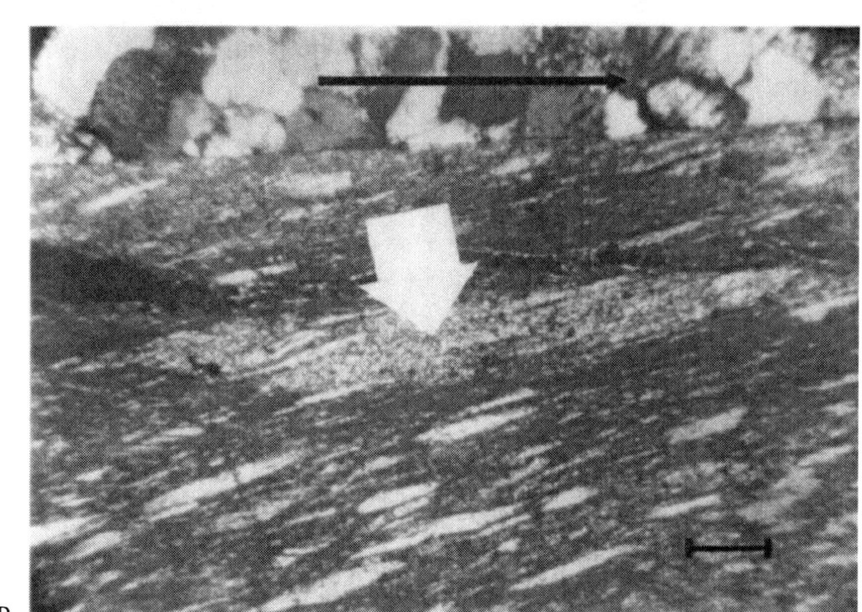
4 B

4 C

## 5. Evolution of Calcite Fabrics in Experimental Shear Zones (III)

M. FRIEDMAN and N. G. HIGGS

(Dextral shear zones oriented horizontally in these photographs)

**A. Ghost structure.** Mosaic of large neoblasts shows ghost structure (*double-ended arrow*) and late, rational twin lamellae (*curved arrow*). Specimen 65, 850 °C, shear strain $\gamma = 5.69$. Scale: 0.1 mm PPL

**B. Ghost structure detail.** Note alignment of impurities trending left to right. Specimen 65, 850 °C, shear strain $\gamma = 5.69$. Scale: 0.025 mm PPL

**C. Indentation at shear zone boundary.** Fine-grained calcite forms indentation of gouge into a depression in the Tennessee Sandstone (*TS*) which forms the wall to the shear zone. Specimen 65, > 600 °C. Scale: 0.035 mm PPL

5 A

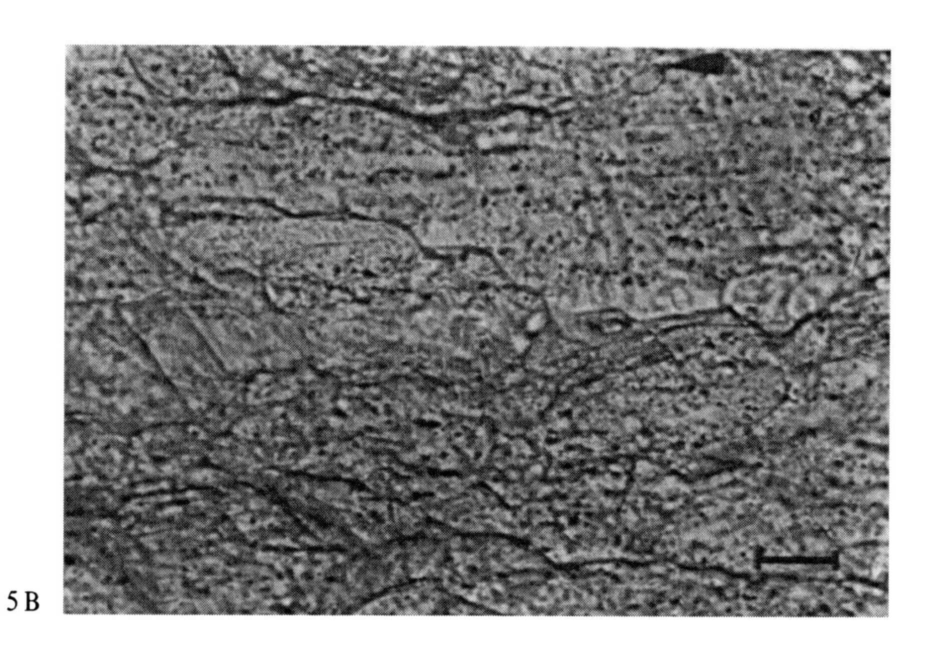
5 B

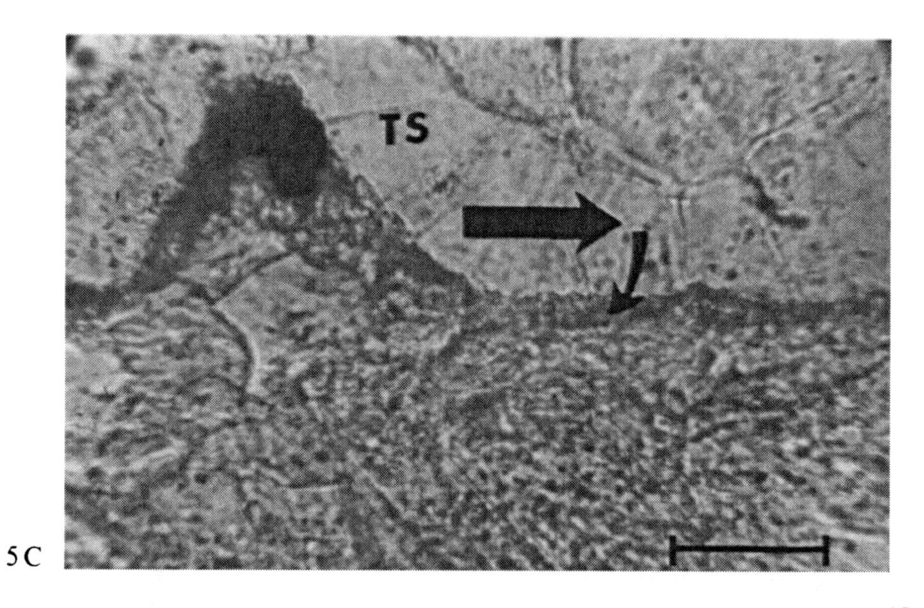

5 C

## 6. Bedding-plane Foliation (I)

R. F. C. Morritt, C. McA. Powell and R. H. Vernon

Siltstone from Late Ordovician Triangle Group, Guineacor Creek, northeast of Taralga, N.S.W., Australia (Scheibner 1973).

The bedding-plane foliation is defined by very elongate biotite flakes statistically parallel to the bedding. The biotite flakes are considered to be detrital because of the following features:

1. Very high length : width ratios, commonly exceeding 20 : 1.
2. Ragged diffuse ends commonly partially crystallized to fine-grained chlorite or "sericite" continuous with the matrix.
3. The length of grains in section is several times larger than the diameter of associated detrital quartz grains, and two orders of magnitude larger than any other associated phyllosilicates for which an authigenic or metamorphic origin can be inferred.
4. The maximum size of biotite flakes in graded beds decreases in sympathy with the decrease in diameter of the detrital quartz grains.
5. The degree of recrystallization of the rock is so low that both rounded and angular shapes of detrital quartz and rare plagioclase grains are clear.

These Ordovician rocks are part of the widespread quartzose flysch succession deposited in the Lachlan Fold Belt (Cas et al. 1980). In the Guineacor Creek area, they have been folded at least twice (Morritt 1979). The first folding in the mid-Devonian produced open, upright folds trending north-northeast, but no axial-surface foliation. The detrital biotite flakes were probably compacted to a bedding-parallel fissility at this time.

---

**A.** Detrital biotite grains statistically parallel to bedding, which is horizontal. Oblique grains have been deformed by shortening perpendicular to the bedding. PPL

**B.** Enlargement of the central part of **A** showing kink bands and chevron folds in biotite grains with (001) oblique to bedding. The chevron-folded grain in the center of the photograph has been shortened by 35% to 40% normal to bedding. PPL

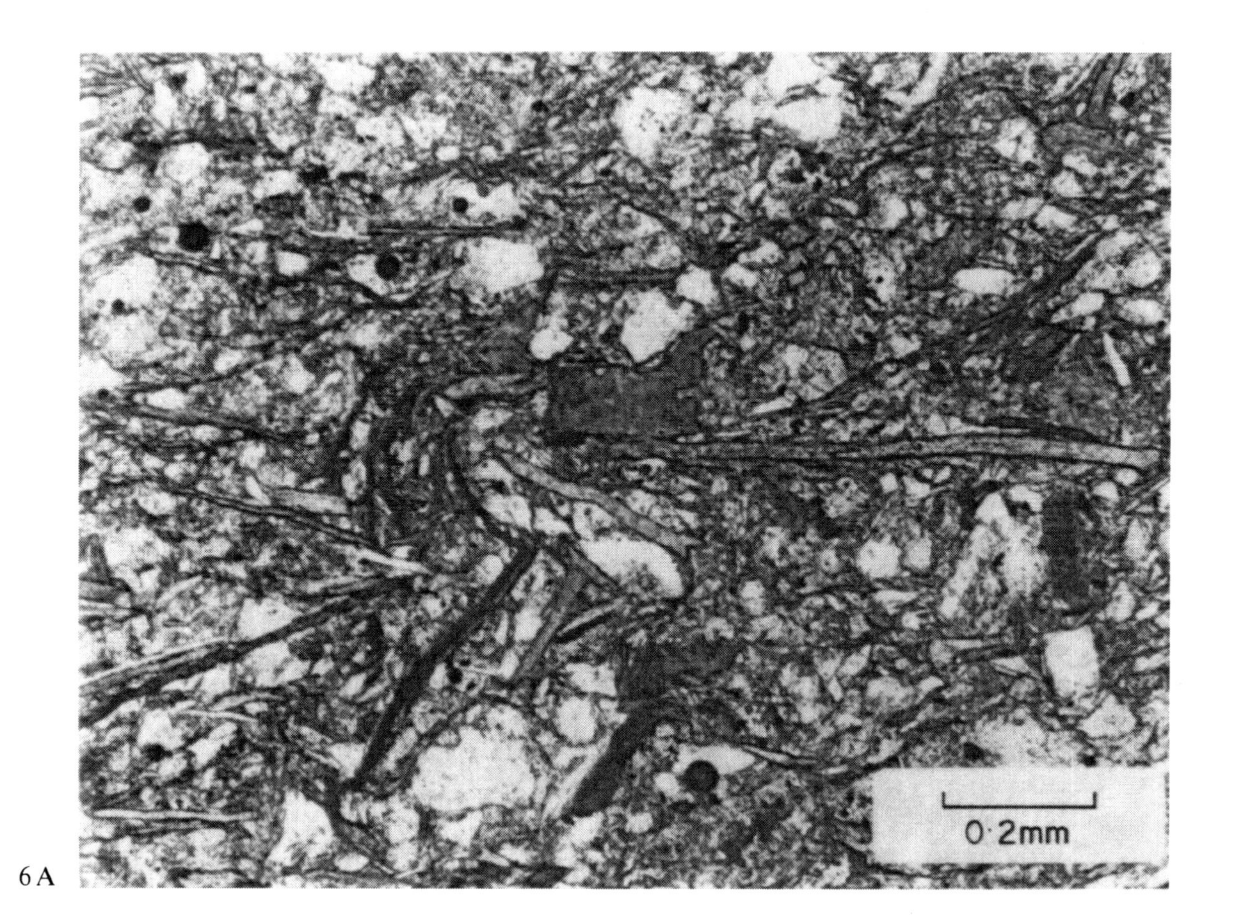

6 A

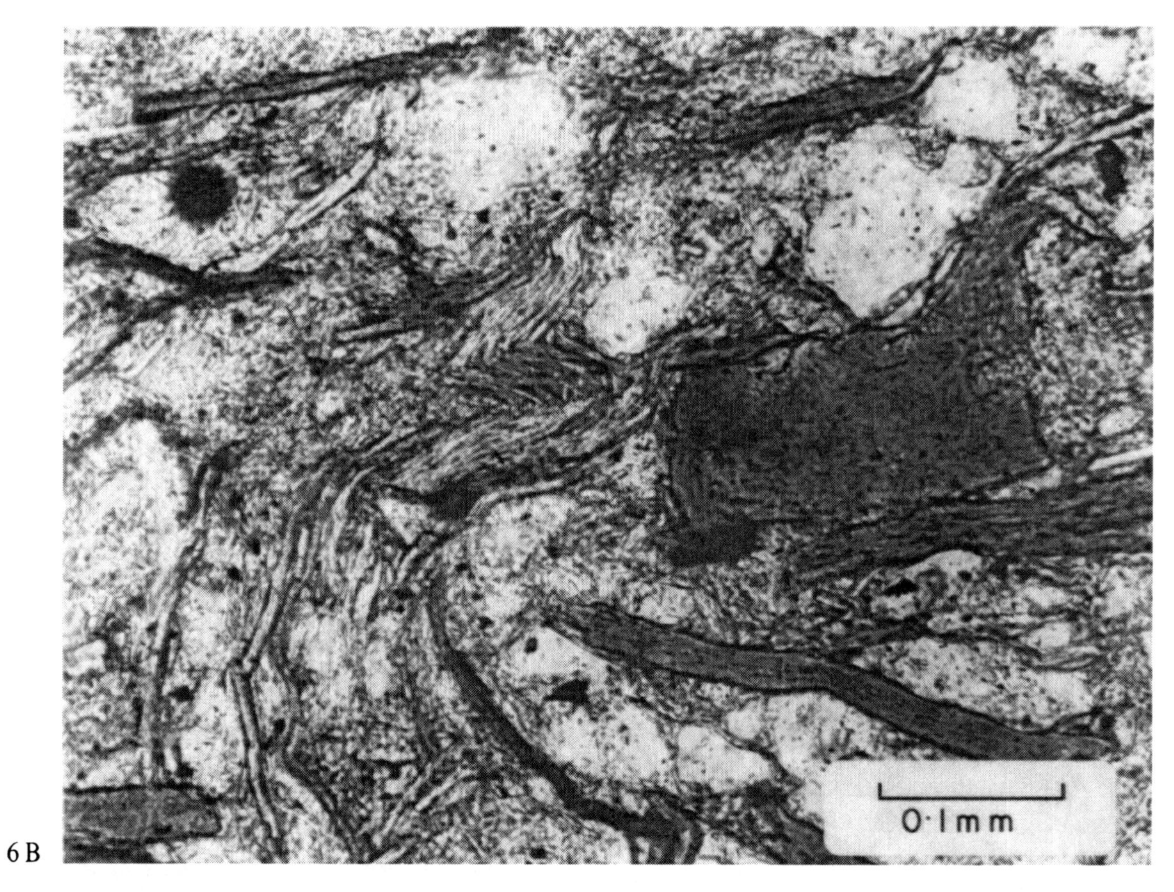

6 B

## 7. Bedding-plane Foliation (II)

R. F. C. Morritt, C. McA. Powell and R. H. Vernon

Detrital biotite in quartzose flysch of the Late Ordovician Triangle Group from Guineacor Creek, northeast of Taralga, N.S.W., Australia (Scheibner 1973) defines a bedding-plane foliation. Biotite grains parallel, or at a low angle to bedding are relatively planar, with only small kinks deforming (001) cleavages (**A**), whereas biotite grains at a high angle to bedding are more highly bent and kinked (**B**), indicating shortening perpendicular to bedding. Whether this shortening is simply sedimentary compaction, or has been enhanced by the first phase of open folding (mid-Devonian), has not been determined, but the bedding-plane foliation is sufficiently well developed to cause the first axial-plane cleavage (Early Carboniferous) to be a crenulation cleavage (cf. Williams 1972).

---

**A.** Elongate detrital biotite grain (length : width = 28 : 1) parallel to bedding. Small kinks at a high angle to the bedding are the only structures deforming the otherwise planar (001) cleavages. PPL

**B.** Deformed detrital biotite grain at a high angle to bedding, which is horizontal. The ends of the biotite grain have been bent and kinked. PPL

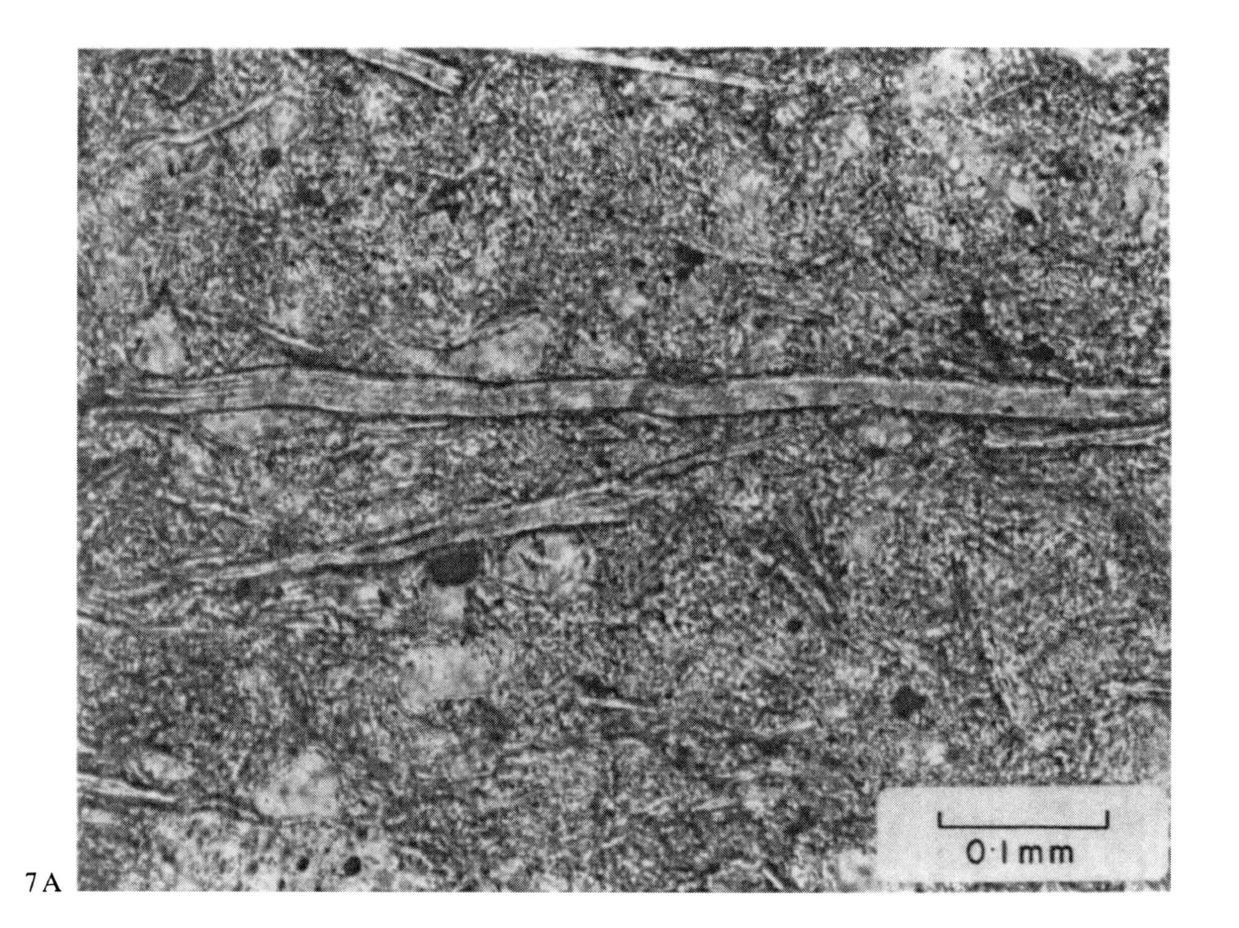

7 A

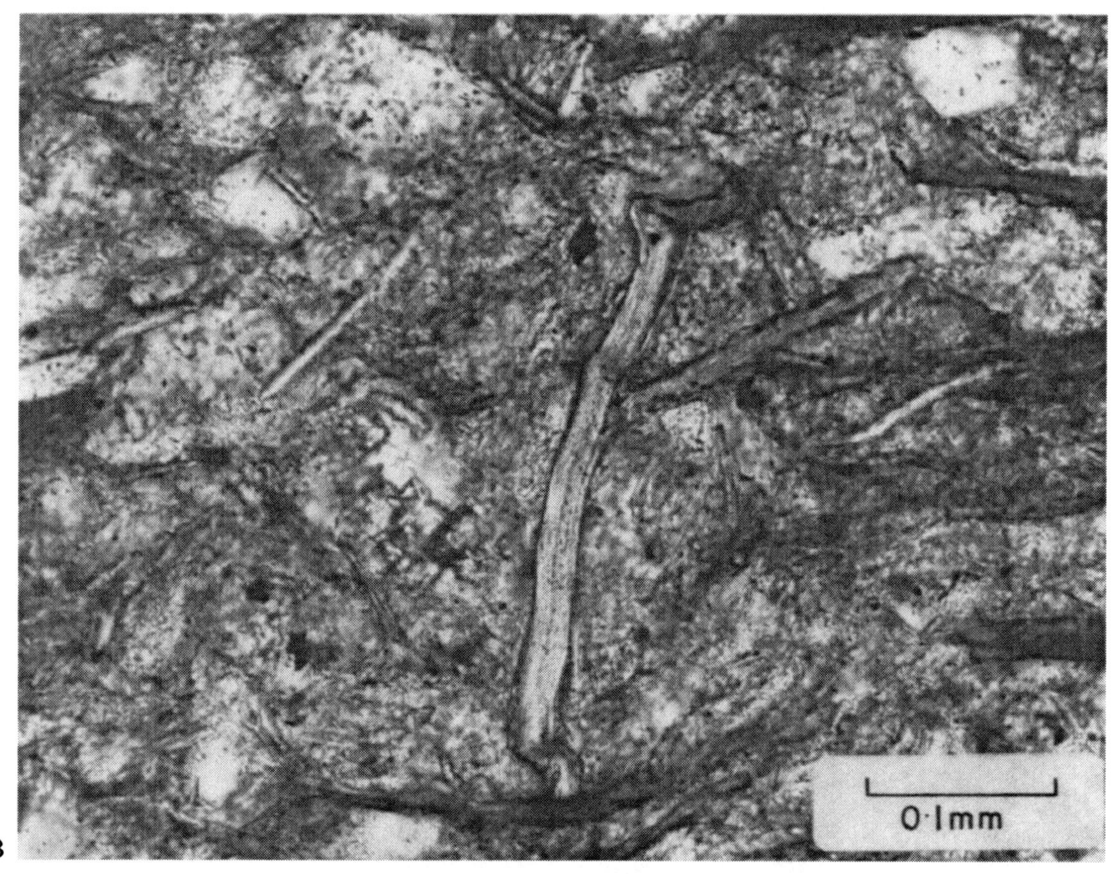

7 B

# Continuous Cleavage

## Section 2: Fine Continuous Cleavage in Rocks Composed Largely of Phyllosilicates

(Plates 8 to 17)

Here ten plates have been brought together to illustrate continuous cleavage in fine-grained materials, composed chiefly of phyllosilicates. Within this group fall rocks which would traditionally have been described as schists, slates or rocks with slaty cleavage. The rocks and experimental specimen in this section developed fabrics which are typically homogeneous at the scale of observation, although porphyroblasts do incidentally occur in one or two cases. The anisotropic fabrics are defined by preferred crystallographic orientation with or without a preferred dimensional orientation. Many other examples of continuous cleavage are to be found, at least locally, in rocks illustrated elsewhere in the book (e.g., Plates 132, 133, 135, 198). There the plates have been selected to emphasize other features shown by the rock.

---

**Plates**

## 8. Penetrative Mica Fabric in a Phyllite

A. J. WATKINSON

This example is from a phyllitic unit within the Maitlen phyllite, a Lower Cambrian formation between the basal quartzite and Middle Cambrian limestone in the Kootenay Arc, NE Washington, U.S.A. The phyllitic cleavage is an $S_1$ cleavage, axial planar to major, overturned to recumbent folds.

The cleavage is constituted as a parallel array of muscovite and chlorite grains with minor biotite. Interstitial lenticular quartz grains have a preferred shape orientation with the long axes parallel to the schistosity. There are also interleaved straight-edged, spindle-like laths of quartz/plagioclase crystal aggregates.

This cleavage is close to an ideal, continuous, non-domainal cleavage and contrasts markedly with the discrete, spaced cleavage in the more quartzitic units of the Maitlen phyllite. NX

8

0·5mm.

# 9. Continuous Cleavage in Fine-grained Rocks

B. M. Bayly

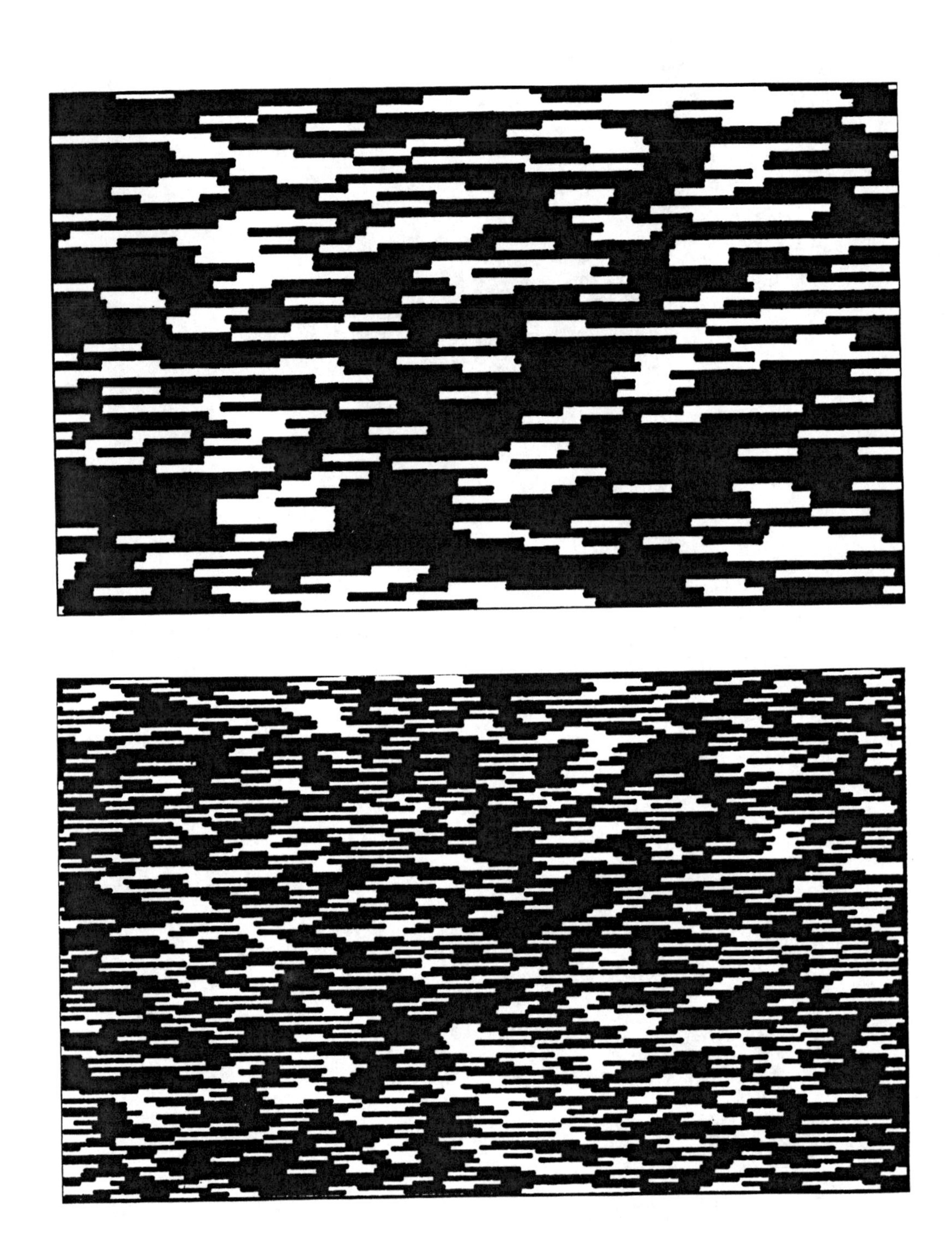

In each example one may ask: is the cleavage continuous or domainal? and one has to consider: if the distribution of grains were indeed random, how much clustering would appear? The diagrams above show textures synthesized by computer, in which uniform elongated grains were colored dark or pale by a random process biassed to give ⅔ dark and ⅓ pale. One concludes that even these rocks, chosen for their homogeneity, are more strongly domainal than random processes would produce.

Details of the rocks are given elsewhere **A**: Plate 12, **B**: Plate 111, **C**: Plate 121. NX

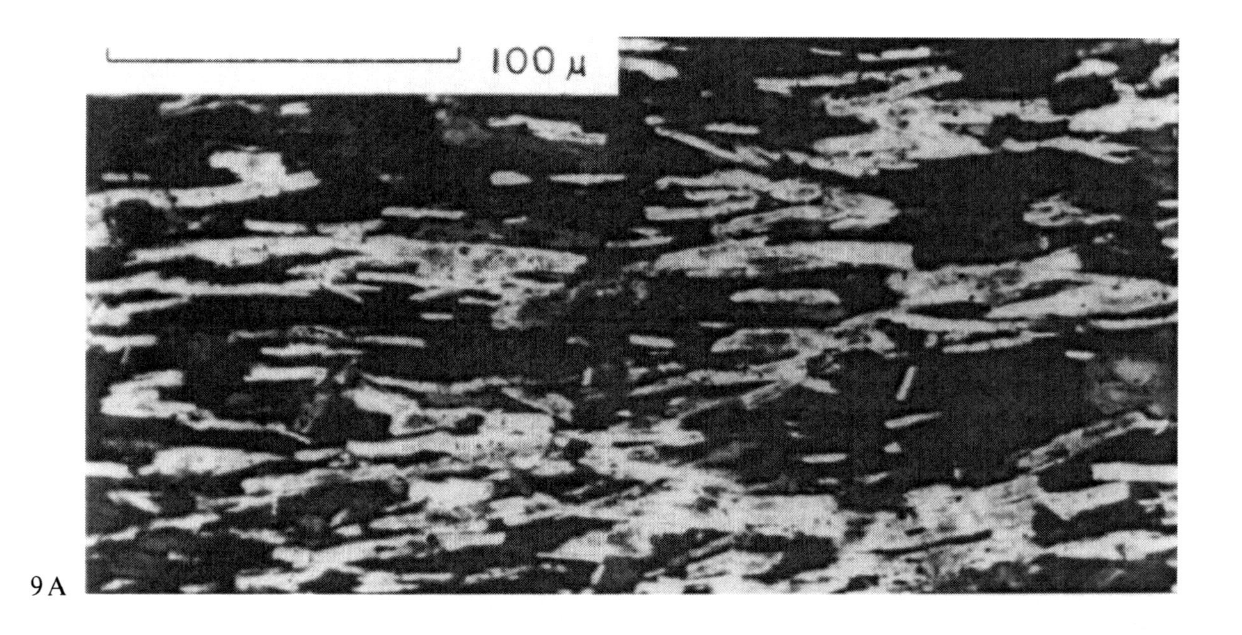

9A

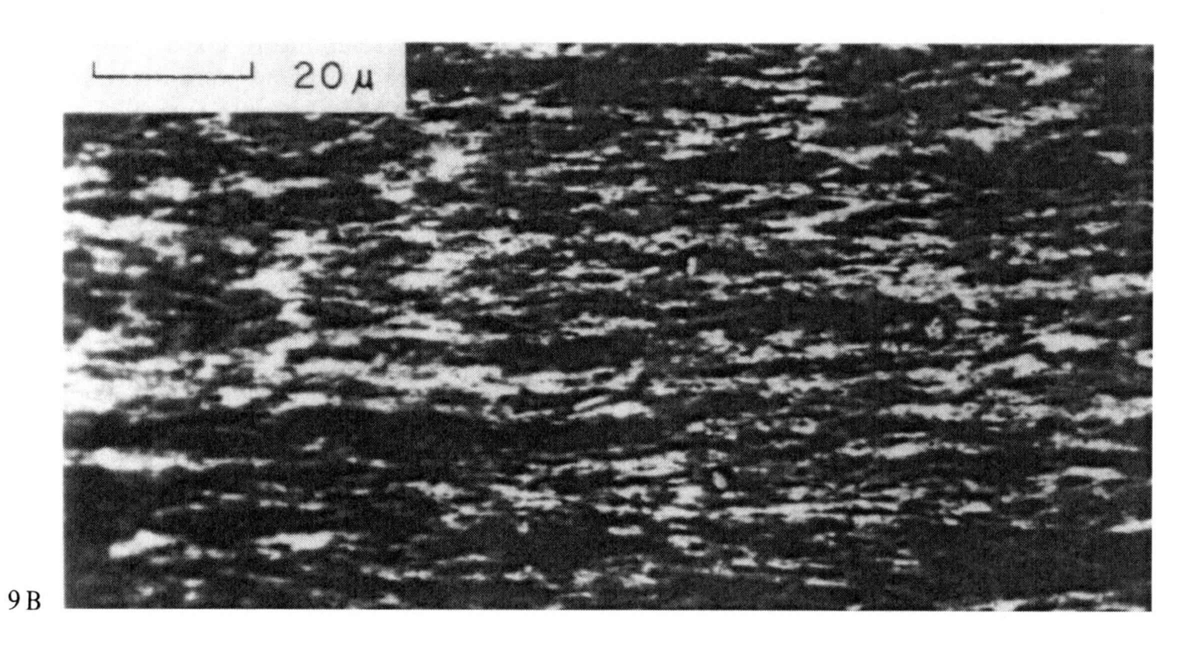

9B

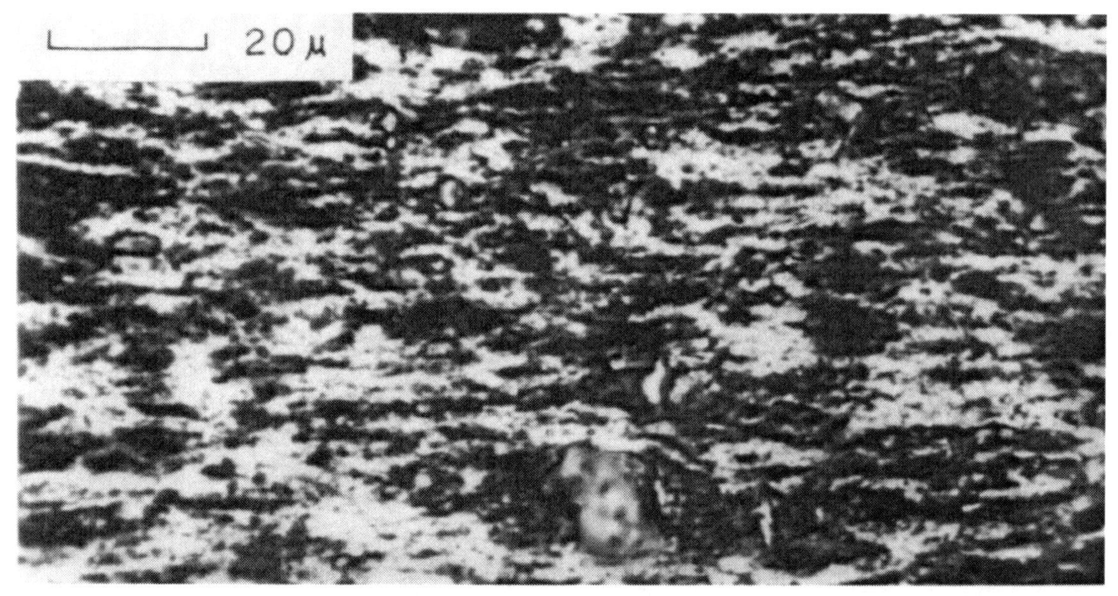

9C

## 10. Continuous Cleavage in Schists (I)

B. M. Bayly

This plate and the following one show different degrees of preferred orientation in what are thought to be typical schists.

---

**A.** The rock is somewhat fissile and breaks into fragments that are not equidimensional; it is definitely foliated but one might hesitate to say that it has cleavage. Many mica flakes are more ragged than euhedral, and preferred orientation is weak.

**B.** Thin euhedral muscovite and more robust biotite flakes have good preferred orientation, but the rock's cleavage is not much better than in (**A**) because the phyllosilicates are dispersed (not in laminae) and are low in abundance (20%–30% of the rock).
Is the cleavage continuous (homogeneous)? The distribution of flakes is *not* homogeneous: muscovite clusters with biotite in a systematic way. However, the biotite flakes are dispersed as well as can be expected at their grain size, so that to call the cleavage continuous would be correct.
The quartz is of smaller grain size than the biotite and this necessarily means that quartz grains are somewhat clustered. Even after admitting that the cleavage is *now* continuous, one might ask if the present texture results from grain growth, from a precursor where fine grains were organized in domains; compare with Plate 11 A that follows. PPL

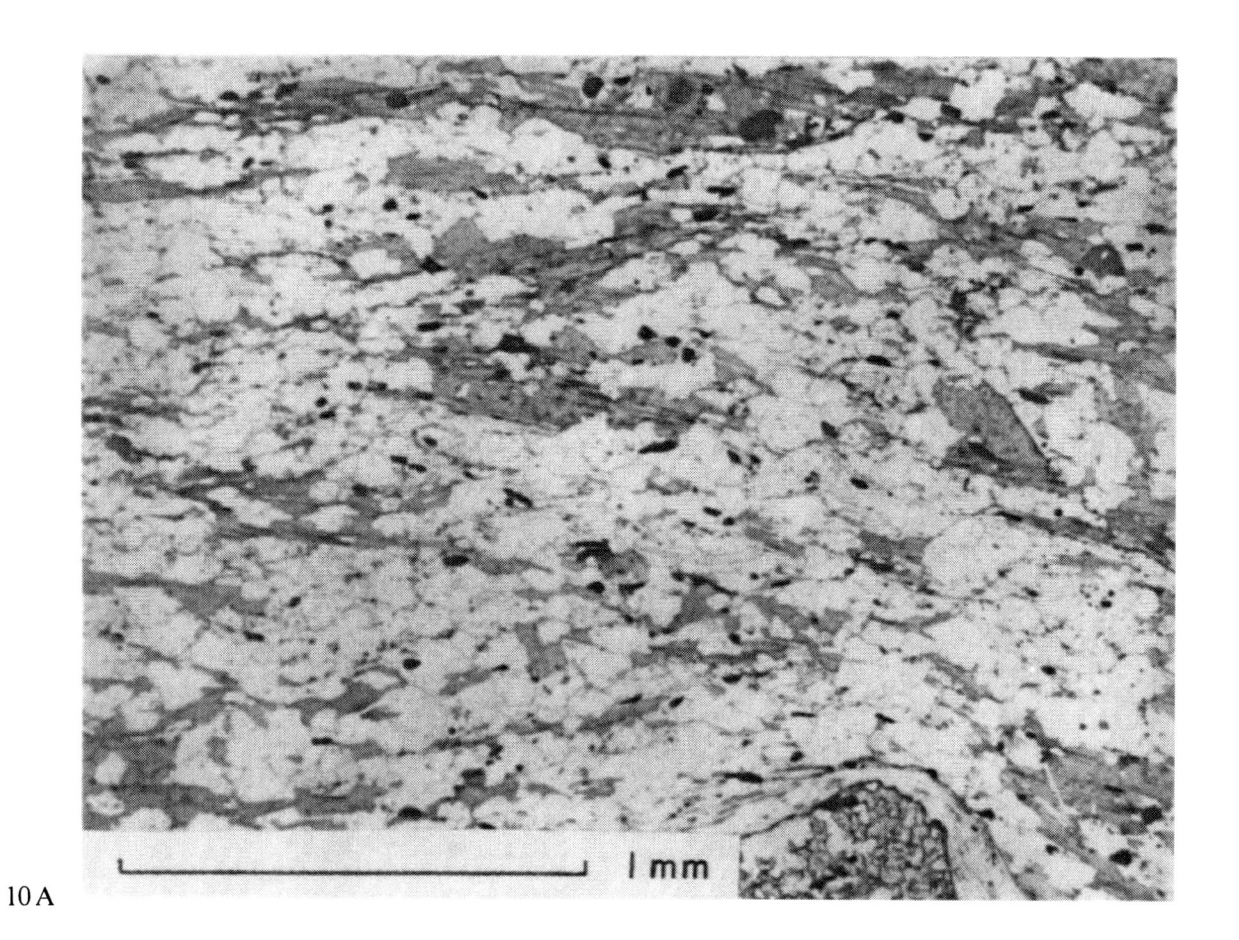


10 A

10 B

# 11. Continuous Cleavage in Schists (II)

B. M. BAYLY

A. The hand-specimen is a homogeneous rock and is tough rather than easily cleaved. As in Plate 10 B, the mica is not organized into laminae well enough to give good cleavage to the rock; the cleavage could be called continuous but poor. However, the mica here shows a slight tendency to be clustered, and in Plate 122 (from the same hand-specimen) the clustering is locally strong. Therefore we should admit the possibility that we have a coarse-grained product from a domainal precursor. This rock could have formed by grain growth, for example from the rock shown in Plate 43. NX

**B, C.** Here the very high degree of preferred orientation is related to the high mica content. Possibilities are: (i) a more fine-grained rock achieved quite good preferred orientation and then grain growth produced the present perfect fit (ii) the laths achieved something like their present size in a less well-organized geometry, with an interstitial mineral present such as quartz; then dissolution of the quartz allowed the flakes to collapse into parallelism, with adjustment of their mutual boundaries as needed.

B: PPL
C: NX

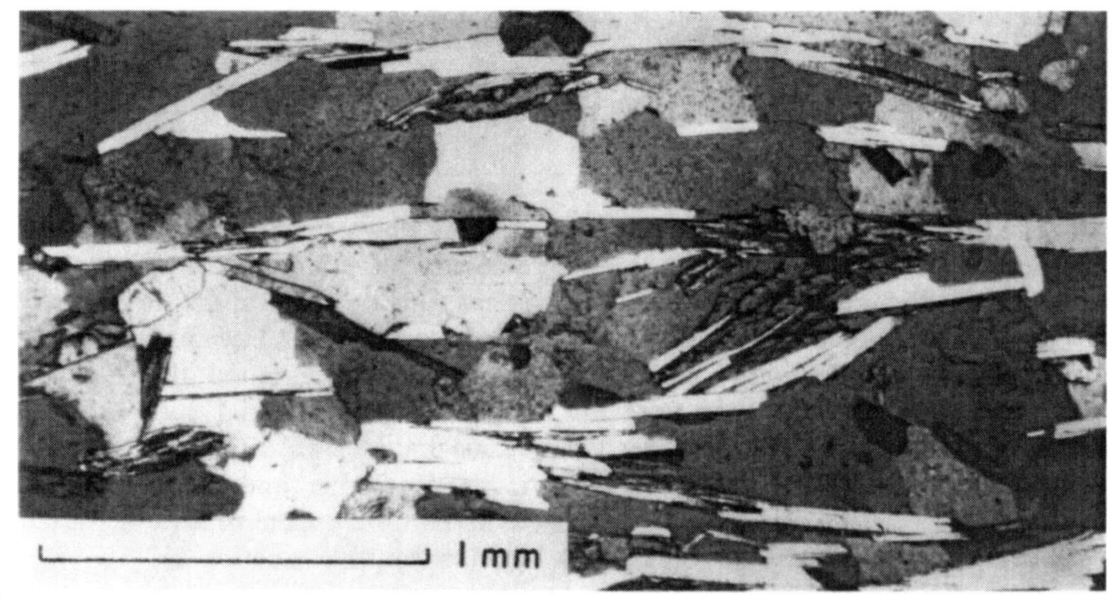

11A

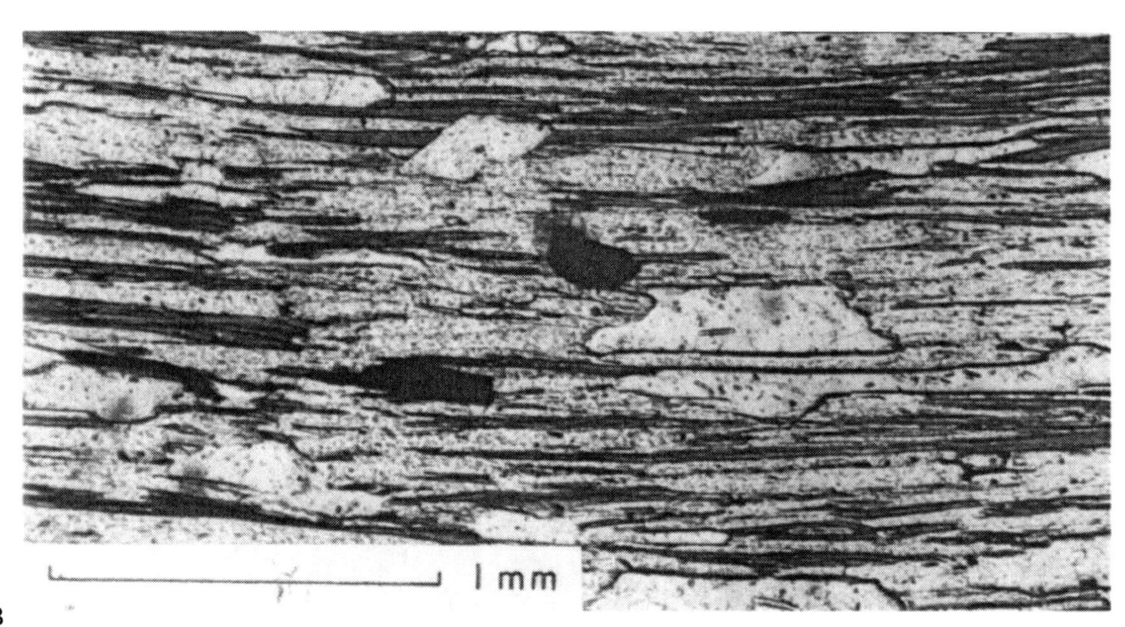

11B

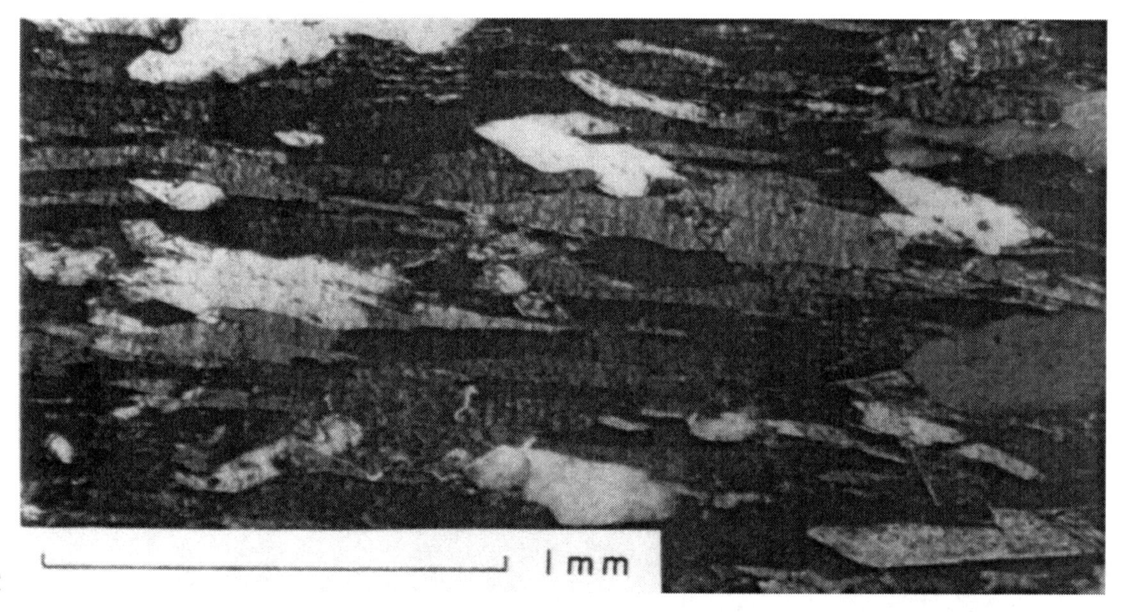

11C

## 12. Continuous Cleavage in Schists (III)

B. M. Bayly

The rock is a layered quartz-muscovite phyllite with megacrysts of biotite and ilmenite. Many biotite flakes are discordant with the foliation and layering, whereas the ilmenite flakes **x** are very well aligned parallel with them. The schistosity and weak cleavage arise from the orientation of the muscovite that pervades the groundmass, shown in greater detail in Plate 9A.

The rock contains dark seams such as the one running down the center of **A** and also shows evidence of migration of quartz in solution (clear rims on some of the megacrysts), but in this instance, the dark seams are believed not to be residues concentrated by preferential dissolution of pale material such as quartz. The seams that are dark in section are rusty-colored in handspecimen, and are probably due more to an oxidizing effect of invading groundwater than to simple dissolution. PPL

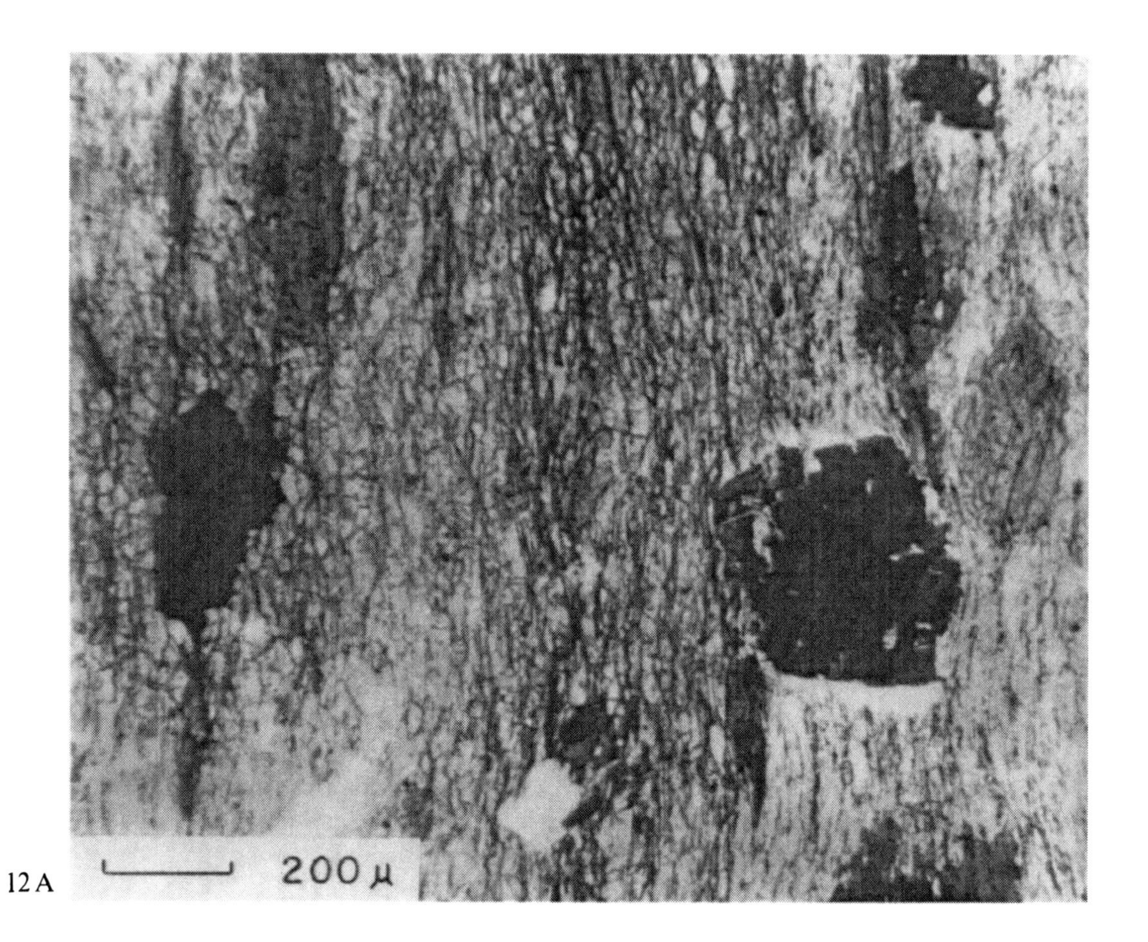


12 A

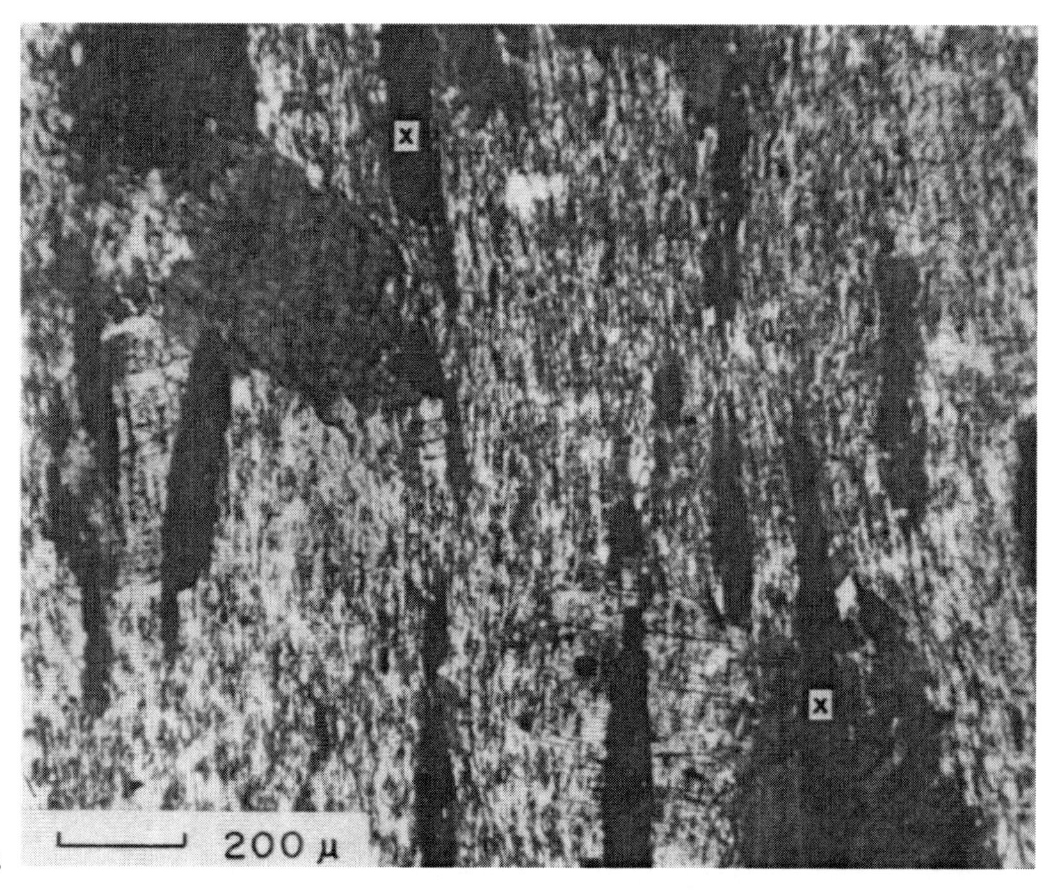


12 B

## 13. Slaty Cleavage, Fleurieu Peninsula, South Australia

N. S. Mancktelow

Photographs from a single thin section (cut perpendicular to cleavage, parallel to lineation) show the biotite fabric typical of slates and phyllites from the low-metamorphic-grade region of Fleurieu Peninsula. This example comes from the hinge of a small fold (wavelength ~ 30 m) near Campbell Creek.

The biotites form two distinct populations (see Mancktelow 1979):

1. "Old Grains" which generally have an irregular shape, are relatively thick perpendicular to (001), do not show strong unimodal crystallographic alignment, and display optical evidence of considerable intracrystalline deformation [undulose extinction, wavy (001)].
2. "New Grains" which have a more regular, characteristic shape [elongate parallel to (001), with long grain boundaries parallel to (001) and rounded ends truncating (001)], show a very strong unimodal crystallographic alignment, and exhibit no optical evidence of significant intracrystalline strain.

---

**A. Overall biotite fabric.** The marked difference in shape and alignment of the two biotite populations is readily apparent. Note in particular the thin elongate shape of one of these populations (the new grains). There is no evidence of any bending of these elongate grains despite the heterogeneity of the fabric and the range of physical properties of the mineral components comprising the rock (quartz, plagioclase, biotite, muscovite). Scale: 0.1 mm PPL

**B. Old and new biotite grains.** Old grains (**O**) with irregular grain boundaries and wavy (001) indicating strong intracrystalline deformation. New grains (**N**) showing their characteristic shape with rounded ends truncating (001), and absence of optical evidence of intracrystalline deformation. Scale: 0.02 mm PPL

**C. Quartz-fiber pressure shadow.** Old biotite grains with (001) at a high angle to the cleavage plane commonly develop quartz-fiber pressure shadows. These biotites may have acted as "hard" grains during the deformation as their orientation effectively precluded mechanical deformation by slip on (001). Scale: 0.05 mm PPL

B and C are reproduced with permission of Elsevier Scientific Publishing Company

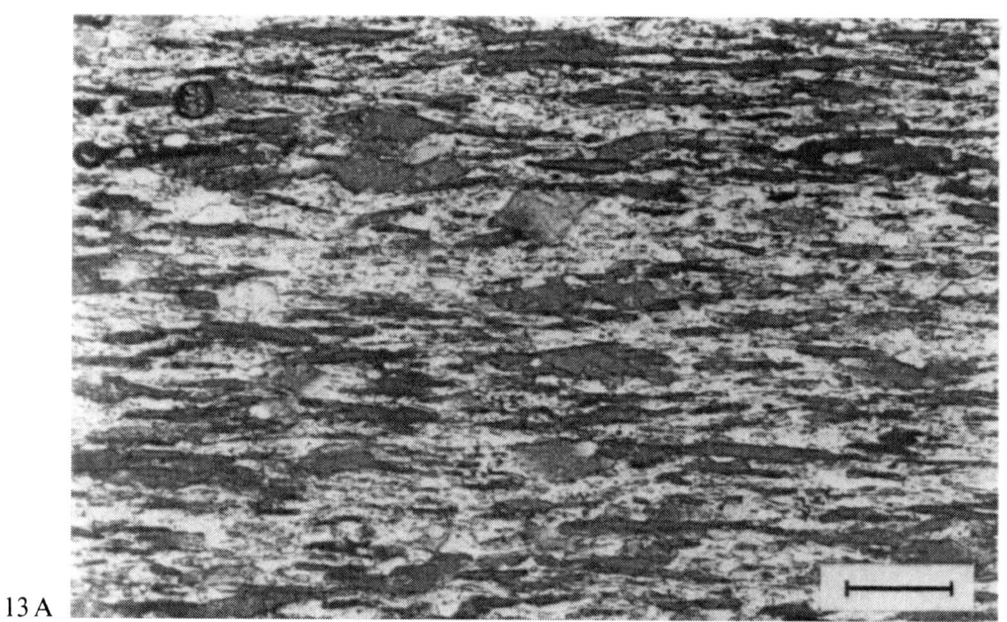

13 A

13 B

13 C

# 14. Fine Structure of Slate

W. DAVIES

Llandeilo. Llechwedd quarry, Blaenau Ffestiniog, Wales, U.K. Old Vein, as worked for top grade roofing slate.

---

**A.** Stereo-pair of fracture surface transverse to cleavage. Planar fabric corresponding to slaty cleavage due to arrangement of clay flakes, mainly chlorite, in subparallel layers separated by lenticles of randomly arranged flakes.

**B.** Stereo-pair of cleavage surface, exhibiting overlapping clay flakes, mainly chlorite.

Scales: 20 $\mu$m, S.E.M.
Specimens coated with gold-palladium 60 nm thick

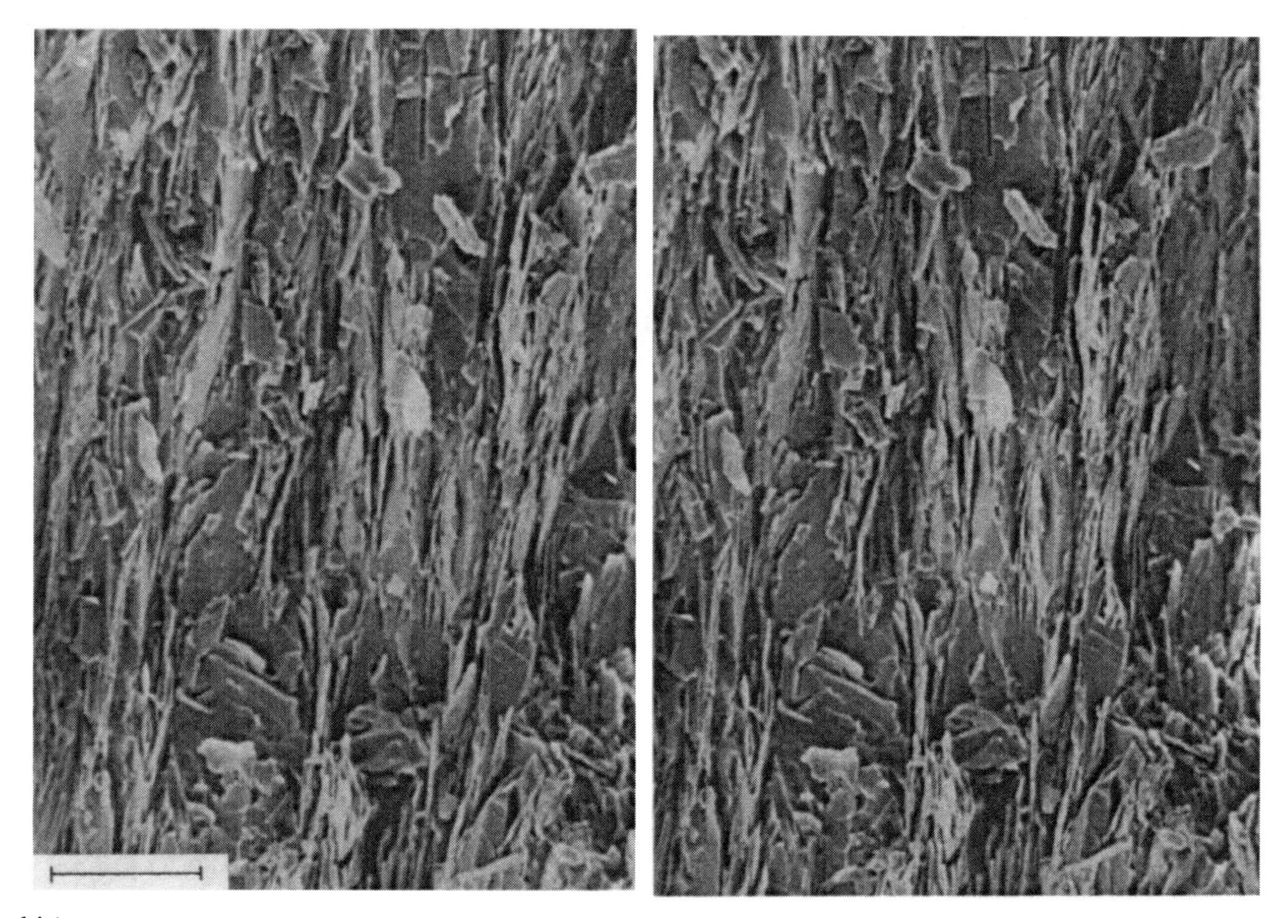

14 A
(pair)

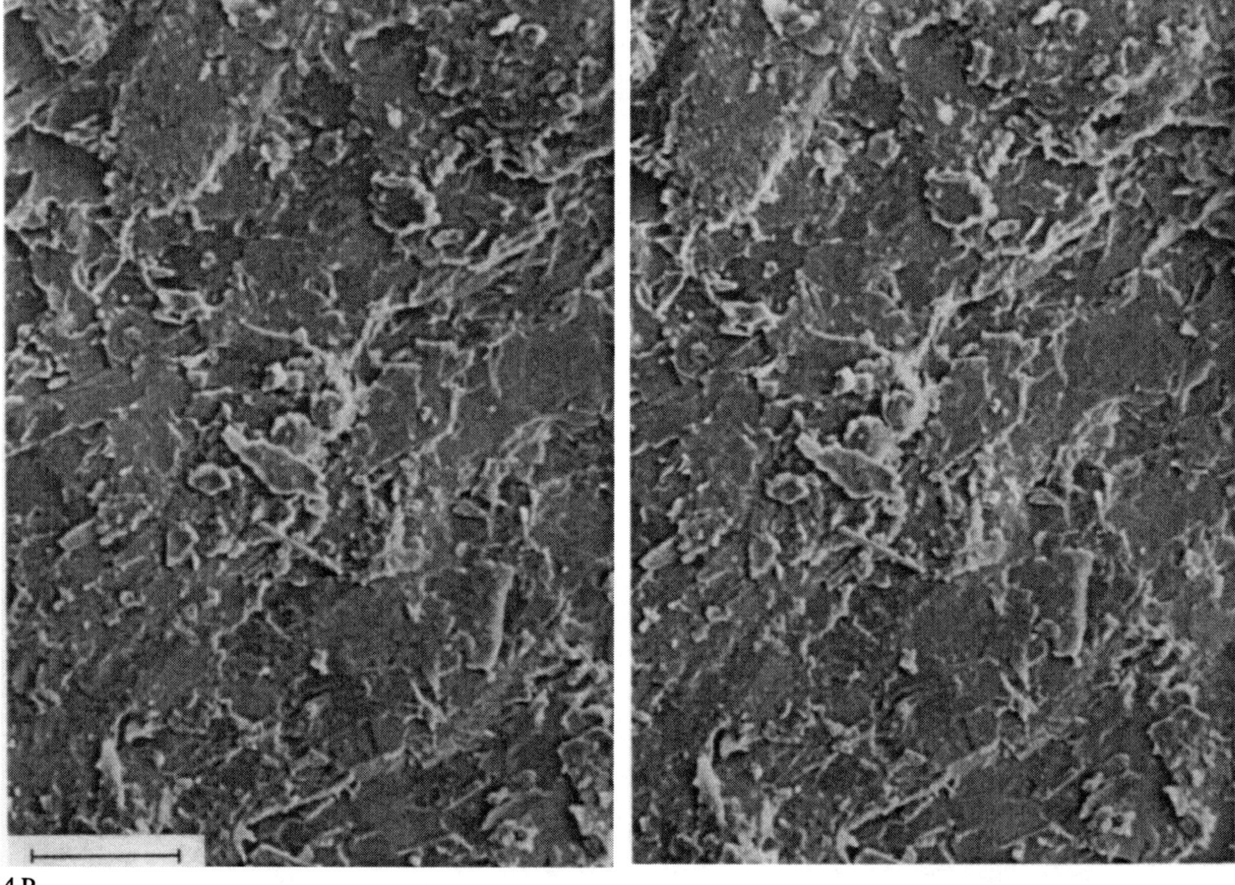

14 B
(pair)

## 15. Cleavage Intensity in Slate

W. Davies

**A. Mudstone devoid of cleavage.** Llandovery, Goginan, Dyfed, Wales, U.K. Clay flakes, mainly chlorite, in completely random arrangement.

**B. Ill-cleaved mudstone.** Llandovery, Cemmaes Road, Powys, Wales, U.K. Mudstone splits into uneven plates, 10 mm or more in thickness and presenting irregular cleavage surfaces. Rudimentary planar structure developed by widely spaced layers of subparallel clay flakes.

**C. Highly cleaved slate.** Murchisoni "shales", Abereiddy Bay, Dyfed, Wales, U.K. Closely spaced layers of subparallel clay flakes, separated by thin lenticles of randomly arranged flakes.

Scales: 20 $\mu$m S.E.M.
Specimens coated with gold-palladium 60 nm thick

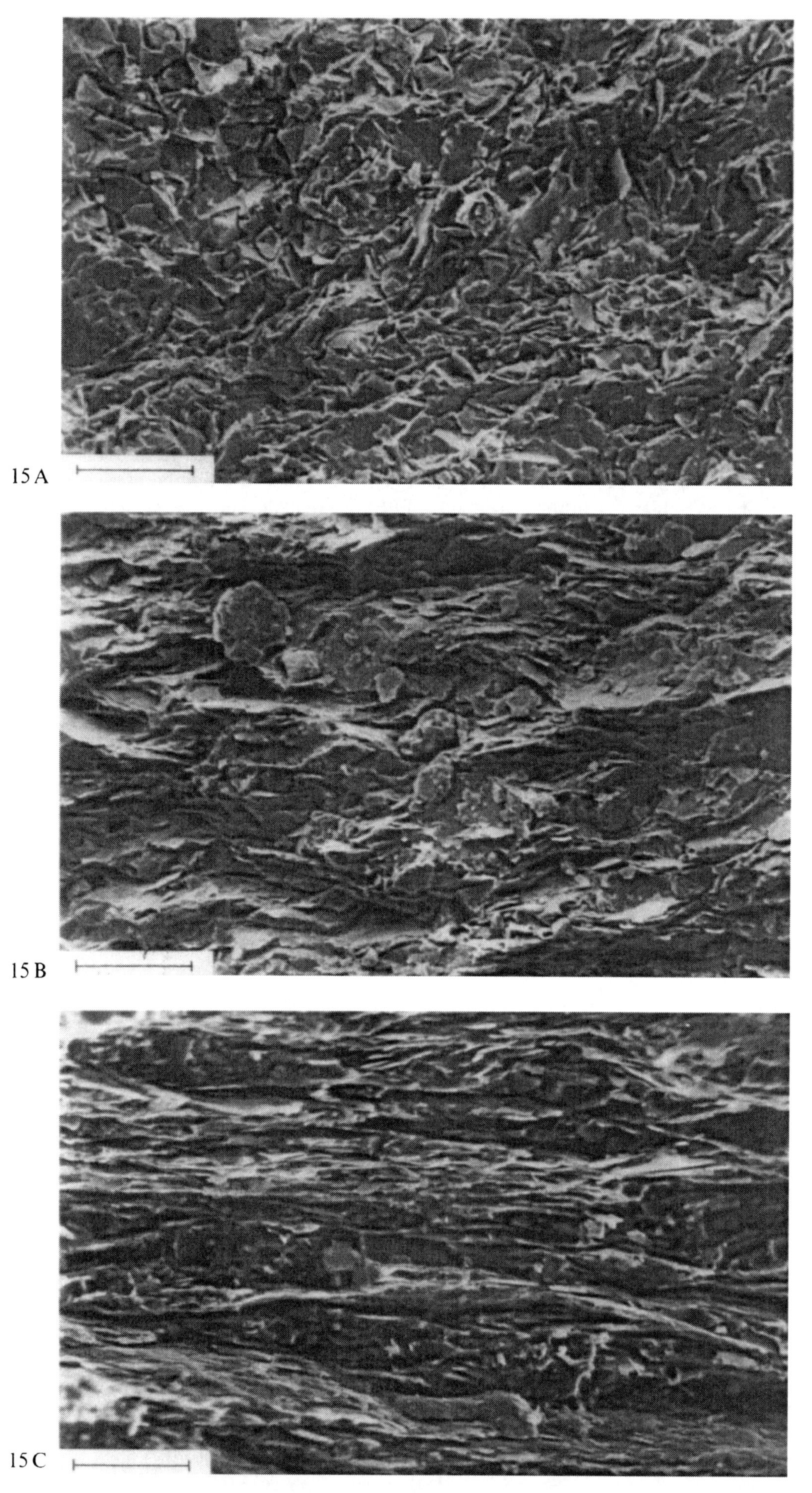
15 A
15 B
15 C

## 16. Two Foliations: Fine Structure

B. Z. LINCOLN

Phyllite from the Middle Ordovician Partridge Formation, collected five miles north of Fairlee, Vermont, U.S.A.

---

**A.** The mineralogical composition and texture of this sample are similar to those of the sample shown in Plates 200 and 214. It has been prepared for transmission electron microscopy by ion bombardment with $Ar^+$ ions (5–6 kV). The *black lines* are a copper grid which was glued to the section for support. *Holes* made by ion thinning are visible between several of the grid bars. The only foliation visible in the photograph is the older, penetrative one. PPL

**B.** Detail of **A**, photographed on a transmission electron microscope. The area photographed is at the edge of the hole indicated by the arrow in the preceding photograph. At this scale, the elongate shape of the phyllosilicate grains is evident. Their long axes are subparallel to the older foliation, which is nearly vertical in **A**. The grain boundaries are straight and the grains are tightly interlocked. Both chlorite and muscovite are present here, aligned approximately parallel to the older foliation. The texture of this phyllite is similar to that of a slate described by OERTEL and PHAKEY (1972). TEM

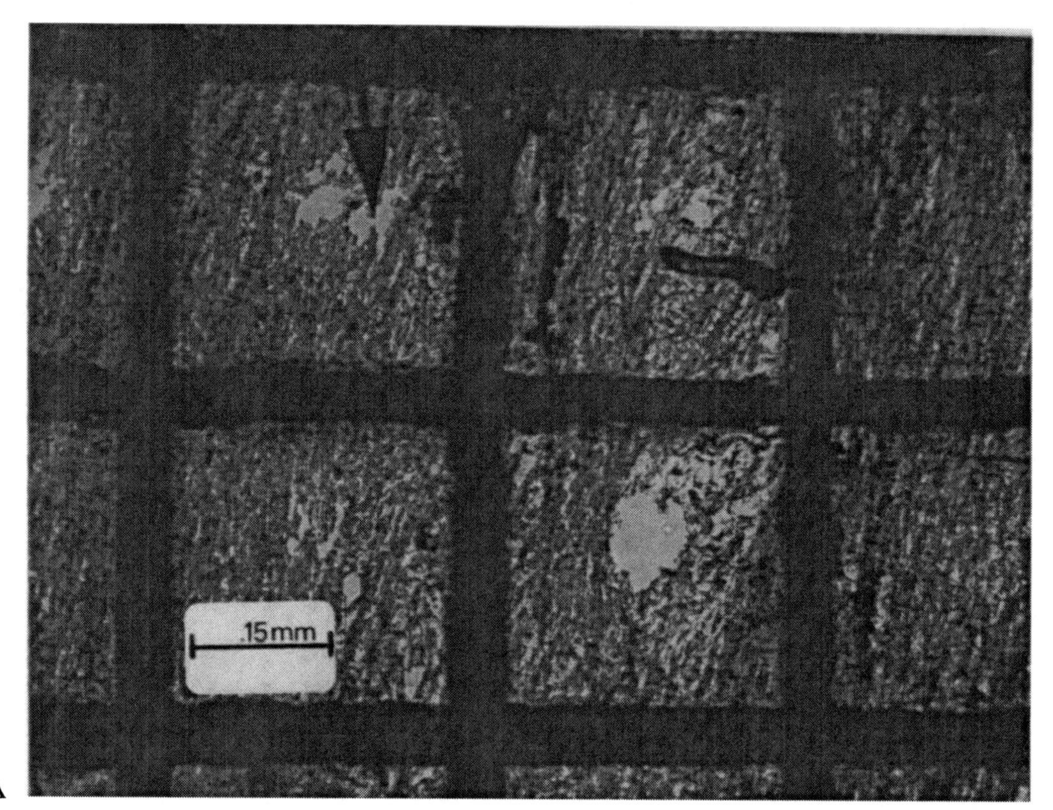

16 A

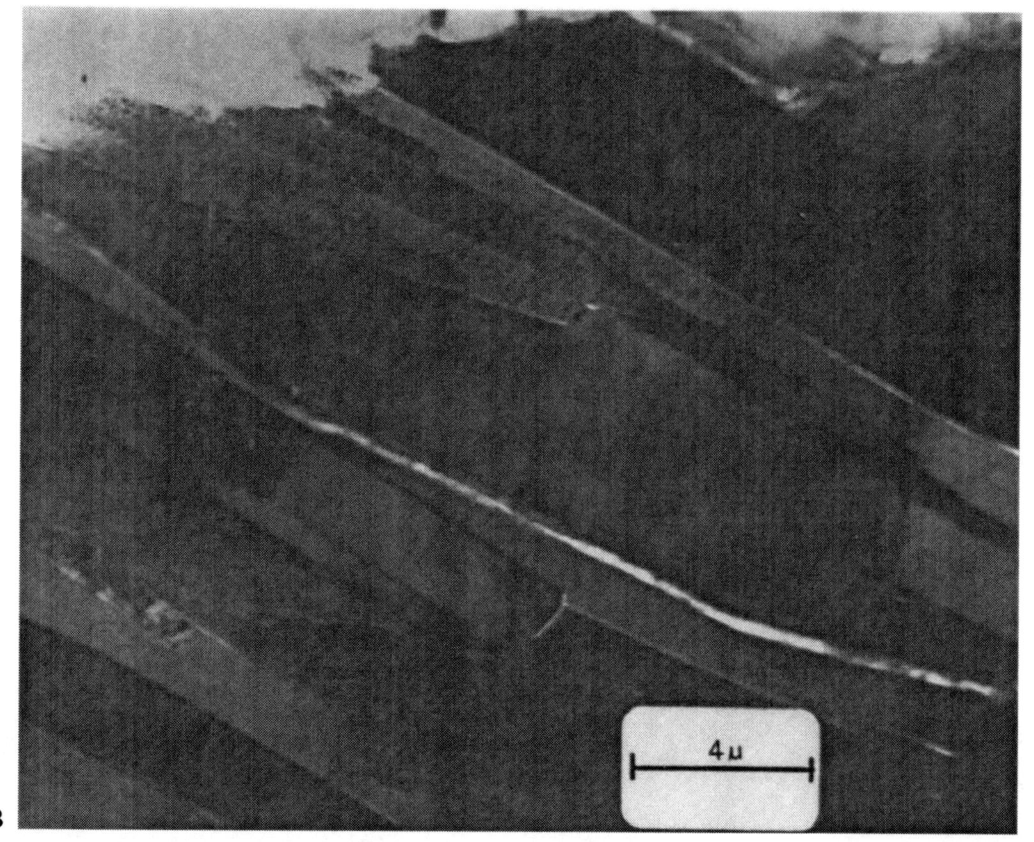

16 B

## 17. Synthetic Axial-Plane Foliation in an Experimentally Folded Salt-mica Specimen

W. D. MEANS

The layered assembly (**A**) comprises two outer layers of aluminum, two thick layers of coarse salt-mica mixture and a thin central layer of more fine-grained mix. The central layer was prepared by dry compaction of a mixture of 75% NaCl and 25% fine mica (*light* and *dark* respectively in **B**). This induced some initial preferred orientation of mica parallel to the "bedding." For the folding experiment the layered assembly was saturated in brine, enclosed in a thick lead jacket, and shortened about 35% at room temperature in five and a half days, using the procedure and apparatus described more fully for specimen 213 in MEANS (1975). This is specimen 219.

In the region covered by the photograph there is a secondary foliation (*vertical*) that makes a small but distinct angle with the "bedding planes" defined by the boundaries of the coarse layers. The *dark ovoid spots* in the coarse layers were circular before folding and provide some information on strains in these layers. PPL

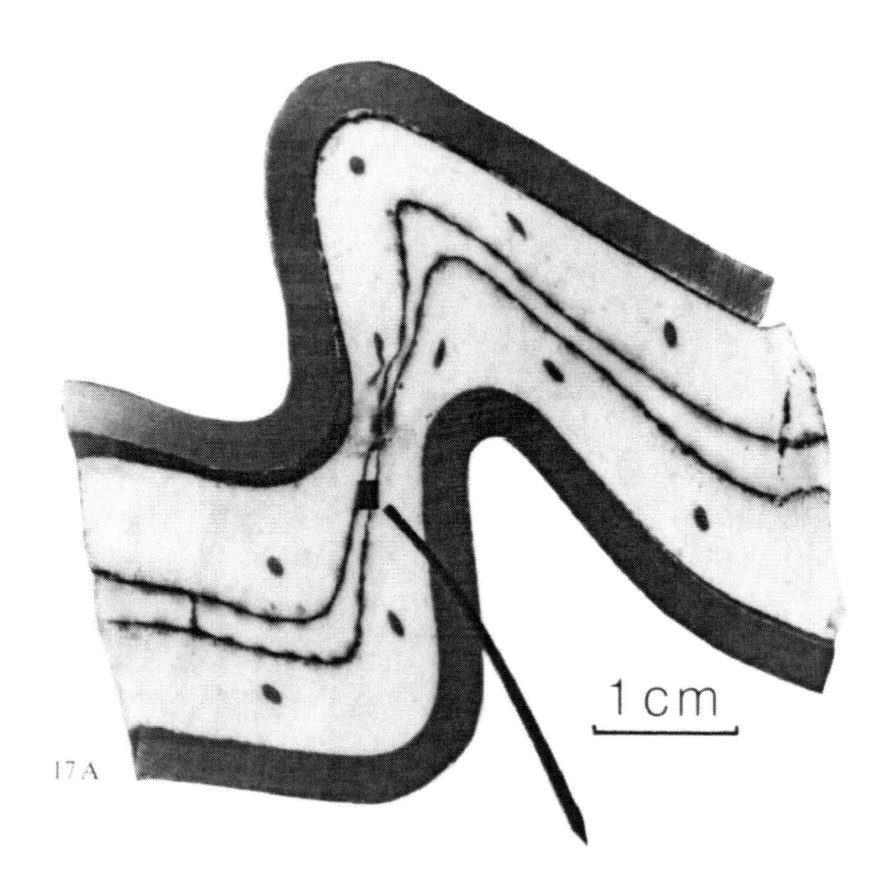


17 A

17 B

# Continuous Cleavage

## Section 3: Continuous Cleavage in Rocks Composed Largely of Non-Phyllosilicate Minerals

(Plates 18 to 23)

Six plates are here drawn together to represent continuous cleavage in aggregates composed of more-or-less equant grains. The range of materials represented is wide: ores, naturally and experimentally deformed ice, meta-igneous rocks and metasediment. The fabrics have developed by an equally wide range of processes: plastic deformation of grains, recovery of grains and particulate flow (intergranular deformation).

Fabrics in ores are also discussed in other sections (Plates 157 to 161) and reference to the index will provide many cross-references to plastic deformation of grains, recovery, and particulate flow.

---

**Plates**

# 18. Fabrics in Ores: Pyrrhotite

K. R. McClay

Pyrrhotite occurs in many deformed sulphide ore bodies (e.g., Rammelsberg, Sullivan, and Mount Isa), where it commonly exhibits a strong shape orientation and also a marked crystallographic preferred orientation. Pyrrhotite deforms by kinking, twinning, and dislocation glide (Atkinson 1974). The deformation characteristics are further complicated by phase transitions in the temperature region of 254° to 308 °C (Craig and Scott 1974). Experimental deformation of pyrrhotite indicates that it can be expected to be a relatively weak mineral during natural deformation (Atkinson 1974, Clark and Kelly 1973) and observations on naturally deformed pyrrhotite support this conclusion. Examples of naturally deformed pyrrhotite are given below.

---

**A. Pyrrhotite.** Mount Isa silver-lead-zinc orebodies, Queensland, Australia.
Bands of fine to medium-grained pyrrhotite occur in shales and siltstones in the stratiform silver-lead-zinc orebodies of Middle Proterozoic age at Mount Isa (Mathias and Clark 1975). Deformation under very low greenschist facies conditions has locally produced a slaty cleavage which is also reflected in the pyrrhotite fabric. Elongate pyrrhotite grains are parallel to the slaty cleavage in the host rocks, which is slightly oblique to the horizontal in the photograph. The pyrrhotite also exhibits a strong preferred orientation. Polished section NX

**B. Pyrrhotite.** Mount Isa silver-lead-zinc orebodies, Queensland, Australia.
Elongate pyrrhotite grains parallel to the schistosity. The internal microstructure is revealed by etching. A few flame-like kinks (*k*) can be observed but in many places the internal microstructure is complicated by exsolution features (*ex*) of monoclinic pyrrhotite. It is probable that the exsolution features are controlled by lattice kinking.
Polished section PPL, etched 50 s with HI

**C. Pyrrhotite.** Pyrrhotite in Ni-Cu ore from Pickwe, Botswana.
Coarse grained (2–3 cm) pyrrhotite occurs in amphibolite facies gneisses. Etching reveals large lattice kinks (*k*) which bend exsolution lamellae of monoclinic pyrrhotite (*e*). Monoclinic pyrrhotite is also exsolved along probable flame-like kinks (*f*). The phase transitions in this deformed pyrrhotite appear to be partly controlled by kinking within the host lattice. The sketch below illustrates some of the features in photograph **C**.
Polished section PPL, etched 50 s with HI

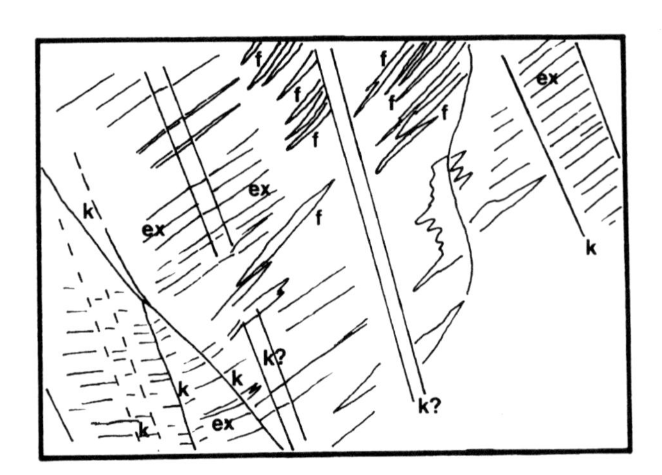


Sketch of photograph **C**: Large lattice kinks (*k*) can be seen bending exsolution lamellae (*ex*) of monoclinic pyrrhotite. Other linear structures along which monoclinic pyrrhotite is exsolved may be relict kink or twin features (*k?*). The flame-like features (*f*) are interpreted as kinks along which monoclinic pyrrhotite has exsolved.

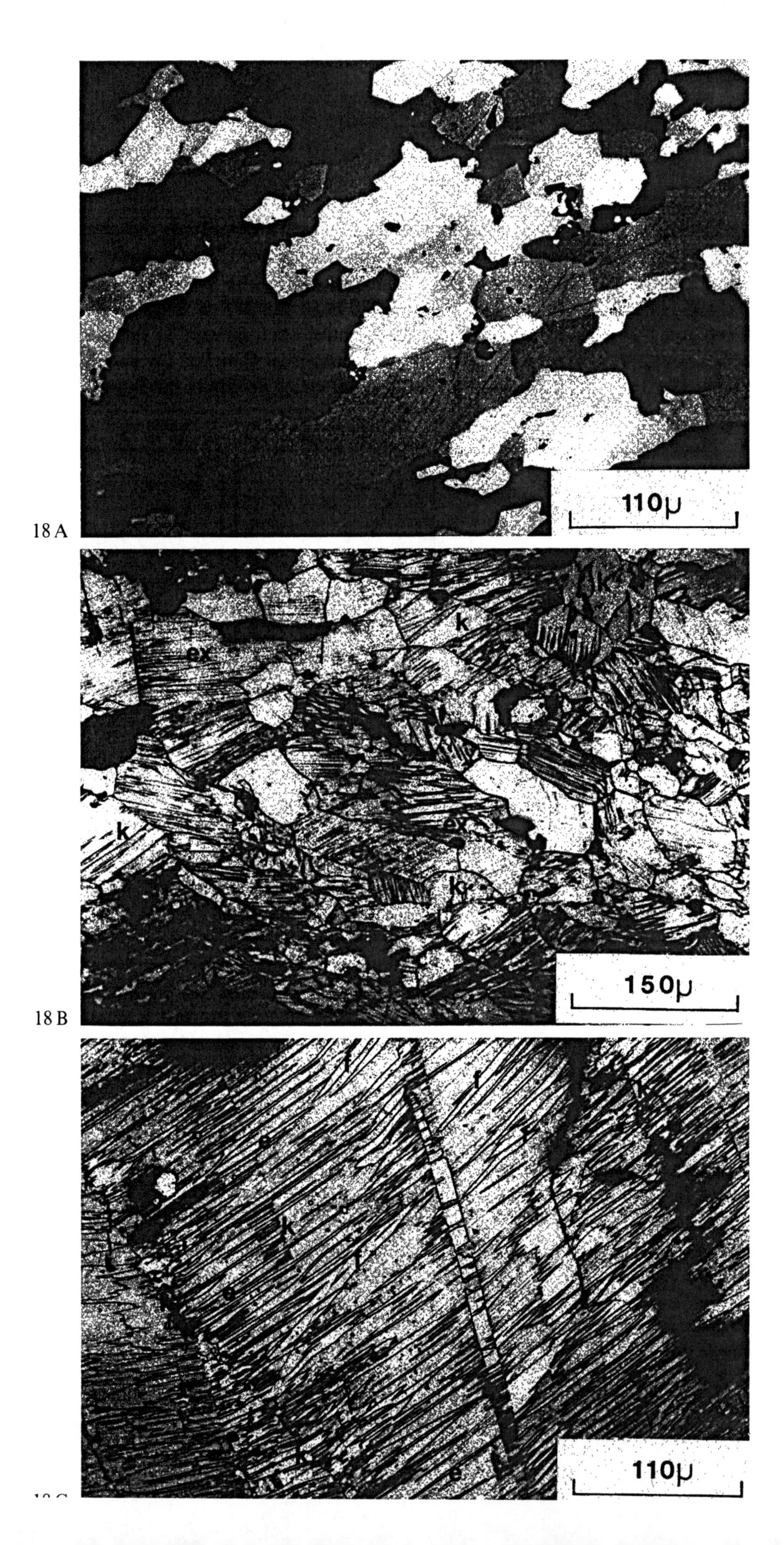

110μ
18 A
ex
k
150μ
18 B
110μ

# 19. Cleavage Structure in Antarctic Ice

D. S. Russell-Head

Elongated air bubbles, aligned in the direction of ice flow, are a commonly observed cleavage-like structure in ice from deformation zones in ice sheets and glaciers. The examples shown here are from the core of a borehole drilled near Casey, East Antarctica. The fluid-filled borehole has been instrumented and horizontal shear rates down to 250 m in the 385 m thick ice sheet are known (Russell-Head and Budd 1981). Samples of the core have been stressed in the laboratory at typical englacial temperatures and stresses. The resulting empirical flow law for ice of differing c-axis preferred orientation has been used to estimate the shear stress profile in the figure below.

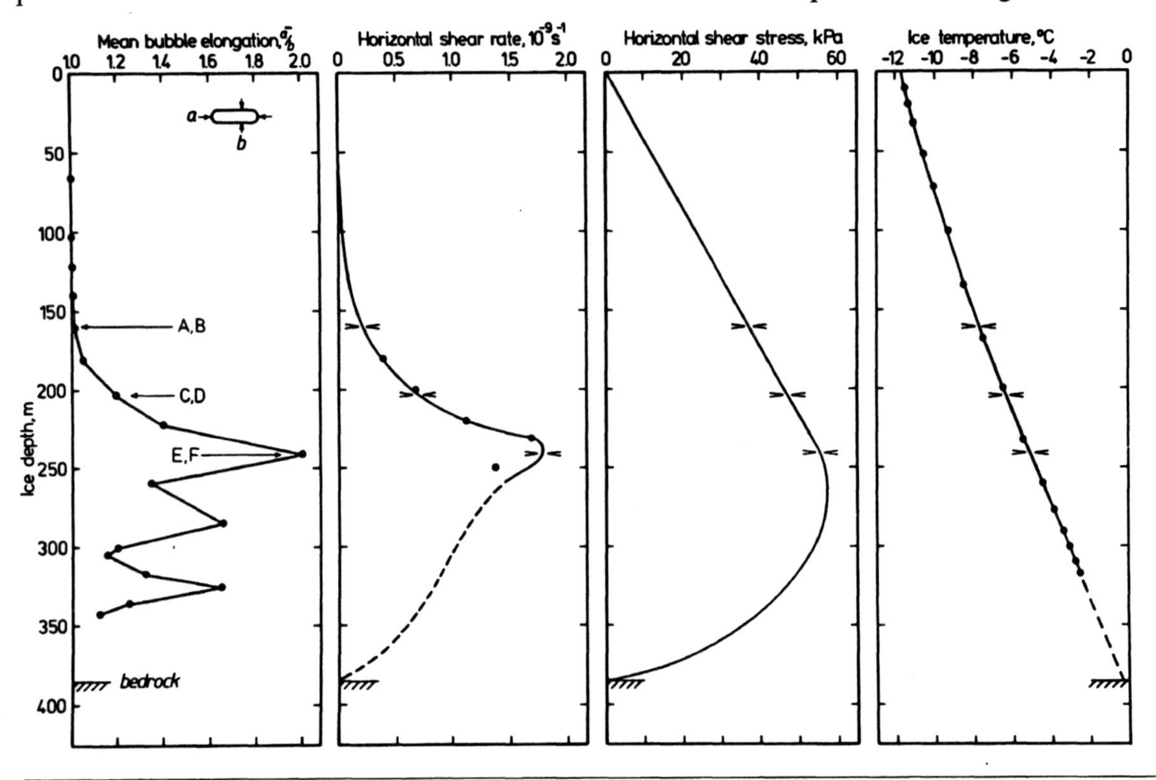


**A. Spherical bubbles.** Sample depth: 160 m. Section thickness: 4.6 mm. This horizontal section contains spherical bubbles with diameters of 0.2 mm to 1.0 mm. Scale: as in (B)

**B. Ice microstructure.** Sample depth: 160 m. The section in **A** has been thinned to show a coarse-grained aggregate with smooth grain boundaries. A subgrain structure is present in many of the grains. 80% of the c-axes lie within 35° of the vertical to the section. NX

**C. Elongated bubbles.** Sample depth: 203 m. Section thickness: 3.2 mm. About half of the bubbles in this horizontal section are elongated in the direction of ice flow. The mean breadth of the elongated bubbles is comparable to the diameter of the larger spherical bubbles. Scale: as in (B)

**D. Ice microstructure.** Sample depth: 203 m. The section in **C** has been thinned to show an interlocking grain structure. The grain size is smaller than in **B** and a subgrain structure is more prevalent. 80% of the c-axes lie within 50° of the vertical to the section. Scale: as in (B), NX

**E. Elongated bubbles.** Sample depth: 241 m. Section thickness: 4.4 mm. Bubbles with elongation ratios > 15 coexist with spherical bubbles in this horizontal section. About 60% of bubbles are elongated. As in **C**, the directions of elongation and flow are parallel. Scale: as in (B)

**F. Ice microstructure.** Sample depth: 241 m. The section in **E** has been thinned to show a highly interlocked grain structure. The grain size is smaller than in **B** or **D** and almost all grains exhibit a well developed subgrain structure. 80% of the c-axes lie within 30° of the vertical to the section. Scale: as in (B), NX

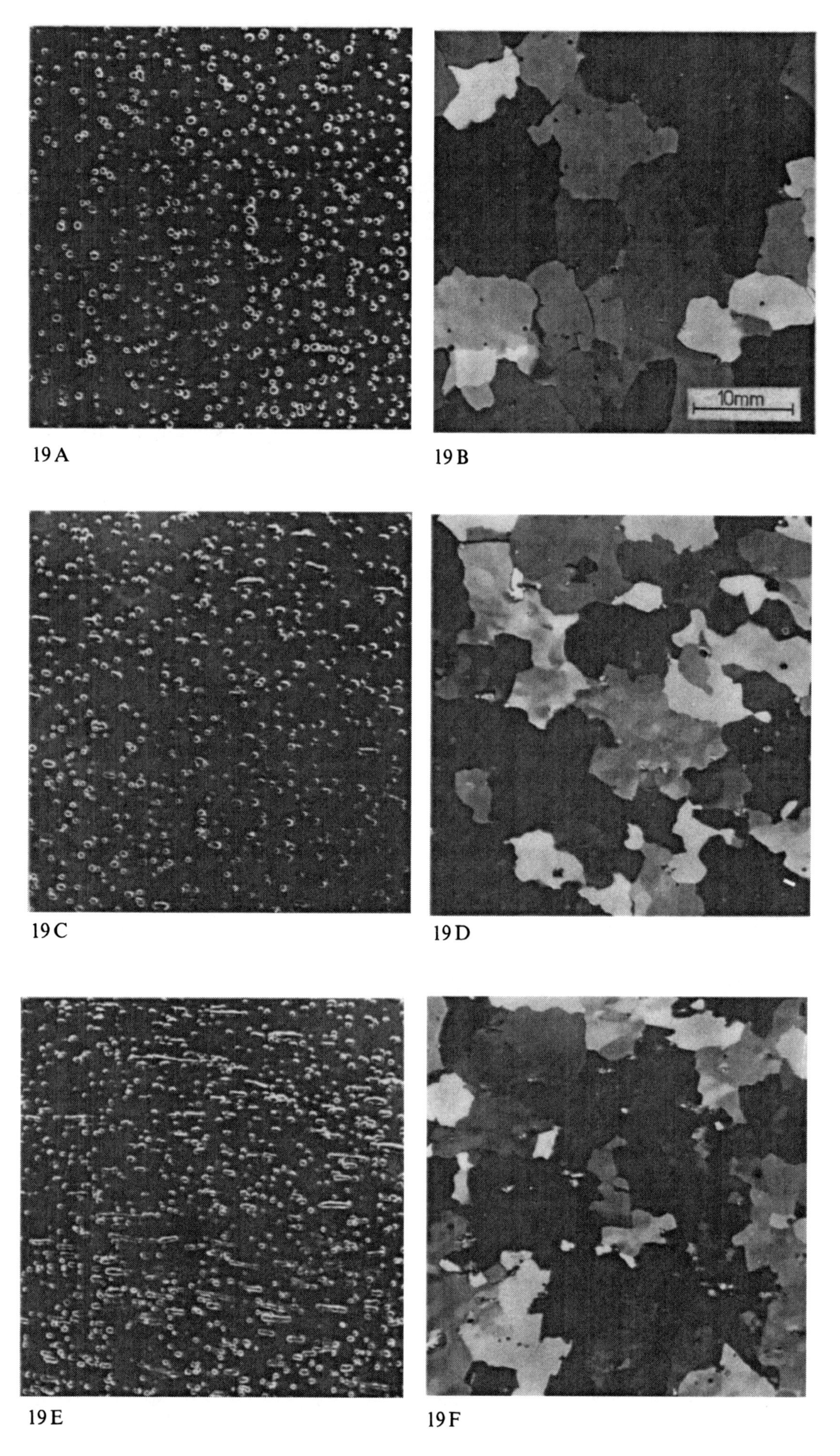


19 A

19 B

19 C

19 D

19 E

19 F

# 20. Cleavage Structure in Experimentally Deformed Ice

C. J. L. WILSON

Ice has been used as an analog for quartz in quartz-rich rocks by deforming an anisotropic sample composed of alternating layers of fine (*F*) grained (1 mm) and medium (*M*) grained (2 mm) polycrystalline ice. Layer boundaries were coated with plastic and adjacent layers were frozen together with a thin film of distilled water (WILSON and RUSSELL-HEAD 1979). Present in the fine-grained layers, and confined to the grain boundaries, was an irregular distribution of small air bubbles of 0.1 to 0.2 mm diameter, at an estimated concentration of 10 $mm^{-3}$. These blocks were shortened in a plane strain apparatus at $-1° \pm 0.05$ °C in a direction parallel to the initial planar layering. After deformation a distinct cleavage-like structure developed; it is defined by an air-bubble elongation in the fine-grained layers. Photographs **A** and **C** are from thick sections taken from the center of two experimentally deformed samples. The direction of shortening in all photographs is north–south.

---

**A. Bubble elongation.** In a sample shortened 30%, buckles are developed in the fine-grained layers *F1* and *F3* with flattening in *F2*. In each fine-grained layer there is a variable air-bubble elongation. In layers *F1* and *F3* a fan-like pattern can be recognized in some hinges. In layer *F2*, where there is essentially no buckling, the air-bubble elongation is perpendicular to the shortening direction.

**B. Bubble elongation and grain microstructure.** Thin section of area illustrated in **A** that shows the relationship between air-bubble elongation and microstructure. In layer *F2* the majority of elongate bubbles are within large recrystallized grains, whereas in the fold hinges in layers *F1* and *F3* there is only partial inclusion, grain size is smaller and the grains contain abundant undulose extinction and subgrain development. The spatially distinct features in photographs **A** and **B** are interpreted as representing areas undergoing different stages in a complex deformation path:

1. The air-bubble elongation in layer *F2* appears to be a result of flattening perpendicular to the direction of shortening.
2. Subgrain development and undulose extinction occur in areas undergoing folding late in the deformation history whereas the rest of the sample has undergone grain growth.
3. In the folded regions the early formed pattern of elongate air-bubbles is rotated into a symmetric fan-like cleavage structure. NX

**C. Cleavage fans.** In a block shortened 36%, large disharmonic sinusoidal folds are developed in the *F* and *M* layers. The axial surface of these folds lies approximately perpendicular to the direction of shortening. The small air bubbles in the *F* layers are elongate and describe a symmetric fan-like pattern. Strain markers inscribed on the outside of the block (WILSON and RUSSELL-HEAD 1979) are also distributed as ellipses about the axial surface of the major folds. Where an original circle lay across the boundary between an *F* and *M* layer it was either (1) significantly elongated in the *F* layer, or (2) displaced parallel to the layer. The thin ice films used to bond the *F* and *M* layers have also been folded partly in conformity with adjacent *F* and *M* layers. However, adjacent to the dilation voids (*A*) the ice films display minor folding where axial surface orientations vary from perpendicular to parallel to the shortening direction. The variable orientation of minor folds is attributed to a component of layer boundary slip. Therefore layer boundary slip, buckling and flattening all contribute to the formation of the symmetric fan of elongate air bubbles that resembles many cleavage fans observed in natural rocks.

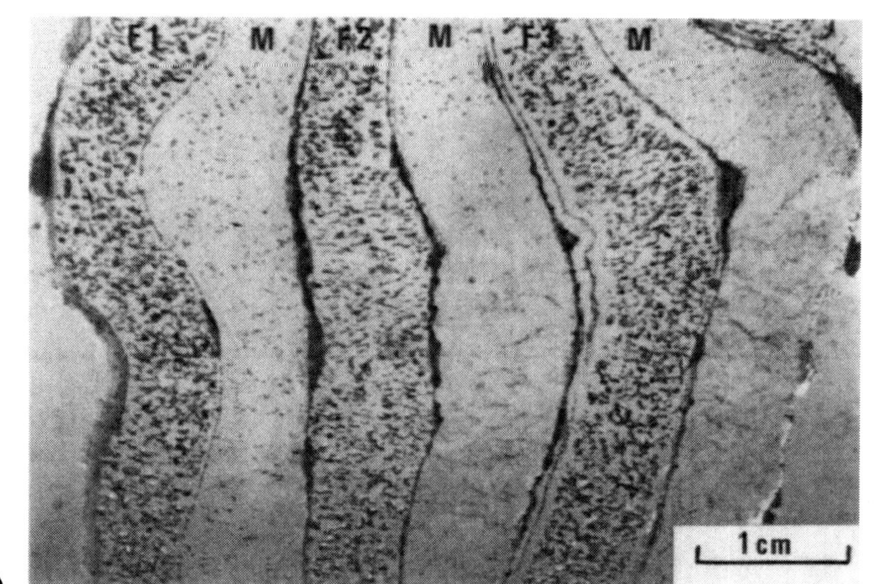

20 A

20 B

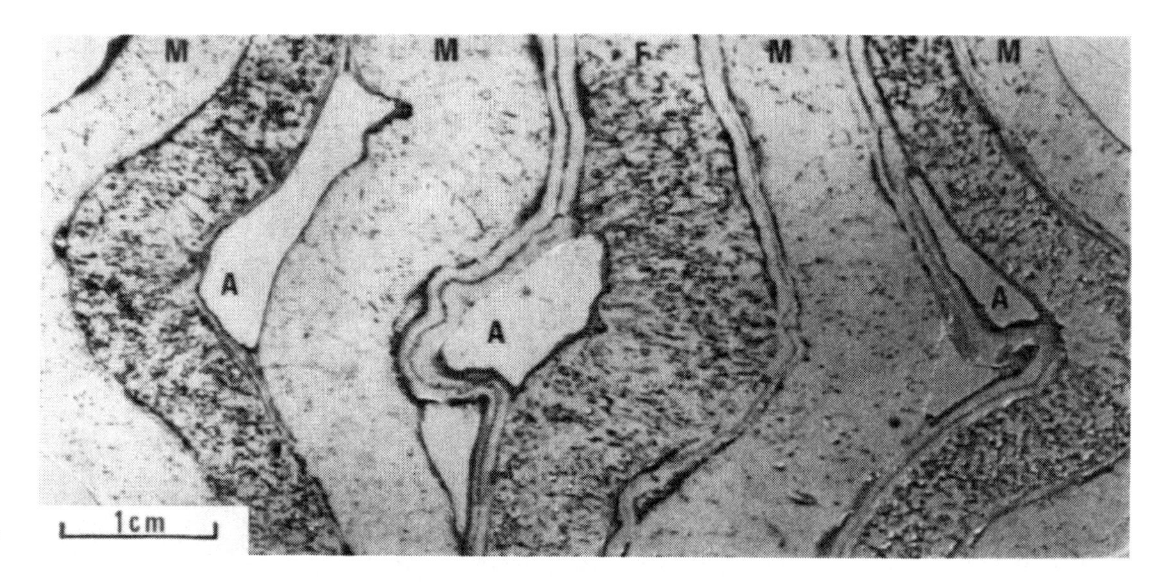

20 C

# 21. Progressive Deformation of Quartz Porphyry (I)

M. M. Kehlenbeck

"Quartz-eye" porphyry from a volcanic unit in the Wabigoon subprovince, Superior Province of Canadian Shield near Thunder Bay, Ontario.
The photographs indicate features seen with progressive deformation from 21 A through to 22 B.

---

**A.** Schistosity develops in matrix of retrogressed feldspar, quartz, epidote, and white mica. "Quartz eyes" are weakly strained: note deformation lamellae in lower large quartz grain. NX

**B.** Undulose extinction of quartz-eye porphyroclasts becomes more noticeable and peripheral subgrains develop. NX

21 A

21 B

# 22. Progressive Deformation of Quartz Porphyry (II)

M. M. Kehlenbeck

**A.** Quartz-eye porphyroclasts are now more elongate and schistosity of matrix does not show strain shadows about them. NX

**B.** Quartz-eye porphyroclasts are now reduced to ribbons whose preferred dimensional orientation reinforces the schistosity. Advanced subgrain formation around edges of the ribbons reduces the contrast between the relict quartz eye and the fine-grained matrix. NX

22 A

0.5 mm

22 B

0.5 mm

# 23. Grain Deformation Versus Particulate Flow: Schistosity Formation in Clastic Sedimentary Rocks

G. J. Borradaile

In the Late Precambrian Dalradian Supergroup of Islay, Scotland, the first schistosity (*S*) is usually well developed in clastic rocks. The following illustrate two modes of the schistosity dependent on composition and grain size variation of the original sediments. The two localities have similar structural settings and occur in the same general area of lower greenschist facies metasediments.

---

**A. Schistose massive sandstone; Port Ellen Phyllite**

**Photograph:** grain size appears small. Schistosity is defined by preferred dimensional orientation of quartz sub-grains formed predominantly on the rims of strained quartz clasts. Relict cores of quartz clasts are also somewhat elongate parallel to schistosity. NX

**Sketch:** same field of view as photograph with feldspar clast (*F*) indicated for correlation. Original outlines of clasts are indicated by dust trails and are sometimes easier to see when the microscope is slightly out-of-focus in PPL. The original clastic outlines are slightly aligned. PPL

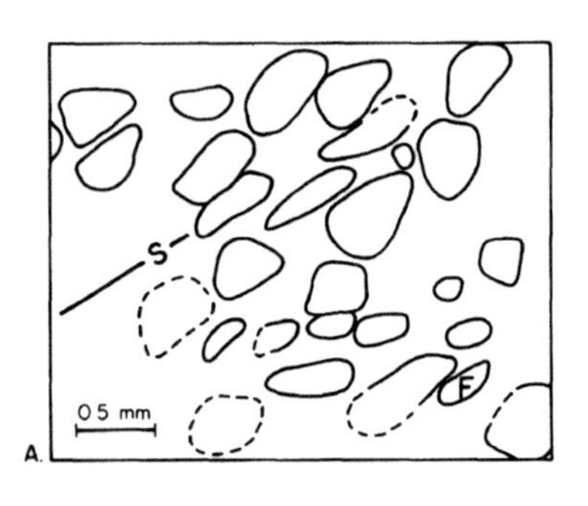


Some grain sliding and rolling is responsible for the formation of schistosity since clastic outlines show a preferred dimensional orientation (*sketch*). However, the tectonic fabric has been considerably enhanced by intracrystalline deformation (producing undulose extinction) and the ensuing recovery which produced slightly elongate aligned quartz sub-grains (*photograph*).
Bulk strain estimates for this rock are:

| $X$ | $Y$ | $Z$ |
|---|---|---|
| $2.13 \pm 0.06$ | $0.90 \pm 0.01$ | $0.52 \pm 0.01$ |

with orientation $XY \simeq S$ (Borradaile 1979a; outcrop *38*).

**B. Schistose, laminated dolomitic siltstones; Dolomitic Formation**

**Photograph:** clastic quartz grains are rarely strained and show little recovery. They are aligned in the fine-grained dolomitic matrix. NX

**Sketch:** prominent bedding lamination (*So*) brought out by the larger quartz clasts. Only the few grains shaded black show undulose extinction or sub-grain formation. PPL

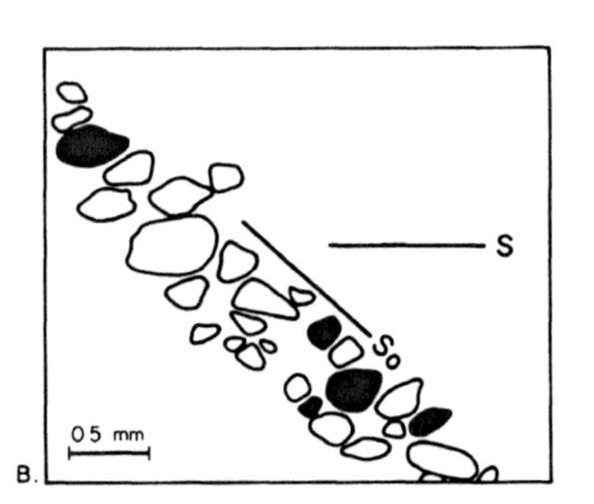


Cleavage is due mainly to the preferred dimensional orientation of quartz clasts which aligned without internal deformation, by rotation through a fine-grained dolomite matrix.

Bulk strain data for this rock are:

| $X$ | $Y$ | $Z$ |
|---|---|---|
| $1.71 \pm 0.16$ | $0.89 \pm 0.06$ | $0.66 \pm 0.08$ |

with orientation $XY \simeq S$ (Borradaile 1979a, outcrop *18B*).

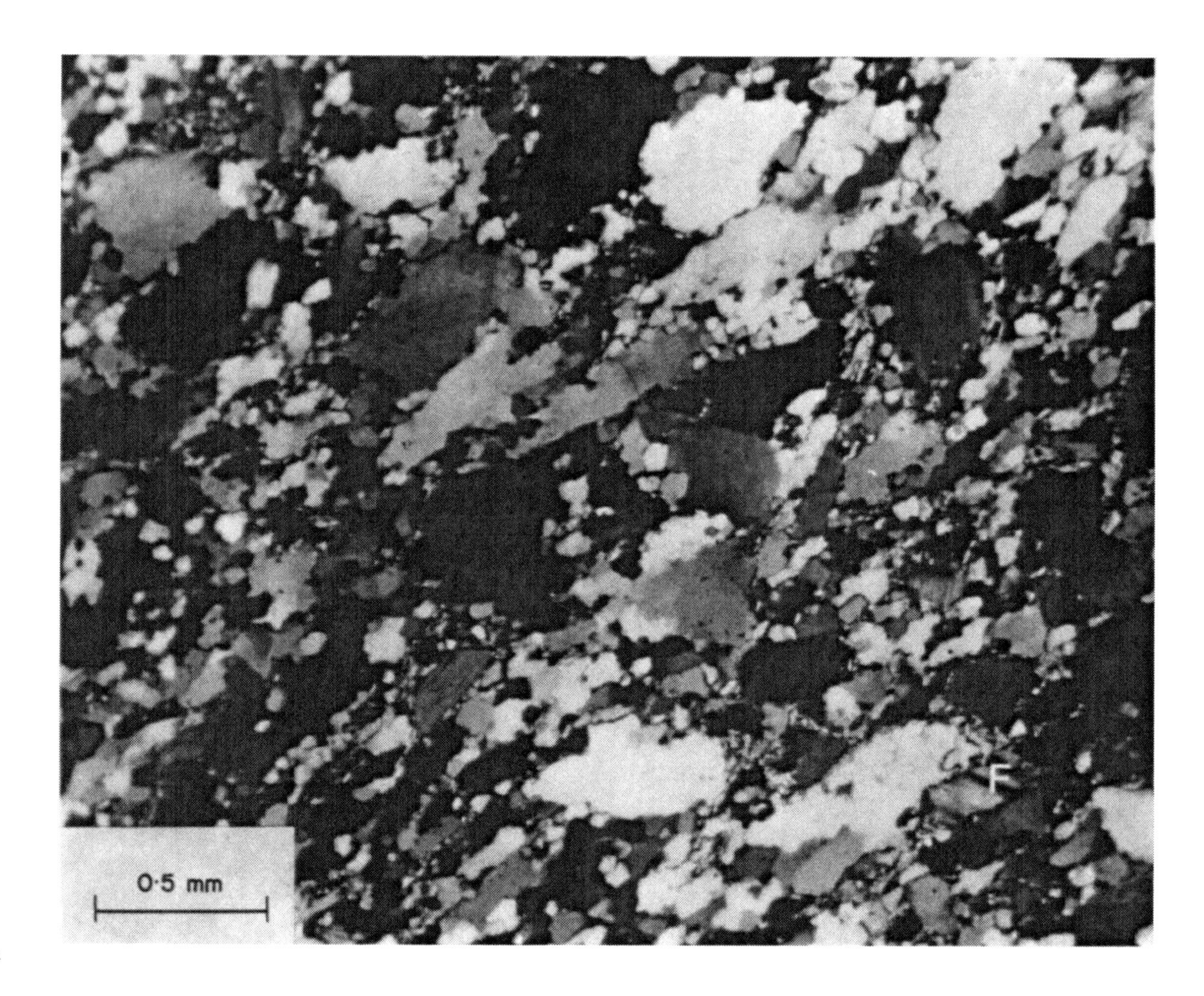


23 A

23 B

## Section 4: Crenulation Cleavage with Gradational Boundaries

(Plates 24 to 43)

Twenty contributions provide examples of this cleavage type. The first two plates (24, 25) illustrate this cleavage type particularly well at the microscopic level. Some authors have emphasized the differentiated nature of these cleavages by calling them spaced cleavage (e.g., Plates 30 to 34) although the microlithons do have diffuse margins on close inspection. This section concludes with Plates 42 and 43 in which the gradual transition to a sharper distinction between adjacent microlithons becomes apparent. Clearer examples of zonal crenulation cleavages with discrete boundaries are dealt with in the following section (Plates 44 to 53).

---

**Plates**

# 24. Crenulation in an Amphibole-Rich Rock

B. M. Bayly

Barnard volcanic member, Ordovician Missisquoi Formation, on Route 9, 1 mile east of Marlborough, Vermont, U.S.A.

Crenulation cleavage normally involves phyllosilicates, but similar effects develop in amphibole-rich rocks. The illustration is a negative print, with the dark hinge regions containing more abundant quartz. The limb regions, paler in the photograph, are dominated by amphibole with minor biotite and epidote, in proportions approximately 4 : 1 : 1. Sodic feldspar might be expected, but if present it is untwinned. It is presumed that the rock was originally a basic intrusion, surface flow or ash unit but because of the thorough metamorphism, this uncertainty cannot be resolved.

A difference from typical phyllosilicate crenulations is in the grain size relations. In Plates 38, 40, 45 and 46 the width of a hinge zone or limb zone is only three to six times the length of a typical flake, whereas in this amphibole rock, the ratio is much greater. It is not clear what fixes the wavelength of crenulations. If a rock that is already somewhat coarse is later crenulated, one can imagine the grain size might set a lower limit to the width of the crenulations that develop; alternatively if a crenulated rock suffers grain growth to coarser grains, one can imagine that the crenulation width might set an upper limit to growth of platy grains. (Garnets are not so restricted and may grow large enough to engulf two or three domains.) We should admit that perhaps many phyllosilicate crenulations develop with, at first, many grain lengths per zone-width; the abundant instances where zone-width is only a few multiples of grain length may result mainly from grain growth occurring after crenulation.

(Ordinary light; rock section placed directly in enlarger)

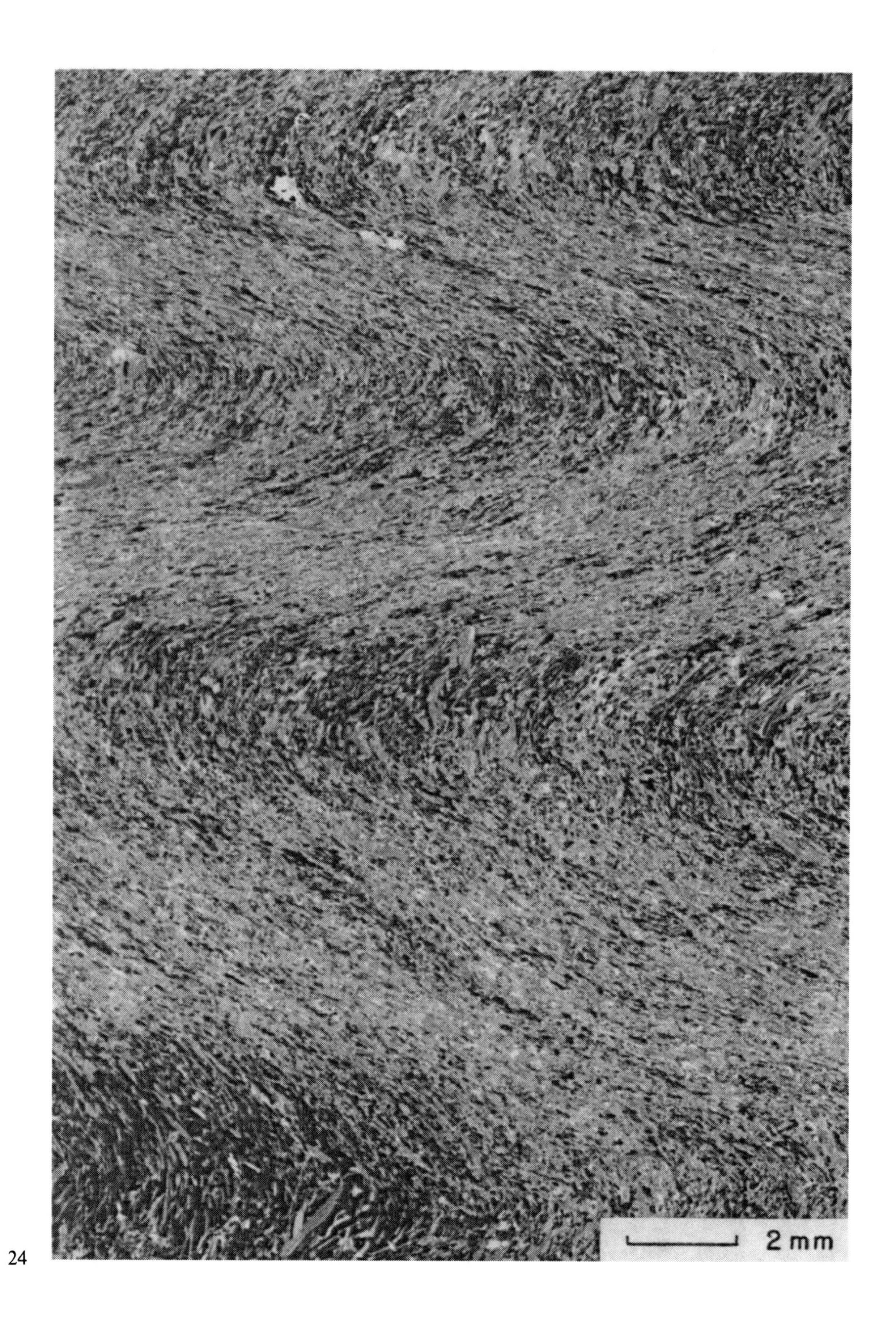


24

## 25. Crenulation Cleavage in Martinsburg Slate, Delaware Water Gap Section, New Jersey, U.S.A.

W. D. MEANS

Characteristic morphology of incipient to moderately developed crenulation cleavage in the Martinsburg slate section described by MAXWELL (1962). The exact locality for this specimen is unknown. Crenulation cleavage dips right. Earlier slaty cleavage is subhorizontal. Domains enriched in layer silicates relative to quartz correspond to darker regions on some of the short limbs of the crenulations. A hand specimen of this rock cleaves more perfectly along the slaty cleavage than along the crenulation cleavage, but the crenulation cleavage interferes sufficiently to prevent this rock from being used for roofing. PPL

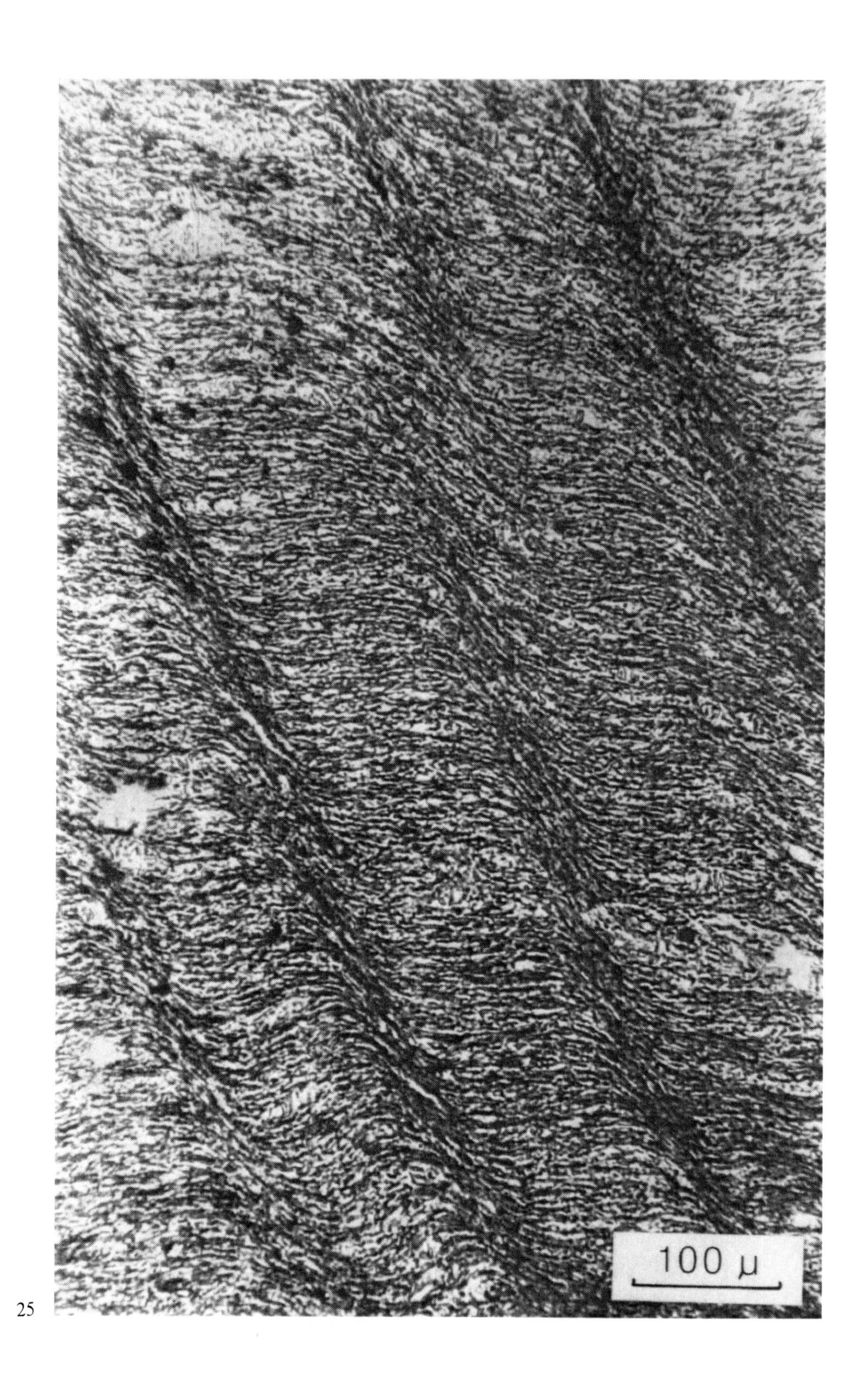


25

## 26. Differentiated Crenulation Cleavage

V. A. WILLIAMS

Crenulated biotite white-mica phyllite, Abercrombie River, northwest of Taralga, N.S.W., Australia (Grid Ref. 233 N 377 E, SCHEIBNER 1973).

The specimen is from a succession of Late Ordovician to Early Silurian quartzites and slates that have been folded three times (WILLIAMS 1979). The first phase of folding occurred prior to the earliest Devonian, and produced east–west trending folds lacking an axial-surface cleavage. The second and third phases of folding occurred after the Early Devonian, probably in the Early Carboniferous, along north–south axial trends with an axial-surface slaty cleavage or schistosity in the second phase, and the crenulation cleavage in the third phase of folding. Metamorphism in the last two deformations was in the greenschist facies.

The crenulation cleavage is defined by alternating quartz-rich and quartz-poor domains in the more pelitic rock units, and is not developed in the psammitic layers. The quartz-rich domains are up to four times wider than the adjacent mica-rich domains, which occupy the shorter limbs of asymmetrical microfolds. The quartz grains are commonly elongate in the direction of the earlier schistosity, although some quartz grains (see *top right-hand corner* of **A**) have more equant rounded sections, possibly reflecting inherited detrital grain shapes.

A possible mechanism for producing the differentiated crenulation cleavage, by solution removal of quartz, is indicated by the following:

1. The much lower abundance of quartz in short limbs of the crenulation microfolds, relative to quartz abundance in the longer fold limbs, in the absence of any other mineralogical differences between these domains.
2. The narrower shape of quartz grains in mica-rich domains (average approximately 0.02 mm wide) relative to quartz grains in quartz-rich domains (0.04 mm wide), coupled with a reasonably uniform length (approximately 0.06 mm) in both domains.

Note that the earlier schistosity in this sample crosscuts the sedimentary layering (horizontal in **A**) at a low angle, and that there is no evidence for an earlier bedding-parallel fissility as described by MORRITT and others from rocks of similar age and lithology near Taralga to the southeast of the Abercrombie River (Plates 6, 7).

---

**A.** Sedimentary layering (inclined gently *upward to the right*), schistosity (*horizontal*), and the differentiated crenulation cleavage. Note the absence of the crenulation cleavage in the psammitic layer at the top. PPL

**B.** Enlargement of the lower left-hand corner of **A**. PPL

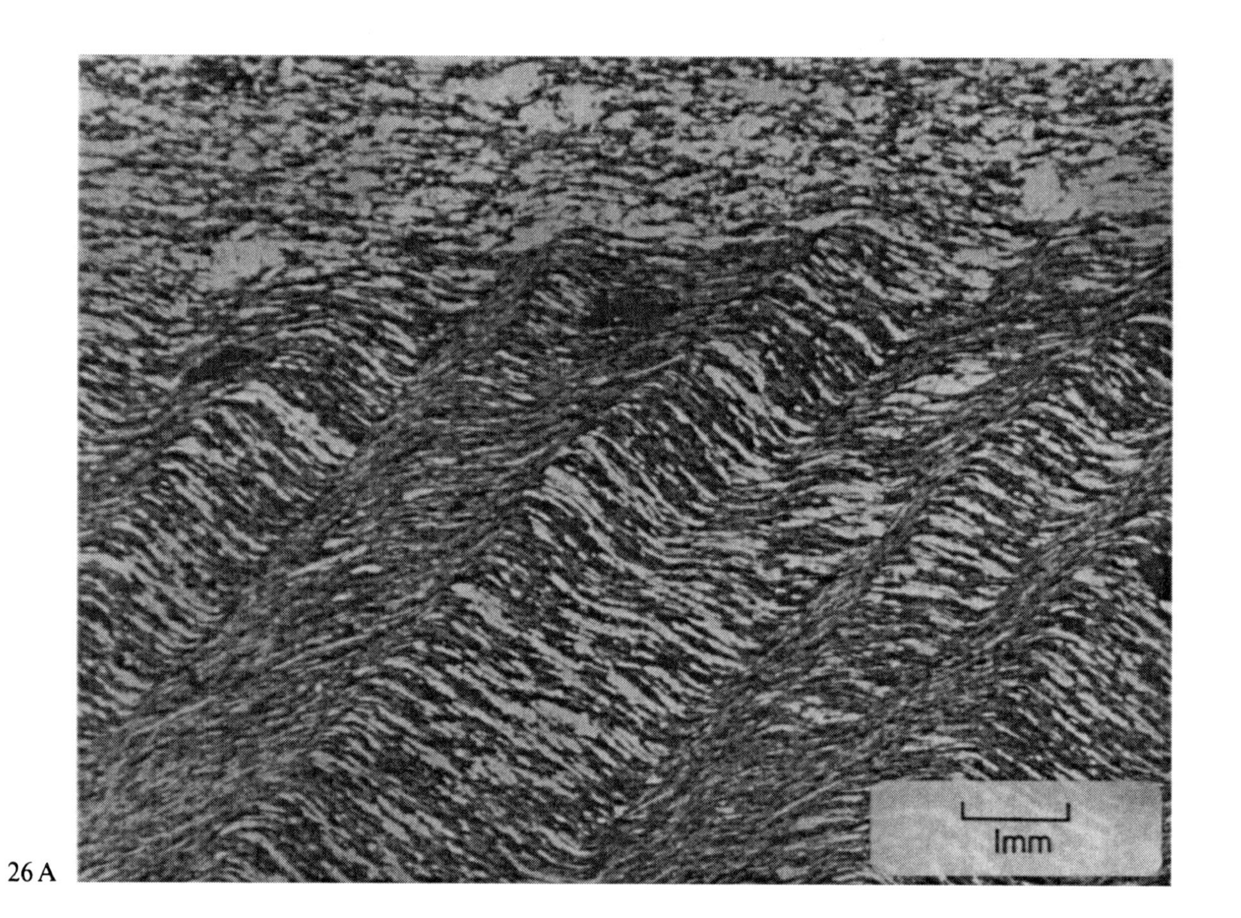

26 A

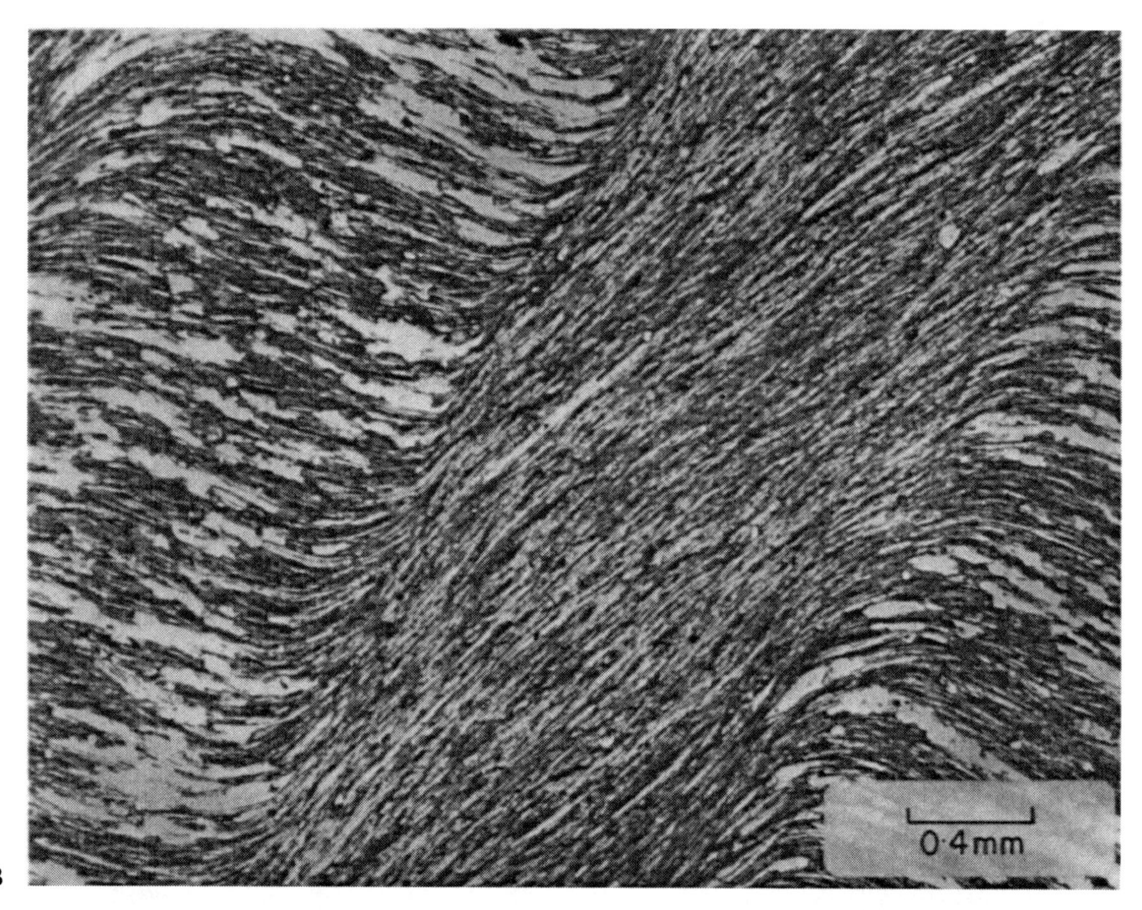

26 B

## 27. Crenulation Cleavage in Rhythmites

O. A. PFIFFNER

Upper Triassic Quartenschiefer at Fuorcla Punteglias (Infrahelvetic complex, eastern Switzerland).

The rhythmites consist of a sequence of alternating light fine-grained dolomitic and dark argillaceous layers, one to a few millimeters thick. The fabric is of post-Early Oligocene/pre-Miocene age (Calanda phase, cf. MILNES and PFIFFNER 1977, PFIFFNER 1977). The Calanda-phase deformation occurred during increasing burial and progressively more ductile behavior of the rocks. Metamorphic mineral assemblages indicate P-T conditions of 3–4 kbar and 350 °C at the end of the Calanda phase (FREY 1978, PFIFFNER 1977). The sample was collected from the normal limb of a Calanda-phase fold with a basement core. The crenulation cleavage developed in the rhythmites is an axial plane cleavage to this fold; it has been named Ausweichungsclivage by A. HEIM (1878, Plate XV, fig. 11). This crenulation cleavage can be traced from the rhythmites down into a dolomite formation, where it becomes a spaced cleavage ("fracture cleavage"), and upwards into slates, where it becomes a slaty cleavage.

The competent dolomite layers accomplish shortening by folding and low angle thrusting ("wedging", e.g., see **A**); folds show a tendency toward class IB (parallel; RAMSAY 1967). The argillaceous incompetent layers accommodate the necessary contact strains by folds of class 3 and secondary wrinkles. The rock cleaves along a set of undulating parallel spaced surfaces, defined by the inflection surfaces of the folds in the dolomite layers. The dolomite layers thin out toward, and become displaced along, these inflection surfaces (e.g., at **B**); but both these features die out with the inflection surface as one traces them up or down (**C** and **D**). The inflection surfaces transfer to new inflection surfaces, probably by means of bedding plane slip (e.g., inflection surface **C**–**B**–**D** is replaced by inflection surface **E**–**F**).

Polished surface of hand specimen

27

# 28. Crenulation Cleavage in Low Grade Metamorphic Rocks

M. R. STEUER

Photographs of cleavage in pelitic rocks of the Taconic allochthon in eastern New York, U.S.A. document the formation of a complex tectonic fabric (crenulation cleavage) in mildly metamorphosed shales and siltstones of the Cambrian Bull Formation. Zonal crenulation cleavage in microlithons is defined primarily by elongation of quartz and phyllosilicates along with minor intergranular residues of insoluble material. Field relations indicate that this early cleavage is axial planar to large recumbent isoclinal folds. A discrete crenulation cleavage cuts the zonal fabrics and is subparallel to axial planes of second-phase folds which fold the early isoclines. Fold axes of both sets are similar in orientation although axial planes are approximately orthogonal. The discrete crenulation cleavage is defined principally by thick accumulations of opaque material with grain elongations being subordinate or absent. Evidence for both cleavages forming via dissolution using the criteria of GRAY (1979) and the inference that recrystallization and grain growth processes were minor contributors to cleavage formation may have implications for cleavage formation in higher grade metamorphic rocks with morphologically identical cleavages.

---

**A. Buckled silt layer.** Cambrian Bull Formation near Salem, New York, U.S.A. The host rock is predominantly shale with thin interbeds of quartzose siltstone, metamorphosed to lower greenschist facies. The E–W cleavage is defined by long axes of quartz and phyllosilicates. N–S cleavage (discrete) is defined by residues of insoluble opaque material. Although the silt layer has experienced $\simeq 40\%$ buckle shortening, dissolution effects are not obvious except at the thinned limb of the fold on the left. PPL

**B. Discrete and zonal crenulations.** Cambrian Bull Formation. From the same hand specimen as **A.** NX

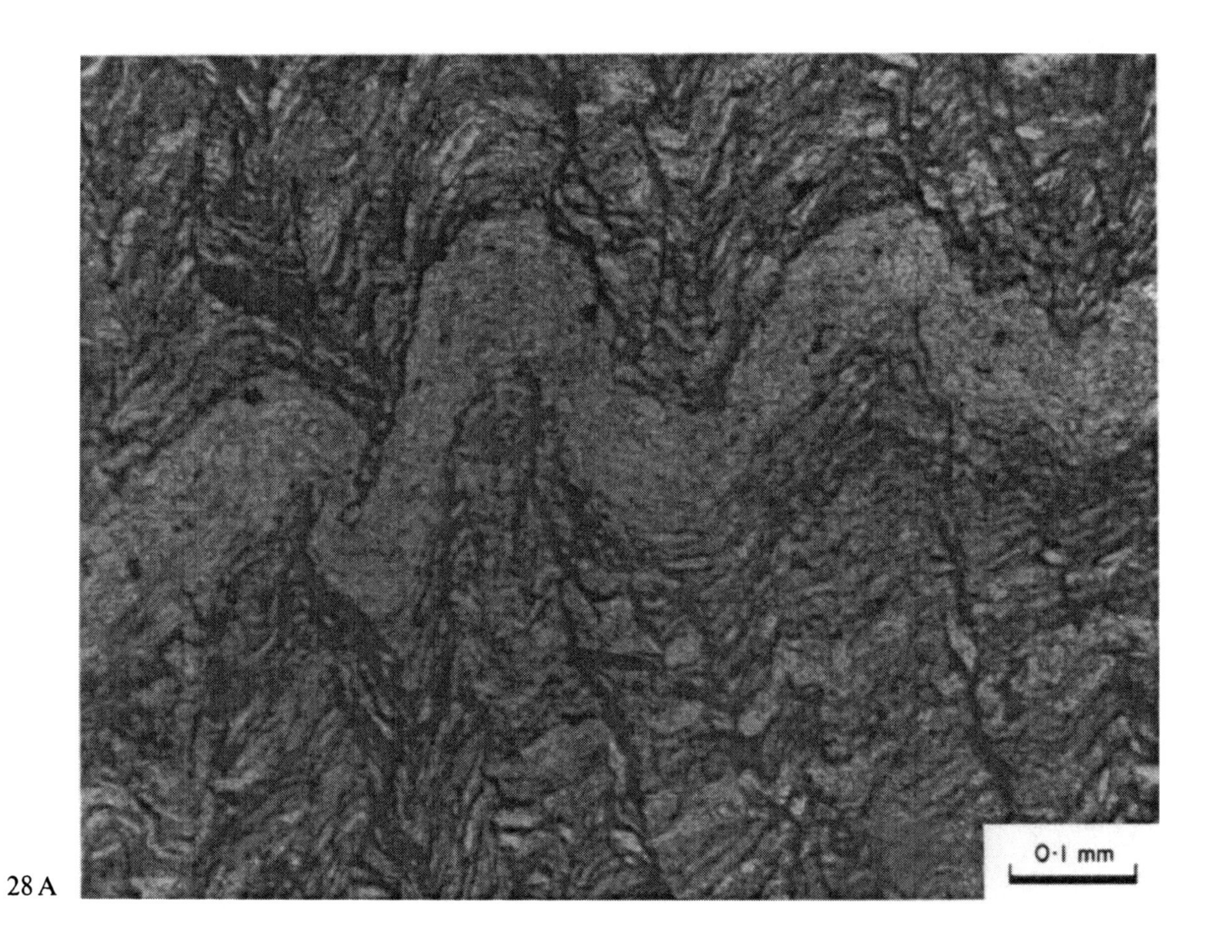

28 A

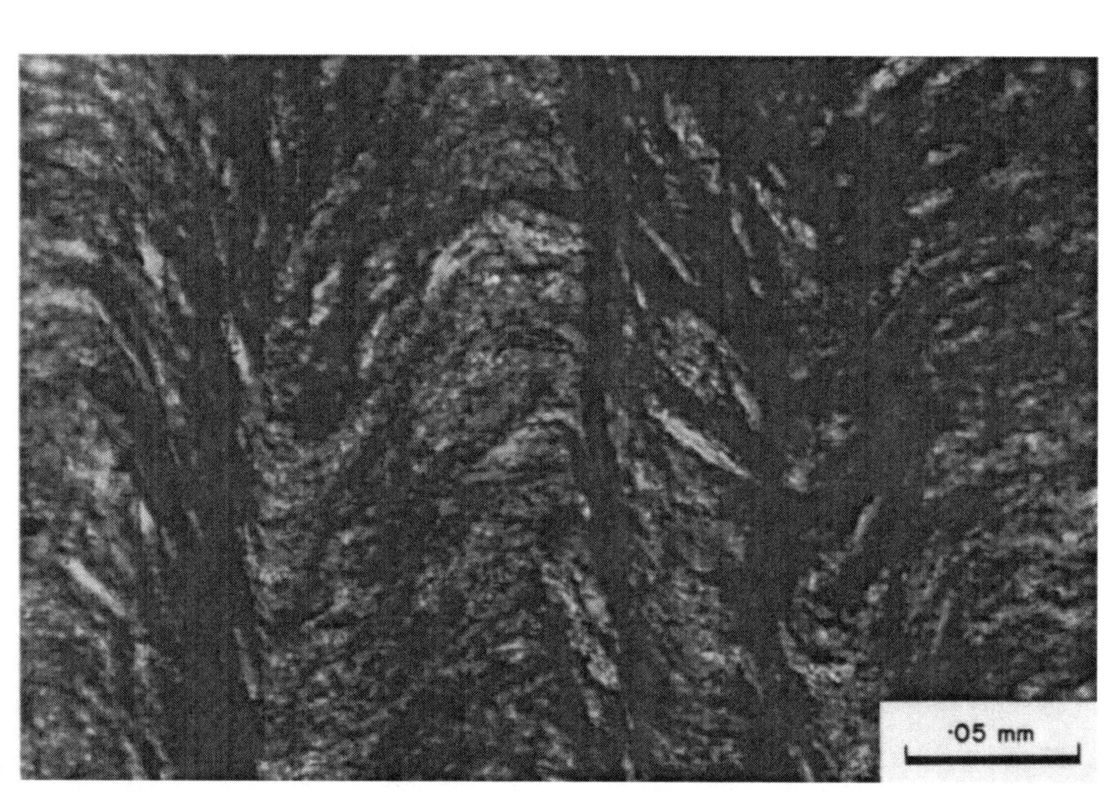

28 B

## 29. Crenulation Cleavage

J. W. Cosgrove

The New Harbour Series rocks of Treaddur Bay, Holy Isle, Anglesey, North Wales, U.K.
The cleavage was formed by the buckling of an earlier crenulation cleavage. Vestiges of this early cleavage [alternations of quartz rich (*light*) and mica rich (*dark*) bands] can still be seen in the hinges of the buckles. The early cleavage was originally at 90° to the later crenulation cleavage but has now been almost completely transposed by the processes of microfolding and metamorphic differentiation. A refolded example is shown in Plate 213.

29

## 30. Low-Grade Dalradian Sediments Carrying Spaced Cleavage: (I) Setting

H. J. Bradbury and A. L. Harris

Dewatering of the Dalradian (Riphean-Cambrian) succession at the onset of the Grampian Orogeny was accompanied by the formation of an extensive pressure solution cleavage system. This is best developed and preserved within greywackes at high levels within the nappe pile at the Highland Border near Dunkeld, Scotland, U.K. (Harris et al. 1976, Bradbury et al. 1979).

---

**A. Downward-facing synform.** Southern Highland Group, Little Glen Shee, Tayside. Large wavelength, first order, $D_1$ synform faces down on the evidence of ripple cross-lamination and grading within psammites. Spaced cleavage within psammite beds forms a strongly convergent cleavage fan by contrast with the dominantly axial planar habit of slaty cleavage within pelites. Greenschist facies. Approximate fold profile section.

**B. Hinge zone of a $D_1$ synform.** Little Glen Shee. Refracted lithons within the transposed semi-pelitic bed (*center*) change their sense of asymmetry around the synform hinge and are continuous with a slaty cleavage towards the original top of a graded unit (*near coin*). The sharp original base of the underlying psammite shows "dip-slip" off-sets along a spaced cleavage.

**C. Spaced cleavage.** Little Glen Shee. Vertical bedding youngs NW (*right* to *left*) on the SE limb of a downward-facing antiform adjacent to the synform in **A** above. (For "facing" see Plates 226, 227.) Variations in cleavage morphology and geometry are similar within each graded unit of which two are shown. The tapering of quartz-dominated lithons from the original base (*right*) toward the pelitic top (*left*) of each graded unit is accompanied by a corresponding increase in the width of the intervening phyllosilicate-rich domains, which often coalesce at the pelitic top. The change in angle between bedding and cleavage from ca. 50° at the base to ca. 20° at the top of each unit defines cleavage refraction arcs which are always convex to antiform axial surfaces. Cleavage orientation is primarily controlled by variations in original clast size and type, while its convexity is dependent on the width of graded units

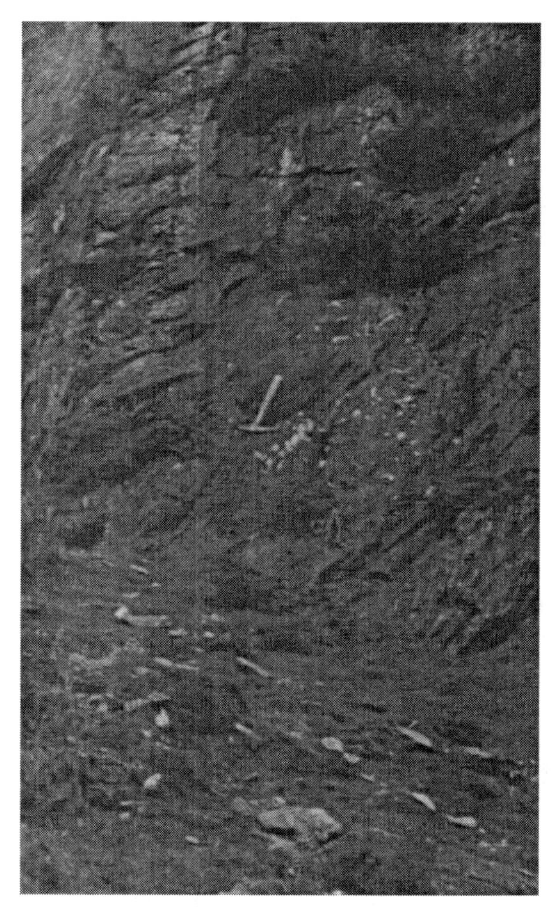

30 A

30 B

30 C

# 31. Low-Grade Dalradian Sediments Carrying Spaced Cleavage: (II) Spaced Cleavage Morphology

H. J. BRADBURY and A. L. HARRIS

**A. Morphological variations in graded units.** Detail of Plate 30. The consistent orientation of bedding laminations (E–W) at the base and top of graded units allows for little rigid-body rotation within individual quartz-dominated lithons (NW–SE) during cleavage formation. On fold limbs, spaced cleavage at the base of graded units (center) has a characteristically asymmetric form in fold profile section. That part of the lithon defined by bedding at the base of graded units is commonly cuspate and convex toward the sandstone in the direction of younging. Lithons are apiculate toward antiform hinges and rounded toward synform hinges. Apices often overlap adjacent lithons and are elongate parallel to the slaty cleavage within the pelitic top of the adjacent graded unit. Though refracted, slaty cleavage is continuous through phyllosilicate-rich domains of the spaced cleavage from the pelitic top of one graded unit to that of the next unit. At fold hinges, lithon "bases" have a symmetrical molar form.

**B. Spaced cleavage variants.** Little Glen Shee. In the lower half of the plate, spaced cleavage (NE–SW) of the type described in **A** above is continuous with a lenticular cleavage similar to that in **C** below.

**C. Lenticular spaced cleavage.** Birnam Grits, Newtyle Hill, Dunkeld, U.K. Cleavage lithons developed within thick beds of poorly graded, immature sandstone commonly have a lenticular habit and often show a marked periodicity of width and length in both sections normal to cleavage. In thin section, the quartz-rich lithons alternate with chlorite-white mica domains against which clastic grains are truncated and within which quartz has undergone grain-shape modification and volume reduction, indicating a dominant pressure solution mechanism (HARRIS et al. 1976). In multiple cleavage folds, a slaty cleavage (ENE–WSW) is developed oblique to the lengths of lithons. In thin section, this comprises dark cleavage folia which anastomose around bearded and sometimes microboudinaged clastic quartz and feldspar. The evidently limited transport of dissolved quartz within lithons contrasts with that inferred from nearby phyllosilicate seams

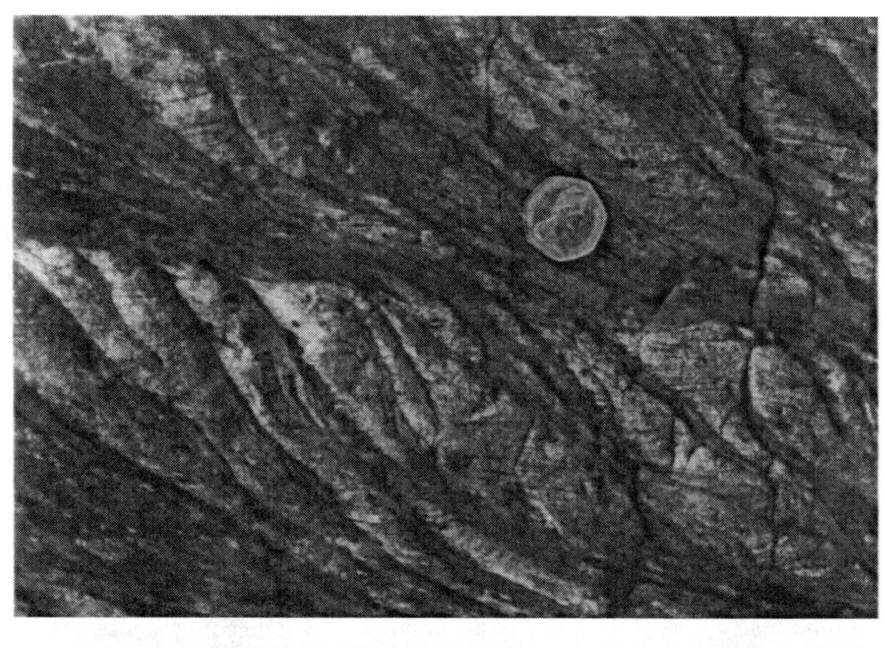

31 A

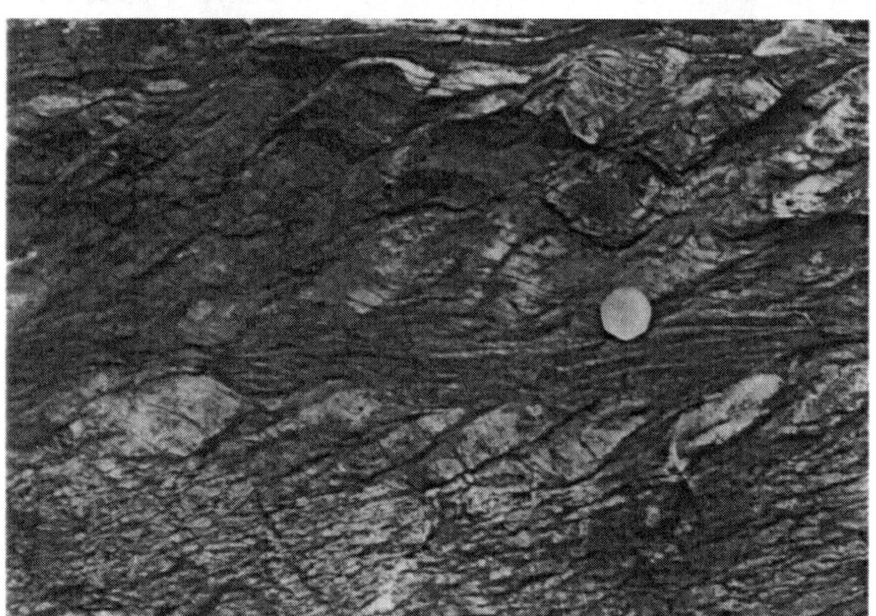

31 B

31 C

## 32. Low-Grade Dalradian Sediments Carrying Spaced Cleavage: (III) The Formation of Spaced Cleavage

H. J. Bradbury and A. L. Harris

**A, B, C. Lithon initiation in bed-delimited spaced cleavage.** Little Glen Shee, Tayside, Scotland, U.K. These photographs illustrate a possible buckle-fold mechanism for the initiation of quartz-dominated lithons of the type described in **A** on the previous page. Bedding laminations within the central semi-pelitic unit in **A** are disposed in asymmetric buckle-folds, the steep and shallow limb zones of which constitute an incipient differentiated layering. In **A**, steep and shallow limb zones can be traced into phyllosilicate-rich and quartz-rich domains respectively of a spaced cleavage developed within adjacent psammites (see also **B** which is an enlargement of part of **A**). Thin sections across close-spaced lithons of the type shown in **A** on the previous page indicate their formation by a combination of buckle-shortening and solution-shortening during rotation within phyllosilicate domains, probably under the type of controls envisaged for the evolution of crenulation cleavages (e.g. Gray and Durney 1979). Intermediate stages of development such as that in **C** show:

1. The obliquity of the slaty-type fabric in phyllosilicate domains to the length of the domain.
2. The continuity of this fabric with bedding in quartz-dominated lithons.
3. The convergence of this fabric and bedding within pelites (cf. Plate 31 A)

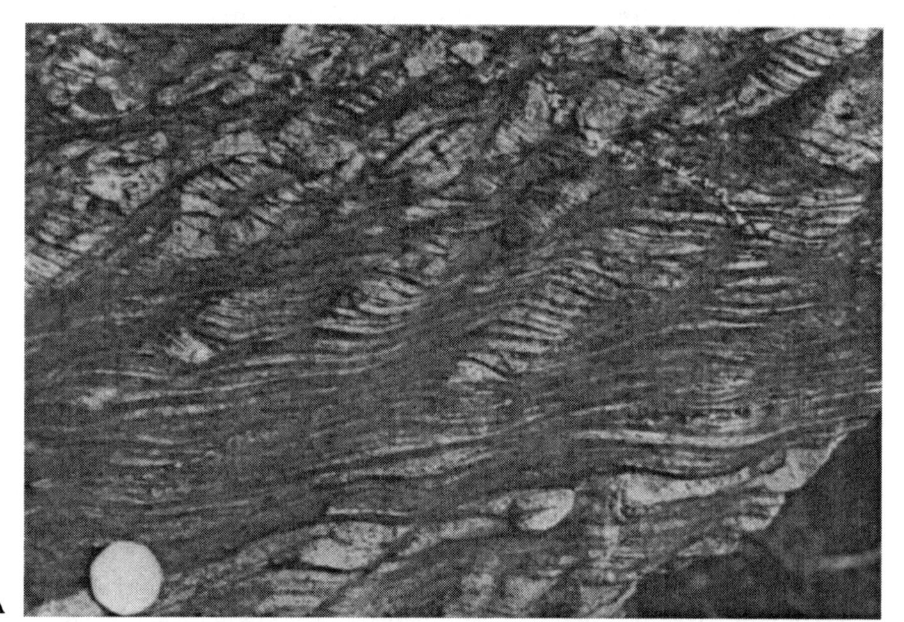

32 A

32 B

32 C

# 33. Low-Grade Dalradian Sediments Carrying Spaced Cleavage: (IV) Deformation of Spaced Cleavage

H. J. BRADBURY and A. L. HARRIS

By contrast with high structural levels in the Tay Nappe, superimposed simple and pure shear strains markedly modified the form and geometry of early pressure solution cleavages at low structural levels during progressive upper greenschist-lower amphibolite facies metamorphism (HARRIS et al. 1976, BRADBURY et al. 1979). The photographs are from the Dunkeld Grits, River Braan, Dunkeld, Scotland, U.K.

---

**A. Shear zone modification of early spaced cleavage.** Early-formed lenticular spaced cleavage is markedly attenuated within narrow shear zones (*top* and *bottom*) separated by augen zones, often en echelon, within a belt of minimum 2 km width at the Scottish Highland Border. Shape fabrics are produced from deformed clastic grain aggregates within shear zones, whereas textures common to quartz-rich domains of the original spaced cleavage are commonly preserved within augen zones, except where a superimposed vertical shortening has produced buckle-folds and a weak slaty fabric from finer-grained matrix minerals (*center*).

**B. Attenuated spaced cleavage.** In spite of the modification of the dimensions of lithons and constituent clastic minerals within shear zones, the original lenticular habit of the cleavage is retained as a braiding. During secondary deformation the originally high angle between bedding and cleavage was considerably reduced to ca. 5°. New zones of preferential solution transfer comparable with phyllosilicate seams of the early cleavage were extremely localized during secondary deformation, at this structural level and under these metamorphic conditions.

**C. Strain bands.** In addition to shear zones and en echelon sets of augen, a sharply defined layering of alternating low and high strain bands is developed during secondary deformation in certain parts of the Braan section

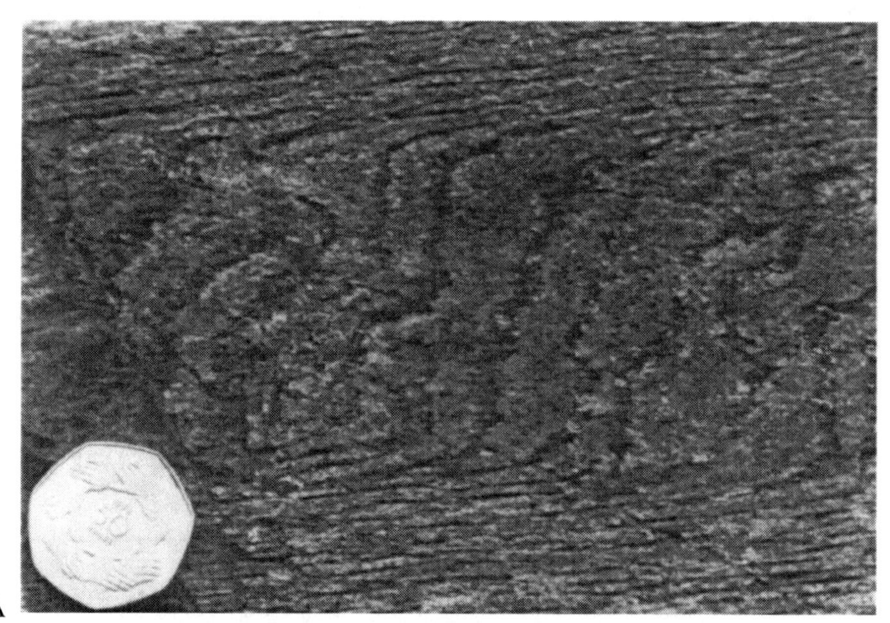
33 A

33 B

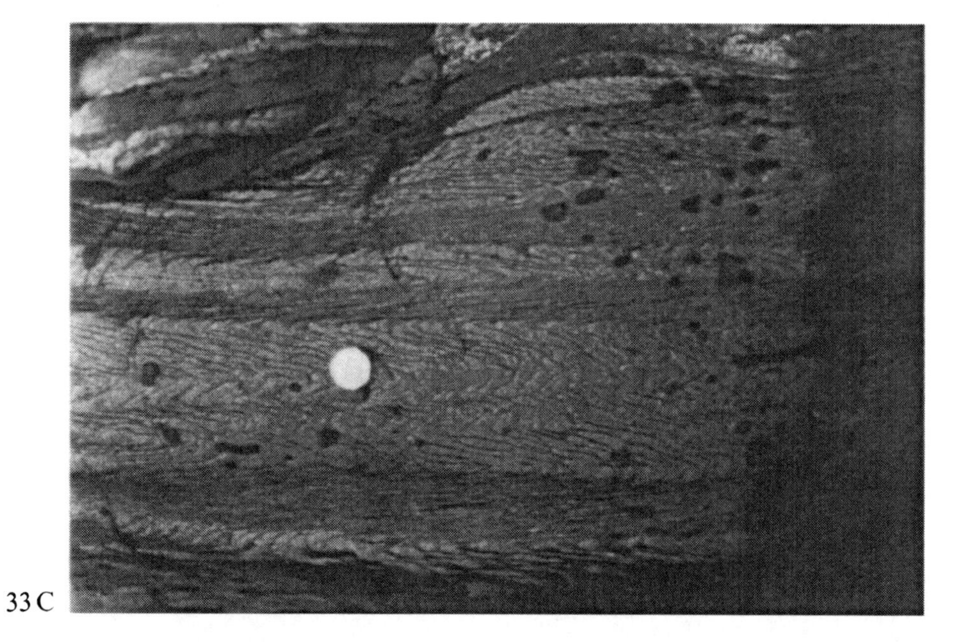
33 C

# 34. Polyphase Deformation of Spaced Cleavage

H. J. BRADBURY and A. L. HARRIS

The Highland Border Steep Belt was formed by late Grampian downbending of previously flat-lying planar fabrics and associated linear elements about the hinge of a major antiform (Highland Border Downbend), probably as a result of basement movement along the ancestral line of the Highland Boundary Fault (BRADBURY et al. 1979). Evidence of the polyphase deformation of early spaced cleavage exists within the hinge region of this major antiform, within the Almandine Zone, at the lowest structural levels exposed within the nappe pile at the Highland Border. The photographs are from Southern Highland Group rocks from Creag Vinean, near Dunkeld, Scotland.

---

**A. Deformation of strain bands.** The variations in the multilayer system produced during the second-phase deformation of the original spaced cleavage into strain bands of the type illustrated in Plate 33 **C**, have controlled the wavelength-amplitude characteristics of the subsequently formed buckle-folds, which are parasitic on the Highland Border Downbend. Crenulation cleavage is only well developed within previously strongly attenuated cleavage stripes.

**B. Foliation-delimited crenulation cleavage.** Crenulation cleavage within the pelite does not significantly penetrate the competent granular sandstone. The sandstone carries the strongly modified spaced cleavage sub-parallel with bedding.

**C. Folding associated with the downbend.** The psammite bed carries the attenuated lenticular spaced cleavage at a discernible angle to bedding. This composite foliation is deformed about open folds parasitic on the Downbend. A bed-delimited crenulation cleavage within the more pelitic unit is developed parallel to the axial plane of the Downbend

34 A

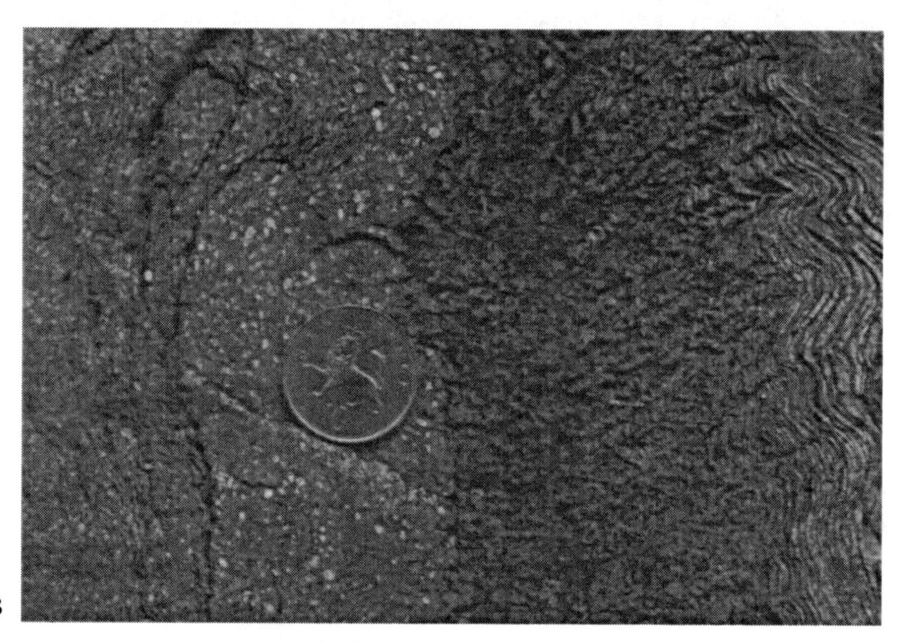

34 B

34 C

# 35. Gradational Crenulation Cleavages in Multilayers

D. R. GRAY

The photographs illustrate the dependence of crenulation cleavage morphology upon layer thicknesses, or more precisely the multilayer thickness ratio ($t_1/t_2$, where $t_1$ is competent layer thickness and $t_2$ is incompetent layer thickness) for regular multilayers consisting of alternating quartz and mica/carbonaceous material layers. All the samples shown are from the Permian Nambucca Slate, N.S.W., Australia. Metamorphic grade is low greenschist facies (LEITCH 1975).

---

**A. Multilayer with $t_1/t_2 < 1$.** The zonal cleavages (vertical) show less thinning of quartzose layers than those in **B** and **C**. Smaller-wavelength crenulations occur in the micaceous layers. PPL

**B. Multilayer with $t_1/t_2 \simeq 1$.** Thinning of both quartzose and micaceous layers occurs along microfold limbs. Well-developed zonal cleavages do not contain quartzose layers. There are no smaller wavelength crenulations in the micaceous layers. PPL

**C. Multilayer with $t_1/t_2 > 1$.** Zonal cleavages are transitional into discrete cleavages along microfold limbs. Quartzose layers are in contact and show thinning and truncation. Some layers now occur as detached segments within hinge zones. PPL

**D. Zonal crenulation cleavage** on the steep limbs of asymmetric microfolds. Cleavage zones contain less quartz and a higher proportion of mica/carbonaceous material. There is thinning and partial truncation of thick quartzose layers whereas thin layers are completely truncated. Layer thinning and cleavage differentiation is due to dissolution and removal of quartz from the microfold limbs (GRAY 1979). PPL

**E. Incipient zonal crenulation cleavage.** Both the micaceous and quartzose layers are thinned in the cleavage zone. Note that the quartzose layers have irregular, interlocking boundaries where they are in contact with adjacent layers. This is due to dissolution and removal of quartz along the layer interfaces as the microfolds develop. PPL

35 A

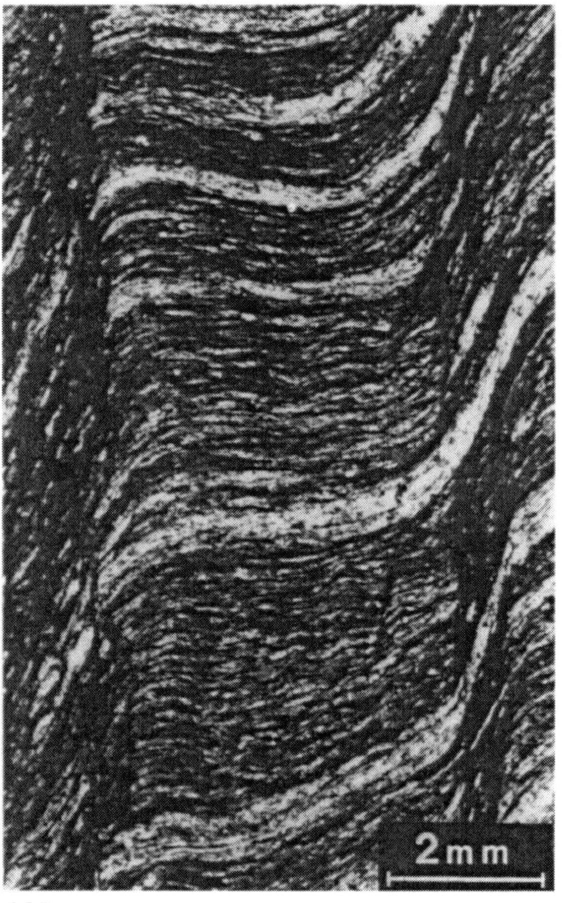

35 D

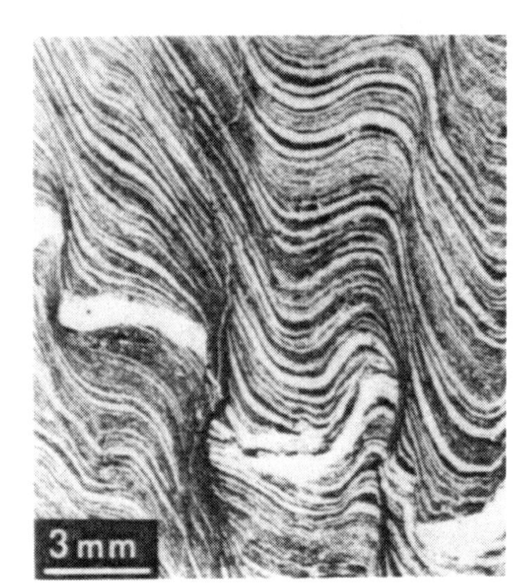

35 B

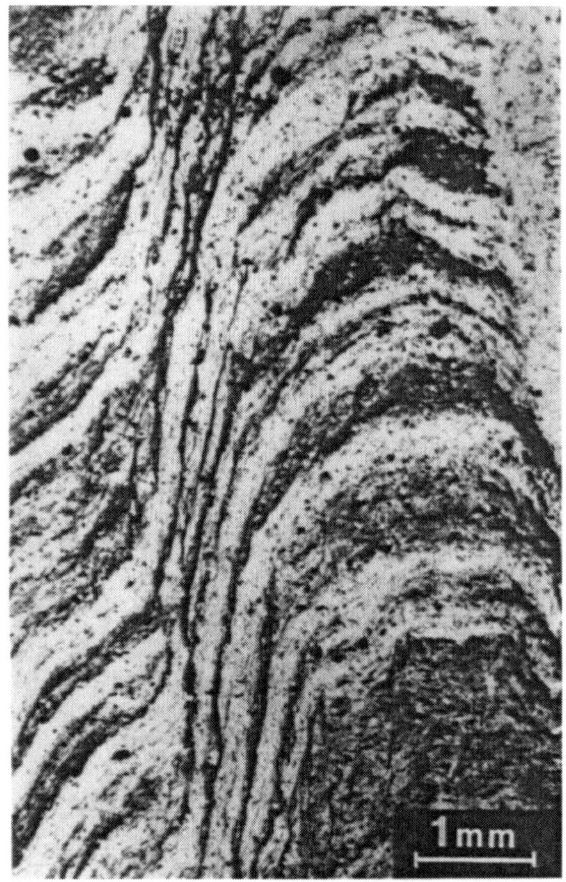

35 E

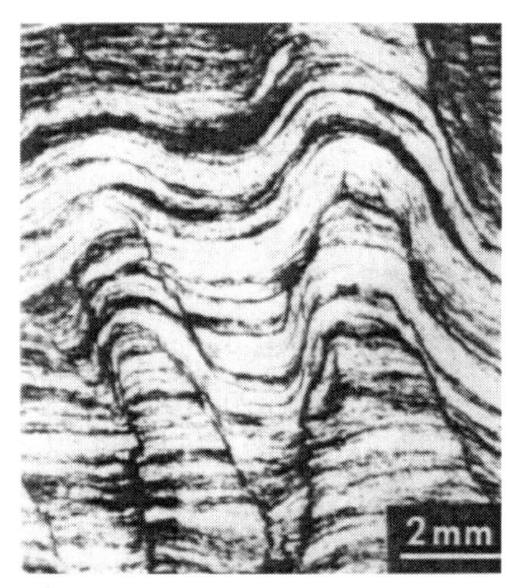

35 C

# 36. Cleavages in Psammitic Rocks

D. R. Gray

**A. Weak rough cleavage.** Undifferentiated Ordovician at Zane Grey Pool, Bermagui, N.S.W., Australia. Weakly developed $S_1$ rough cleavage (*vertical*) in singly-deformed wackestone. Cleavages are short, discontinuous seams that wrap around detrital quartz and feldspar grains. *Seams* represent solution surfaces where phyllosilicates have been passively concentrated due to removal of quartz and feldspar. PPL

**B. Strong rough cleavage.** Undifferentiated Ordovician at Zane Grey Pool, Bermagui, N.S.W., Australia. Well-developed $S_1$ rough cleavage seams (*vertical*) in silty sandstone. Cleavages are more closely spaced, more continuous, thicker seams that anastomose between larger grains to produce a trapezoidal appearance in the fabric. Shape-modified detrital grains have either planar truncated boundaries or irregular sutured boundaries. Mica beards have developed on larger grains. Cleavage development is due to dissolution and removal of quartz and feldspar from the cleavage accompanied by metamorphic grain growth of phyllosilicates. PPL

**C. Intersecting rough cleavages.** Silurian Chesleigh Formation at Sofala, N.S.W., Australia. Poly-deformed psammite with intersecting $S_1$ rough cleavage (*oblique*) and $S_2$ rough cleavage (*vertical*). The $S_1$ cleavage is transitional into a slaty cleavage in the adjacent slate, whereas $S_2$ cleavage is transitional into a discrete crenulation cleavage (see fig. 15, Gray 1978). PPL

**D. Incipient zonal crenulation cleavage.** Undifferentiated Silurian, Limerick, N.S.W., Australia. Poly-deformed psammite with a zonal cleavage defined by accentuated $S_1$ rough cleavage traces. The accentuated traces are thicker and darker and are arranged en echelon along the steep limb of a microfold. Cleavage development is due to further dissolution along the existing rough cleavages. PPL

36 A

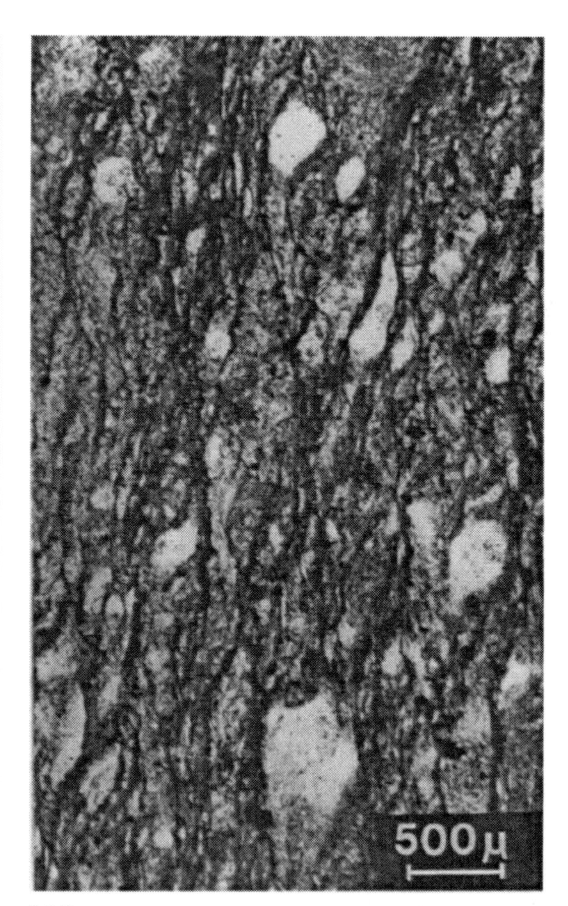

36 B

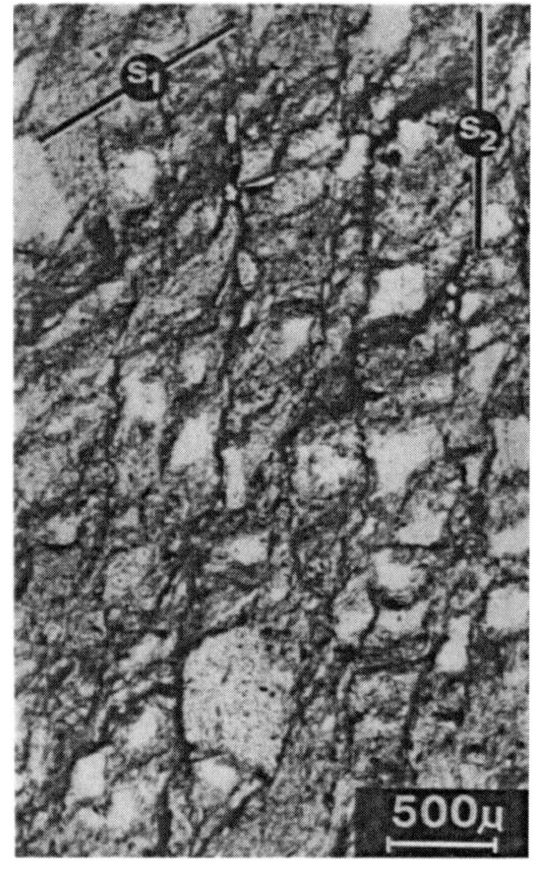

36 C

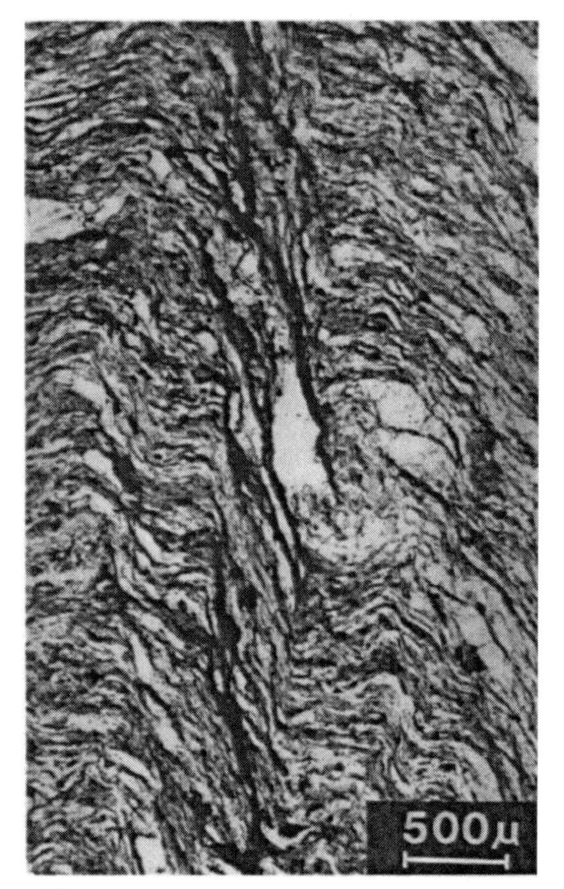

36 D

## 37. Cleavages in a Poly-deformed Schist

D. R. GRAY

Ordovician Abercrombie Beds, near Bigga, N.S.W., Australia. These photographs show variations in morphology of three cleavages which have developed in a biotite-schist (biotite-quartz-albite-muscovite assemblage).

---

**A. Mesoscopic structural relationships.**
Four foliations are present:
$S_0$ (bedding): defined by alternating silty and pelitic laminae (oblique). Note the grading in the silt lamina.
$S_1$ (slaty cleavage): defined by a dimensional preferred orientation of mica, quartz and albite (see E).
$S_2$ (crenulation cleavage): defined by a striped layering (*subhorizontal*) with zonal cleavage morphology.
$S_3$ (crenulation cleavage): defined by dark, subplanar traces (*vertical*) with both discrete and zonal morphology. PPL

**B. Intersecting cleavages.** Zonal crenulation cleavage ($S_2$) re-crenulated and truncated by another zonal cleavage ($S_3$). Note the progressive decrease in leucocratic minerals from the $S_2$ to the $S_3$ cleavage. PPL

**C. Zonal crenulation cleavage ($S_2$).** Domainal fabric (*striped layering*) showing the progressive grain size reduction and removal of quartz and albite necessary to produce the $S_2$ cleavage. PPL

**D. Discrete crenulation cleavage ($S_3$).** Discrete morphology of $S_3$ cleavage within a wide pelitic zone. Note the distinct $S_1$ fabric and the restricted development of $S_3$ to the steep limbs of the asymmetric microfolds. PPL

**E. Refracted crenulation cleavage ($S_3$).** Zonal crenulation cleavage ($S_3$) refracted through a silty layer. Note the slight discordance between $S_0$ and $S_1$ at the contact between the siltstone and the pelite. PPL

37 A

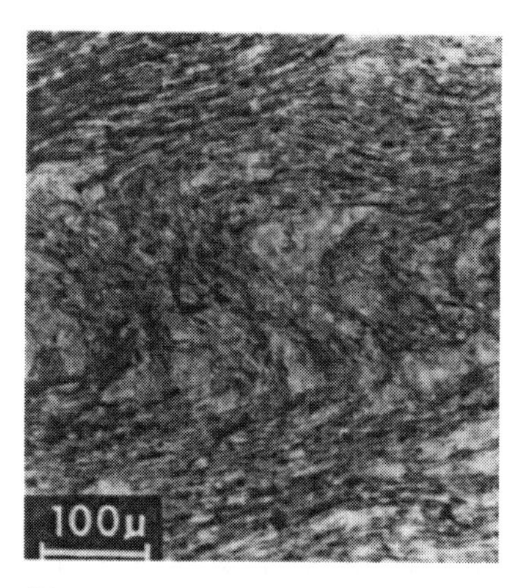

37 C

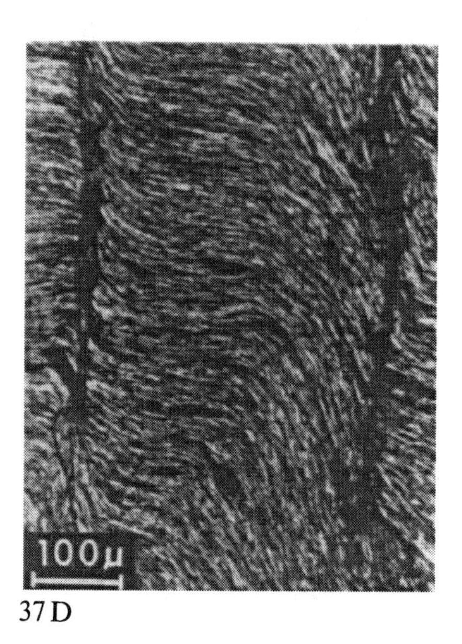

37 D

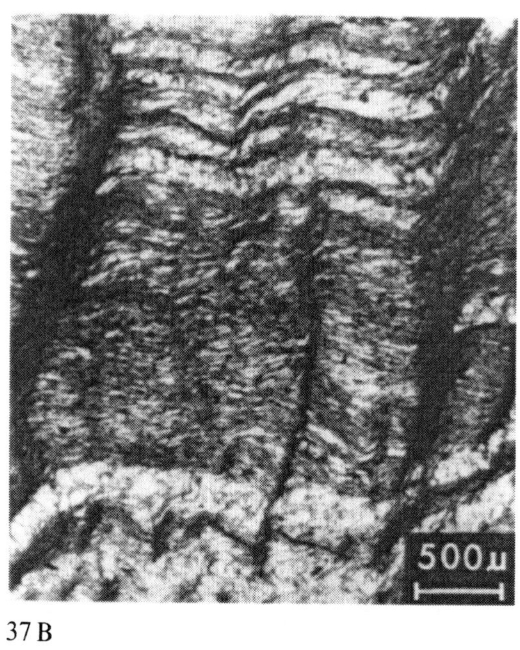

37 B

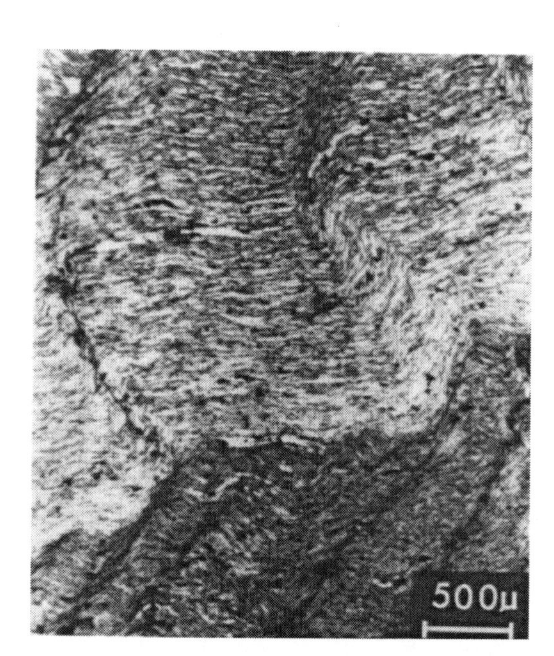

37 E

## 38. Gradational Crenulation Cleavages in Schist

D. R. GRAY

The specimen shown in the photographs is a fine-grained biotite schist from the Proterozoic Sturtian Tapley Hill Formation at Greenock, S.A., Australia. (Specimen by courtesy of Dr. K. Mills.) Metamorphic grade is within the biotite zone of the greenschist facies, typified by biotite-quartz-muscovite assemblages in pelites.

Crenulation cleavage development, shown by progressive fabric-modification from **C** to **D** to **B**, is accompanied by changes in the grain shape of quartz and albite grains, and changes in mineralogical proportions (i.e., there is a progressive decrease in proportion of quartz and albite in the cleavage relative to the host rock). The differentiation is due primarily to dissolution of quartz and albite along the limbs of developing microfolds accompanied by diffusional transfer of dissolved material to the fold hinges where it precipitated as overgrowths on existing grains. The degree of differentiation is dependent on the tightness of the associated microfolding (GRAY 1979, figs. 1 and 2). Crystallization of mica in the cleavage is unlikely since there are no visible changes in the size and shape of micas around the microfolds. Furthermore, the micas in the hinges are bent and in some places kinked, indicating that no recovery or recrystallization of mica has occurred. Quartz and albite grains in developing cleavage zones show sutured and corroded boundaries typical of a dissolution-deformation mechanism.

---

**A. Zonal crenulation cleavage.** Incipient zonal cleavage along the short limbs of asymmetric microfolds. The cleavage zones are noticeably darker and contain a lower proportion of quartz and albite than the host rock. PPL

**B. Domainal fabric** with alternating zonal crenulation cleavage and "quartzose" microlithons. The cleavage corresponds to the steep limbs of asymmetric microfolds. Note that the quartz/albite grains in the microlithon are larger and more equidimensional than grains in the preexisting fabric (C) reflecting deposition of dissolved material in the microlithon zones. The microlithon pictured has a "quartz"/mica ratio of 60:40 (which is higher in quartz than the pre-existing fabric) and a mean "quartz" aspect ratio (l/w) of 1.8. PPL

**C. Incipient zonal crenulation cleavage** along the limb of a microfold. The incipient cleavage has a higher proportion of mica relative to quartz/albite and a difference in quartz/albite grain shape compared with the pre-existing fabric shown in the hinge zones (microlithons). The microlithons have a "quartz"/mica ratio of 54:46 and a mean "quartz" aspect ratio (l/w) of 2.2, whereas the cleavage has a "quartz"/mica ratio of 42:58 and a mean "quartz" aspect ratio (l/w) of 3.0. PPL

**D. Zonal crenulation cleavage** with a marked increase in mica relative to quartz/albite. Note the thin lenticular, inequant habit of relict quartz/albite grains in the cleavage. The cleavage has a "quartz"/mica ratio of 15:85 and a mean "quartz" aspect ratio (l/w) of 6.0. PPL

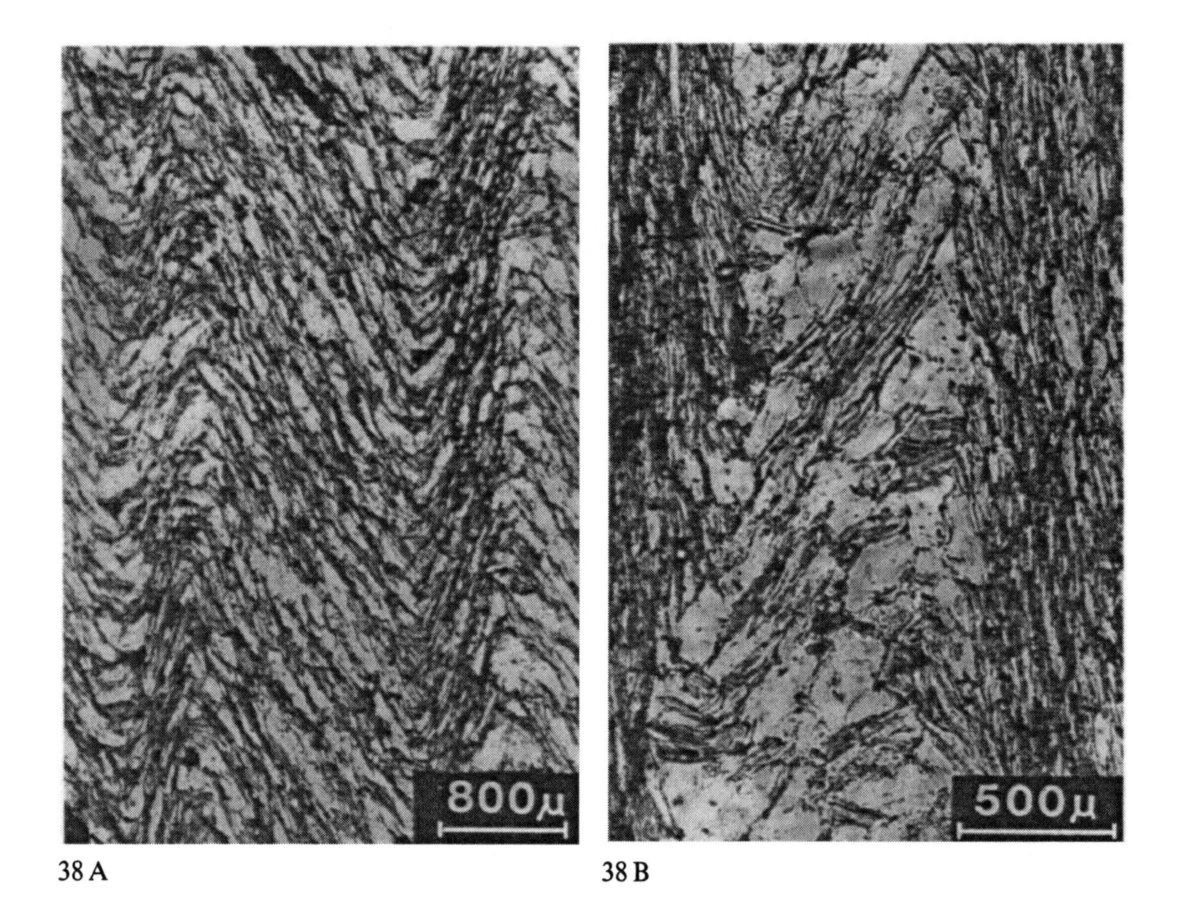


38 A 38 B

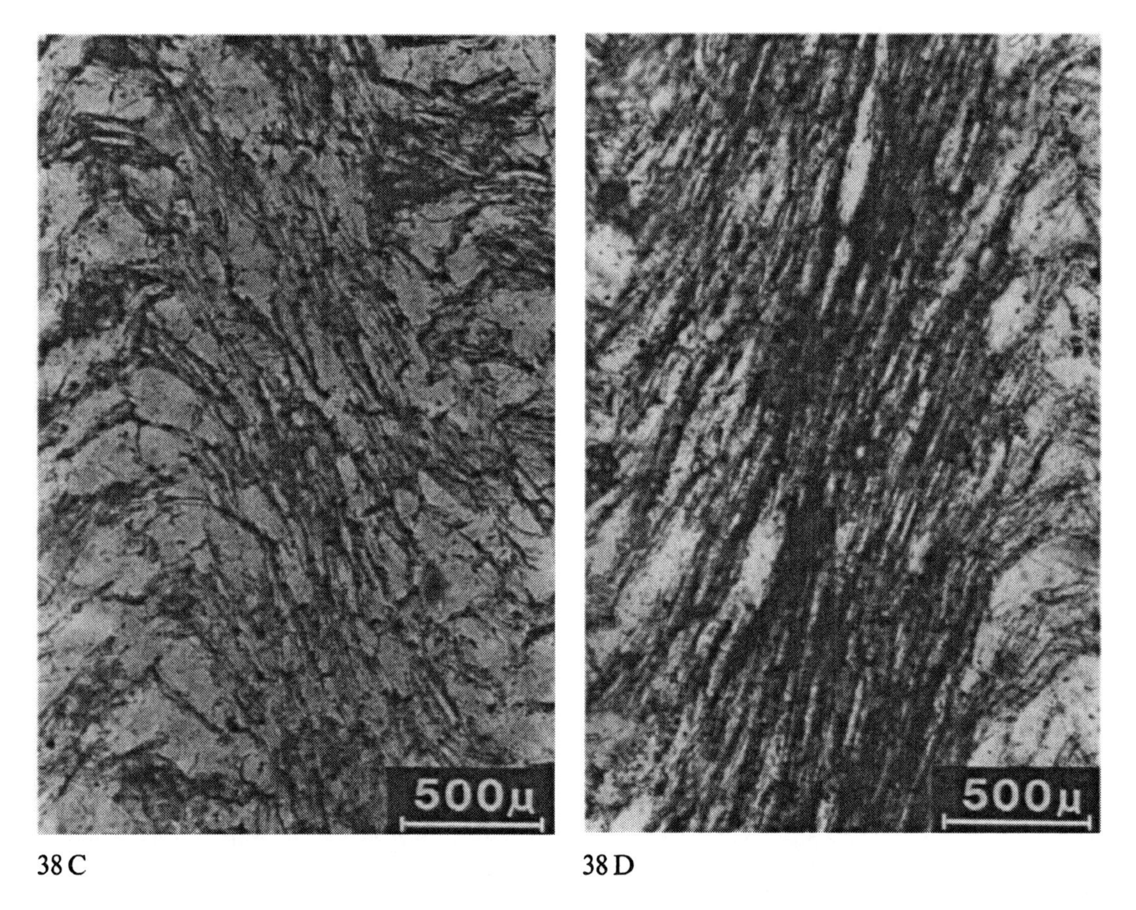


38 C 38 D

# 39. Spatial Variations in a Crenulation Cleavage: Microfaults and Crenulations

R. J. HOLCOMBE

The photographs here and in Plate 46 show morphological variations in the second-generation schistosity, $S_2$, in the Ocoee Gorge/Ducktown area of southeastern Tennessee, U.S.A. The sections shown are all from psammopelites of the Upper Precambrian Ocoee Series. The principal variations between samples are metamorphic grade and intensity of the second deformation, which has no effect at low chlorite grade but produces macroscopic folds near the staurolite isograd at Ducktown.

---

A. **Microfaults,** Walden Creek Group, middle chlorite grade, Ocoee Gorge. Penetrative microfaults and minor shears define the lowest intensity occurrence of the $S_2$ in the area. Scale: 1 mm PPL

B. **Crenulation bands.** Great Smoky Group, upper chlorite grade, Ocoee Gorge. Branching and anastomosing normal micro-kinks (DEWEY 1965) with irregular spacing characterize $S_2$ at this point. The *dark lines* marking the apparent orientation of $S_1$ within the bands are actually secondary crenulations and the true $S_1$ orientation is considerably steeper (see sketch). These secondary crenulations are interpreted to be second-order structures akin to Riedel shears rather than an overprinted fabric. The crenulation bands are slightly depleted in quartz, albite and carbonate relative to the rest of the rock (shown by microprobe analysis). Scale: 1 mm PPL

C. **Coherent crenulations.** Great Smoky Group, biotite grade, Ocoee Gorge. The $S_2$ foliation is defined by strongly differentiated crenulations in which the folded $S_1$ is continuous across the $F_2$ microfolds. This section is from the hinge of a small $F_2$ fold. Scale: 1 mm PPL

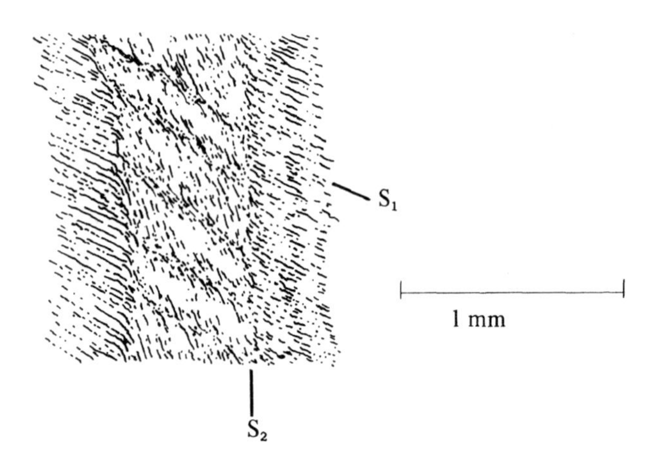

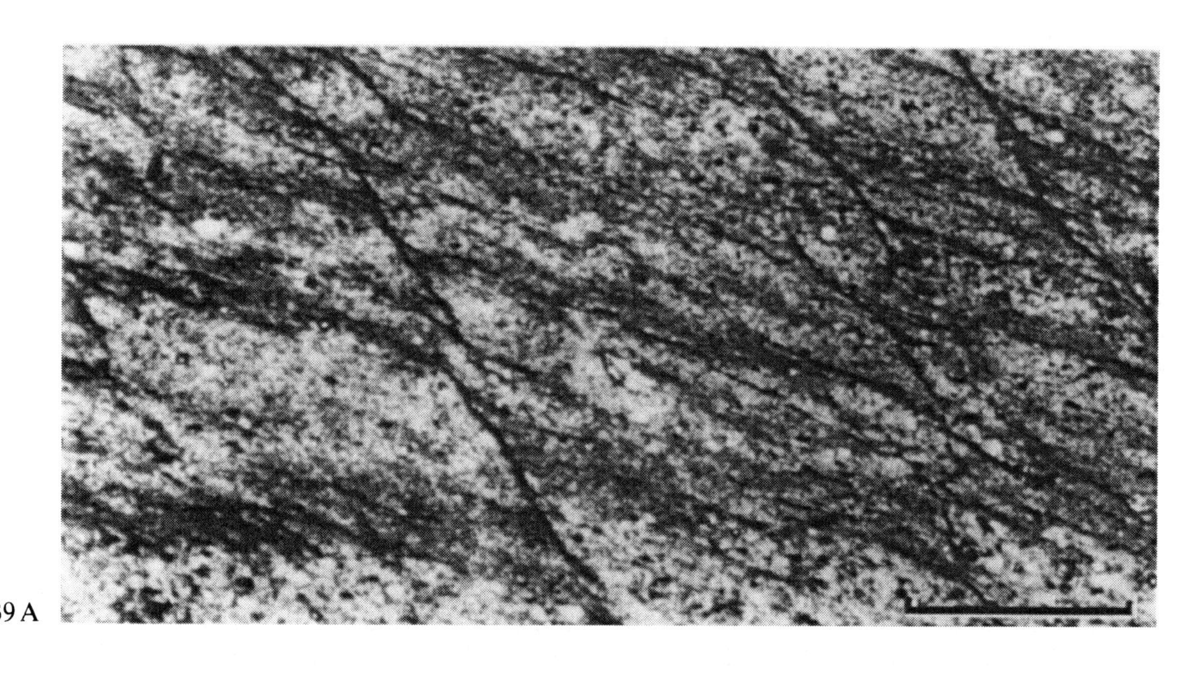
39 A

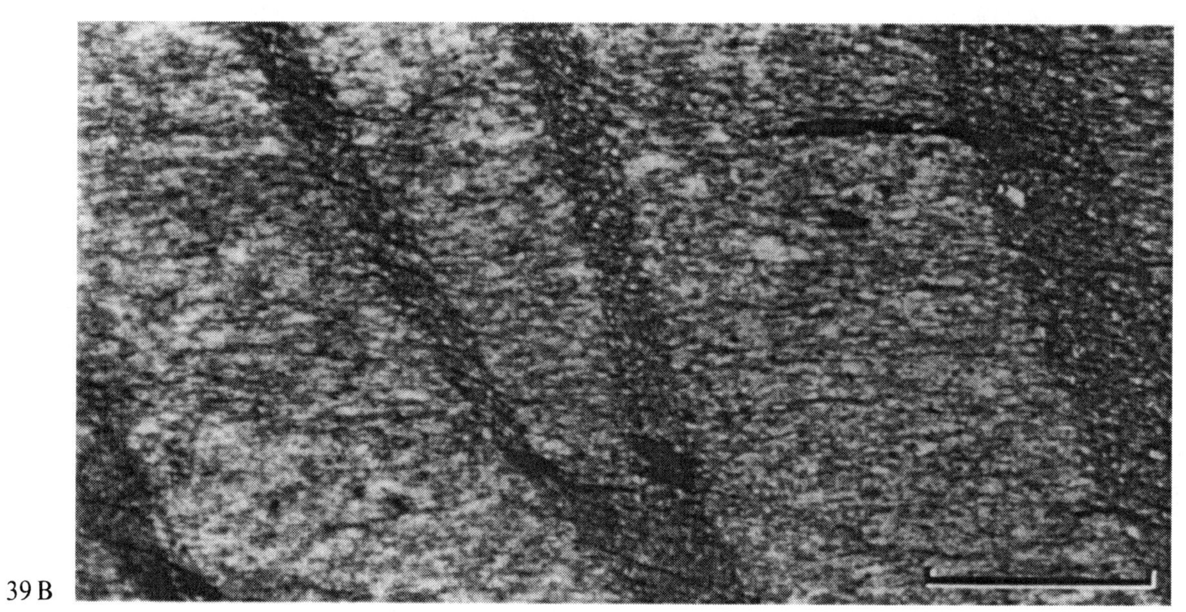
39 B

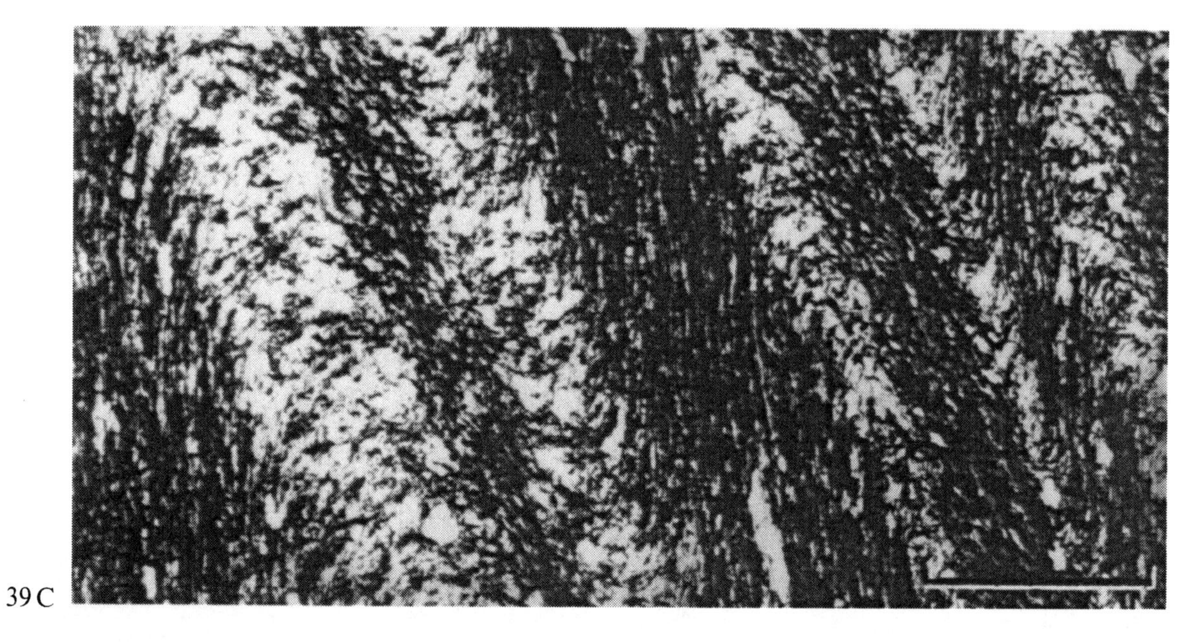
39 C

## 40. Segregated Crenulation Cleavage

C. McA. Powell and D. Martin

Himalayan Central Gneiss, about 0.8 km NE of Sissu, Chandra Valley (Powell and Conaghan 1973a)

A segregated crenulation cleavage in a biotite-muscovite-quartz-feldspar schist developed during $D_2$ accompanying Neogene overthrusting along the Main Central Thrust (Powell and Conaghan 1973b). Small-scale folds to which the crenulation cleavage is axial planar have subhorizontal axes trending WNW–ESE and axial surfaces dipping less than 10° to the southwest. The cleavage folded is $S_1$ of Powell and Conaghan (1973a), and, in places, $S_1$ is itself a segregated crenulation cleavage.

In the photograph, $S_1$ is folded into an S-fold running from the *upper right to lower left*. The segregated $S_2$ crenulation cleavage, subhorizontal in outcrop, has much higher concentrations of mica on the microfold limbs than in the hinges, and the quartz and feldspar grains are fewer and have reduced cross-sectional areas in the limbs compared with the fold hinges. Muscovite increases in modal percentage and cross-sectional area with increasing tightness of microfold profile, especially on the limbs of the microfolds.

The segregated cleavage has presumably developed by selective removal of quartz and feldspar from the fold limbs in the manner suggested by Marlow and Etheridge (1977, fig. 10). Rb-Sr dating using a biotite-whole rock pair from a sample 200 m from this outcrop gives a biotite age of 19 million years (Powell et al. 1979, sample 73–561).

40

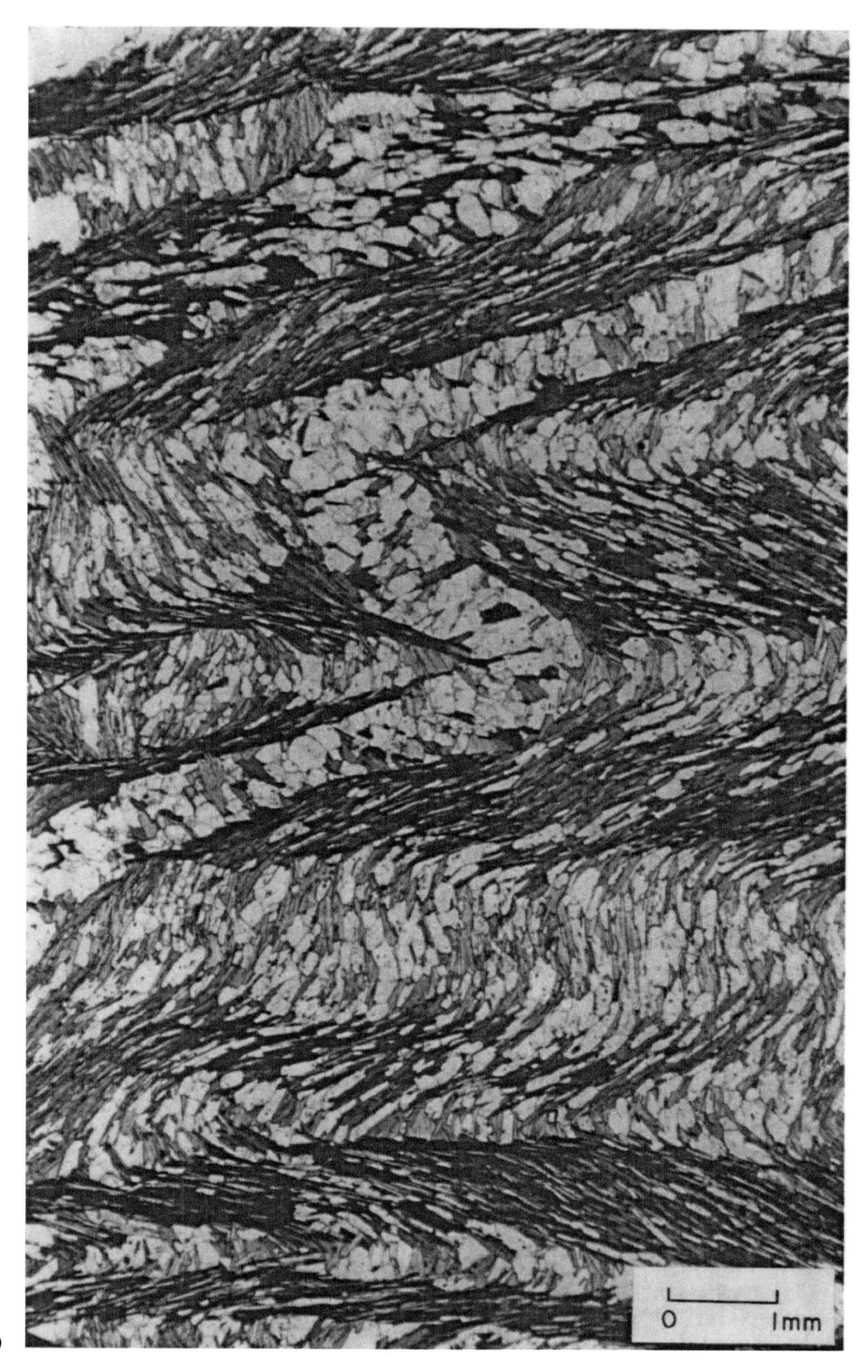

# 41. Crenulation Cleavage and the Buckling of Anisotropic Materials

J. W. COSGROVE

It has been shown both experimentally and theoretically that the geometry of structures that develop in a mechanically anisotropic material when it is compressed depends upon the degree of anisotropy of the material and the orientation of the principal compression with respect to the axes of symmetry of the anisotropy (COSGROVE 1976). The figure shows how the geometry varies with these two parameters.

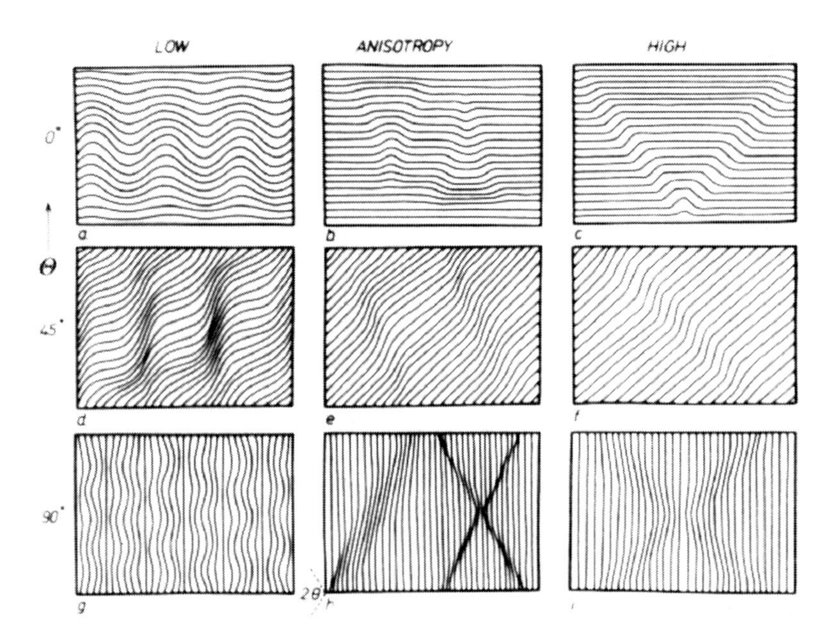


If it is assumed that the lines on these diagrams represent a mineral fabric (i.e., a natural example of an anisotropic material) then $\Theta$ is the angle between the mineral fabric and the principal compression prior to the formation of the structures. Mineral migration in response to stress gradients established during the development of these structures often results in the formation of discrete zones or planes of weakness. These planes are intimately associated with the structure and constitute a crenulation cleavage.

It is interesting to compare the structures shown in the diagram with structures developed in certain natural anisotropic materials; cf. **a, b,** and **d** of the diagram with the photographs **A, B** and **C.**

---

**A.** Symmetric crenulation in a quartz-muscovite schist from the Lukmanier pass, Switzerland. PPL

**B.** Asymmetric folds and an embryonic crenulation cleavage in a quartz rich multilayer, Treaddur Bay, Anglesey, North Wales, U.K.

**C.** A fold with gently divergent axial planes (RHS of photograph) developed in a multilayer made up of quartz rich (*light*) and mica rich (*dark*) layers. The removal of quartz from the limbs of the fold by the processes of pressure solution and mineral migration forms a coarse crenulation cleavage.

**D.** Crenulation cleavage in Devonian slate of the Mosel valley, Berncastle, West Germany. The crenulating of the mineral fabric is by normal kinking and the discrete cleavage planes are caused by the removal of quartz from the normal kink bands by pressure solution. PPL

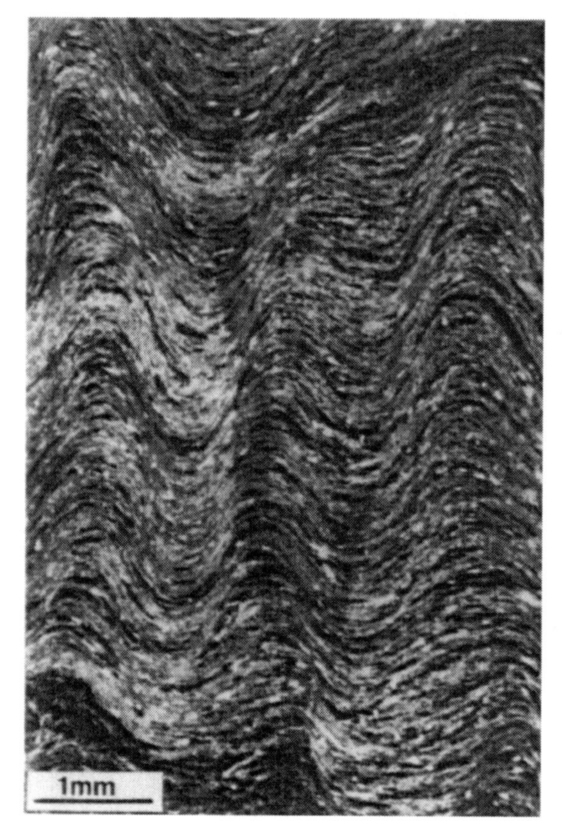


41 A

41 C

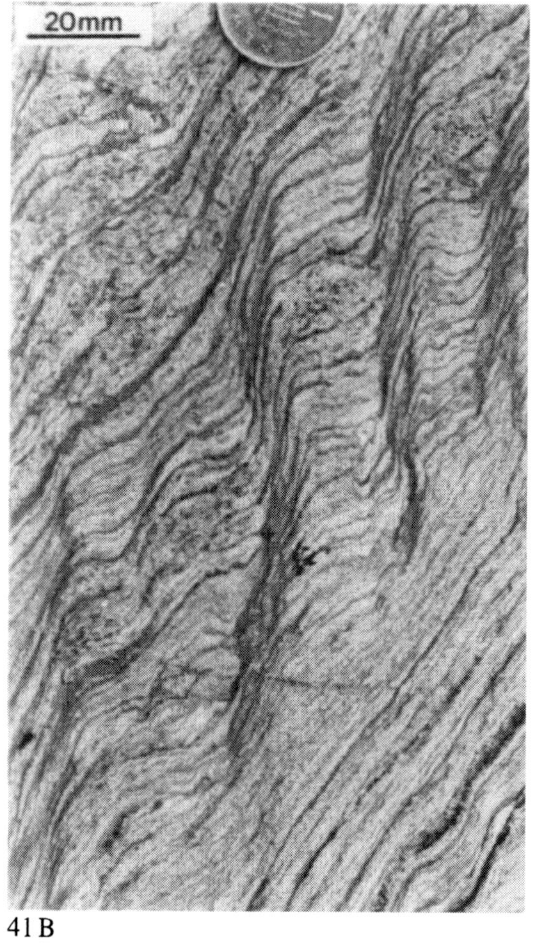


41 B

41 D

## 42. Crenulation with Cleavage Discontinuities (I)

D. PUPPOLO and B. M. BAYLY

Ordovician Moretown Formation at West Cummington, Mass., U.S.A., roadcut on Route 9 beside Westfield River (PUPPOLO 1979, HATCH 1975). The original sediment has been metamorphosed to a quartz-mica rock with prograde garnet and extensive retrograde development of chlorite. An earlier layering runs *from bottom left to top right* and is cut by a cleavage, *vertical* in the picture. The complex history of the region is given by HATCH. Here we draw attention to three features:

1. The two populations of phyllosilicates, "old grains" and larger new chlorite flakes. The latter appear mainly on steep crenulation limbs, but also sporadically away from them.
2. The thin lines, white in the plate and black in the actual rock, that cut across the texture and are true parting surfaces.
3. The en echelon geometry of these lines and the occasional bridges across them, reminiscent of maps of the San Andreas fault system, the Dead Sea rift, etc. (see sketch).

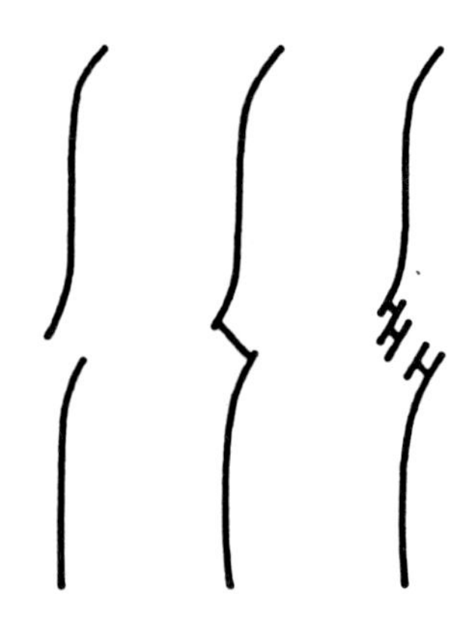

(caption continues on Plate 43)

Negative print; rock section placed directly in enlarger.

42

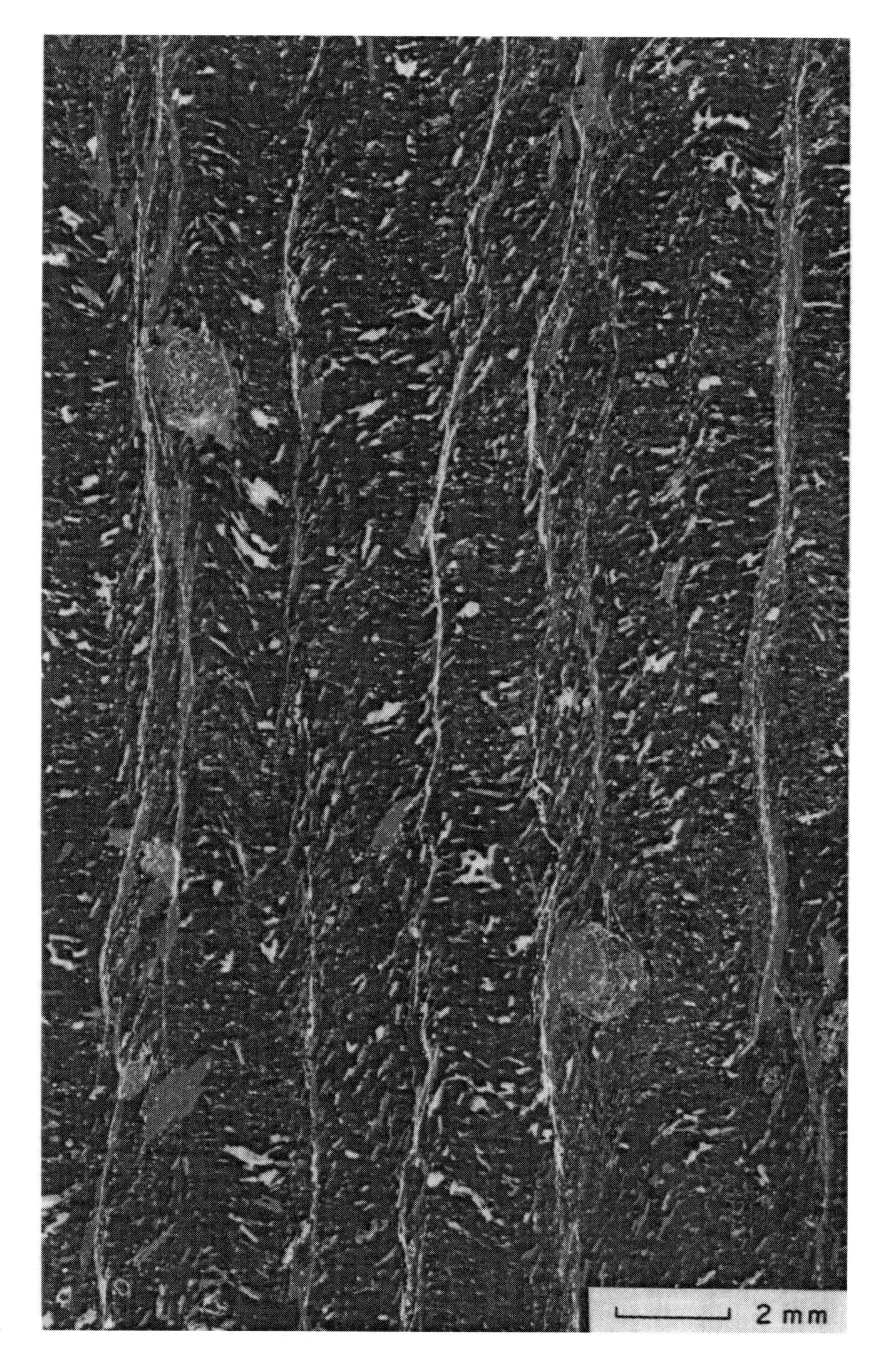

## 43. Crenulation with Cleavage Discontinuities (II)

D. Puppolo and B. M. Bayly

(Caption continues from Plate 42; the two samples are about 1 m apart.)

The nature and origin of the parting surfaces are illustrated again opposite. Here the actual partings are less well developed and less conspicuous; the rock has more zones of finite width that are just steep limbs of crenulations, and parting develops mainly on such steep limbs in apparently a rather arbitrary way. A recognizable lithologic unit cuts across two phyllosilicate-rich zones at **AA**; it seems that these zones do not have particularly large amounts of shear strain on them at least in this plane of section. Truncated megacrysts at **B** and **C** suggest that some dissolution has been going on.

(Plates 44 and 221 are from the same outcrop)

Negative print; rock section placed directly in enlarger.

43

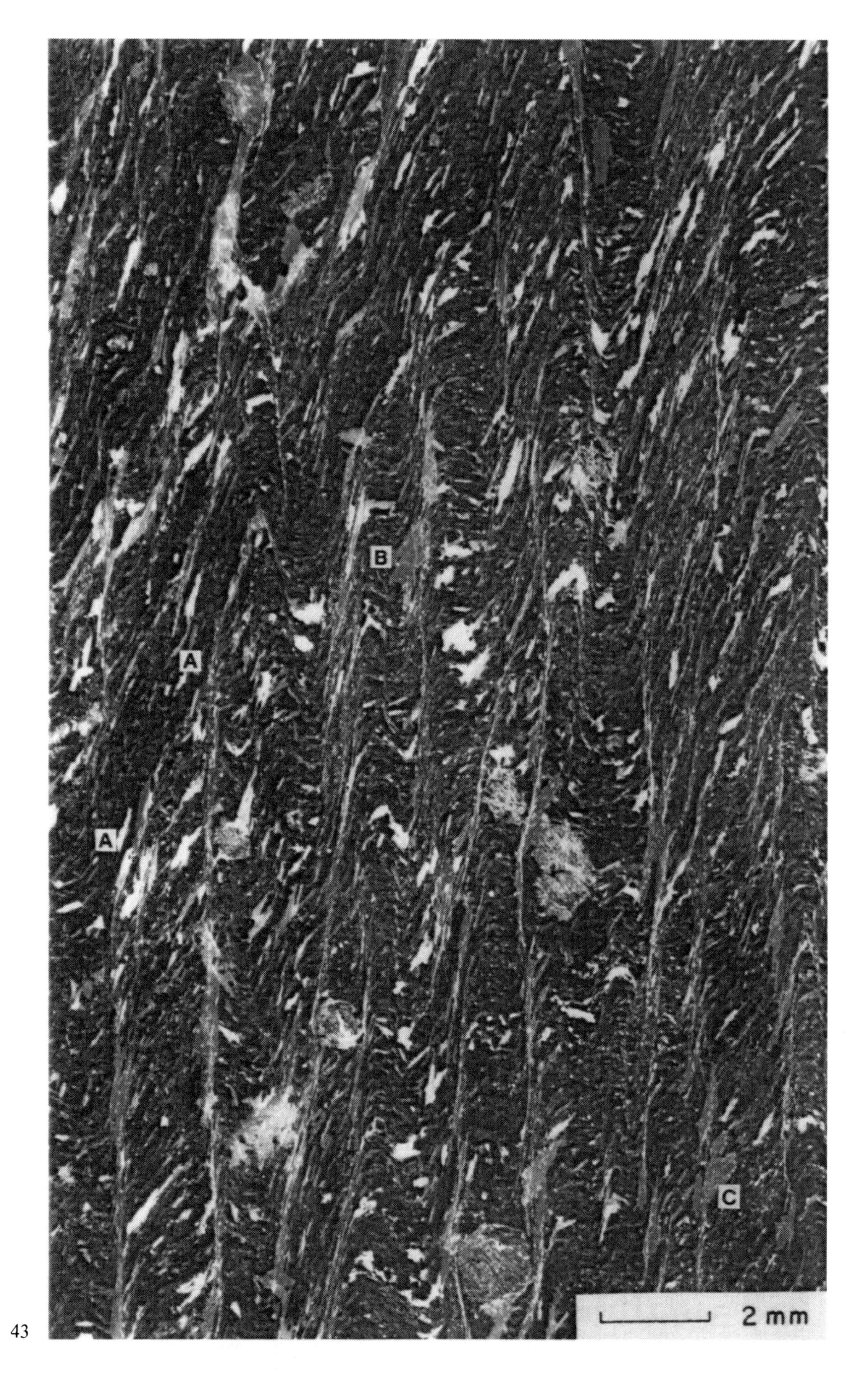

# Spaced Cleavage

## Section 5: Zonal Crenulation Cleavage with Discrete Boundaries

(Plates 44 to 53)

Ten plates are here brought together to illustrate this cleavage type. The first two plates characterize the type: crenulated zones and other zones abut sharply. While these features are common at the microscopic scale in crenulated rocks, an analogous morphology is also revealed in the fine structure of slates studied by electron microscope (Plates 48 to 52).

---

### Plates

# 44. Continuous and Discontinuous Domain Boundaries

D. PUPPOLO and B. M. BAYLY

Ordovician Moretown Formation at West Cummington, Mass. U.S.A., roadcut on Route 9 beside Westfield River (PUPPOLO 1979, HATCH 1975). The original sediment has been metamorphosed to a quartz-mica rock with prograde garnet and extensive retrograde development of chlorite.

The purpose of the plate is to show the variable geometry of the boundary between cleavage zone and microlithon. For the most part, foliation is continuous across the boundary, curving smoothly to its new orientation as at **A**. Locally, however, there is discontinuity like that in Plate 45 (see **B, C** and **D**), and we may speculate about why the discontinuity develops.

Suppose that for accidental reasons the geometry of a microlithon is as shown (**Fig. 1**). Further shortening perpendicular to the axial plane can be accommodated at *E* by clockwise rotation and at *F* by counterclockwise rotation without shortening of mica sheaves. At *G*, however, the sheaves are parallel to the supposed shortening; they cannot accommodate by rotation and they resist change of length. Hence they prod into the neighboring cleavage zones with more force than along *E* or *F*, and create more angular profiles (**Fig. 2**).

Returning to the plate, we find that sites **B**, **C** and **D** do seem to fall at transitions from left-leaning slopes to right-leaning slopes. All this suggests that despite its appearance, the cleavage domain in Plate 45 may be a crenulation limb; the angular discontinuity may be due basically to a parallelism of microlithon sheaves with shortening direction; there may have been a small amount of dissolution or slip at the discontinuity but the essential effect could perhaps be the same as at **B, C** and **D** here, developed to a more striking extent.

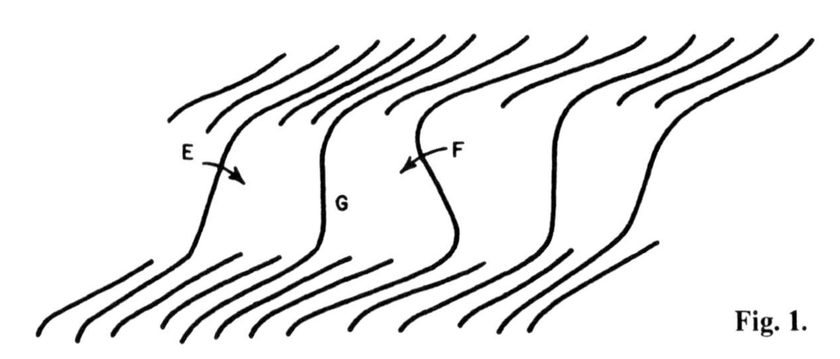


Fig. 1.

Angular discontinuity is enhanced by grain growth: megacrysts **H** and **J** have more discontinuous angular relations to the cleavage zones below them than the finer-grained groundmass does on either side.

Negative print; rock section placed directly in enlarger.

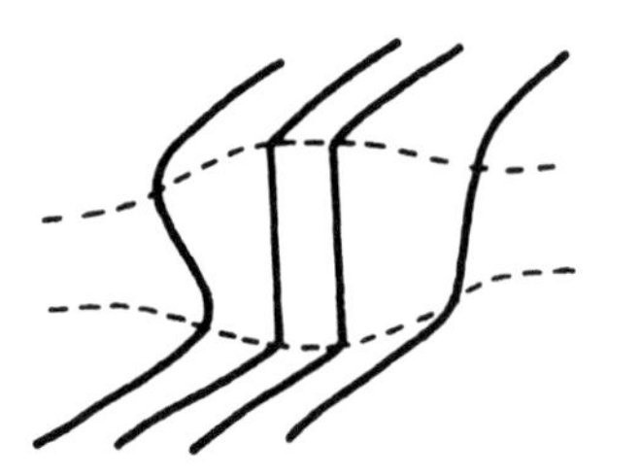

Fig. 2.

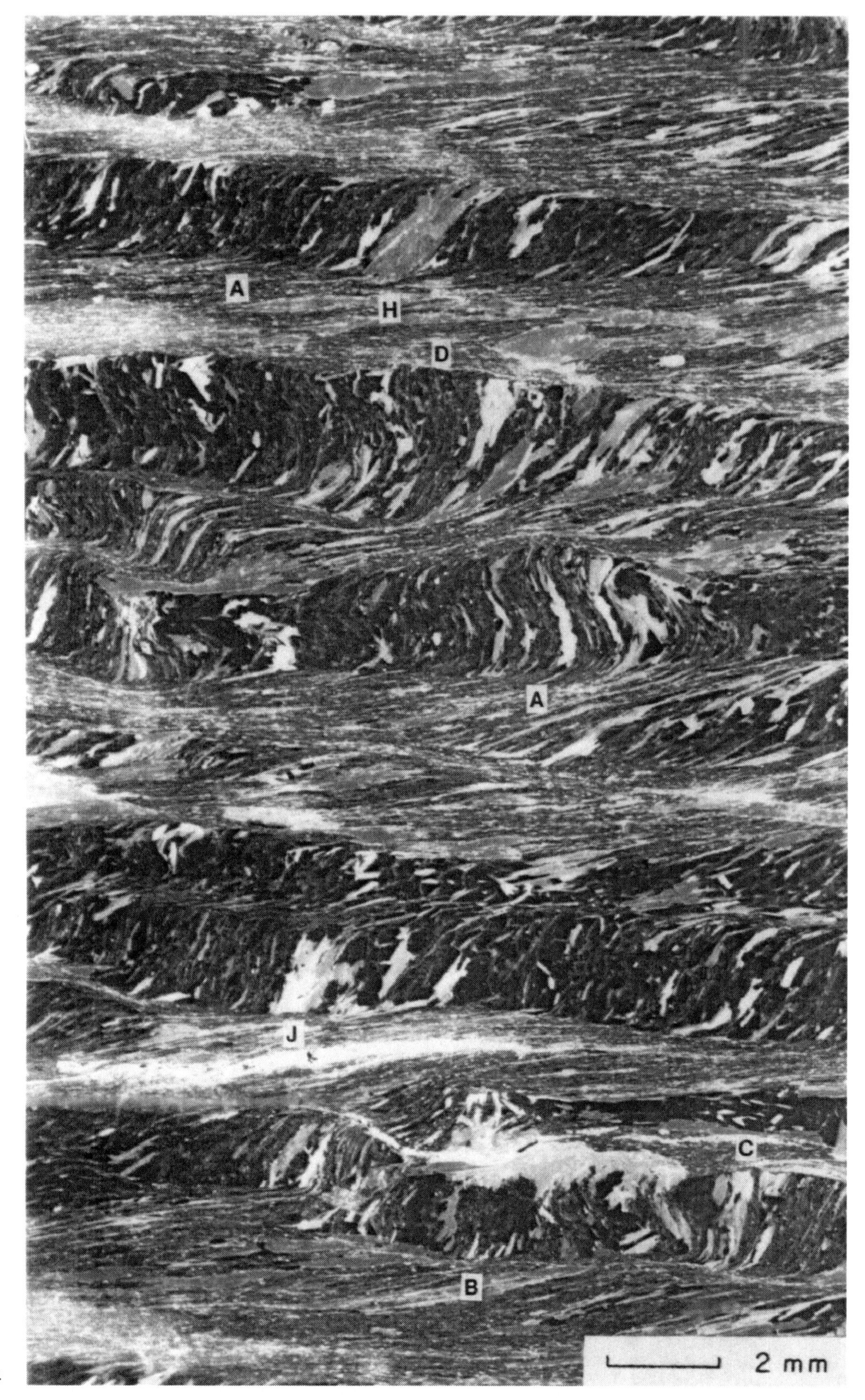

44

## 45. Cleavage Domain in Crenulated Chlorite Schist, Chester, Vermont, U.S.A.

W. D. MEANS

This rock is made up entirely of a dark green chlorite. An earlier foliation has been crenulated and is interrupted by cleavage domains such as the one shown that are parallel to the axial planes of the crenulations. The cleavage domain consists of chlorite optically indistinguishable from the chlorite outside the cleavage domain. Notable and somewhat unfamiliar features are: failure of many of the cleavage domains to follow the limbs of crenulations, and sharp discontinuities in basal plane orientations across the boundaries of the cleavage domain.

NX

45

# 46. Spatial Variations in a Crenulation Cleavage Above Garnet Grade

R. J. HOLCOMBE

These crenulation-cleavage morphologies are typical of all the psammopelites above the garnet isograd in the Ocoee Gorge-Ducktown area of southeastern Tennessee, U.S.A. (see also Plate 39). The principal control on whether the crenulations evolve into a gneissic layering or not seems less dependent on metamorphic grade and probably is mainly a function of composition (and perhaps of deformation intensity).

---

**A. and B. Incoherent crenulations.** Great Smoky Group, garnet grade, Ducktown. This is the dominant type of $S_2$ morphology in the amphibolite facies psammopelites. Remnant hinges of crenulations are defined by muscovite orientations in the quartz-rich domains. Muscovites in the mica-rich domains are parallel to the domain boundaries. There is no continuity of the crenulated $S_1$ fabric across the structure. Biotite (*mid-grey*) occurs preferentially in the quartz-rich domains where it forms irregularly shaped grains filling the space between curved muscovite trains. Quartz grains in the mica-rich domains are very much smaller than in the quartz-rich domains. Scale: 1 mm PPL

**C. Gneissic layering.** Great Smoky Group, staurolite grade, Ducktown. Alternating quartz-rich and mica-rich layers define the $S_2$ foliation. Rare micas or mica trains crossing the quartz-rich layers at very acute angles are the only remnants of the crenulations. Scale: 1 mm PPL

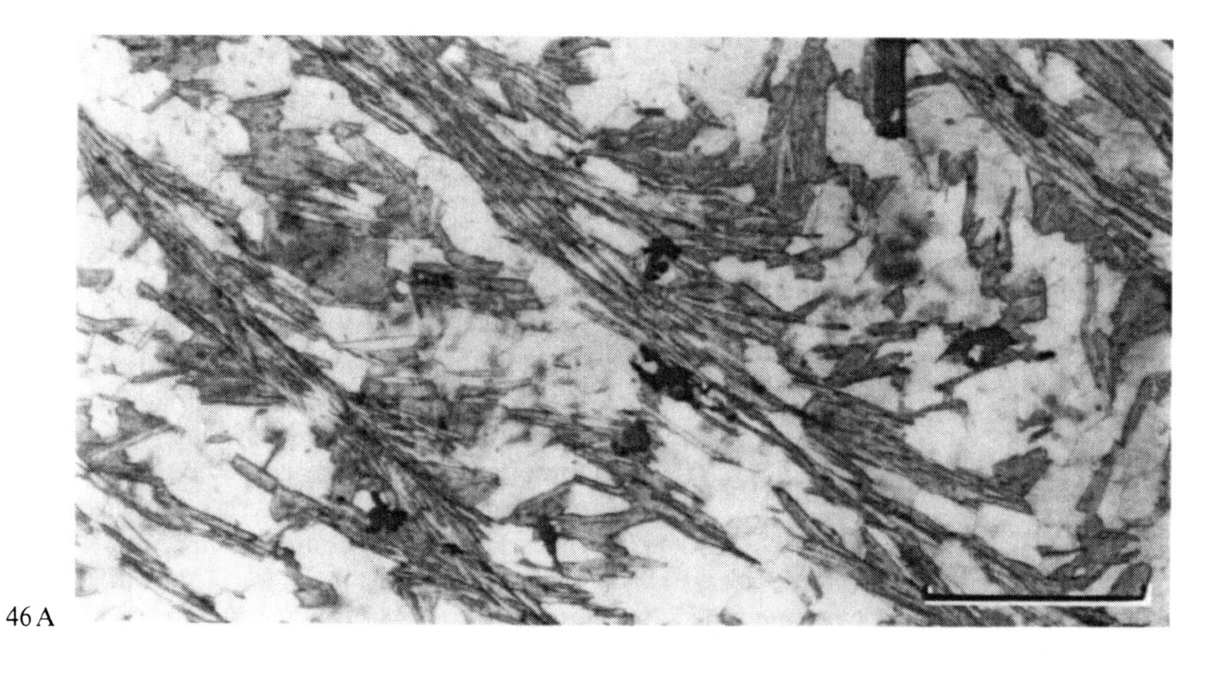
46 A

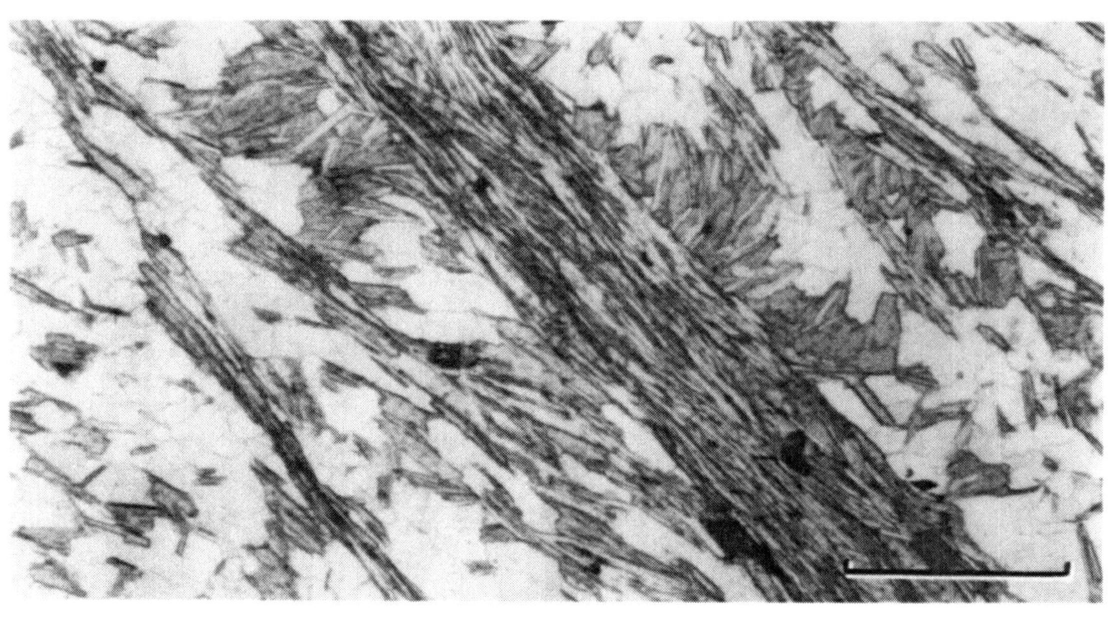
46 B

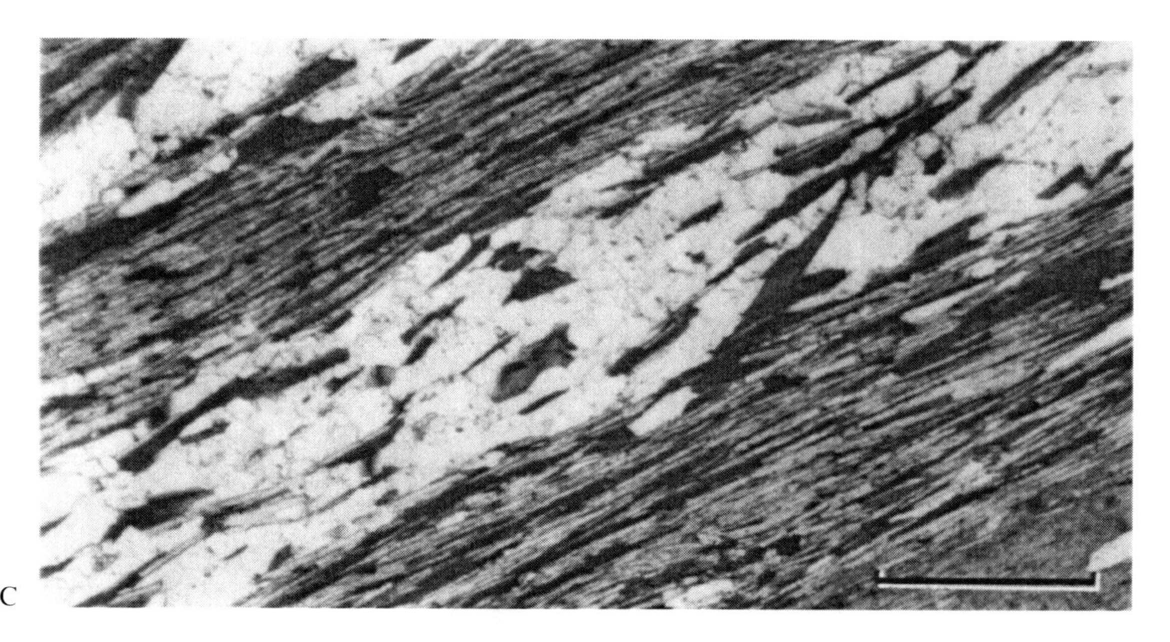
46 C

# 47. Microfabric of Slates from the Rheinische Schiefergebirge: Without Syntectonic Recrystallization

K. Weber

Grain fabric studies on very low grade metamorphic slates from the Rheinische Schiefergebirge reveal two types of "slaty" cleavage:

i. First cleavage without syntectonic phyllosilicate recrystallization (this plate).
ii. First cleavage with syntectonic phyllosilicate recrystallization increasing in keeping with increasing metamorphism (Plate 48).

First cleavage without syntectonic recrystallization is of the crenulation cleavage type. The crenulation fabric shows orthorhombic symmetry in fold-hinges and monoclinic symmetry in fold limbs. In the fold hinges, limbs of symmetrical microfolds form the cleavage planes. The limbs of the asymmetrical microfolds which are forming the cleavage fabric in a macroscopic fold limb display different deformation paths. One set of limbs defines the cleavage planes. With increasing deformation, this set of limbs is more and more rotated into a cleavage-parallel attitude. The other set of limbs forms the cleavage lamellae and its deformation is essentially coaxial. S.E.M.

47

## 48. Microfabric of Slates from the Rheinische Schiefergebirge: with Syntectonic Recrystallization

K. Weber

With increasing metamorphism the crenulation fabric is modified by solution of grains along cleavage planes and precipitation of material within the cleavage lamellae. The previous plate shows this stage of development of the fabric in a Carboniferous slate from the northern part of the Rheinische Schiefergebirge (Madfeld Sheet). With further increase of metamorphism the crenulation fabric is gradually obliterated and the deformation path more and more modified by crystallization and recrystallization processes. This plate shows the first cleavage fabric in a syntectonically, strongly recrystallized Eifel slate from the Rheinische Schiefergebirge (Eversberg Sheet). Note that contacts between the phyllosilicates which define cleavage planes and those which define cleavage lamellae are very sharp. They may be interpreted as solution contacts. At this stage of syntectonic metamorphism, solution processes play an important part in the deformation history. The fabric results from progressive solution, crystallization and recrystallization superimposed on bending, kinking, and internal rotation achieved by slip parallel to the basal planes of phyllosilicates.

From a mechanical point of view, cleavage formation is basically a crenulation process. With increasing metamorphism the deformation path is modified by chemical componental movement, leading to the formation of a syncrystalline first cleavage which may be called "crystallization cleavage" as opposed to "crenulation cleavage". S.E.M.

---

Reproduced with permission of Schweizerbart'sche Verlagsbuchhandlung from Geologisches Jahrbuch

48

# 49. Microstructure of a Spaced Cleavage

R. J. KNIPE and S. H. WHITE

The variation in cleavage development around folds at Rhosneigr, N. Wales, U.K., allows the microstructures present at different stages of cleavage development to be studied. The rocks contain more than 70% phyllosilicates and the metamorphic grade is lower greenschist. Analysis of the microstructures reveals that slaty cleavage development is preceded by heterogeneous crenulation development of the bedding fabric. Thus mechanical rotation of existing phyllosilicates toward the cleavage orientation is the important reorientation mechanism. The rotation is accompanied by grain boundary sliding, crystallization of new grains in dilation sites and possibly by pressure solution processes. Microstructures in areas with well-developed cleavage lamellae indicate that concentrated deformation (bending and kinking) at cleavage lamellae borders followed by the syntectonic growth of new grains accounts for the lateral extension of cleavage lamellae (KNIPE and WHITE 1977, WHITE and KNIPE 1978.) Metamorphic reactions are concentrated along the cleavage lamellae borders. These give rise to new grain compositions in the cleavage lamellae (KNIPE 1979).

---

**A.** The Rhosneigr slate illustrating the anastomosing nature of the spaced cleavage lamellae. The slate is composed primarily of phyllosilicates with chlorite and quartz concentrated in lenticular domains (**ld**) between the cleavage lamellae. Micas occur preferentially in the cleavage domains. PPL

**B.** The detailed microstructures of the well-developed spaced cleavage shown in **A**. Note the variation in the internal structure of cleavage lamellae with width. The thin cleavage lamella (**a**) is composed of fine-grained phyllosilicates while the wide lamella (**b**) contains large grains. The region between these cleavage lamellae is composed of phyllosilicates with their (001) planes at a high angle to the cleavage direction. These grains exhibit varying amounts of deformation (compare grains **1**, **2**, and **3**). Quartz (**q**) is also present in the area between the cleavage lamellae and the control of its shape by adjacent phyllosilicates together with its low dislocation density suggests that the quartz may be filling in an extension site. T.E.M.

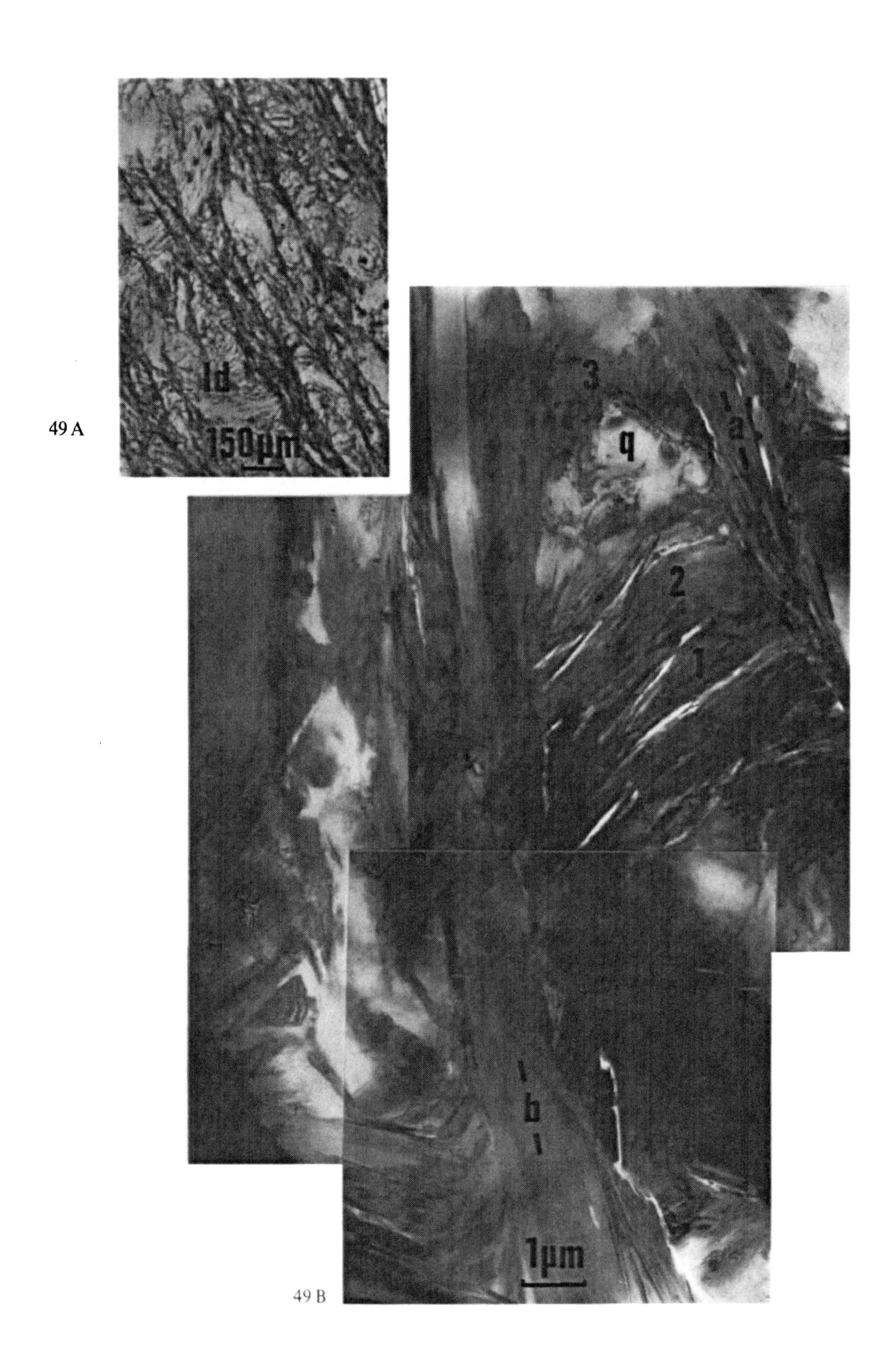

49 A

49 B

# 50. Initial Cleavage Lamella

S. H. WHITE and R. J. KNIPE

Ordovician slate, Rhosneigr, N. Wales, U.K. (KNIPE and WHITE 1977; WHITE and KNIPE 1978). Electron micrographs and line diagram of a common type of initial cleavage lamella showing its relationship to a crenulation. The constituent grains of the initial cleavage lamella are oriented slightly obliquely to the lamella border (*clb*). The cleavage lamella can be traced back to one limb of a crenulation (*pcr*). With increasing distance from this crenulation, i.e., as the lamella border becomes more oblique to the average bedding direction (*b*), fewer phyllosilicates can be traced through the lamella (*t*); most grains located outside the lamella abut against the cleavage lamella borders. Quartz (*dotted*) becomes abundant in possible extension sites and occasionally phyllosilicate grains parallel to the cleavage border are bounded by phyllosilicates with their (001) planes at high angles to the cleavages (*o*). Note also the double hinge ($h_1$ and $h_2$) in grain $X$. This type of initial cleavage lamella is preferentially developed on the limbs of microfolds with a wavelength of approximately 30 $\mu$. The interpretation of these microstructures is discussed fully by KNIPE and WHITE (1977).

T.E.M. 1000 kV

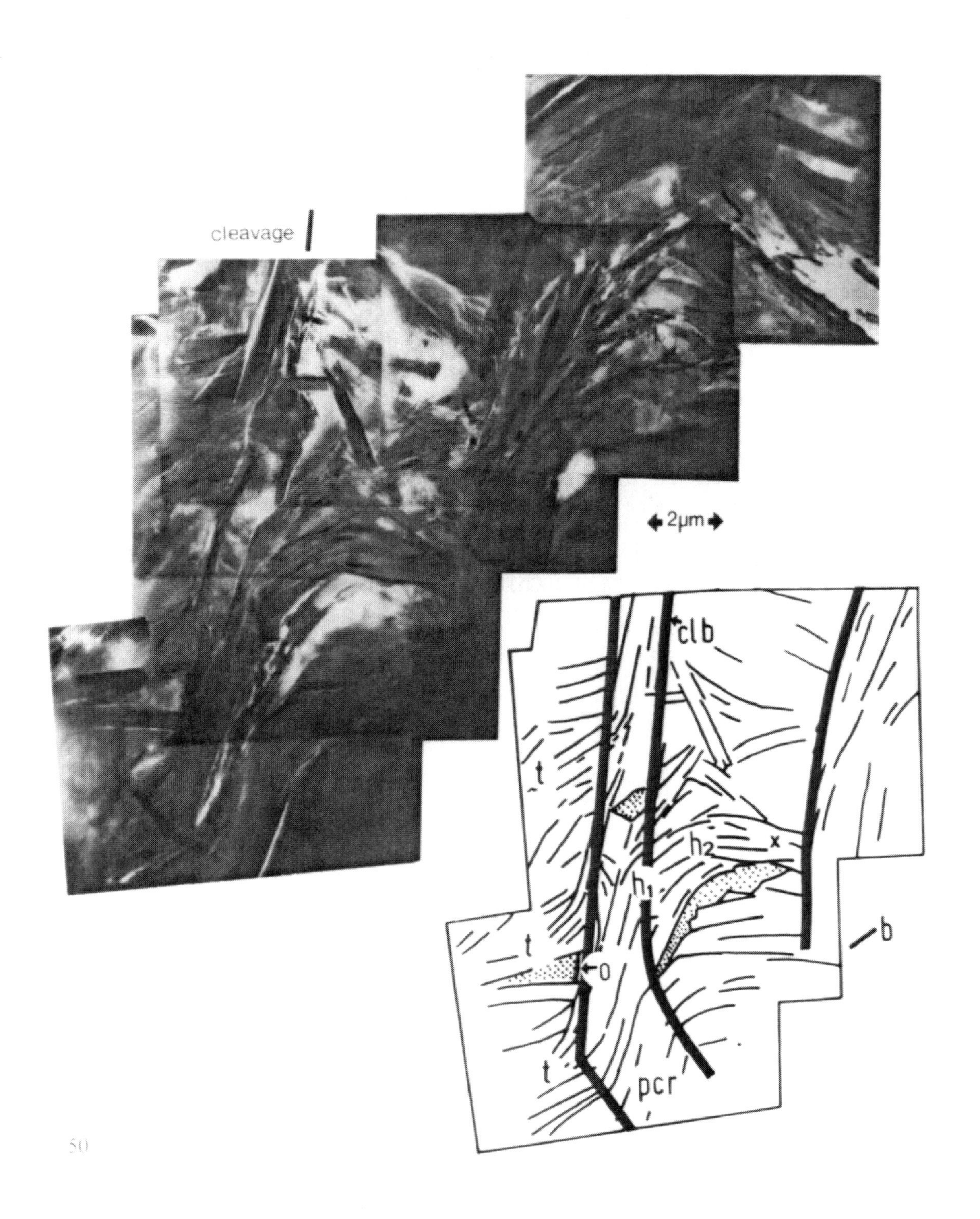
cleavage
2μm
clb
t
h2
x
h1
b
t
o
t
pcr
50

# 51. Cleavage Lamella Border Microstructures

R. J. Knipe and S. H. White

Ordovician slate, Rhosneigr, N. Wales, U.K. (Knipe and White 1977; White and Knipe 1978)

**A.** Triangular area of quartz (Q) containing inclusions located between bent phyllosilicates. This is interpreted as evidence for the crystallization of quartz between cleavage lamellae (Cl) in dilation sites generated during deformation. T.E.M. 1000 kV

**B.** Variation in grain size across a well developed cleavage lamella (CI) from an extremely fine-grained border (f) to a coarse-grained lamella core (c). The preferred orientation of phyllosilicates parallel to the cleavage is better developed in the lamella core. T.E.M. 1000 kV

**C.** Concentrated bending of phyllosilicates at the border of a well developed cleavage lamella. The phyllosilicates do not bend into parallelism with the cleavage orientation (Cl) and are separated from the cleavage lamella by a zone of fine-grained material (f). The fine-grained zone is interpreted as developing from the breakdown of the highly deformed phyllosilicates. It is suggested that new strain-free grains grow from this fine-grained zone and that the lateral migration of the deformation zone and the fine-grained zone allows the lateral extension of the cleavage lamellae. T.E.M. 1000 kV

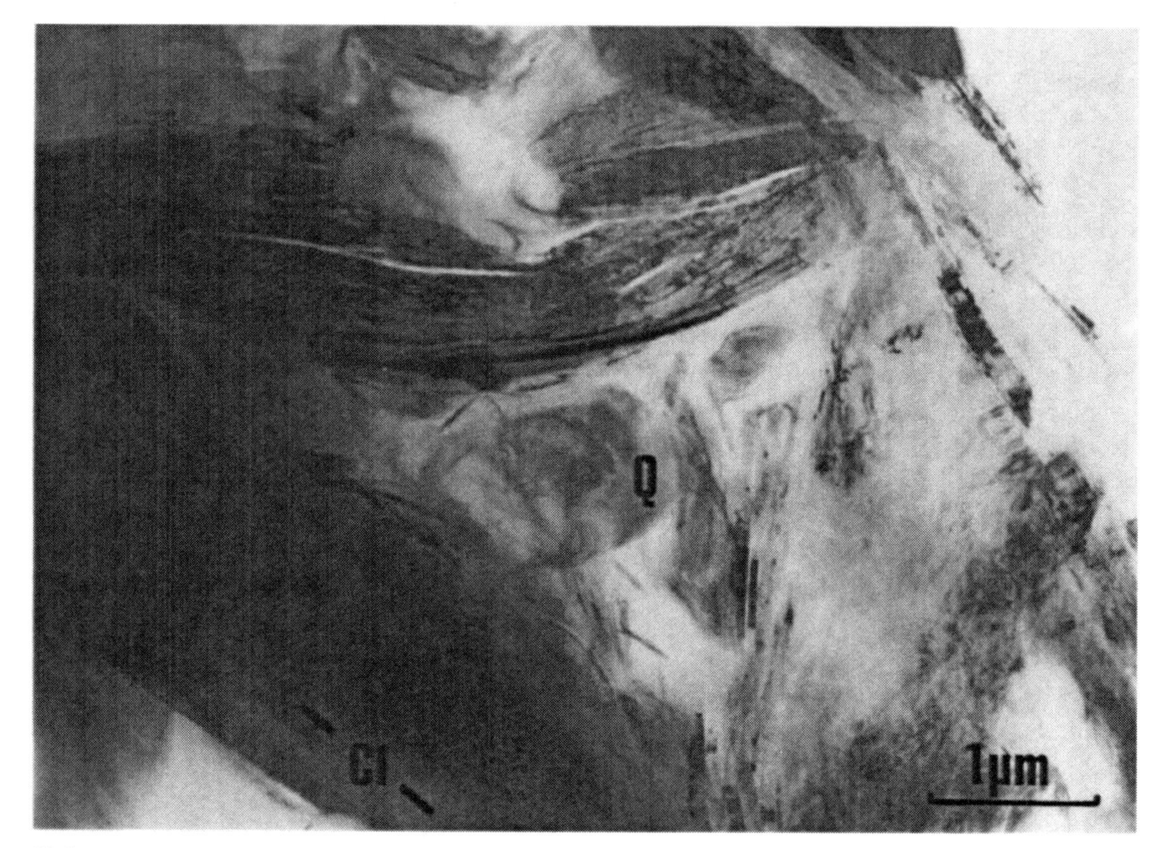


51 A

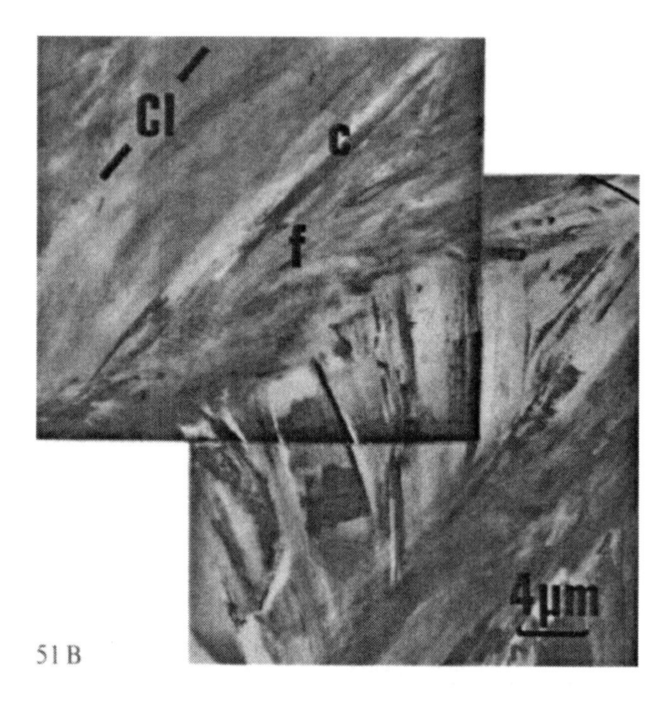


51 B

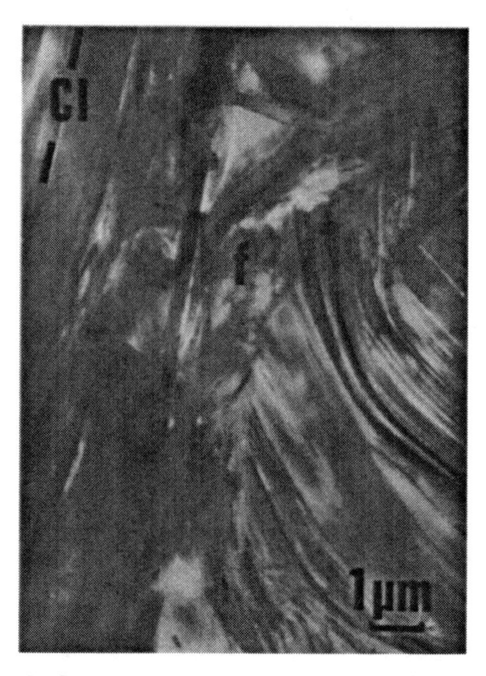


51 C

## 52. Cleavage Lamella Border Structures and Microchemistry

R. J. KNIPE

Ordovician Slate, Rhosneigr, N. Wales, U.K. KNIPE 1979). High-resolution electron micrograph lattice images reveal the structure of a cleavage lamella border. The finely spaced lines within grains are images of the basal (001) lattice planes. The adjacent cleavage orientation is indicated by **Cl**. Note:

1. the extremely fine grain size at the border (area **A'**)
2. the large number of low-angle grain boundaries (**G**)
3. the interweaving texture (**W**) of grains.

The *triangular plots* (i) to (iv) in the diagram illustrate the chemistry of micas present in the Rhosneigr Slate. Grains located in three microstructural sites, i.e., between cleavage lamellae (*ii*), in cleavage lamellae (*iii*), and along fine-grained cleavage lamella borders (*iv*), have been analyzed. (*i*) illustrates the compositional range of individual grains and the fields where different micas plot.

$A = Al_2O_3/SiO_2$, $K = K_2O/SiO_2$,
$F = Fe_2O_3 + MgO/SiO_2$, $A + K + F = 100$.

The analyses were conducted with a scanning transmission electron microscope using probe sizes of approximately 400 Å. The results reveal that micas between the cleavage are illitic, while phengites appear only at the cleavage lamellae borders and within the cleavage lamellae. These chemical variations illustrate that realignment of phyllosilicates during cleavage development is accompanied by the growth of new grains with compositions stable at higher grades. Chlorite grains also show variations in compositions between these microstructural sites. The growth of new grains can be related to both recrystallization and crystallization processes which are controlled by deformation. (See KNIPE 1979).

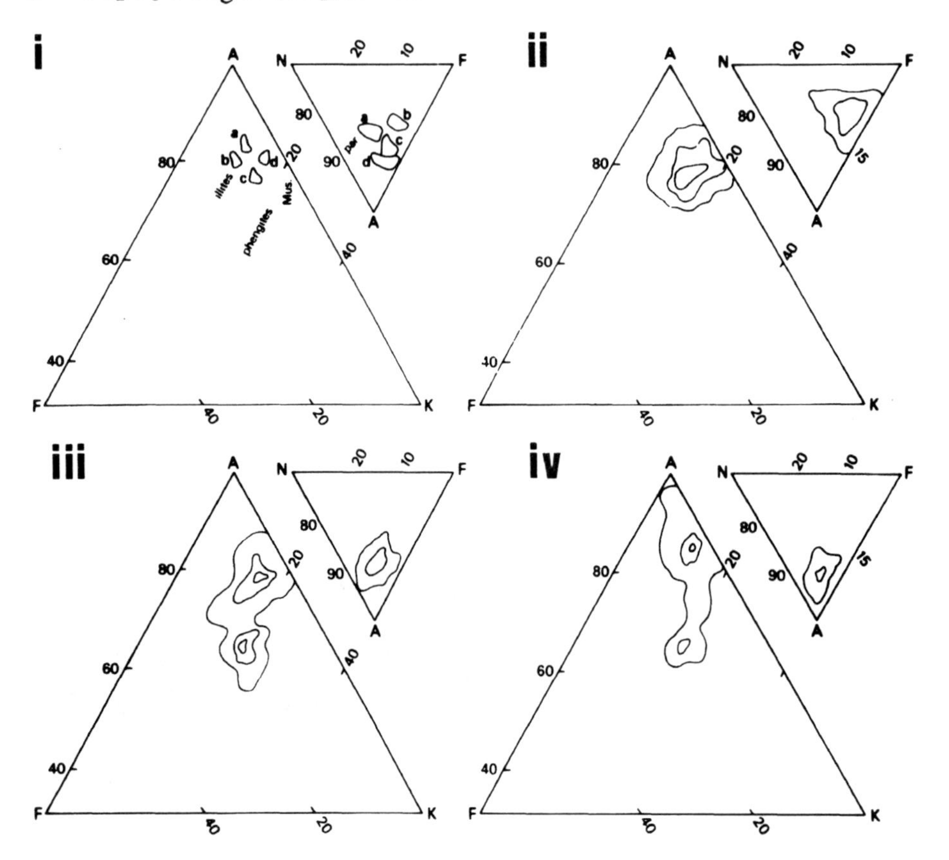

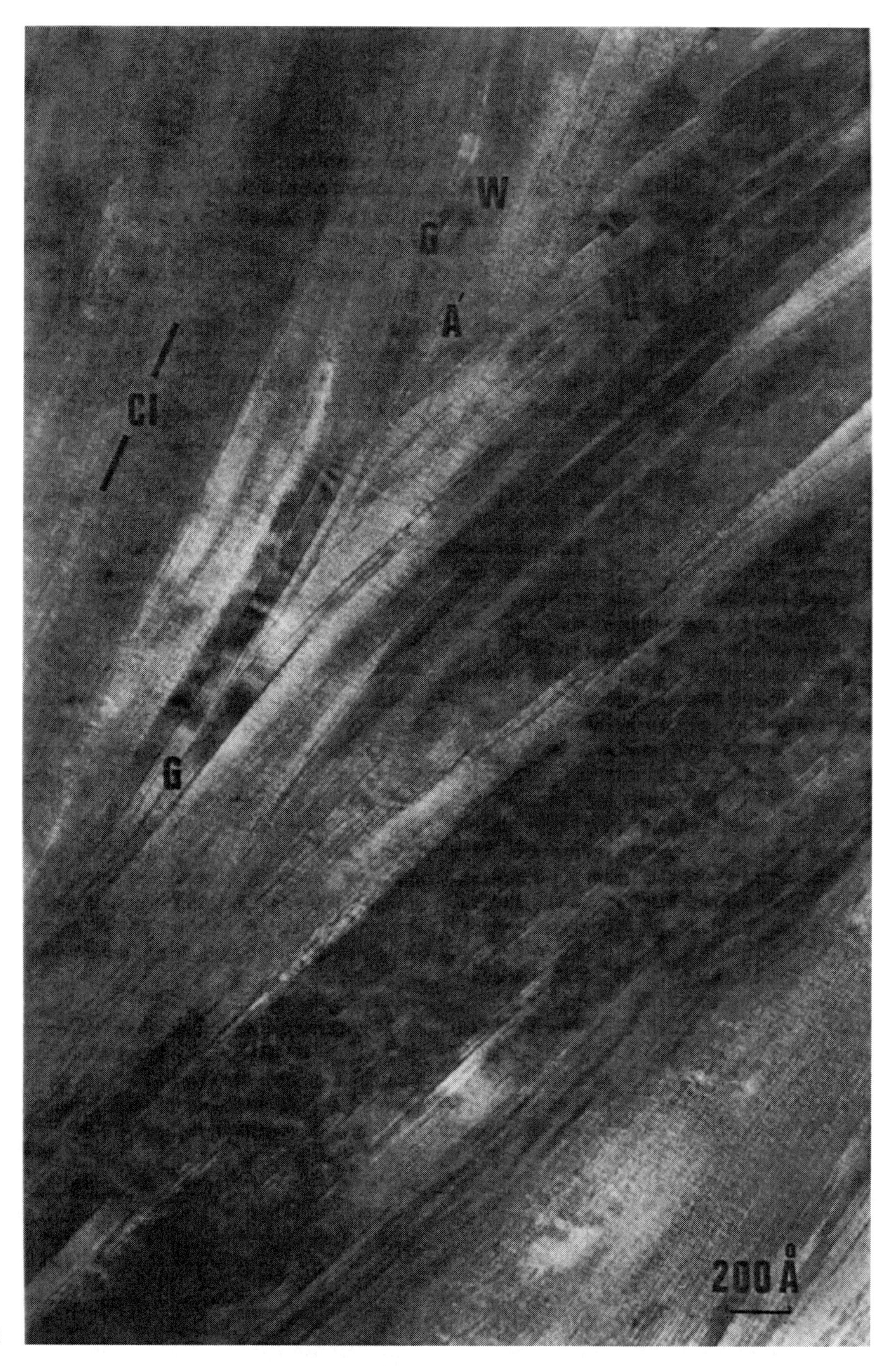

52

# 53. Morphological Alteration of Crenulation Cleavage with Progressive Metamorphism

J. W. GRANATH

Below the andalusite-K feldspar isograd in the Cooma complex, New South Wales, Australia, the predominant foliation in metapelitic rocks is a differentiated crenulation cleavage that is also common, but more sparsely developed, in metapsammitic rocks. Muscovite and biotite combine to form the mica domains of these crenulation cleavages, biotite predominating in the psammitic rock types and muscovite concentrated in the mica domains. At these low grades of metamorphism ilmenite is an important accessory phase, comprising perhaps 3 to 5 modal percentage of the rocks. The photographs illustrate a metamorphic foliation in psammitic paragneisses above the isograd that apparently was derived from the crenulation cleavage (C) during progressive metamorphism. In locally used terminology, this paragneiss is often called a "corduroy gneiss" in reference to the ribbed appearance the layering imparts to the outcrop.

---

**A. Mesoscale character of corduroy layering.** Ordovician Binjura beds at Spring Creek, near Cooma, N.S.W., Australia (JOPLIN 1942). The persistent, anastomosing layering in the metapsammitic unit (**ps**) is a compositionally defined metamorphic layering axial planar to a major upright anticline in the graded metaturbidites of the Cooma complex. This layering contrasts with a foliation vaguely discernable in thin sections from the metapelitic unit (**pe**), which appears to be relict quartz microlithons of a crenulation cleavage now obscured by the porphyroblastic growth of andalusite and K feldspar. More prominent in the pelites are discontinuous granitic veins (**g**). Compositionally differentiated crenulation cleavages, described above and illustrated in **C**, are the strongest element of the structural fabric of metasediments below the andalusite-K feldspar isograd, less than 1 km from this outcrop.

**B. Microscale character of corduroy layering.** Photograph from metapsammitic rock in **A**. The metamorphic layering is defined by the high concentration of red-brown biotite ($TiO_2 = 2–3$ wt. %; GRANATH 1976) and poikiloblastic andalusite (**a**) relative to the intervening quartzofeldspathic microlithons. The muscovite content of this rock is much reduced relative to comparable psammites at lower grades of metamorphism, and that present here may be secondary. Ilmenite is completely absent. Because K feldspar is also absent, the following oxidation reaction is suggested:

$$2\,FeTiO_3 + 2 \text{ phengitic muscovite} + 16 \text{ Ti-poor biotite}$$
$$= Al_2SiO_5 + 18 \text{ Ti-rich biotite} + \tfrac{1}{2}\,O_2,$$

which can be balanced using microprobe analyses from appropriately representative rocks of the Cooma complex. In the more muscovite-rich pelites, K feldspar as well as andalusite and Ti-rich biotite was produced. Note the poor preferred orientation of grains in the biotite-andalusite layer and the statistical parallelism of mica grains in the quartzofeldspathic layers to the domain boundary. This microstructure is thought to have replaced that of a typical crenulation cleavage with the loss of the microfold waveform defined by the orientation of mica grains in the older schistosity. PPL

**C. Microstructure of crenulation-cleaved precursor to B.** Early schistosity ($S_1$) is crenulated with development of mica domain ($S_2$) in long limbs of the microfolds. Note muscovite in mica domains, where ilmenite is also concentrated. PPL

53 A

53 C
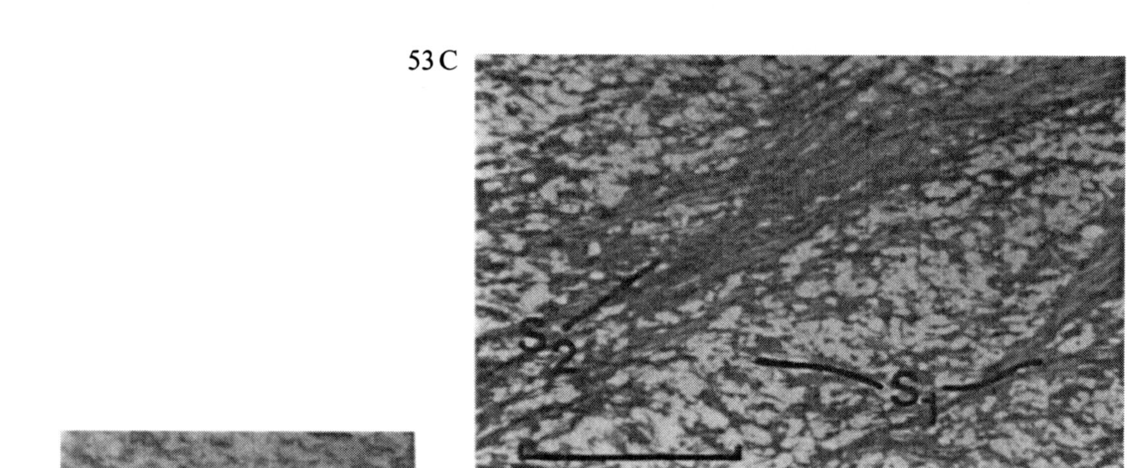


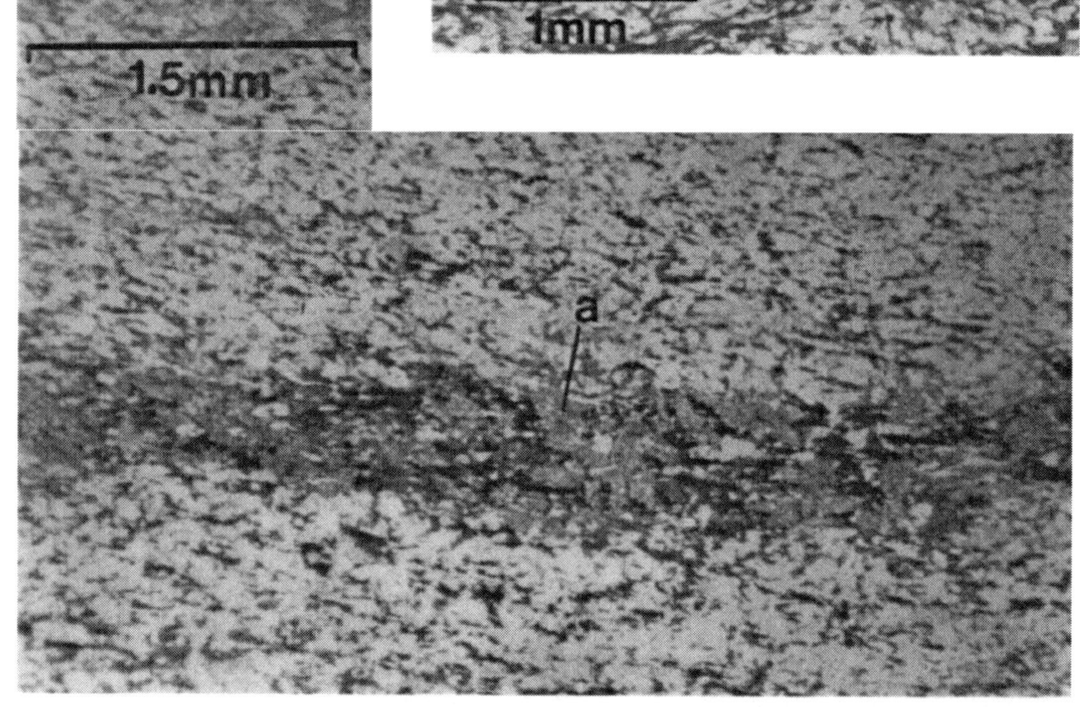

53 B

# Spaced Cleavage

## Section 6: Crenulations Bounded by Cracks

(Plates 54 to 64)

Eleven contributions form this section. The first illustrates the transition from a zonal crenulation cleavage (as per Section 5) to one with discrete cracks between the crenulations. Other contributions draw attention to possible movements on or across the discrete cracks and solution removal ("pressure solution") is frequently referred to in this context.

---

### Plates

# 54. Discrete Versus Zonal Crenulation Cleavage: Attitude of the Earlier Fabric

G. J. Borradaile

This sample is from Late Precambrian rocks of the Dalradian Supergroup near Kintra on the Mull of Oa, Islay, Scotland. The principal structural elements are a crenulation cleavage ($S_2$), earlier penetrative, continuous cleavage ($S_1$) and quartz-calcite veins that were originally planar but which formed at a consistently oblique angle to $S_1$ (see sketch of central part of photograph). In the ensuing discussion reference will be made to *folds* of the *quartz veins,* in locating features of interest.

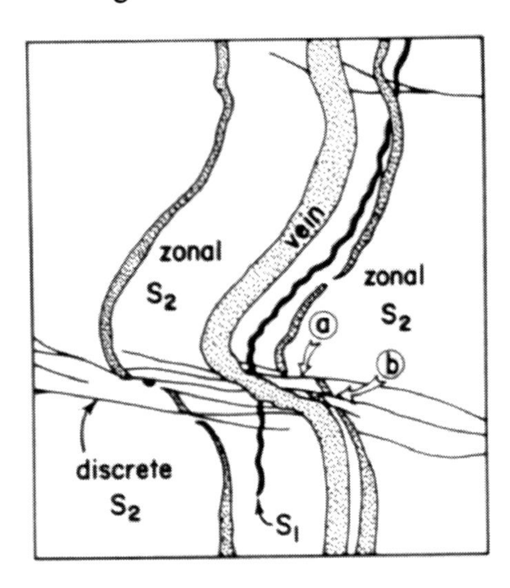


In the regions of the long limbs of the folds $S_2$ is a zonal crenulation cleavage. On the short limbs of the folds $S_2$ is a discrete crenulation cleavage. This change in morphology appears only to be influenced by the orientation of the earlier anisotropy (also see Cosgrove, Plate 41). On the short limbs of the folds the $S_1S_2$ angle is about 80° whereas on the long limbs of the folds that angle is only 50°.

Several other features of interest can be seen in the discrete crenulation cleavage surfaces. These include terminations, bifurcations, dip-slip offsets (*a* in figure) and reverse offsets (*b* in figure) of veins. The dip-slip offsets of veins may be interpreted either as shortening features (see Plate 117) or as small shear faults. The reverse offsets (*b*) can be interpreted only as shear faults and, in this instance, they cannot be interpreted as complete fold-pairs which have been completely removed by "pressure solution" or diffusive mass transfer (e.g., see Plate 113).

These features indicate that the discrete or zonal character of the crenulation cleavage is due to the attitude of the pre-existing anisotropy. Both shortening across $S_2$ and shear parallel to $S_2$ has taken place. The $S_2$ surfaces cannot therefore be parallel to the trace of the principal (XY) plane of total strain in the plane of the photograph. PPL

54

# 55. Discrete Crenulation Cleavage: Evidence for both Shortening and Shear

G. J. Borradaile

Late Precambrian Dalradian Supergroup; Tayvallich Slates, Loch Trallaig near Oban, Scotland. The crenulation cleavage ($S_2$), the second tectonic fabric in this area, crenulates both the pre-existing continuous fine cleavage $S_1$ of the pelite and the sedimentary laminations ($S_0$) of silty material. The sedimentary laminations are vertical in the photographs. The following petrographic observations indicate that the cleavage may not be simply interpreted as the result of a single mechanism or a simple strain history.

---

**A.** Detail of location outlined in (**B**). Partly crossed nicols.

**B.** Overview. Position of enlargement (**A**) located by box. PPL

### Observations

1. $S_2$ cleavage surfaces: enriched in phyllosilicates and opaques, quartz absent or present only in grains of small cross-sectional area and very high aspect ratio (= length/width).
2. $S_2$ cleavage surfaces: refract, anastomose and bifurcate (see **B**).
3. $S_2$ cleavage surfaces sharply truncate clastic grains.
4. Silt laminae "stepped" at termination of $S_2$ cleavage surfaces (see **A**).
5. In the "stepped" regions the siltstones have quartz grains of slightly higher aspect ratio than elsewhere in silt laminae; also quartz grains here show slight undulose extinction.

### Implications

The thinner, "stepped" portions of the silt laminae indicate that there has been some shearing parallel to the $S_2$ cleavage surfaces. The weak undulose extinction and modest grain shaping in these parts of the silt laminae suggest the shear displacement has there been achieved by plastic granular deformation and some intergranular flow.
In the phyllitic portions of the rock the cleavage surfaces are associated with reduction in quartz content, reduction in grain size of quartz, increase in aspect ratio of quartz and the $S_2$ surfaces may sharply terminate clasts in a secondary fashion. Diffusion of quartz, requiring the presence of a fluid phase at this metamorphic grade, would account for these features.
This crenulation cleavage shows evidence of shear parallel with it and shortening (by solution removal) perpendicular to it in different parts of the specimen. The different components of movement could have occurred simultaneously if the $S_2$ cleavage developed at an angle to the principal stresses. Alternatively, the different components may have developed at different times in a noncoaxial strain history during which the cleavage was sometimes oblique to the principal stresses (allowing shear) and sometimes nearly parallel to principal stresses (allowing shortening by solution removal)

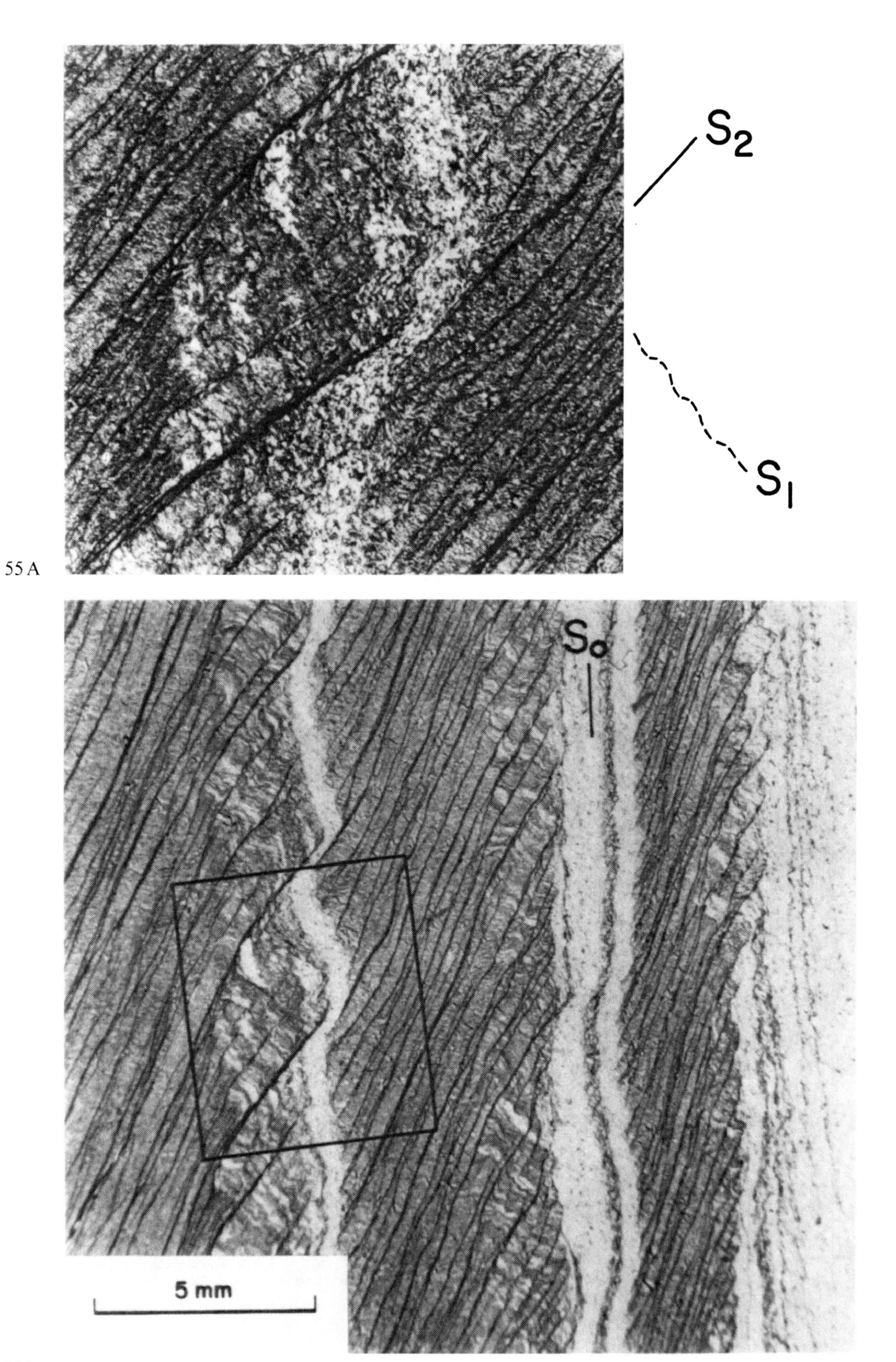


55 A

55 B

## 56. Axial Plane Cleavage and Stylolites

P. F. WILLIAMS

From the closure of a fold in phyllite from the Pyrenees. Layer silicates and opaque material are concentrated in both the stylolites and along the cleavage septae. The two structures differ only in size and planarity. The stylolites have a typical dentate form and are mostly much narrower than the planar cleavage septae. There is a complete transition from one structure to the other. The stylolites occur only in the hinge of folds, as shown here, and close examination reveals no evidence of shear strain parallel to the stylolite. The cleavage septae on the other hand occur predominantly on the limbs of the folds and coincide with the limbs of microfolds indicating that there has been shear strain parallel to the septae.

It is suggested that the structures are contemporary and that the difference in morphology is due to a difference in their straining histories. The stylolites developed in a coaxial strain environment, in fold closures, where there was no shear strain parallel to their enveloping surface. The cleavage septae developed in a noncoaxial strain environment where shear parallel to the seams prevented them from developing a dentate form. Note that some cleavage lamellae develop by coalescence of arrays of small seams that are parallel to the folded surface. Photograph **B** is a detail of the central part of **A**. NX

56 A

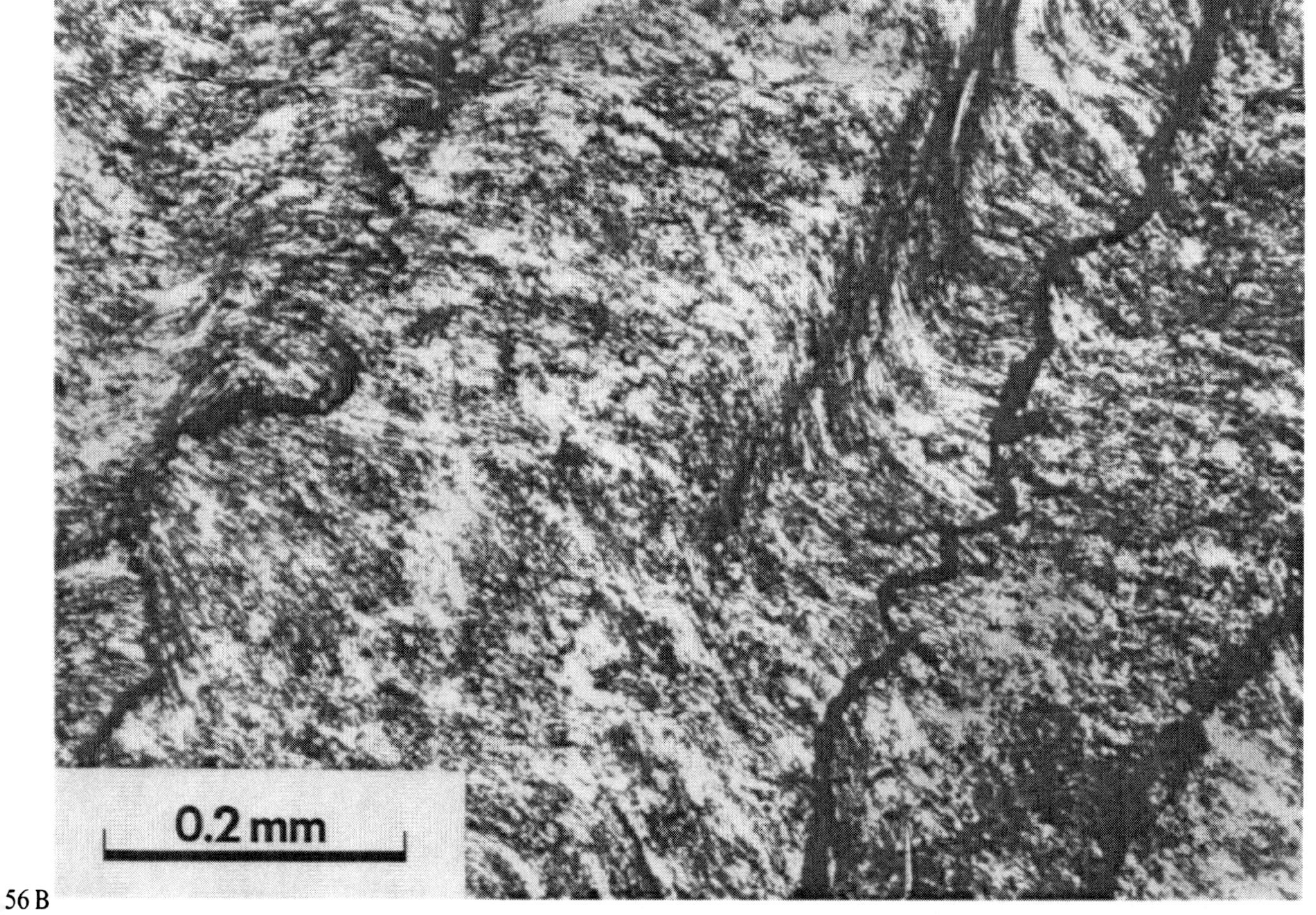


56 B

# 57. Discrete Crenulation Cleavages in Slates

D. R. Gray

**A. Discrete cleavage traces.** Silurian Captains Flat Formation, north of Captains Flat, N.S.W., Australia. The discrete cleavages (*vertical*) are dark, undulating and irregular, anastomosing, spaced discontinuities which truncate the weakly crenulated pre-existing slaty fabric (*horizontal*). The original fabric is either bent, sharply deflected, or undeflected adjacent the cleavage, but is clearly truncated by the cleavage. Segments of rock adjacent individual discrete cleavages show mutual interpenetration suggesting that there has been no significant movement along the cleavages. PPL

**B. Discrete cleavage traces.** Silurian Blowering Beds at Blowering Dam, N.S.W., Australia. Smooth and undulating discrete crenulation cleavages (*vertical*) associated with asymmetric microfolding in a slate fabric. The cleavages coincide with the short steep limbs of the microfolds. Their spacing and position in the slate is dependent on the microfolding. PPL

**C. Apparent offset of layering.** Undifferentiated Ordovician at Limerick, N.S.W., Australia. Crenulated lithological layering (*horizontal*) truncated and partially offset along discrete cleavages (*vertical*). The discrete cleavages are transitional with microfolds along their lengths and they coincide with the steep limbs of the crenulations. Where the layering shows maximum offset the cleavage is most distinct; it has a thin, dark, discrete "seam-like" form. Such offsets are attributed to dissolution along the limbs of microfolds. PPL

**D. Microstructure of discrete cleavages.** Undifferentiated Ordovician, Gundagai, N.S.W., Australia. The discrete cleavages (*vertical*) are anastomosing, subplanar films of dimensionally oriented white mica interspersed with minor opaques and iron oxides. The micas, irrespective of cleavage orientation and irregularities in the trace, are as near to parallelism with the cleavage trace as can be determined with a petrological microscope. This microstructure is incompatible with a brittle-deformation origin for these cleavages. PPL

---

A, C, D are figs. 2, 3 and 4 from Gray 1977b, reproduced with permission of the University of Chicago Press

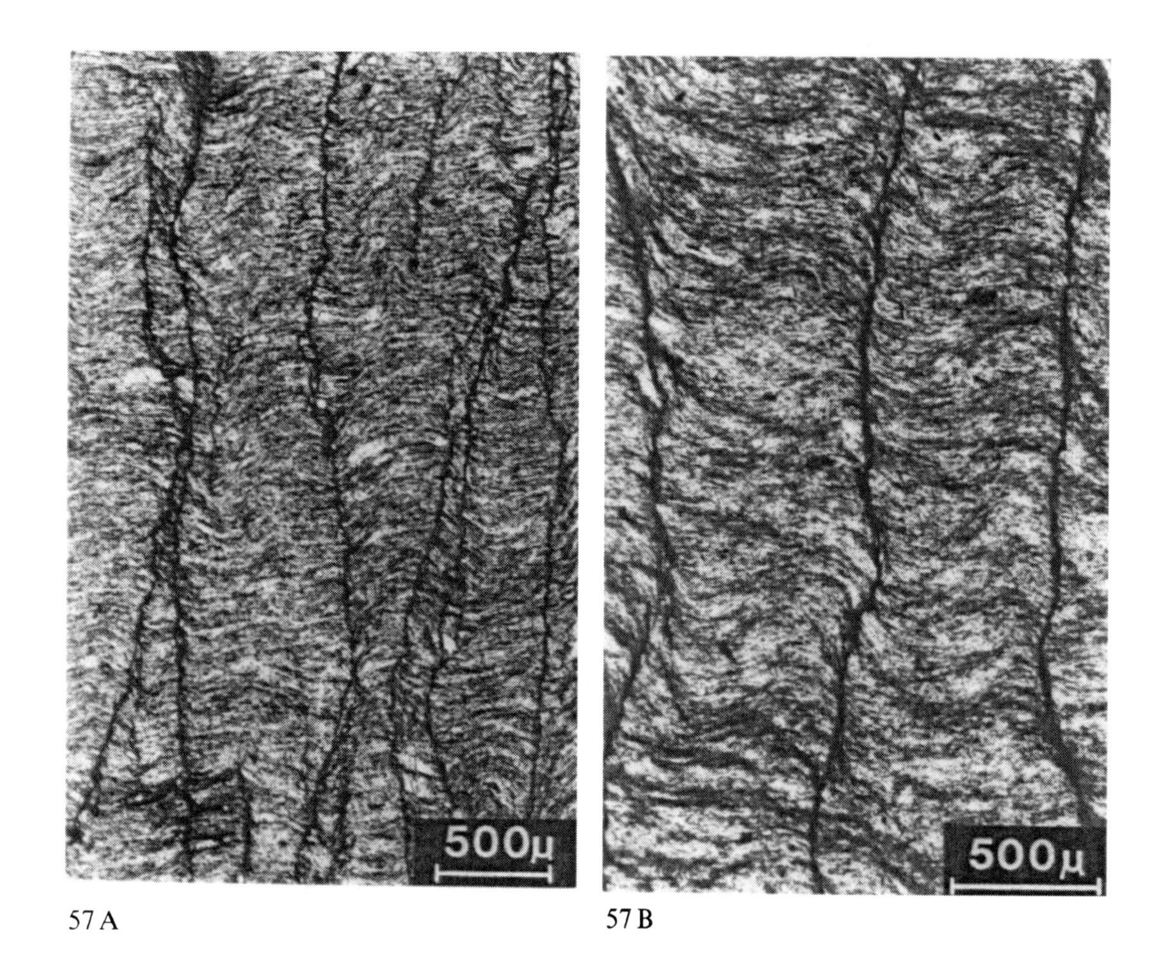


57 A

57 B

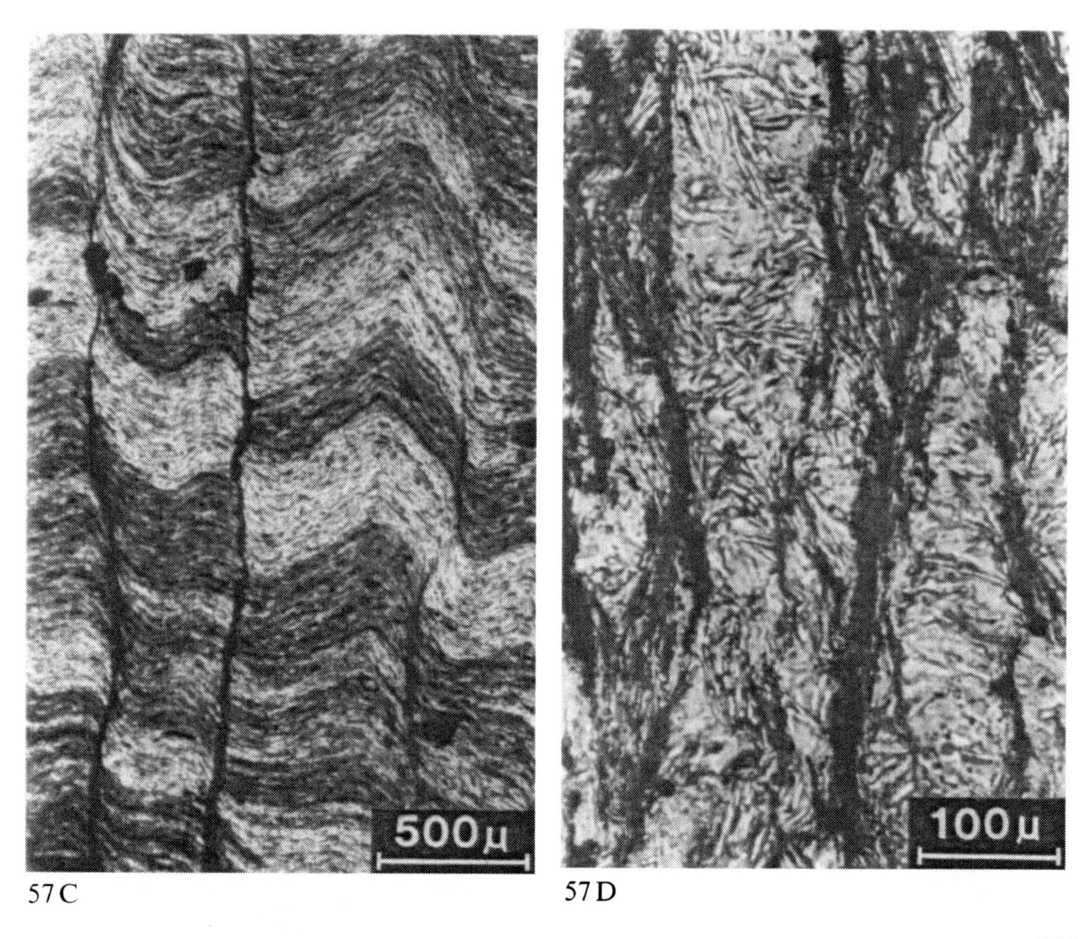


57 C

57 D

## 58. Crenulation Cleavage in Calcareous Slates

D. A. C. Gardner

Calcareous slates of the Upper Cambrian, Upper Chancellor Formation exposed near Golden, British Columbia in the Western Main Ranges, Southern Canadian Rocky Mountains, exhibit a late, locally developed crenulation cleavage which is formed by a combination of asymmetric microfolding (crenulation) of penetrative slaty cleavage with attendant alignment of white mica on limbs of crenulations, and local concentrated pressure solution. The cleavage is localized on overturned limbs of tight folds where bedding and the slaty cleavage (the same cleavage as illustrated in Plate 105) have been rotated toward the horizontal and subsequently have been refolded into broad open folds with a wavelength of 0.5 to 1.0 m. The crenulation cleavage is subparallel with the axial surfaces of these late broad folds.

---

**A. Crenulation.** Slaty cleavage (*SC*) and bedding (*B*) are folded into asymmetric crenulations; a weak cleavage is formed by mica alignment along limbs of crenulations. PPL

**B. Solution zone and crenulation cleavage.** This photograph of the same sample as **A**, above, illustrates a dark, through-going surface (*S*) oriented subparallel with the axial surfaces of crenulations, where clear offset of slaty cleavage and compositional layers defining bedding, and accumulation of insoluble residues suggest a pressure solution origin. Similar, approximately regularly spaced zones at 2 to 3 mm intervals occur throughout the sample. PPL

58 A

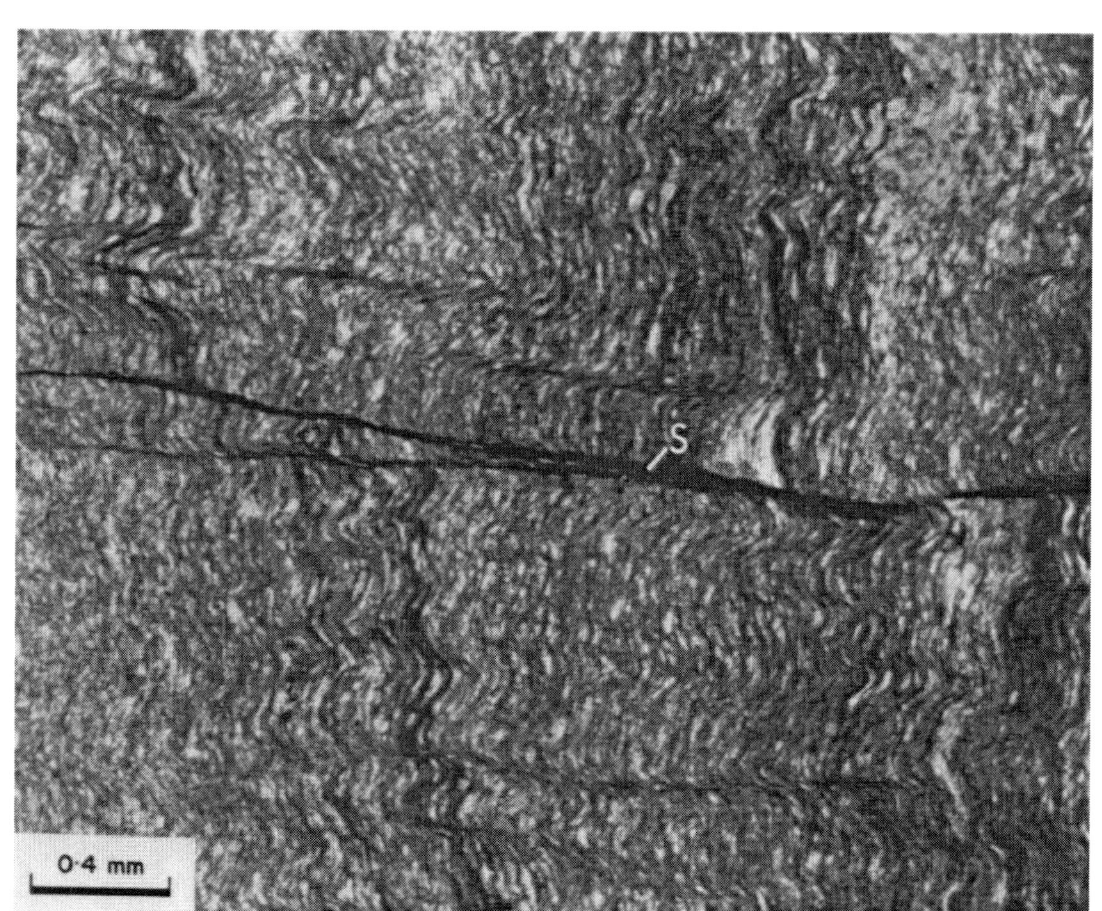

58 B

# 59. Crenulated Fibrous Vein

C. McA. Powell and R. F. C. Morritt

A crenulated fibrous vein from Late Ordovician quartzose flysch in Guineacor Creek, northeast of Taralga, N.S.W., Australia (Scheibner 1973). The fibrous vein is formed by intergrowth of an opaque mineral (hydrated iron oxide, probably goethite) with quartz and "sericite". Originally, the fiber axes were 40° to 50° oblique to the vein walls. The vein has subsequently been folded and partly removed by solution along discrete crenulation-cleavage surfaces during the Early Carboniferous folding (Powell et al. 1977). (Also see Plate 113.) The microfabric of the very fine grained slate that is host to the vein is domainal, with dark cleavage folia marking discrete crenulation-cleavage surfaces, and open crenulation microfolds occurring in the microlithons.

---

**A. Overview.** The fibrous vein is truncated at the discrete crenulation-cleavage domains. Assuming that the vein was initially planar, and that subsequent deformation was a combination of buckling with no volume loss in the plane of the section, followed by solution removal of parts of the rock along the discrete crenulation-cleavage surfaces in a direction normal to the cleavage, the strain in the section can be calculated. In this example, there has been a total shortening of 67% normal to cleavage, an elongation of 30% parallel to cleavage, and a volume loss of 57%. Of the shortening strain, 35% was by buckling and 65% by dissolution. PPL

**B. Detail.** The central portion of **A** is enlarged showing the nature of the fibrous intergrowth and a median line of transparent and small opaque grains within the vein. Note the crenulation microfabric within the microlithons in the host slate, and the abrupt truncation of this microfabric and the folded vein at the discrete crenulation surfaces. PPL

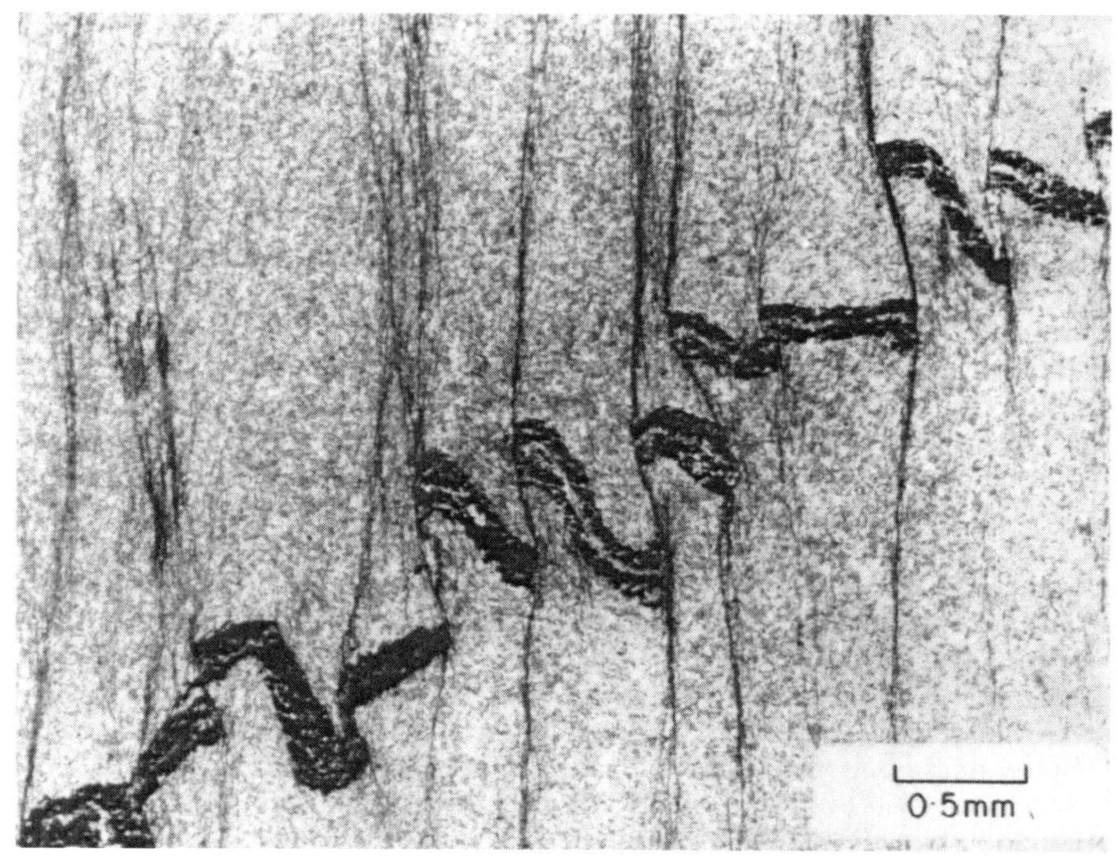


59 A

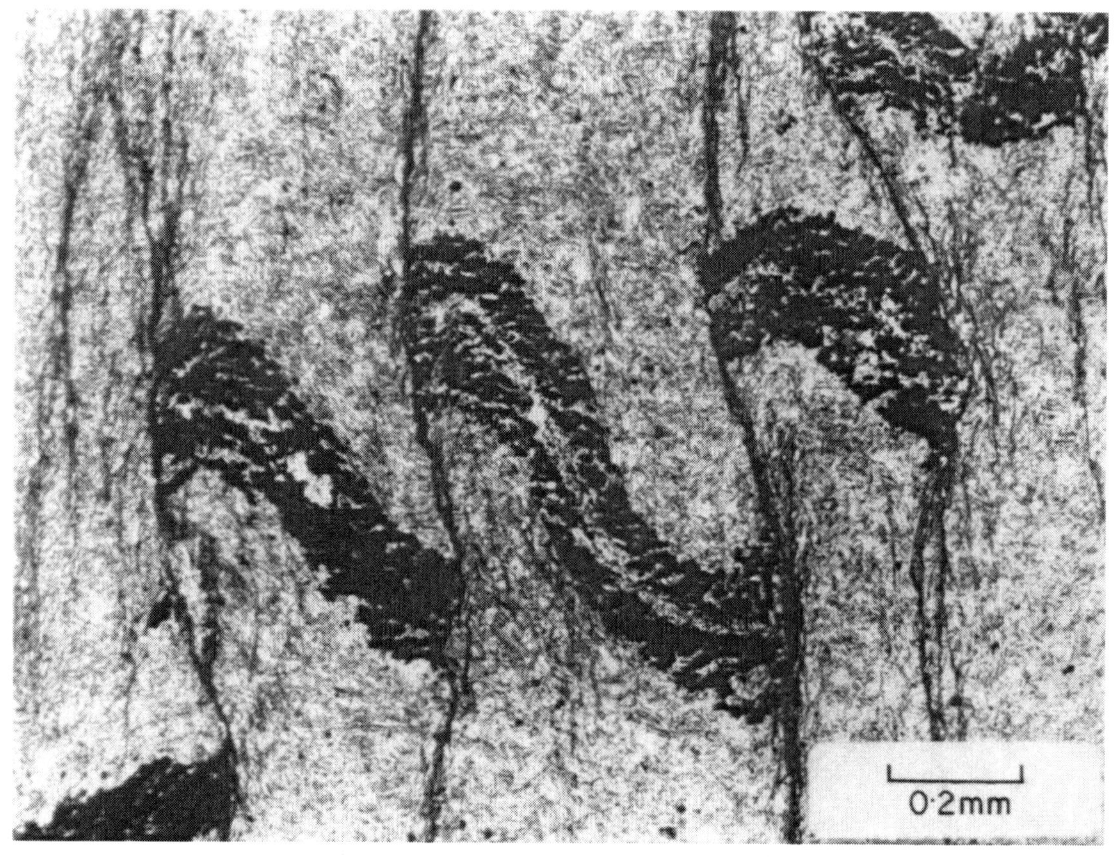


59 B

# 60. Crenulation Morphology of a First-generation Cleavage

C. McA. Powell, R. F. C. Morritt, and R. H. Vernon

The importance of crenulations in the development of slaty cleavages has been brought to light by the discovery that some slaty cleavage forms by progressive strain from early crenulations (Williams 1972, White and Knipe 1978). In this example from the Late Ordovician Triangle Group in Guineacor Creek, northeast of Taralga, N.S.W., Australia (Scheibner 1973), the first secondary foliation, which appears slaty in hand specimen, is seen to be a fine-scale crenulation cleavage in thin section. The structural history of the rocks involved at least two periods of folding. The first folding, in the mid-Devonian, produced open, upright folds trending north–northeast, but without axial-surface foliation. Detrital biotite was probably compacted to form a bedding-parallel foliation at this time (see Plates 6 and 7). The second folding, part of the meridional Early Carboniferous folding (Powell et al. 1977), impressed tight, upright, chevron folds with wavelengths varying from cm to km, and formed a crenulation cleavage parallel to the axial surface. This crenulation cleavage was the first cleavage oblique to bedding, and has similar structural style to that reported in rocks of comparable age and facies from Bermagui, southern N.S.W. (Williams 1972). At Bermagui the crenulation cleavage is more coarsely developed, particularly in psammitic rocks, where mineralogical differentiation in the crenulation microfolds gives the rocks a striped appearance. In the Guineacor Creek area, the size of the crenulations is smaller, but transitions from zonal crenulation cleavage, through discrete crenulation cleavage, to a disjunctive slaty cleavage can be seen.

---

**A. Refraction of crenulation cleavage.** The cleavage refracts across a silty band, which grades upward from a sharp base. The cleavage in the pelite (lower part of photograph) is zonal, and passes abruptly into more widely spaced, discrete crenulation cleavage at the base of the silty bed. This discrete crenulation cleavage curves upward as the detrital grain size decreases, and passes continuously into closely spaced zonal crenulation cleavage in the more pelitic part of the bed. PPL

**B. Detail.** Crenulation cleavage detail in a pelitic band. The dark folded vein is composed mainly of opaque minerals, and has been shortened normal to the cleavage by both buckling and solution removal. Both discrete and zonal cleavage morphologies are present, the discrete crenulation cleavages corresponding to zones of solution removal of the vein. Assuming that the vein was originally planar, an estimate of the strain in this section is a shortening of 65% normal to the cleavage, of which one-third to a half was accomplished by solution removal. PPL

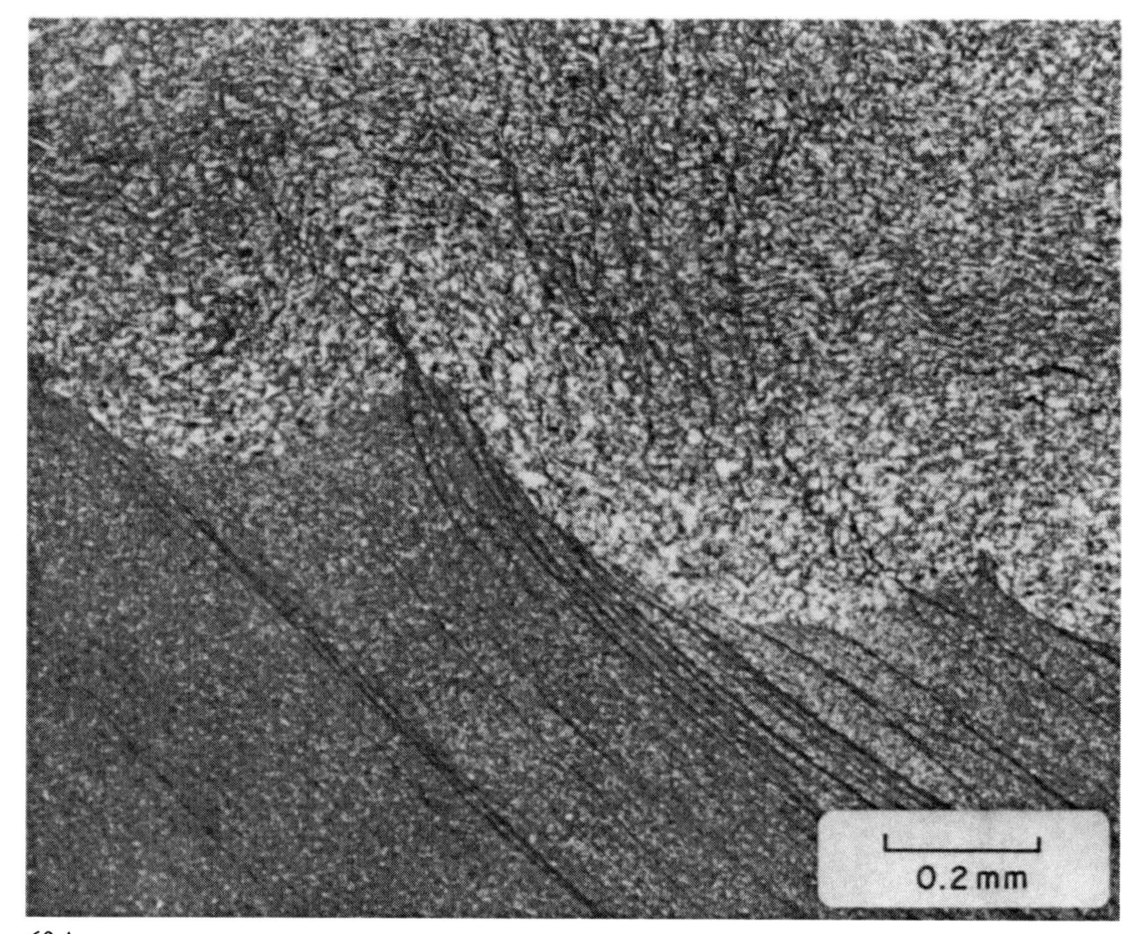


60 A

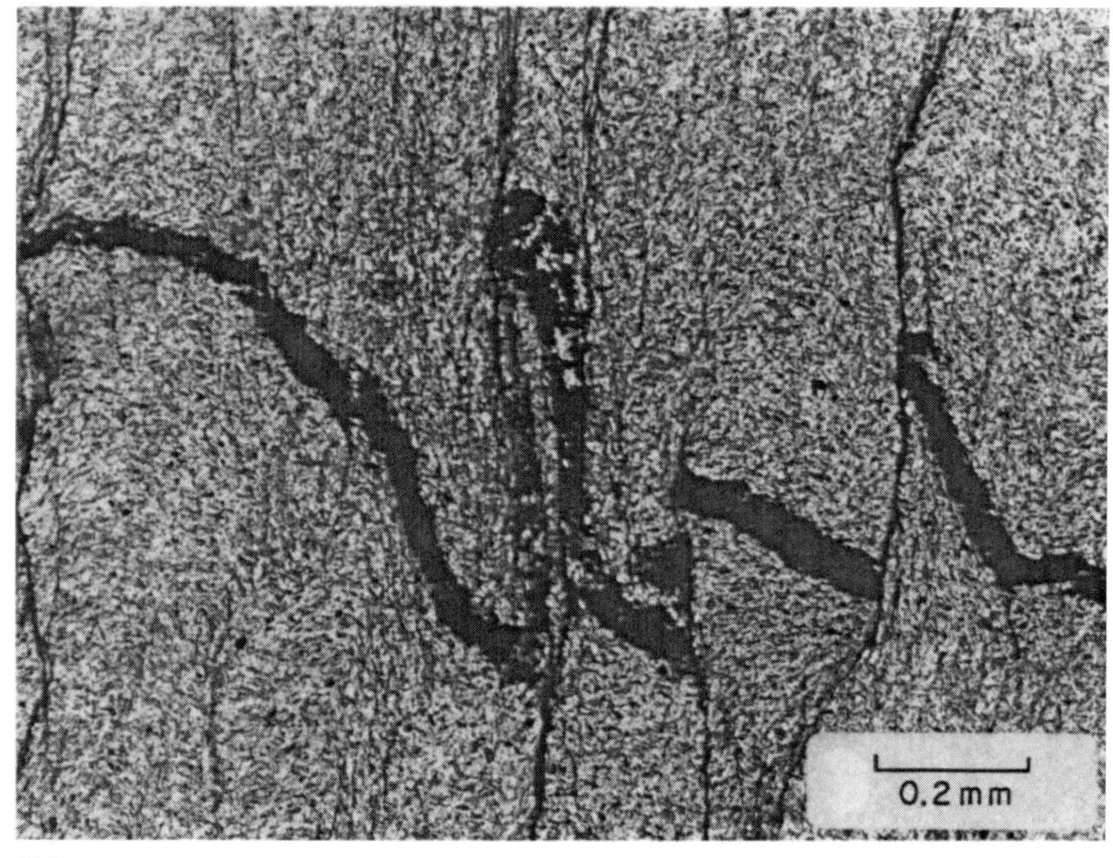


60 B

# 61. Crenulation Cleavages in Low Grade Metamorphic Rocks

M. R. Steuer

See introductory remarks for Plate 28.

**A. Dissolution of silt bed at discrete cleavage.** Cambrian Bull Formation. Apparent offset of silt beds due to dissolution and removal of material along N–S discrete cleavage in siltstone. Recrystallization effects are not apparent and metamorphic grade is presumed to be very low. N–S cleavage, although generally planar in three dimensions, is locally anastomosing. PPL

**B. Apparent offsets.** Cambrian Bull Formation. Host rock is dark shale with siltstone laminae. Discrete cleavage is NE–SW and can be seen continuing across the dark shale. Apparent offsets (see Plate 117) in the silt are viewed as dissolution features representing approximately 15% rock volume loss using layer reconstructions after Gray (1979). Dissolution effects are more predominant in fold hinges and preferred beds, complicating bulk rock loss calculations. PPL

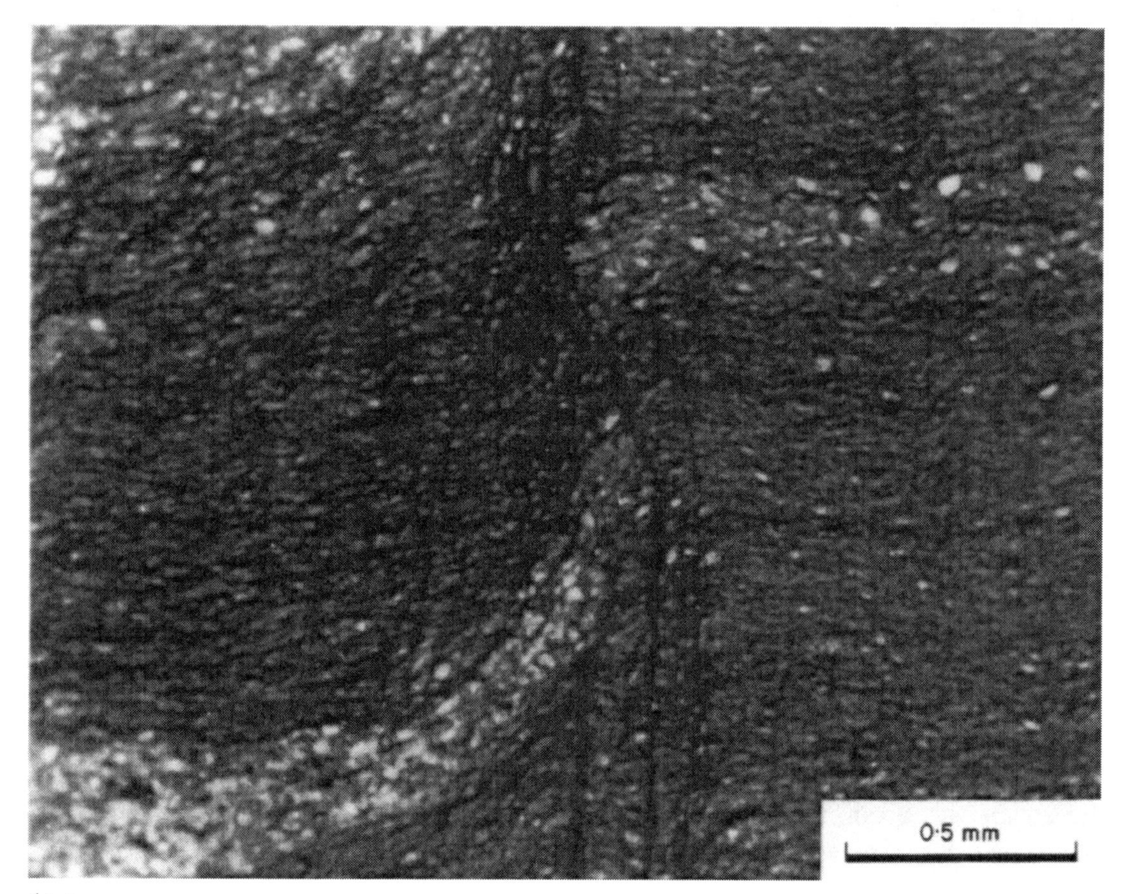

61 A

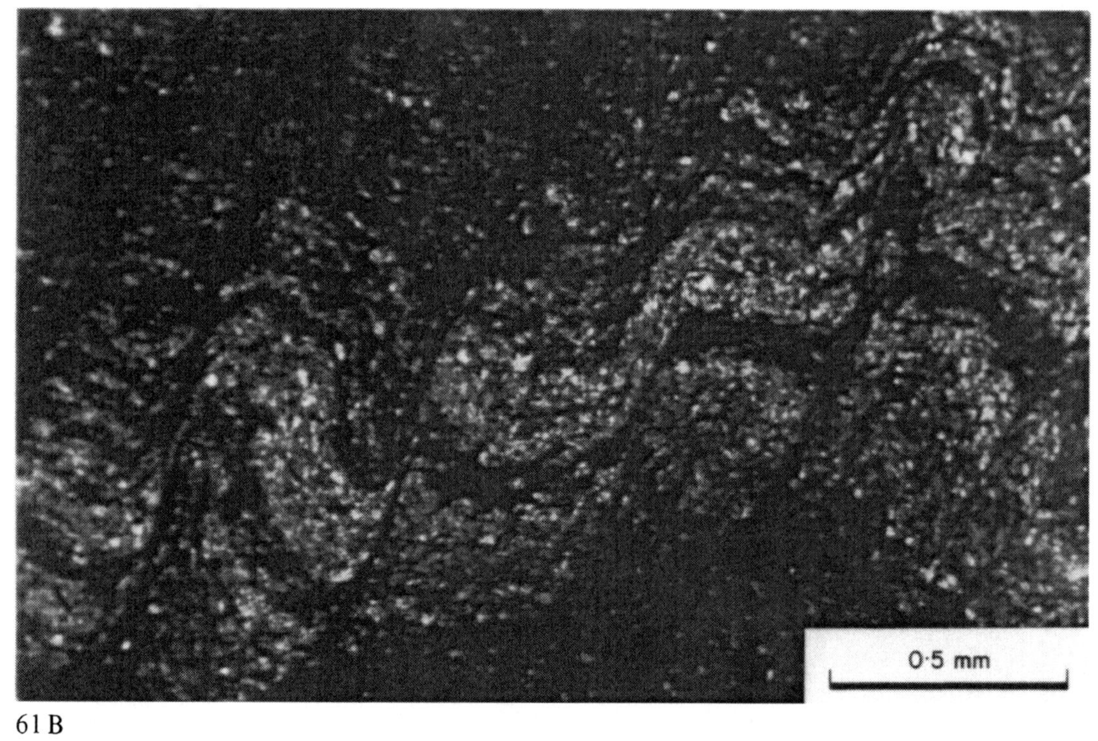

61 B

## 62. Crenulation and Dissolution in Shales

L. Solé-Sugrañes

In Billings (Dutchess Co., New York, U.S.A.) crenulation cleavage is superimposed on well-developed slaty cleavage in shales of the Normanskill Formation (Fisher et al. 1970). Slaty cleavage is defined by the complete orientation of detrital phyllosilicates and neoformation of very small flakes of chlorite and white mica growing parallel to the cleavage surfaces. Slaty cleavage and metamorphism developed during the main deformation episode of the Taconian orogeny (Ordovician), but crenulation could be of late Taconian age or more probably of Acadian age (Devonian).

Development of crenulation cleavage starts by microfolding the previous slaty cleavage. The shape and symmetry of these microfolds is controlled both by the viscosity contrast between the cleavage surfaces and the orientation of the shortening direction relative to these surfaces (Cosgrove 1976, Solé-Sugrañes 1976). Dissolution surfaces are preferentially superimposed on the short limbs of microfolds.

In the quartz-rich layers the wavelength of the folds is several times longer than in the mica-rich layers. The viscosity contrast between them was about 35 : 1 during the development of slaty cleavage. This was estimated from the wavelength of isolated single-layer microfolds (Sherwin and Chapple 1968). The amount of shortening due to crenulation is about 30%. In the most competent layers the strain was accommodated by folding but in the less competent material it was accommodated by dissolution and minor microfolding.

---

**A. Crenulation cleavage.** Defined by microfolding of previous slaty cleavage and stylolite-like dissolution surfaces in Normanskill shales. Dissolution surfaces grow on the short limbs of asymmetrical microfolds but they can cut across some axial planes. PPL

**B. Detail of A.** Showing the aspect of the dissolution surfaces on their progressive development on the short limbs of microfolds in places with marked viscosity contrast. PPL

**C. Lithological control of crenulation.** Dissolution surfaces and microfolds develop in mica-rich layers (*right part*) but they do not penetrate quartz-rich layers (*left part*). Spacing of the dissolution surfaces probably depends on the wavelength of the microfolds that in turn is controlled by the thickness and viscosity contrast of the layers. PPL

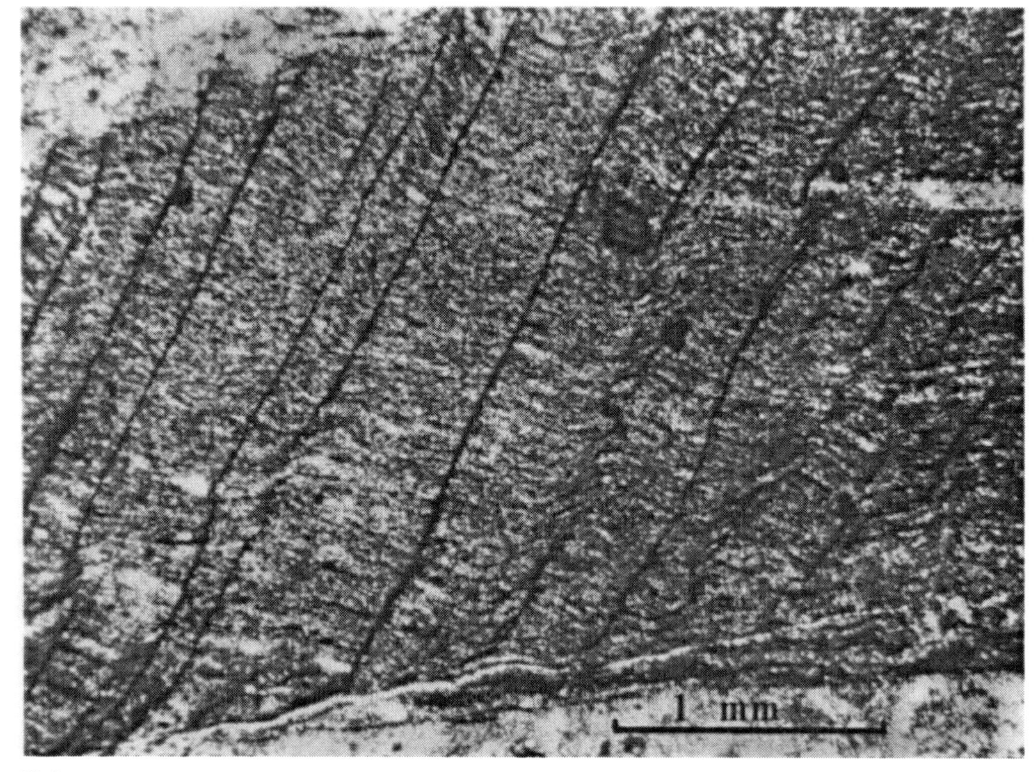

62 A

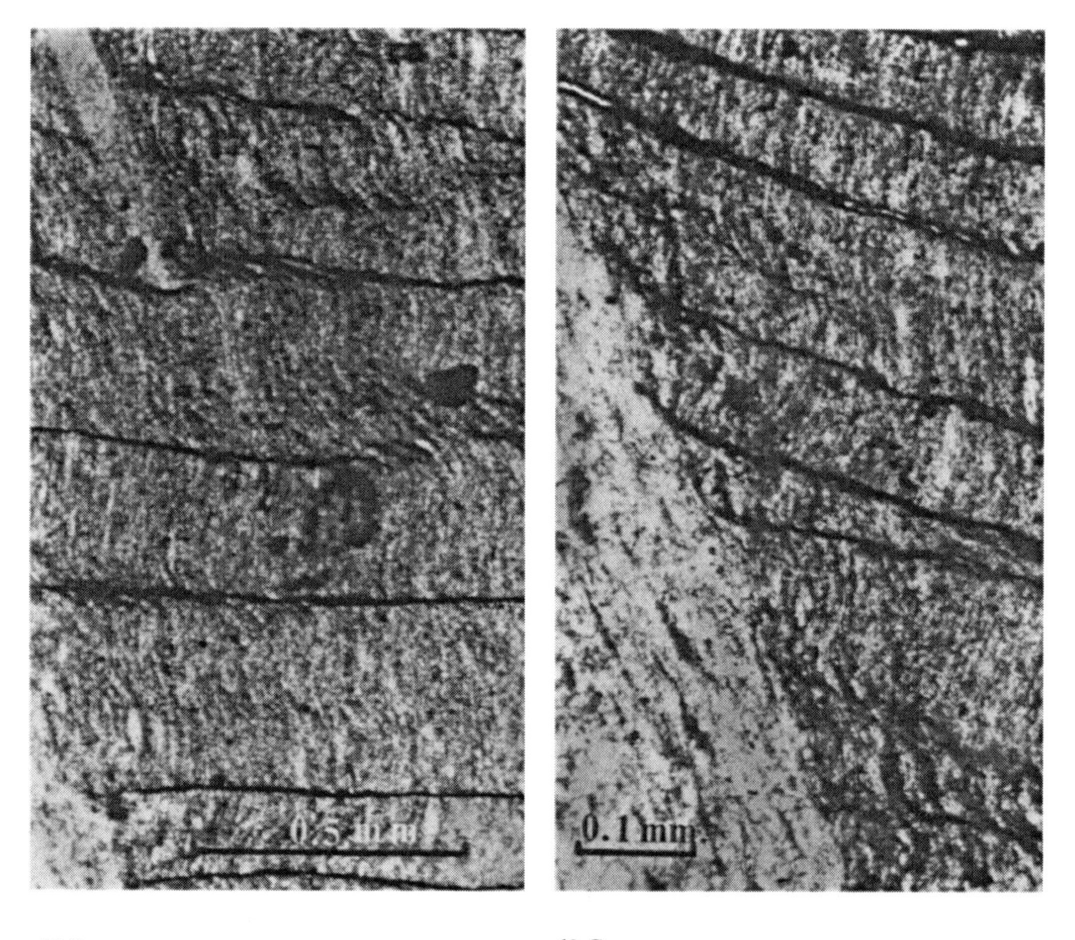

62 B

62 C

# 63. Deformation of Slaty Cleavage, New Jersey, U.S.A.

J. C. MAXWELL

Slaty cleavage has been broadly arched and also locally sheared and sharply folded, with accompanying local development of spaced cleavage.

---

**A. Single shear zone deforming slaty cleavage.** Martinsburg slate in road cut about 2 miles south of Columbia, N.J. along U.S. 46. At hammerhead, dextral shear, approximately parallel to bedding, sharply flexed the slaty cleavage.

**B. Spaced cleavage cutting folded slaty cleavage.** Lackawanna Railroad right-of-way, just south of Ramseysburg and about 3.5 miles south of Columbia, N.J. Sketch of outcrop is given in MAXWELL 1962, fig. 8.

**C. Detail of B.** A cleavage with spacing one mm or less (*vertical*) cuts the well developed slaty cleavage (*horizontal*). Scale: 1 mm PPL

63 A

63 B

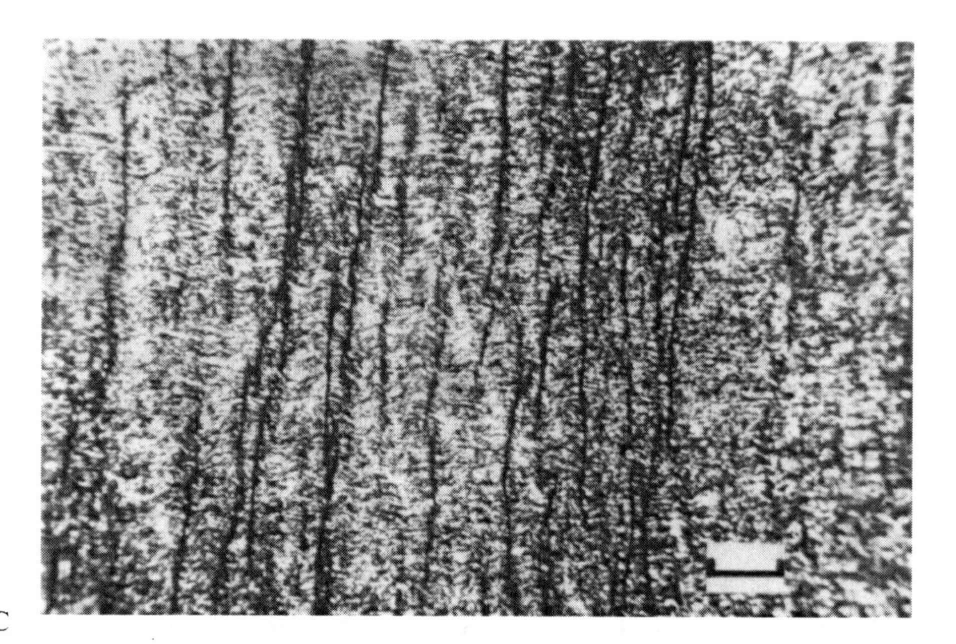

63 C

# 64. Spaced Cleavage in the Miette Group, Lake Louise, Alberta, Canada

J. H. SPANG, A. E. OLDERSHAW, E. D. GHENT, and M. Z. STOUT

Samples are from an outcrop of calcareous, sandy and silty argillite in the Upper Proterozoic Miette Group, approximately 2 km NW of Lake Louise in the Southern Rocky Mountains, Alberta, Canada. In outcrop the cleavage is axial planar within a cylindrical syncline. The lower-hemisphere equal-area projection shows the poles to layering (*solid circles*), poles to cleavages (*solid triangles*), calculated best-fit great circle and the pole to the great circle (*solid star*). The fold hinge plunges 5° on a bearing of 153°. Based on X-ray diffraction patterns the rock is predominantly composed of quartz, chlorite, sodium-bearing muscovite, calcite, and tourmaline.

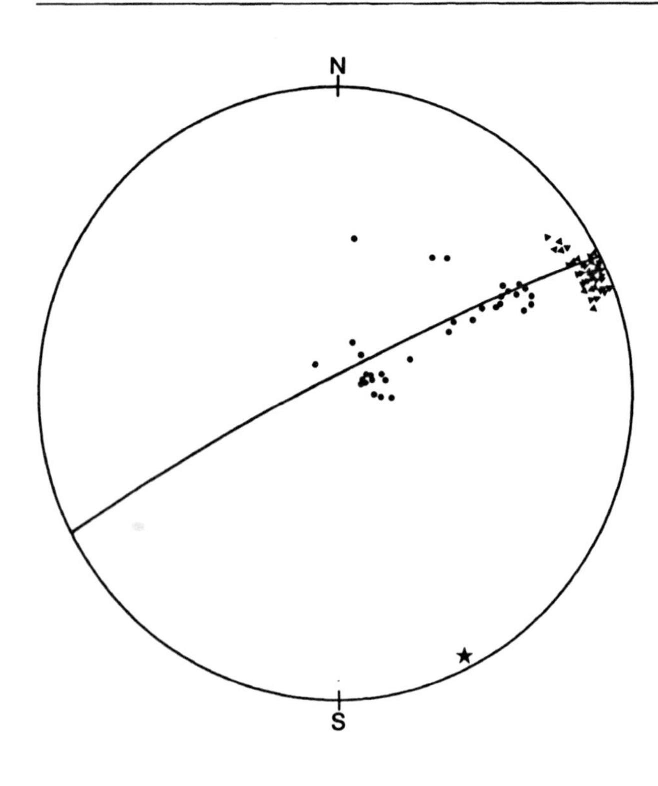


**A. Spaced cleavage.** The compositional layering (*Comp Layer*) is due to variations in the ratio of quartz to layer silicates. Scale: 1 mm PPL

**B. Chlorite porphyroblasts.** The chlorite porphyroblasts (**CL**) generally have the trace of their (001) planes at a high angle to the rock cleavage and range up to 0.15 mm in maximum dimension. The matrix consists of complex intergrowths of muscovite, chlorite, calcite and pyrite. The porphyroblasts occasionally appear to be truncated against the cleavage planes. Some of the chlorite porphyroblasts are composite, containing interleaved muscovite. Scale: 0.1 mm PPL

64 A

64 B

## Section 7: Disjunctive Cleavage Defined by Simple Cracks

(Plates 65 to 77)

Many of the thirteen examples illustrated here would have formerly been described as "fracture cleavage". More cautious descriptions are generally offered here. The examples are from limestones (Plates 65 to 69) and sandstones (Plates 70 and 71) though microfaults in shales, silts and mylonites are also shown to have the frequency of occurrence and splitting-character of cleavage. The offset produced on some crack-like cleavages is attributed to shear movements in some cases and in other cases to dip-slip offsets produced by "pressure solution" removal of material (see also Plate 117).

---

**Plates**

# 65. Spaced Cleavage of Solution Origin in the Scaglia Rossa, Italy

W. ALVAREZ and T. ENGELDER

Thin section evidence for solution origin of spaced cleavage in the Scaglia Rossa is presented on Plate 81. Additional evidence for this interpretation is provided by field observations:

1. Insoluble chert nodules are imbricated, showing tens of percentage shortening (**A**).
2. Cleavage is deflected away from strain shadows protected by chert nodules toward places where the nodules have been telescoped (**A**).
3. Bedding is offset at bedding-cleavage intersections in the direction that would be produced by dissolution and removal of limestone along the cleavage surfaces (**B**).

---

**A. Spaced cleavage in the upper chert-bearing member of the Scaglia Rossa.** Eocene, section between Maesta Confibio and Badia di Sidria (43°27′N, 12°45′E). Spaced cleavage in the Scaglia Rossa dips steeply *to the right* in this photograph. Bedding dips gently *to the left* and is marked by a horizon of chert nodules with dark cores and lighter rims. Origin of the cleavage through dissolution is shown by imbrication of the chert nodules and by convergence of cleavage surfaces toward points where the nodules have been telescoped. Scale: 15 cm rule

**B. Spaced cleavage in the lower chert-bearing member of the Scaglia Rossa.** Upper Cretaceous, 4 km ESE of Fiastra on the road to Bolognola (43°0.5′N, 13°11′E). View of a bedding plane showing offset produced by dissolution along spaced cleavage surfaces. The mechanism for producing this kind of offset bedding is shown by GROSHONG (1975, fig. 4 B) and by ALVAREZ et al. (1976, fig. 5) and by Plate 117 in this volume. Scale: hammer

65 A

65 B

## 66. Solution Cleavage and Estimates of Shortening, Umbrian Appenines

W. ALVAREZ and T. ENGELDER

Within the general category of solution cleavage, subdivisions of *weak, moderate, strong,* and *very strong* cleavage have been proposed (ALVAREZ et al. 1978). The table shows the attributes of the four types. Typical examples of the degrees of intensity are shown in the plate. *On the right hand side the photographs show the plan view of bedding while on the left hand side the view is perpendicular to bedding.* Telescoped chert horizons allow the shortening to be established and the correlation with the intensity of the fabric is listed below.

| | |
|---|---|
| Undeformed (**A, F**): | Zero shortening |
| Weak cleavage (**B, G**): | 0%– 4% shortening |
| Moderate cleavage (**C, H**): | 4%–25% shortening |
| Strong cleavage (**D, I**): | 25%–35% shortening |
| Very strong cleavage (**E, J**): | > 35% shortening |

The authors note two possible paths of cleavage development. Firstly, a single weak cleavage develops and becomes progressively stronger, remaining statistically perpendicular to the direction of maximum shortening all the time. Secondly, cleavage is initiated on two sets of planes at 60° and 120°, with the direction of maximum shortening bisecting the obtuse angle. While cleavage is weak, irregularities mask the difference between these two paths; at moderate intensity, the two paths are most readily distinguished; at strong intensity, the 60° paired cleavages interfere and the geometry approaches the same pattern as a single cleavage perpendicular to shortening (Plates **C**, and especially **H**, show this transition); at very strong intensity, the two paths are wholly indistinguishable.

| Characteristics | Average Spacing [a] | Designation |
|---|---|---|
| Surfaces generally toothed, stylolitic: their preferred orientation weak | > 5 cm | Weak (**B, G**) |
| Surfaces discrete (note 60° angles in **H**); toothed character less marked, preferred orientation stronger | 1–5 cm | Moderate (**C, H**) |
| Surfaces in places wispy and anastomosing, in places concentrated into major surfaces; toothed surfaces rare | 0.5–1 cm | Strong (**D, I**) |
| Surfaces sigmoidal; abundant calcite veins perpendicular to cleavage | < 0.5 cm | Very strong (**E, J**) |

[a] Spacing is measured perpendicular to cleavage without regard to bedding or the cleavage/bedding angle. In the Umbrian Appenines, the characteristics are so recognizable that there is almost never disagreement between field geologists over the cleavage type present in a given outcrop. However, the spacings were measured in nearly pure limestones or the least marly parts of impure beds. It is possible that in marls, the various qualitative attributes might correlate with different average spacings

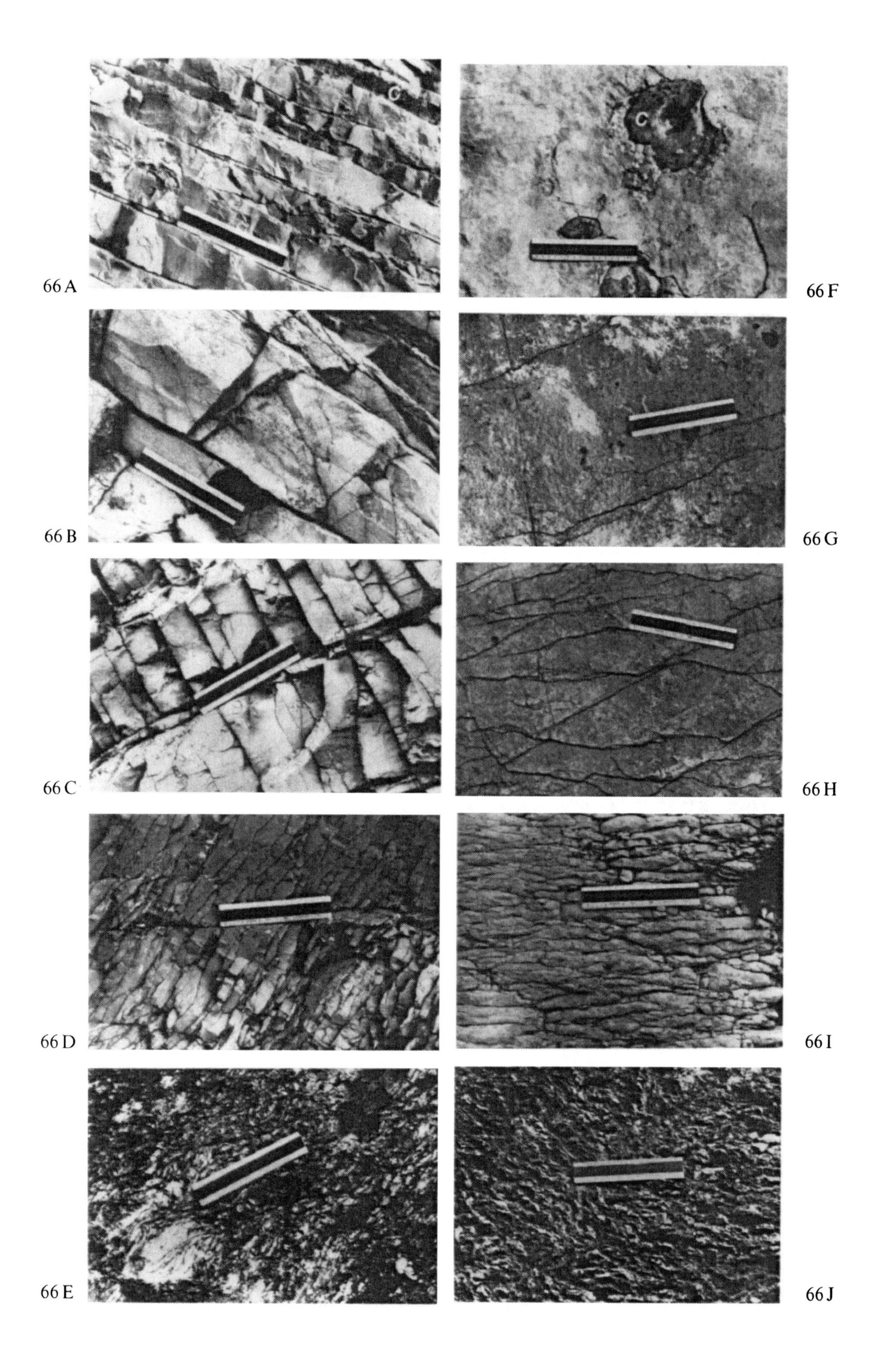

66 A
66 F
66 B
66 G
66 C
66 H
66 D
66 I
66 E
66 J

# 67. Fractures and Solution Cleavage (I)

S. A. Sansone

Solution cleavage, due to dissolution of calcite in limestones, accounts for a minimum of 16% layer-parallel volume loss at this locality. The estimate is made by measuring the following; relative concentration of insoluble material in the cleavage seam ($C_1$) and outside the seam ($C_0$); cleavage seam thickness (T); and cleavage spacing (S). The following equation gives the amount of bulk strain due to localized solution removal:

$$\varepsilon = \frac{S}{S + T\left(\frac{C_1}{C_0} - 1\right)} - 1 .$$

From a search in the immediate area for the missing or dissolved calcite it is concluded that most, if not all, of the calcite was transported out of the system. There is evidence that solution cleavage is nucleating on or following a pre-existing fracture pattern. The photographs are of two different views and scales of the solution cleavage.

---

**A. Configuration of solution cleavage in bedding plane.** Helderbergian Lower Thacher at Rosendale, New York, U.S.A. A photograph of a stone rubbing, made by laying rice paper on the bedding plane and rubbing the protruding cleavage with graphite, displays the configuration of the cleavage. The cleavage seams appear as connected straight line segments that at intervals bifurcate to give the anastomosing pattern. This anastomosing, bifurcating pattern suggests the cleavage is following a fracture or joint set.

**B. Configuration of solution cleavage normal to bedding and cleavage.** Helderbergian Lower Thacher at Rosendale, New York, U.S.A. Note how the cleavage seams zig-zag. The zig-zagging is believed to be derived from a fracture set that has been preserved during the dissolutional event that formed the cleavage. Note how the cleavage seams thin in the upper layers, indicating a lesser amount of bulk strain due to dissolution

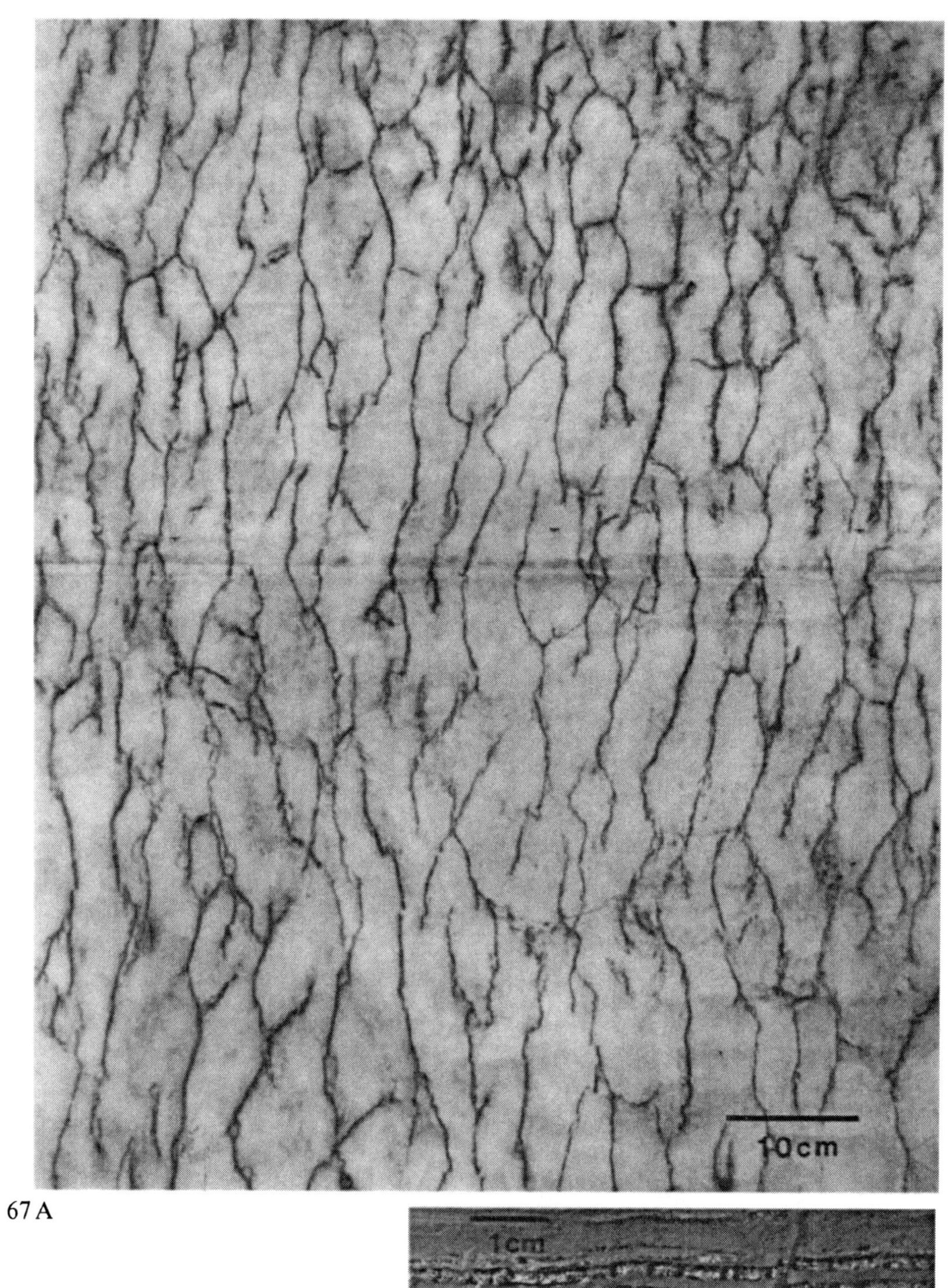


67 A

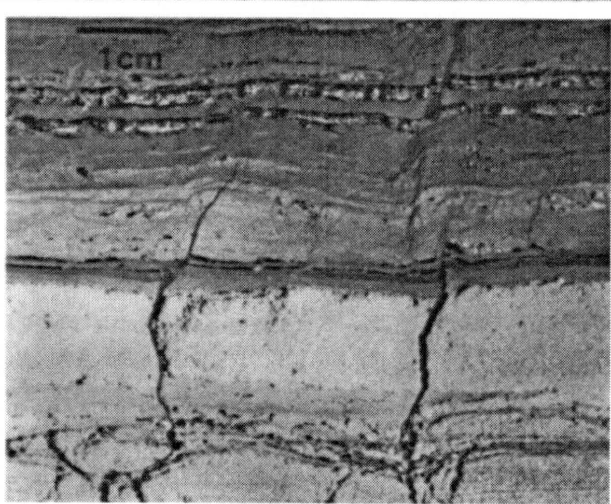


67 B

# 68. Fractures and Solution Cleavage (II)

S. A. Sansone

**Solution cleavage in three dimensions.** Helderbergian Lower Thacher at Rosendale, New York, U.S.A. The photograph is of a three-dimensional model developed from a 2500 $cm^3$ block of Lower Thacher limestone. The Thacher is a fine-grained, thin-bedded limestone. The rock was cut, normal to bedding and cleavage, into nine slabs and each slab was etched in 10% HCl for 30 min. The cleavage, standing out in high positive relief, was traced onto sheets of Plexiglas and placed in serial section. The print is in four shades of black and grey to provide the three-dimensional effect. The *black lines* represent the outline of the cleavage closest to the viewer and the lighter shades are deeper, or further down, in the specimen. View is normal to cleavage and bedding. Note how the cleavage seams zig-zag as if they follow a closely spaced fracture pattern that predates the cleavage. The cleavage bifurcates down and back into the photograph, and is essentially unconnected to adjacent cleavage seams.

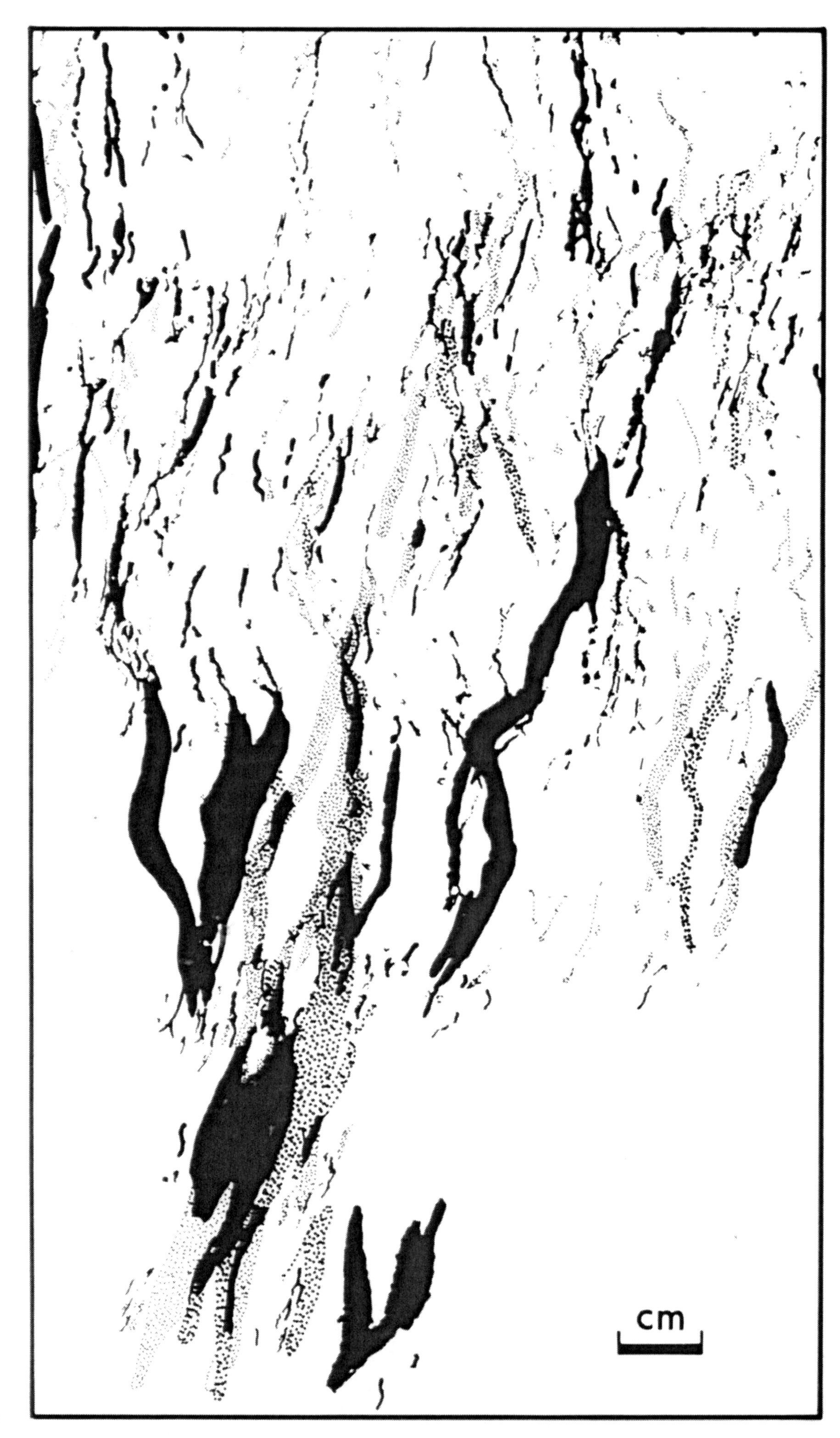
cm

# 69. Neotectonic Fracture Cleavage in Weakly Deformed Pliocene Chalks

P. L. HANCOCK

A Late Pliocene fracture cleavage cuts weakly deformed Pliocene chalky limestones within the Comiso Fault Zone on the northwest margin of the Iblean (Ragusa) platform, a tectonic unit external to the Alpine chain (CAIRE 1973, MASCLE 1974). Relationships at Punta Branco Grando, where the structures are well exposed, are schematically illustrated in the figure. The globigerinid-rich Pliocene chalks (Trubi) are involved in a gentle anticline cut by a nearly vertical, curviplanar fracture cleavage. The chalks are unconformably overlain by uncleaved early Quaternary calcarenites. Mid to late Quaternary normal faults downthrowing to the northwest displace the Pliocene and Quaternary rocks, and kink-bands parallel to the faults deform the cleavage. The relationships are shown in the sketch profile below.

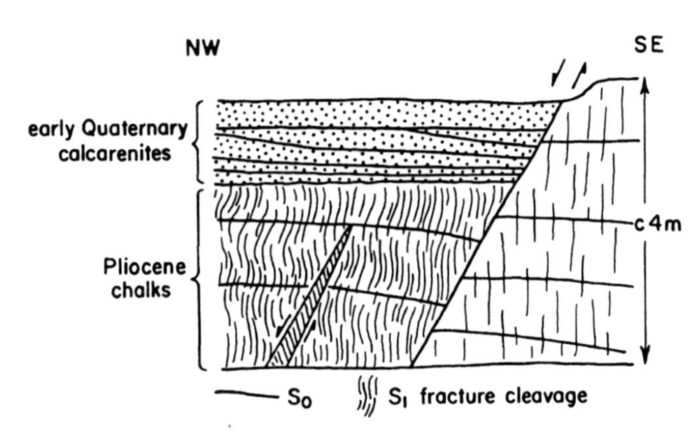


**A. Fracture cleavage cutting Pliocene chalks.** Punta Branco Grande, 14 km SSW of Vittoria, Sicily. The curviplanar, nearly vertical fracture cleavage surfaces cut chalks dipping gently to the left.
Scale: 1 m rule

**B. Quaternary kink-band deforming fracture cleavage.** Punta Branco Grande, 14 km SSW of Vittoria, Sicily. Nearly vertical fracture cleavage surfaces cutting Pliocene chalks are deflected in a negative (reverse) kink-band parallel to nearby Quaternary normal faults.
Scale: 1 m rule

**C. Fracture cleavage surface cutting a globigerinid test in Pliocene chalk.** Punta Branco Grande, 14 km SSW of Vittoria, Sicily. The photograph shows an unfilled extension fracture from the cleavage set illustrated (**A**). The fracture cuts a globigerinid test and offsets its outer wall by dilation. The thin section was prepared after impregnation with Araldite to prevent the rock splitting apart on the cleavage plane during specimen preparation.
PPL

69 A

69 B

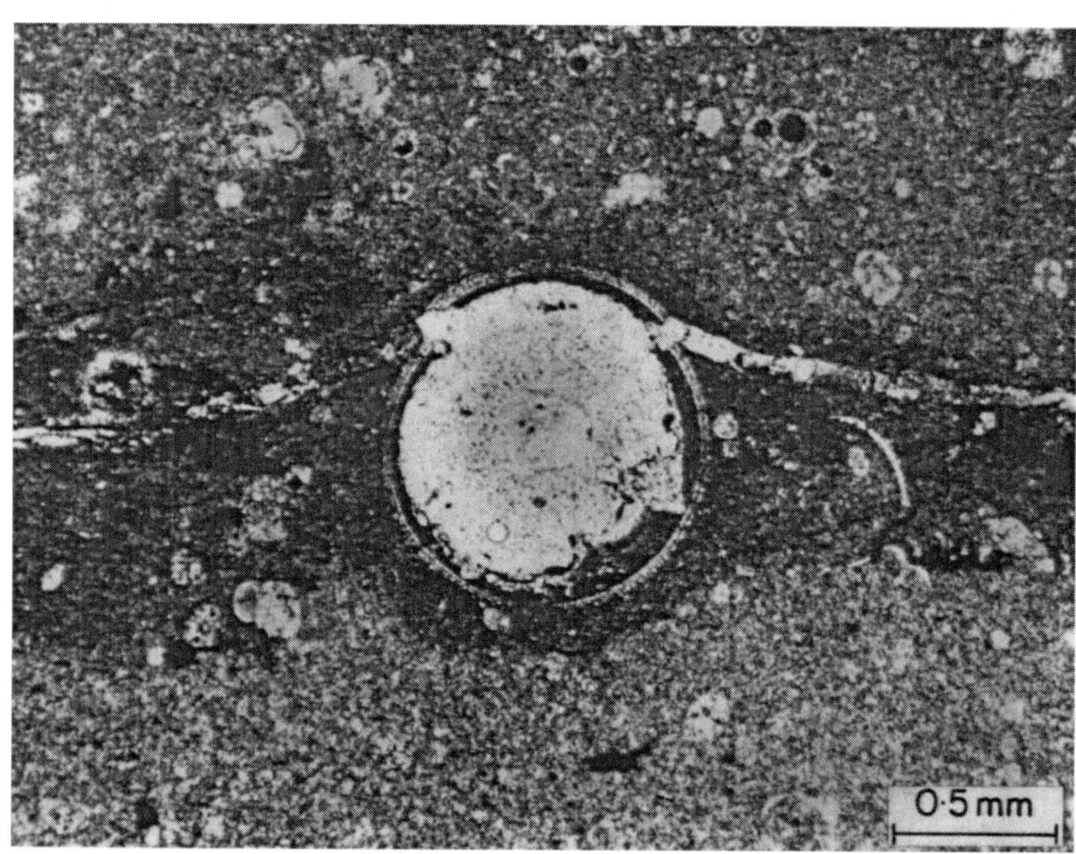

69 C

# 70. Distinction Between Cleavage and Joints Using Fracture Separation

P. L. HANCOCK

Although most cleavages are related to processes such as recrystallization, grain rotation, pressure solution or crenulation, some sets of closely spaced fractures form a structure which is commonly called fracture cleavage. To specify the meaning of "closely spaced" an arbitrary maximum fracture separation may be selected, or, as PRICE and HANCOCK (1972) propose, a maximum separation of, say 5 cm, may be combined with a separation/layer-thickness (s/t) aspect ratio not exceeding 0.05 (Fig. **1**).

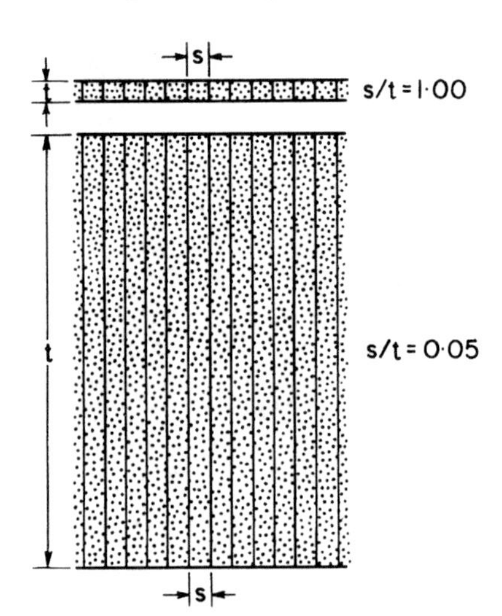


Fig. 1.

The illustrated examples are extension fractures cutting Oligocene molassic sandstones of the Campodarbe Formation within the Jaca Basin of the South Pyrenean Zone (PUIGDEFÁBREGAS 1975), their geometrical relationships to the late Oligocene folds being shown in Fig. **2**.

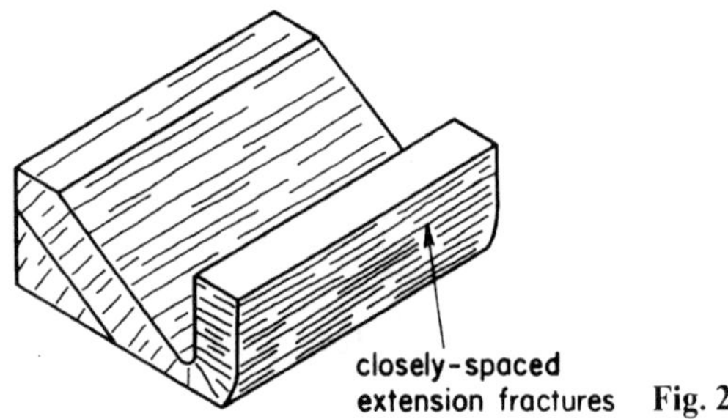


Fig. 2.

**A. Joints cutting sandstones.** Oligocene, about 1.0 km NE of Villalangua, Huesca Province, Spain. Nearly horizontal fractures cut vertical sandstones and their intersections on bedding planes form a nearly horizontal lineation. The average separation of fractures is 22 cm and the average s/t ratio is 0.60, and thus the fractures belong to a joint set.

Scale: Rock face about 5 m high.

**B. Fracture cleavage cutting sandstones.** Oligocene, about 1.5 km ESE of Villalangua, Huesca Province, Spain. A composite, nearly vertical bed of sandstone is cut by fractures inclined gently to the left. The average fracture separation is 4.3 cm and the s/t ratio is less than 0.03, and hence the fractures define a cleavage.

Scale: 30 cm hammershaft

**C. Extension fracture cutting sandstone.** Oligocene, about 1.5 km ESE of Villalangua, Huesca Province, Spain. The irregular, matching walls of the fracture, together with the absence of evidence for shear or pressure solution indicate that the fracture is of extensile origin.

NX

70 A

70 B

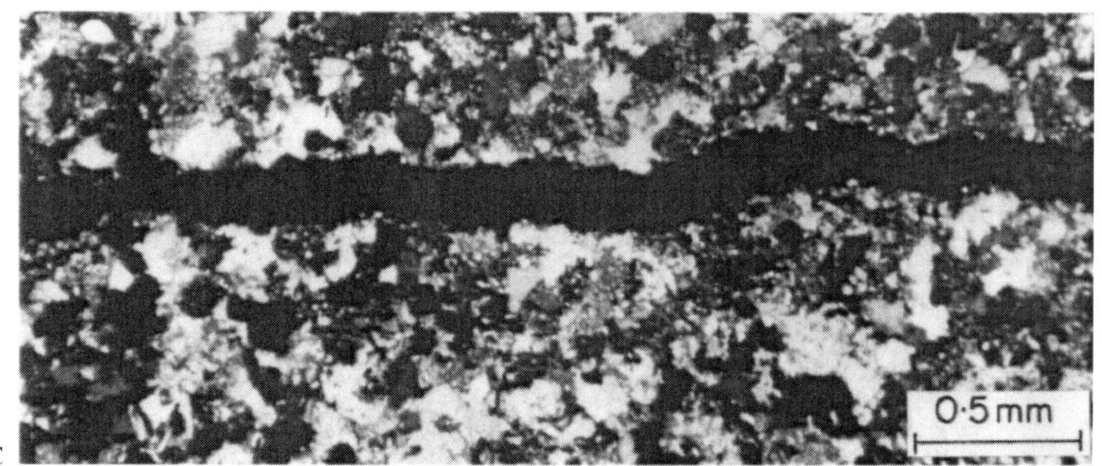


70 C

# 71. Distinction Between Cleavage and Joints Using Morphology

P. L. Hancock

When deciding whether a set of surfaces defines a cleavage an observer is likely to take into account the separation of surfaces, and he may also allow for the influence of a separation/layer thickness aspect ratio (see Plate 70). The morphology of the surfaces is also significant because it reflects the origin of the cleavage. These examples occur in sandstones within the Milford Haven Group of the Old Red Sandstone (Siluro-Devonian) in the Variscan fold belt of southwest Wales, U.K. (Allen and Williams 1978, Hancock 1973).

---

**A. Pressure-solution cleavage surfaces and close joints.** Milford Haven Group, Marloes, Dyfed, Wales. The bedding plane supporting the 1 m rule is crossed by the irregular traces of pressure-solution cleavage surfaces which form a lineation bisecting the obtuse angle between conjugate shear zones (Fig. **1**). The nearer bed is cut by smooth planar surfaces. On the basis of their morphology most geologists would call these surfaces joints despite their average separation being less than that between the pressure solution cleavage surfaces.

**C. Pressure-solution cleavage surface cutting sandstone.** Milford Haven Group, Freshwater West, Dyfed, Wales. Thin section of the cleavage visible in **B**. The cleavage is seen to be a reflection of narrow domains comprising concentrations of insoluble residues and inequant quartz grains with their long axes parallel to the domain. A calcite veinlet is emplaced within the domain. Hematite cement renders much of the slide opaque. PPL

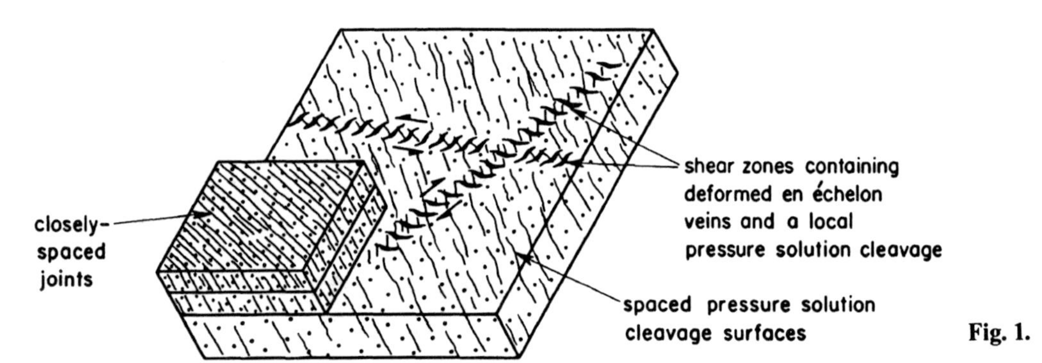


**Fig. 1.**

**B. Pressure-solution cleavage surfaces and conjugate joints.** Milford Haven Group, Freshwater West, Dyfed, Wales. The irregular lineation, formed by the intersection of pressure-solution cleavage surfaces with bedding, bisects the obtuse angle between conjugate shear joints (Fig. **2**). Scale: 25 cm rule

**D. Joint plane cutting sandstone.** Milford Haven Group, near Marloes, Dyfed, Wales. The photograph illustrates the morphology of one of the joint planes visible in the upper bed of sandstone shown in **A**. PPL

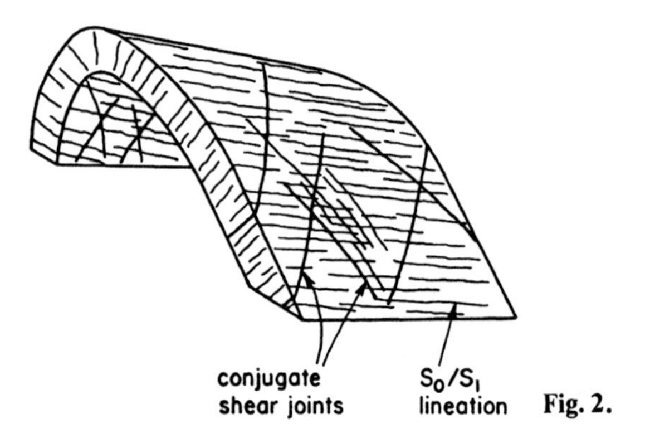


**Fig. 2.**

71 A

71 B

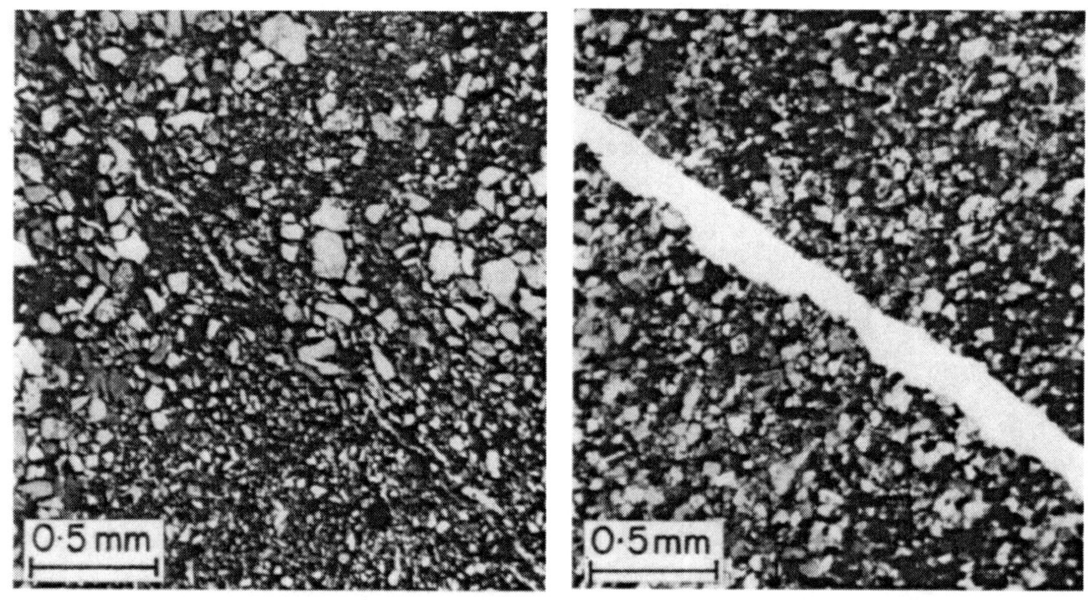

71 C, D

## 72. Reticulate Cleavage in the Goonoo Goonoo Mudstone, Australia

K. A. W. Crook

Reticulate cleavage (Crook 1964) is a type of disjunctive cleavage which is typically developed in weakly deformed mudstones and muddy limestones. It consists of quasi-planar fractures which are penetrative on an outcrop scale (**A**) but not on the scale of a hand specimen. The fractures are commonly 0.5–1.25 cm apart and tend to anastomose in profile view (**B**). The fractures also tend to anastomose in the direction of strike of the cleavage (**C**). The cleavage surfaces are therefore irregular on a small scale. They are not slickensided and to the naked eye neither recrystallization nor the growth of new minerals is visible. On an outcrop or larger scale, however, the cleavage surfaces are statistically planar, fanning about the axial surfaces of folds which they may be used to define.

In the Devono-Carboniferous mudstones of the Tamworth Trough, Australia, where reticulate cleavage was first recognized, the cleavage is widespread. Its degree of development varies regionally, sympathetically with the interlimb angles of associated major folds. Where it is incipiently developed, reticulate cleavage is expressed as a weak ordering in the orientation of the fractures that result from weathering. It may therefore be easily overlooked or dismissed as being of no regional significance. Stronger development of the cleavage enhances the quasiplanar fractures until they are at least as prominent as the bedding fissility.

Several observations suggest that reticulate cleavage develops in response to bulk strain in the whole sequence attendant on folding. The rocks that display this cleavage are in part flyschoid (**A**), but the form of the cleavage does not vary with changes in the thickness or apparent competence of interbedded siltstone and fine sandstone units. Typically these coarser grained units lack cleavage, and are transected by widely spaced joints. Slickensided bedding planes are very rare, suggesting that relative movement between competent beds was not a factor in cleavage formation. Furthermore Crook (1964, Plate 3) found that irregular intraformational mudstone blocks included in well-indurated rudites displayed a reticulate cleavage of the same morphology and orientation as that in adjacent mudstones.

---

**A. Reticulate cleavage** in flyschoid mudstone with siltstone interbeds (Crook 1964). Goonoo Goonoo Mudstone, Spring Creek, Montary, south of Tamworth, N.S.W., Australia.

**B. Anastomosing fracture pattern** typical of profile views of reticulate cleavage (Crook 1964).
Scale: 20 cm

**C. Reticulate cleavage as seen on bedding plane** in mudstone (Crook 1964). The fractures anastomose along the strike direction of the cleavage. Goonoo Goonoo Mudstone, Marangaroo, south of Tamworth, N.S.W.

72 A

72 B

72 C

# 73. Anastomosing Cleavage: Lenticular Configuration

G. J. BORRADAILE

Silurian sandstones and siltstones, Petit Rocher Sud, New Brunswick, Canada (BORRADAILE 1978).
Spaced, disjunctive cleavage divides beds into lenticular microlithons of less deformed material by anastomosing in three dimensions.

---

**A.** Plan view of sand bed also shown in **B**. Disjunctive cleavage surfaces anastomose subparallel to rule. Lenticular microlithons in between cleavage surfaces are deformed so as to produce spoon-shaped synforms. Scale: cm

**B.** End elevation of sand bed illustrated in **A**, together with underlying more fine-grained bed. Bedding is horizontal. Cleavage surfaces are more closely spaced in the underlying fine-grained bed. In the coarser sand bed the surfaces divide the bed into lenses which are about the same shape as in plan view (**A**). Scale: cm

73 A

73 B

# 74. Closely Spaced Antithetic Shear Surfaces in Siltstone

J. G. DENNIS

Here is an example of a fabric which does *not* constitute cleavage. It consists of closely spaced mini-fractures in siltstone of the Miocene Monterey Formation in San Pedro, California.

These fracture surfaces in unmetamorphosed siltstone are due entirely to soft-sediment simple shear parallel to bedding (see diagram below). Hence they are secondary shear surfaces. Criteria which distinguish them from cleavage include: (1) complete lack of grain fabric orientation, and (2) short principal axes of finite strain at an acute angle to the fractures; hence, presumed greatest compressive stress at an acute angle to the fractures throughout the deformation process.

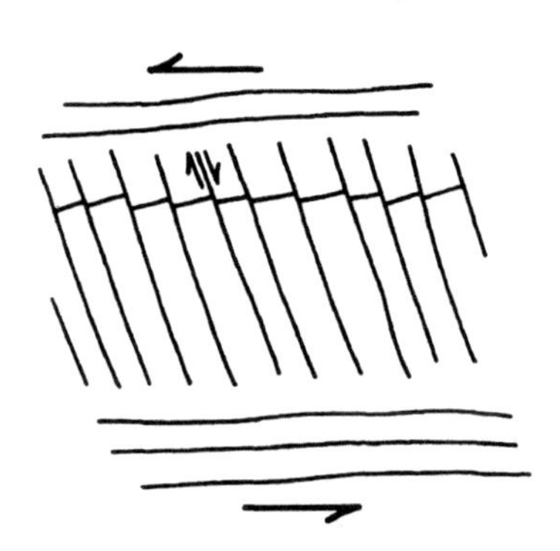

74

# 75. Microfaults and Mylonitic Schistosity

G. J. Borradaile

See also Plates 77, 153 B, 154

The photographs are of a major, steep-dipping transcurrent displacement zone known as the Quetico Fault in the Superior Province of the Canadian Shield. The locations are at Dog Lake near Thunder Bay, NW Ontario, but the structures typify the northern side of the east–west Quetico Fault along at least 200 km of its length. The photographs are of horizontal surfaces, viewed down to the north. The sketch summarizes the gradation that exists between the micro-faulted gneisses on the north side of the fault zone (**A**) and the lenticular mylonite fabric in the fault zone about 1 km further south (**B**). The microfaults suggest a dextral displacement for the Quetico Fault zone.

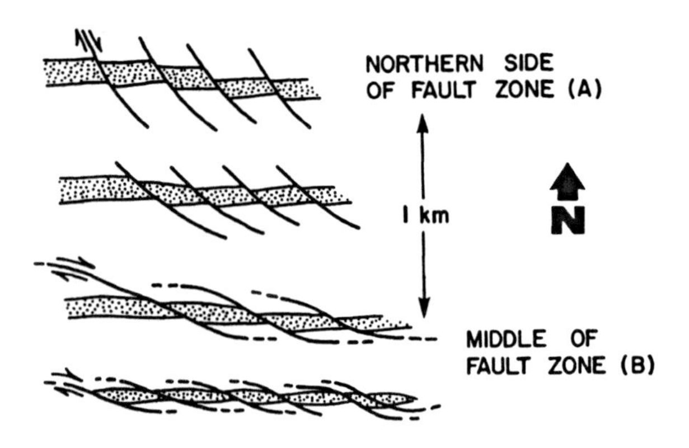


**A. Northern margin of the Quetico Fault zone.** Banded migmatitic gneisses near the displacement zone show microfaults with dextral displacement. The layering in the gneisses is subparallel to the Quetico Fault zone, so that the microfaults are oblique to the Quetico Fault zone.

**B. In the Quetico Fault zone.** The gneisses show a mylonitic schistosity parallel to layering which is defined by trains of subgrains around porphyroclasts and by quartz ribbons. The oblique dextral microfaults are still present, however, and now form a significant component of the fabric, imposing a lenticular aspect on the bands across which they cut. Where the microfaults curve parallel to layering they disappear into grain boundaries. This suggests that the dextral slip on microfaults is taken up along grain boundaries in directions parallel to the banding and therefore also parallel to the Quetico Fault zone which is parallel to the layering

75 A

75 B

# 76. Folding and Cleavage in Martinsburg Slate, Northern New Jersey, U.S.A.

J. C. Maxwell

Major folds have wavelengths of hundreds to thousands of meters and are asymmetric. Structures indicating compression parallel to bedding are characteristic of the short, steep northwest-facing limb, while the long, more gently dipping southeast-facing limb may show extensional features.

---

**A, B. Buckling of calcareous siltstone bed in vertical limb of major fold.** Martinsburg Slate exposure along U.S. 46, 1.5 miles northwest of Belvedere, N.J. Slaty cleavage is accentuated so that bedding is obscured within the concave side of folds, but bedding is visible along straight and convex portions of the folded calcareous siltstone. Note well-preserved vertical bedding in slate away from the crumpled siltstone, and spaced cleavage and calcite-filled tension fractures in the latter. Approximately 40% shortening of the slate matrix is required, presumably by dewatering.

**C, D. Deformation style of thin silt beds.** In slate, a few feet away from thicker beds such as those in **A** and **B.** Note uniform tilt of silt beds between bounding shears which are themselves slaty cleavage planes, indistinguishable in thin section from remainder of slate fabric. This fabric recalls glide packets in deformed metal crystals. The amount of shortening is comparable to that provided by buckling of thicker layers

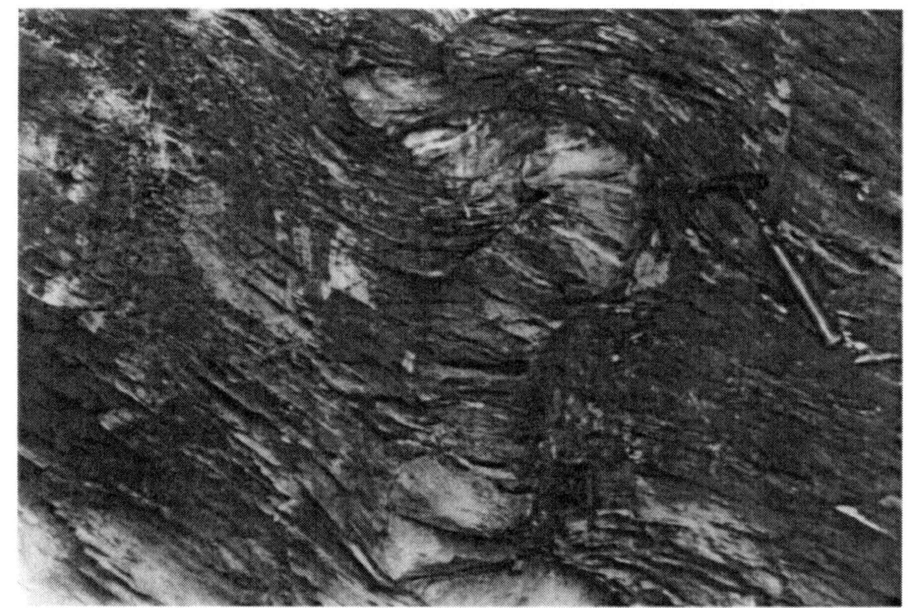
76 A

76 B

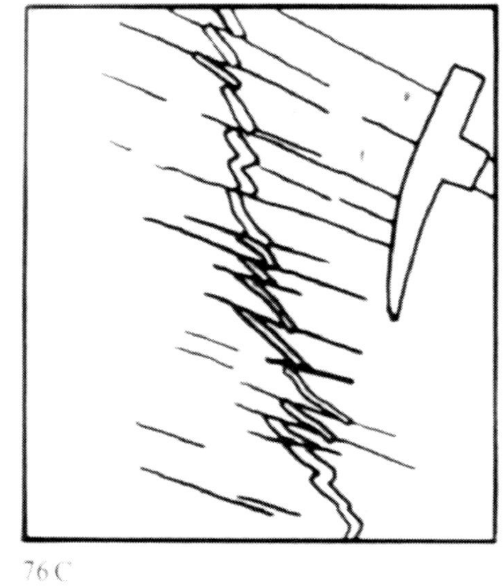
76 C

76 D

# 77. Dilation Along Grain Boundaries

G. J. Borradaile

Annealed mylonite in Archean rocks, Little Turtle Lake near Mine Centre, Northwest Ontario, Canada. (See also Plates 99 and 100.) Part of the Quetico Fault (see Plate 75).

**A. Overview.** The specimen is taken from the flank of an isoclinical fold in a mylonite zone. The quartz texture has been annealed to give triple junctions of quartz grains ("foam texture" of Vernon 1976). Certain crudely aligned grain boundaries define surfaces of dislocation and of small relative movements of parts of the aggregate. These surfaces appear as darkened and thickened grain boundaries (*arrowed*). NX

**B. Detail.** The darkened grain boundaries along which movement is believed to have occurred are indicated as *black* in the sketch of the photograph:

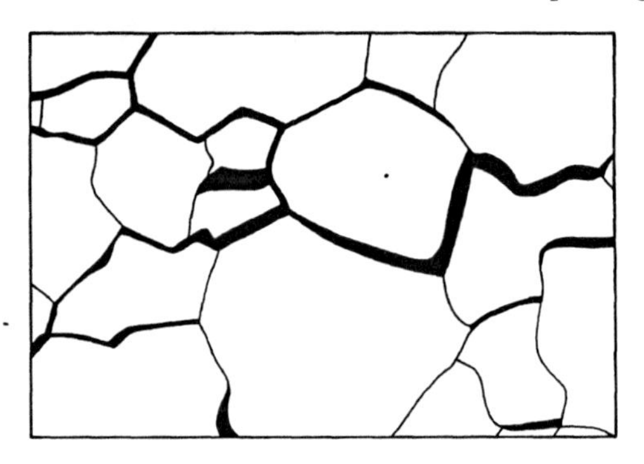

In the photograph it may be seen that these grain boundaries are crowded with small fluid inclusions which can just be resolved in the photograph (*arrow*). These surfaces may represent dilatant zones formed during particulate flow of the aggregate. NX

77 A

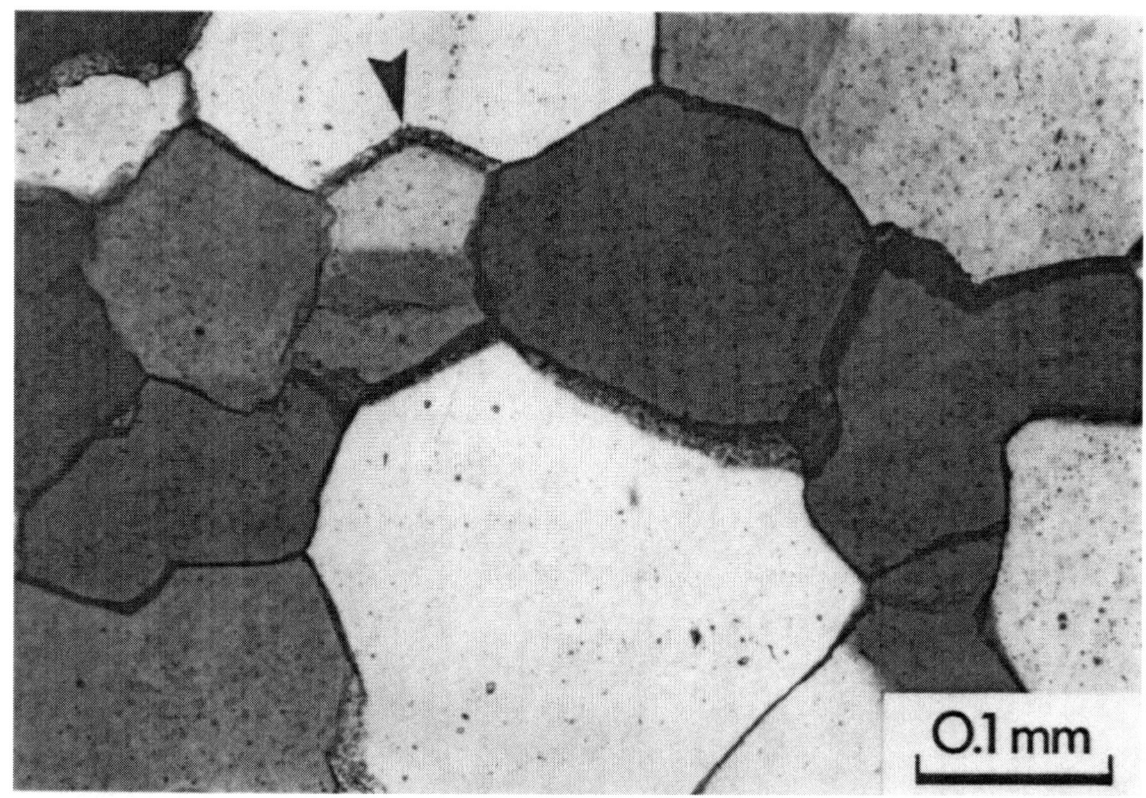


77 B

## Section 8: Disjunctive Cleavage Defined by Wiggly Cracks or Seams

(Plates 78 to 98)

The spaced surfaces defining a disjunctive cleavage here differ from those of the previous section in that they show appreciable width and they are not planar. In some cases they may be better described as seams. In the 21 contributions in this section "pressure-solution" is commonly referred to: the "seams" defining cleavage being interpreted as stylolitic residues in some cases. Truncated grains or fossils (e.g., Plates 87, 93–96) and overgrowths on grains (e.g., Plates 85, 87) and the presence of veins (e.g., Plate 98) may support the pressure-solution hypothesis. Understandably, the rock types involved are largely limestones and sandstones.

---

**Plates**

## 78. Incipient Cleavage in Fine Sandstone, Center White Creek, New York, U.S.A.

W. D. MEANS

The dark folia are mica-rich films. The coarse grains are quartz. The rock exhibits inferior cleavage in hand specimen.

The sample is from the Austin Glen member of the Normanskill formation of middle Ordovician age, at Stop 6 on a field trip led by D. B. POTTER and M. A. LANE (1969). The locality is near the extreme western edge of the Taconic allochthon.

The outcrop from which the sample came contains large recumbent folds, with the cleavage shown here approximately parallel to the axial planes. The position of this sample relative to limbs or hinges of folds is not known, however. The metamorphism is chlorite grade. PPL

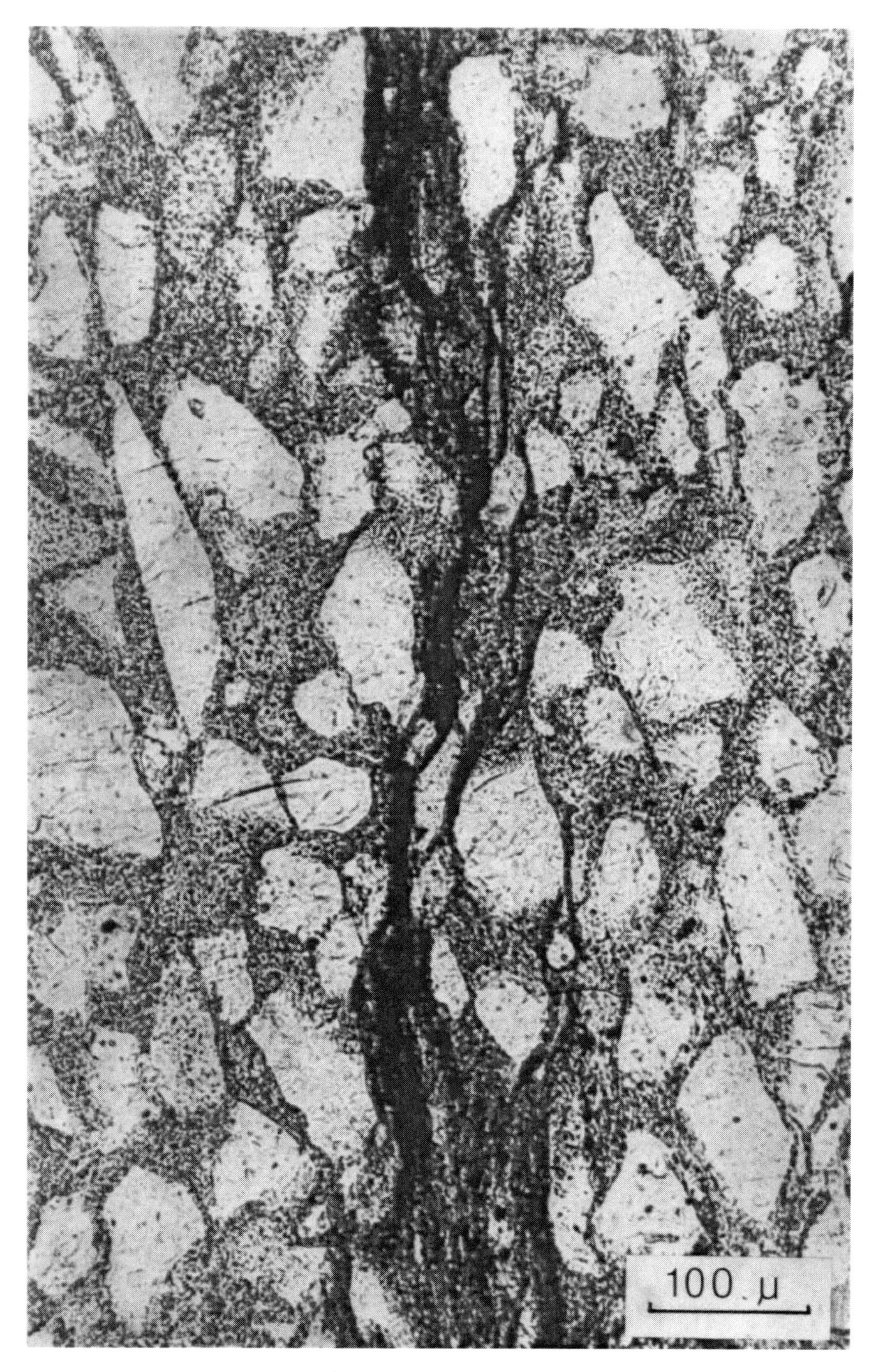


78

# 79. Stylolitic Solution Cleavage From Silurian-Devonian Limestones, Pennsylvania

T. ENGELDER and W. ALVAREZ

These specimens are from the Valley and Ridge Province of the Folded Appalachians. They come from the upper of the two packets of carbonates (Cambrian-Ordovician and Silurian-Devonian) that characterize the Paleozoic stratigraphy of central Pennsylvania, U.S.A.
Sample localities are in an area of major folds with gently dipping limbs. Shortening on the scale of the outcrop was accomplished by solution of material along stylolite surfaces. The general alignment of the stylolite surfaces is roughly parallel to the axial surfaces of the major folds, i.e., strike N 65°–70° E, dip vertical. (The stylolite surfaces are roughly perpendicular to bedding, and correspond to the "Horizontal-Stylolithen" of German authors, e.g., PLESSMANN 1972, where directional reference is made to the orientation of the pits and columns on the stylolite, not to the orientation of the surface).

---

**A. Stylolite formed during shortening parallel to bedding.** Upper Silurian Keyser-Tonoloway Limestone; quarry on Pa 335, 0.5 mile east of American Legion Post, Mifflintown, Pa. A significant but unknown amount of shortening has taken place by removal of limestone at the stylolite, producing a thick selvage of insoluble residues (*white* in photograph). Subsequent pull-apart of the suture is indicated by the presence of sparry calcite (*black*) flanking the clay selvage. Note the rigid-body rotation of block *p* with respect to block *q*. At *r*, two stylolite branches are present, and bedding in the small fragment remaining between them shows a rotation of about 60°.
Scale: 2 mm; Unpolarized light, negative print.

**B. Stylolite formed during shortening parallel to bedding.** Lower Devonian Helderberg Group; quarry just south of a sawmill on Pa 235, 2 miles south of McAlisterville, Pa. General features are similar to **A**. In addition, note that in most of the photograph, pull-apart of the suture is represented entirely by sparry calcite (*black* in photo) filling an extension vein adjacent to the clay selvage (*white*), whereas at *s* less sparry calcite is present next to the clay selvage, and extension occurred through the development of many fine calcite-filled veinlets (*black*).
Scale: 2 mm; Unpolarized light, negative print

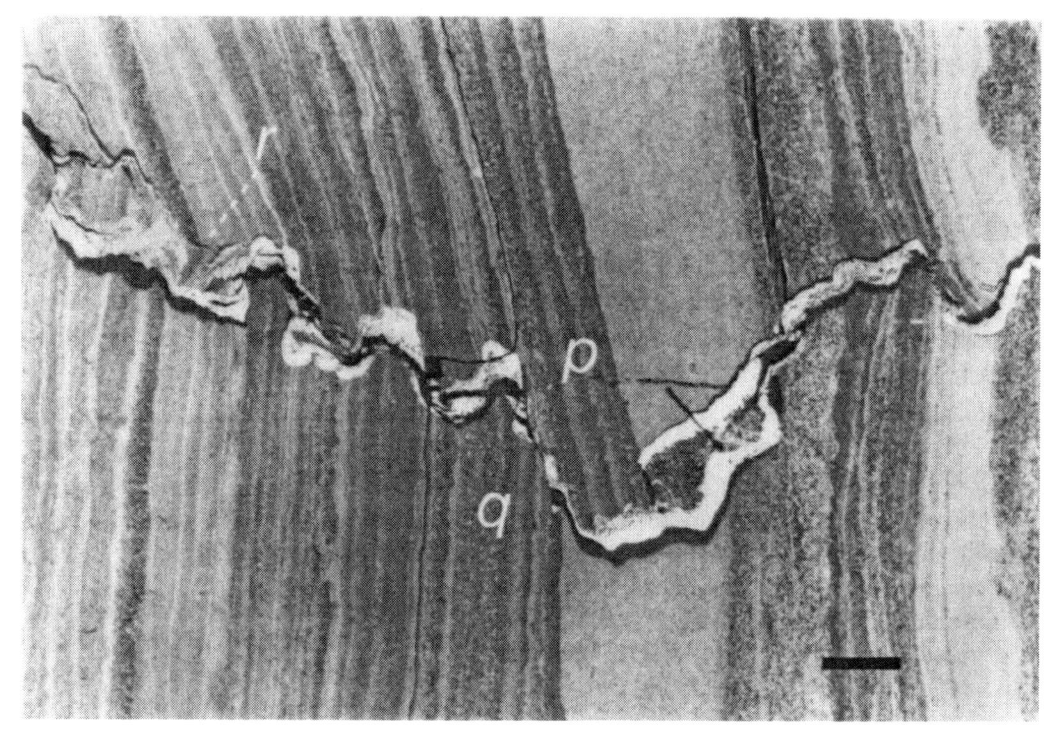


79 A

79 B

# 80. Pressure Dissolution Deformation and Pre-cleavage Calcite Veins

T. O. WRIGHT, G. GILL, and L. B. PLATT

Pre-cleavage calcite veins in the limy shales of the lower part of the Martinsburg Formation at Camp Hill, Pennsylvania, U.S.A., show differential pressure dissolution from the surrounding matrix. These are photographs in plain light of thin sections cut perpendicular to bedding. Cleavage is parallel to bedding. The pre-cleavage calcite veins have been able to resist the shortening of the surrounding limy shale, although the calcite in the veins is twinned.

---

**A. Unbroken pre-cleavage vein.** The large vein in the center of the photograph acted as a buttress against the cleavage shortening normal to the bed. The vein shows minor crumpling along the edges – probably due to shortening by twinning. The bed is 40% thicker at the vein than it is half way between the vein and the next one (20 cm away). Note the thinner post-cleavage vein to the right. PPL

**B. Broken pre-cleavage vein.** This sample was collected from the same bed 20 cm from the previous subject. Here the vein resisted shortening but broke near the center. The fracture became the locus of pressure dissolution of the vein itself and of the surrounding limy shale. Shortening is indicated by the concentration of insoluble opaque material on either side of the break. Similar concentrations of insolubles can be seen associated with the ends of smaller pre-cleavage veins. PPL

**C. Enlargement of broken pre-cleavage vein.** The pattern of insoluble residues shows maximum loss of rock adjacent to the vein. The light-colored area adjacent to the break on the left of the vein is a void; it was filled with black insoluble residues when cut, but some was lost during grinding of the section. The light lamination band above the break (0.5 mm) shows that the break decoupled the bed-wide buttressing by the vein into two segments. Note the post-cleavage veins in the photograph; they cut across the fabric and are not themselves distorted. PPL

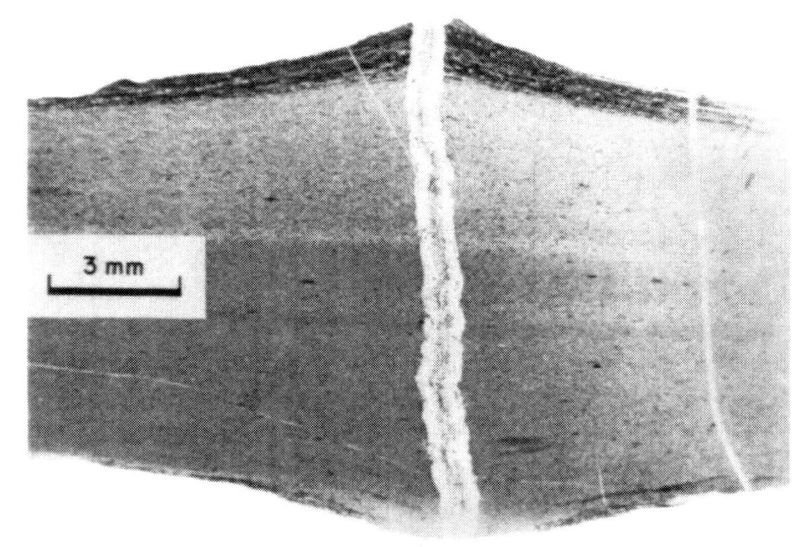


80 A

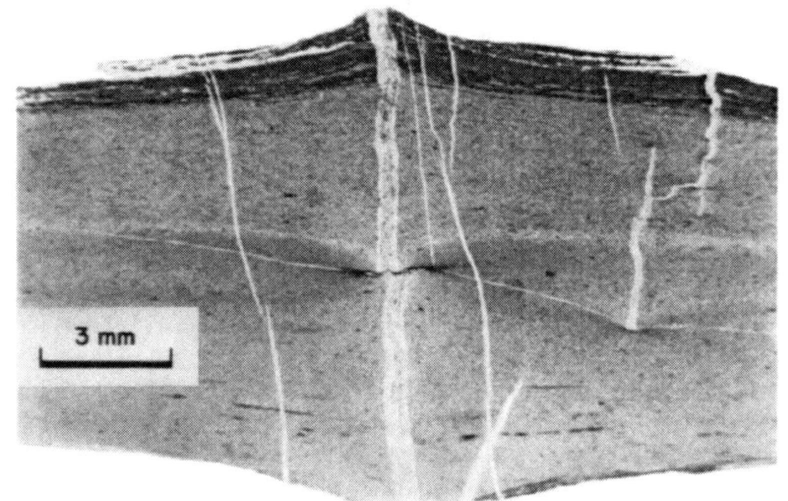


80 B

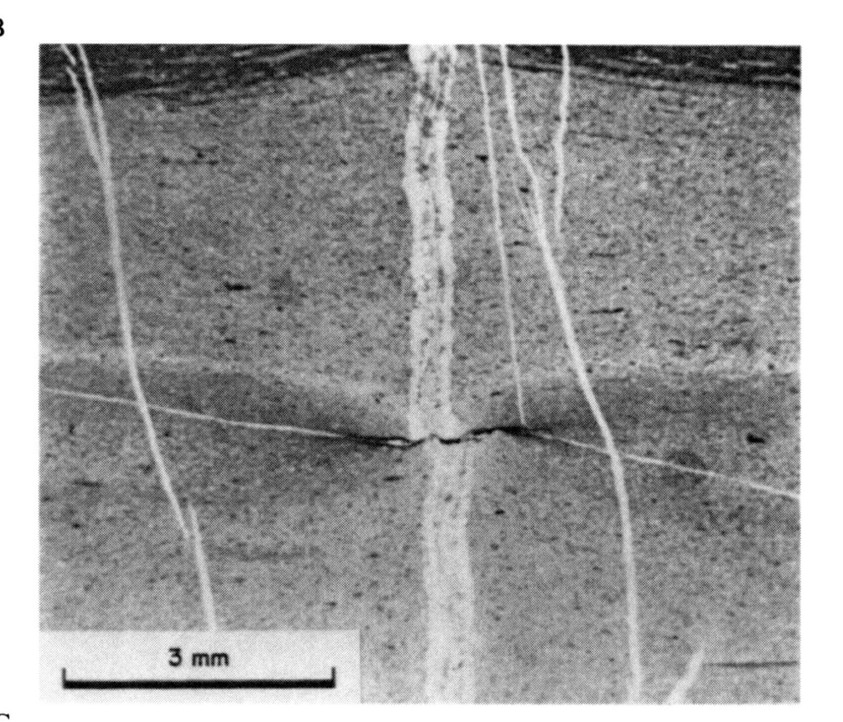


80 C

# 81. Spaced Cleavage of Solution Origin in the Scaglia Rossa, Italy

W. ALVAREZ and T. ENGELDER

The pink, pelagic, foram-coccolith limestone of the Scaglia Rossa represents about 50 my (Cenomanian to Middle Eocene) in the Umbrian sequence of northern Peninsular Italy. The Umbrian Apennines are characterized by open folds with wavelengths of 5–10 km and amplitudes of a few km. Spaced cleavage is common on the flanks of some of these folds.

The Scaglia Rossa is divided into a lower chert-bearing member, a middle chert-free member, and an upper chert-bearing member (ARTHUR and FISCHER 1977). Spaced cleavage is best developed in the upper member and the upper part of the middle member, while the lower parts of the formation often contain more widely-spaced stylolitic solution cleavage. However, these shortening features are not present in all outcrop areas.

The spaced cleavage in the Scaglia Rossa is interpreted as a solution cleavage formed during bedding-parallel shortening of the unit. Thin section evidence supporting this interpretation is provided by (1) the presence of clay selvages (insoluble residues) along the cleavage surfaces, and (2) truncation of foraminifera at the cleavage.

A more complete discussion of the cleavage in the Scaglia Rossa is given by ALVAREZ et al. (1976).

---

**A. Stylolitic solution cleavage in the lower, chert-bearing member of the Scaglia Rossa.** Upper Cretaceous, section at the Cimitero di Moria (43°30.5′N, 12°35′E) in the Umbrian Apennines. The stylolite surface, marked by a selvage of insoluble residues (*white* in photograph) runs in a generally horizontal direction, and is irregularly serrated because of the presence of pits and columns on the surface. Bedding is roughly vertical and shows up as a weak compositional banding (the limestone has been strongly bioturbated). This stylolite is similar in origin and orientation to the spaced solution cleavage higher in the formation.

Scale: 1 mm; Unpolarized light, negative print

**B. Solution cleavage in the middle, chert-free member of the Scaglia Rossa.** Paleocene (?), section between Maesta Confibio and Badia di Sidria (43°27′N, 12°45′E). Bedding is roughly vertical, spaced cleavage roughly horizontal. Wispy white cleavage is due to accumulation of insoluble residues; coarse black cleavage marked by voids where clay residues have been lost during preparation.

Scale: 3 mm; Light as above

**C. Spaced cleavage in the upper, chert-bearing member of the Scaglia Rossa.** Eocene, locality as in **B**. Bedding dips toward left, cleavage horizontal. Wispy, white accumulations of insoluble residues mark both the cleavage and the bedding orientations. (For a discussion of stylolitic pseudobedding in this unit see ARTHUR and FISCHER 1977.) Calcite filling (*black*) on major cleavage planes indicates subsequent extension. Scale: 1 mm; Light as above

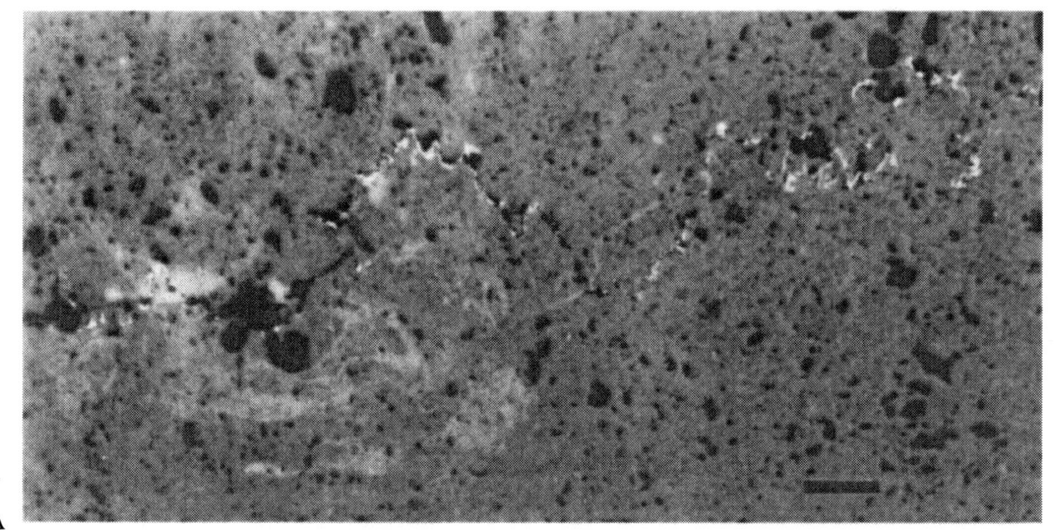

81 A

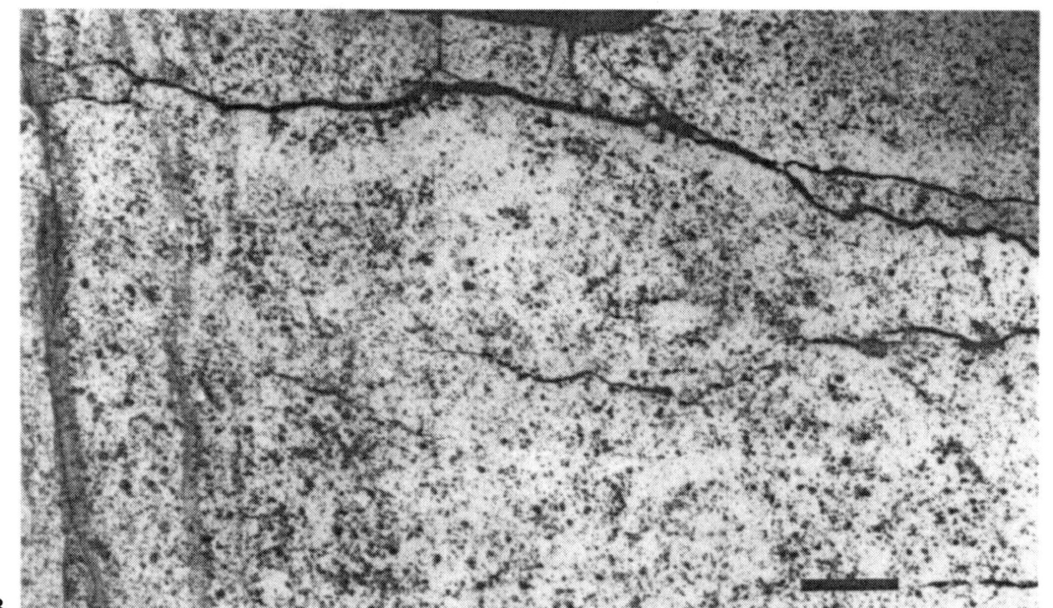

81 B

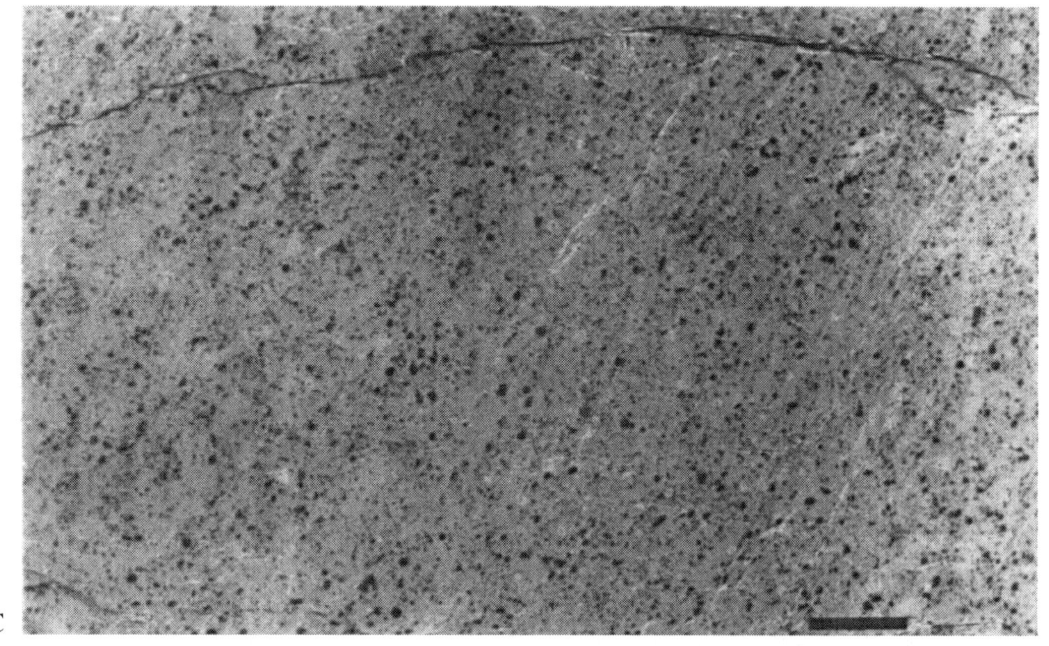

81 C

## 82. Deformation Fabrics in Chalks (I)

Y. Mimran

**A. Closely spaced, steeply northwards-dipping bedding surfaces and subvertical cleavage.** Upper Cretaceous, Upper Chalk at Corfe Castle, Dorset, U.K. (Mimran 1977). The outcrop is located at the steep northern limb of an east–west trending asymmetric anticline, and the view is westward. The close spacing is due to horizontal compression subperpendicular to bedding, and involved more than 80% volume reduction by compaction and large-scale losses of calcium carbonate. The volume losses were determined by the relative enrichments in immobile components (such as iron, magnesium, titanium, clay minerals in general, etc.).

**B. Solution seams surrounding flint nodules.** Upper Cretaceous, Upper Chalk, Lulworth Cove, Dorset (Mimran 1977). The dark clay seams are explained by the existence of local high stresses in the matrix surrounding the strain-resistant nodules. In order to accommodate the bulk volume losses induced by deformation, the matrix surrounding strain-resistant particles had to undergo dissolution.

**C. Deformed burrows.** Upper Cretaceous, Upper Chalk, Lulworth Cove, Dorset (Mimran 1976). The rock is a very hard light grey chalk dipping 80° to the north. The cross sections of the burrows on a horizontal surface shown in the photograph are elliptical. The long axes of the ellipses are subparallel to the strike (E–W) and the axial ratios range from 2 : 1 to 5 : 1. These elliptical sections are not true strain ellipses as they were derived from noncircular sections of circular cylinders (Ramsay 1967). However, the shape ratio 2 : 1 is undoubtedly less than the ellipticity of the strain ellipse on the examined section and therefore shortening of at least 30% must have occurred sub-perpendicular to the bedding

82 A

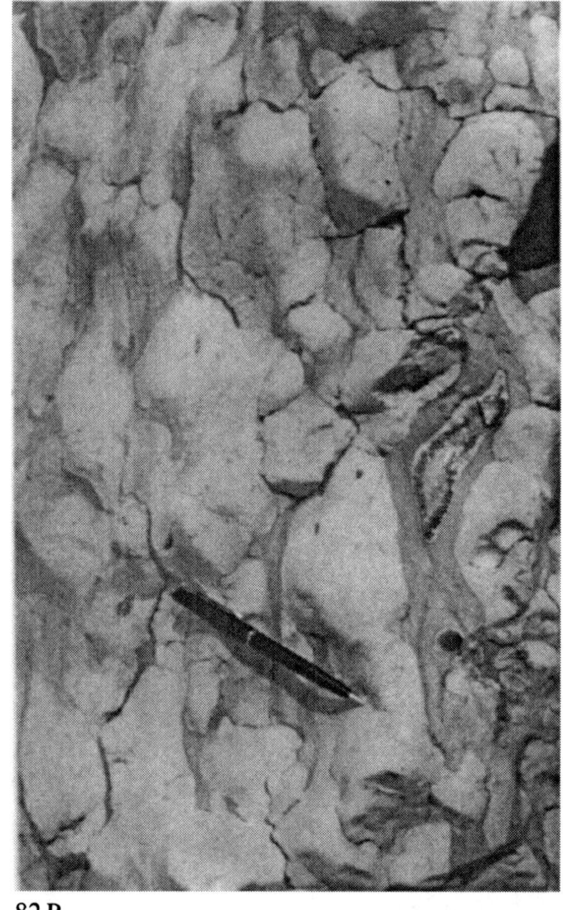

82 B

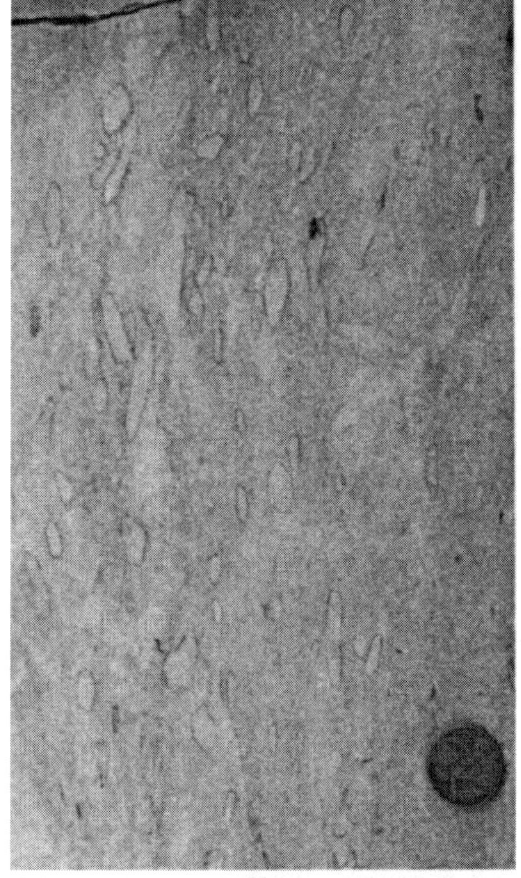

82 C

# 83. Deformation Fabrics in Chalks (II)

Y. MIMRAN

**A. Advanced deformation of chalk.** Upper Cretaceous, Upper Chalk, Isle of Wight, England (MIMRAN 1977). Abundant interpenetration of grains due to grain-to-grain pressure solution. The "free" grain surfaces underwent rounding due to their exposure to pore fluids during deformation (confining-pressure solution). This type of deformation followed a previous stage of purely mechanical compaction with almost no pressure solution involved. The density of the chalk at that previous stage increased from an average of 1.7 g $cm^{-3}$ to an average of 2.2 g $cm^{-3}$. The total volume loss by both pure compaction and extensive dissolution and migration of calcium carbonate exceeds 80%. The volume loss was computed by the relative enrichments in immobile components such as iron, magnesium, titanium, clay minerals in general, etc. Scale: 6 μ, S.E.M.

**B. Deformed calcispheres.** Upper Cretaceous, Upper Chalk, Dorset, England (MIMRAN 1976). The deformed fossils present preferred orientation of long axes. The axial ratios measured on calcispheres are about 1:3 although the bulk strain of the rock is much larger, as deformation involved volume reduction due to pressure solution and migration of calcium carbonate. The calcispheres are relatively strain-resistant particles and therefore the strain is heterogeneous.

Polished etched surface. Scale: 16 μ, S.E.M.

**C. Heterogeneous deformation in chalk.** Upper Cretaceous, Upper Chalk, Dorset (MIMRAN 1976). Highly deformed matrix developing "wavy" fabric surrounding strain-resistant calcispheres.

Scale: 32 μ, S.E.M.

---

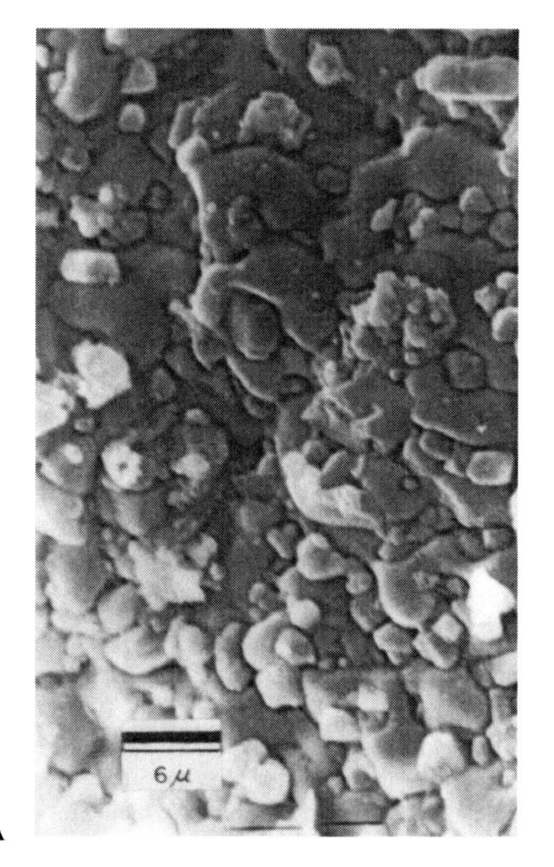


83 A

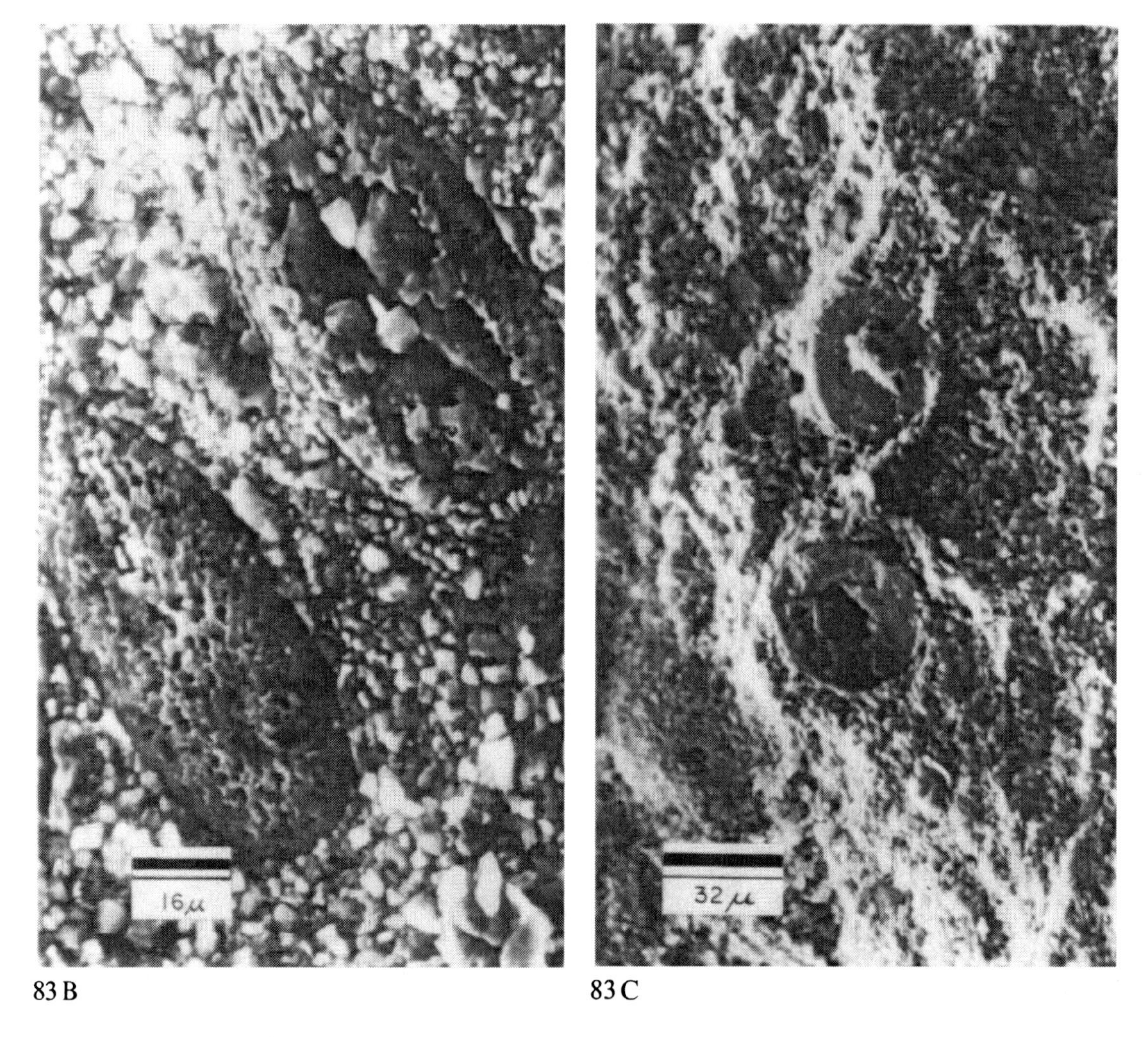


83 B

83 C

# 84. Progressive Development of Cleavage in Mature Arenite

C. A. BOULTER

From the Precambrian of Frankland, SW Tasmania, Australia. The fabrics described here were produced in an initial stage of recumbent folding under chlorite-grade greenschist metamorphism. The rocks are pure quartz arenites with less than 1% argillaceous component. A complete sequence of fabric development can be traced from undeformed examples with clearly defined detrital grains and quartz-cement overgrowths to totally recrystallized quartzites. The first expression of a tectonic fabric in the field is a weak dimensional preferred orientation of new grains formed by the recrystallization of the diagenetic overgrowths. The detrital grains display only slightly undulose extinction and minimal dimensional preferred orientation. Well defined planes of weak cohesion are apparent where about 30% shortening across the cleavage has occurred and flattening of detrital grains enhances the recrystallized overgrowths. The quartz arenites are grain-supported, and hence stresses would have been transmitted through the framework of detrital grains.

---

**A. Recrystallized diagenetic overgrowth.** There is slight strain – measured as zero by the Elliott (1970) method – of the detrital grains, which are not recrystallized. Because the overgrowth was deposited from aqueous solution, it would have had a higher OH-bond content than the detrital grains, and thus be hydrolytically weakened during deformation. This specimen shows patchy recrystallization of overgrowths during deformation, with perhaps 50% being unaffected. MARJORIBANKS (1976) and DONALDSON and OJAKANGAS (1977) figure similar features, but were not able to exclude other mechanisms of formation. In a deformed grain-supported arenite with less than 10% shortening in Z, quartz-mica beards, with perfectly aligned mica, give a preferred dimensional orientation, which represents an incipient cleavage. Mica preferred orientation is clearly not due to rotation of mica platelets. NX

**B. Quartz beards.** The $D_1$ fabric is a strong quartz-beard alignment with minor amounts of mica. The dark material in the beard is derived from iron oxides that both coat the detrital grains and form alternating layers with quartz as pre-tectonic pore-space infill. Quartz in the beard shows marked dimensional preferred orientation, whereas detrital-grain alignment is poorly developed. During $D_1$ the rock was shortened by only 19% in Z (+4% Y, +19% X), as determined by the Elliott method in areas showing little or no later overprint. The effects of a second deformation which involved mass transfer have been heterogeneously imposed obliquely on the first fabric. The restricted field of view in this photograph does not show the angular difference between $S_1$ and $S_2$. PPL

**C. Diagenetic overgrowth.** Minor amounts of mica are present, but not in sufficient quantities to control quartz-grain shape. $D_1$ strain could not be measured in this specimen because of reasonably strong $D_2$ fabric development, but $D_1$ strain is estimated to be similar to that seen in **B.** NX

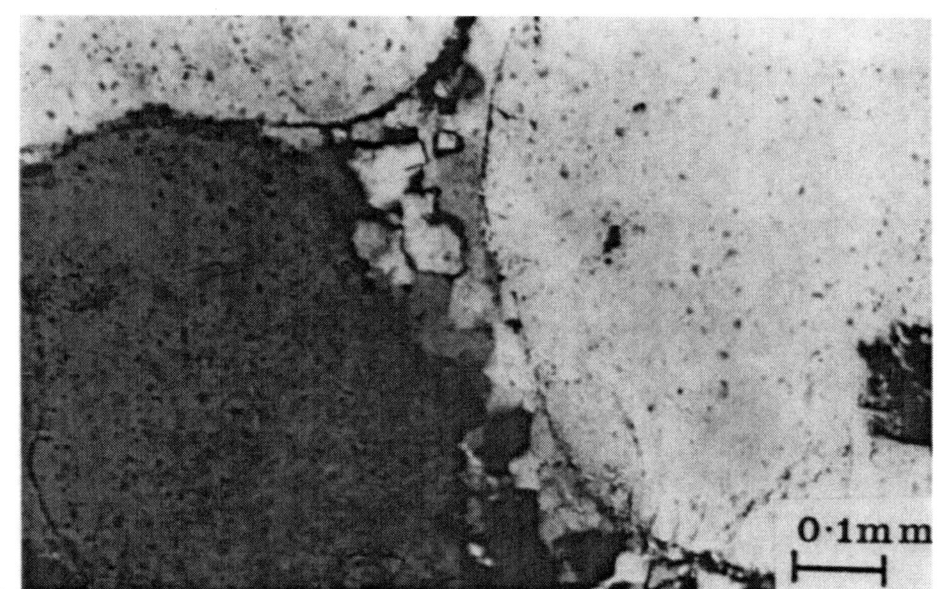


84 A

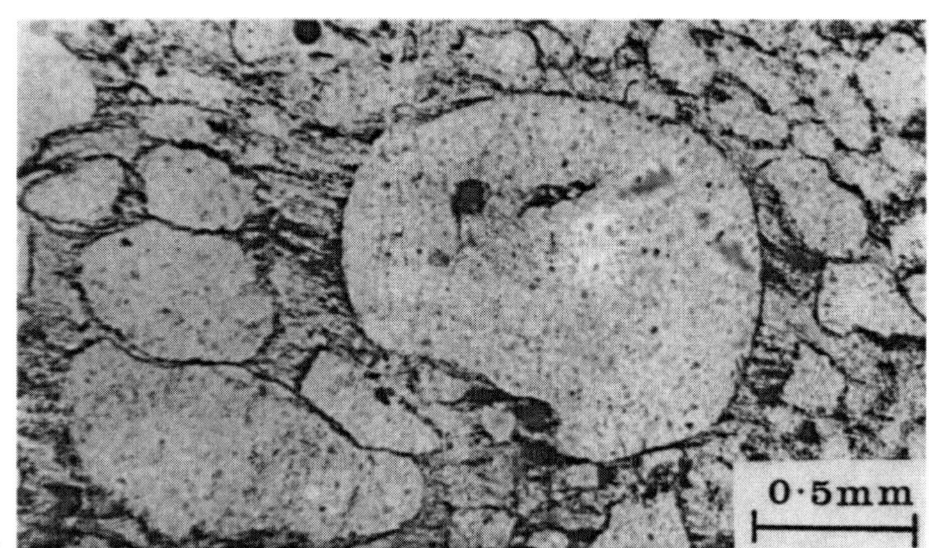


84 B

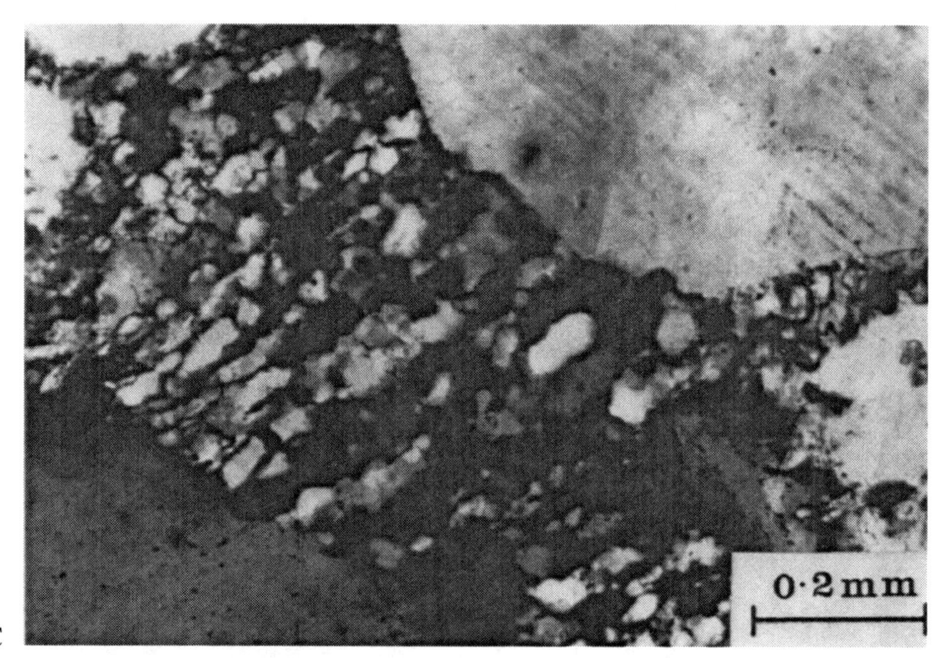


84 C

# 85. Quartz Overgrowths and Pressure Solution

R. B. DE BOER and P. J. C. NAGTEGAAL

**A. Quartz overgrowth next to pressure solution contacts.** *ps* pressure solution contact; *o* quartz overgrowth. The texture is the result of uniaxial compaction of a sample of pure quartz sand (median size 170 $\mu$), interlayered with thin clay layers (not shown in photograph).

Experimental conditions: temperature 340 °C; effective pressure increased by 6 to 7 atm/day from 20 atm to 710 atm and was then kept constant for 159 days. Initial porosity 42%; ultimate porosity 30.9%; pore pressure 160 atm; initial porewater composition 1 M NaCl solution (DE BOER et al. 1977, DE BOER 1975). NX

**B. Large quartz overgrowth.** *ps* pressure solution contact; *o* quartz overgrowth. For experimental conditions see **A** (DE BOER 1977). NX

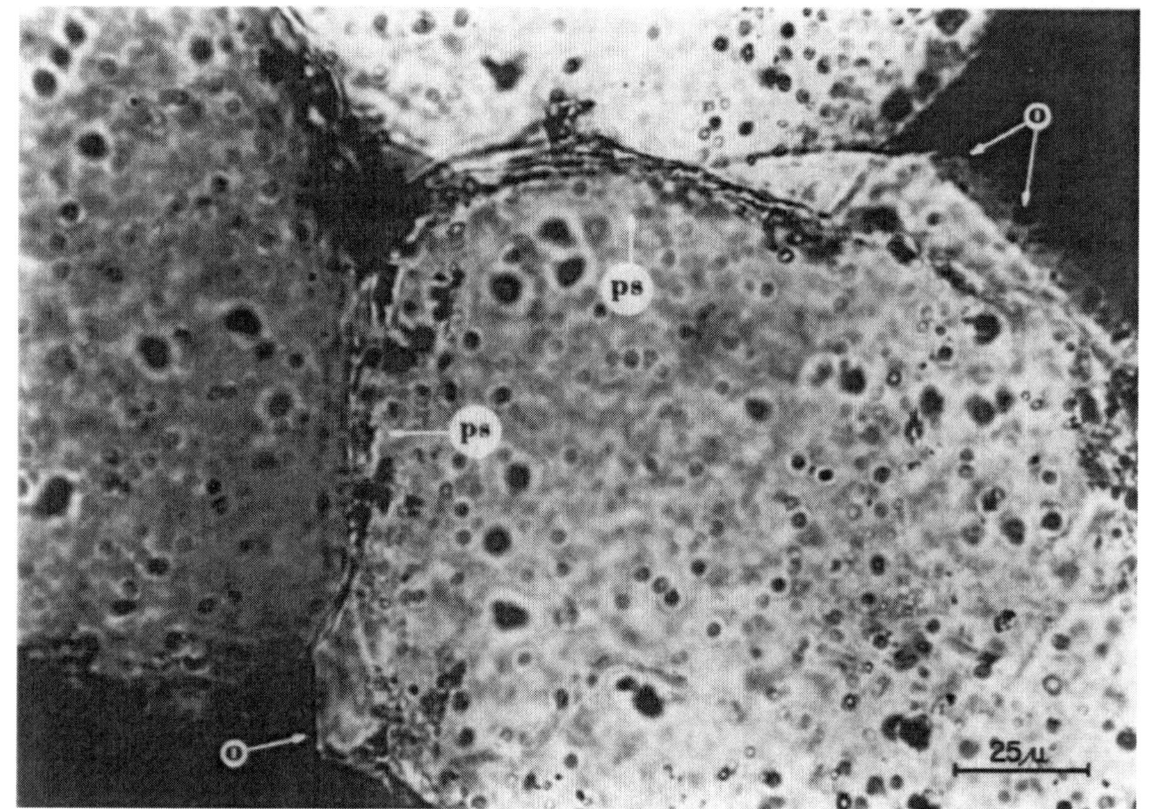

85 A

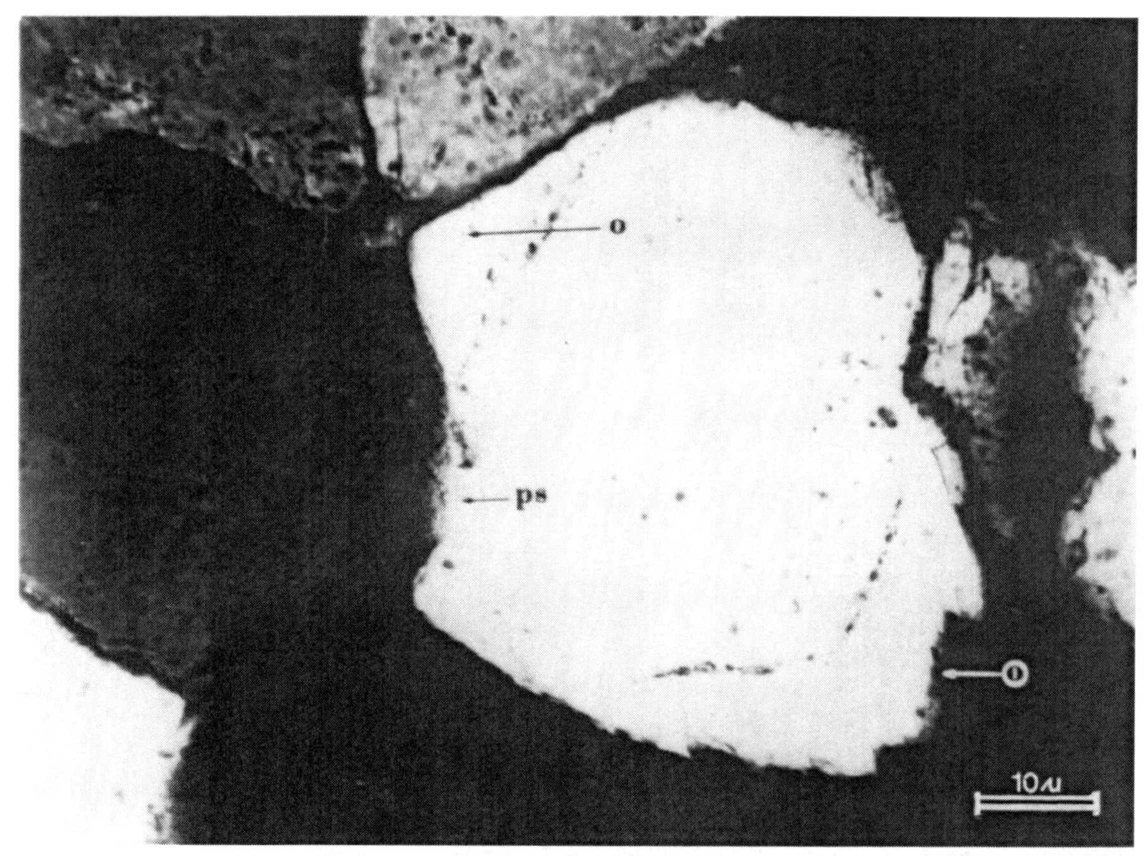

85 B

# 86. Pressure Solution in Sandstone

R. B. DE BOER and P. J. C. NAGTEGAAL

**A. Pressure solution in experimental sample.** *ps* pressure solution contacts; *o* quartz overgrowth. The texture is the result of uniaxial compaction of a sample of pure quartz sand (median size 170 $\mu$), interlayered with thin clay layers (not shown in photograph).

Experimental conditions: temperature 340 °C; effective pressure increased by 6 to 7 atm/day from 20 atm to 710 atm and was then kept constant for 159 days. Initial porosity 42%; ultimate porosity 30.9%; pore pressure 160 atm; initial porewater composition 1 M NaCl solution (DE BOER 1975, DE BOER et al. 1977). NX

**B. Pressure solution in deeply buried sandstone.** Venezuela, Lower C Sands, depth 4250 m. *ps* pressure solution contact; *o* quartz overgrowth. (DE BOER et al. 1977). NX

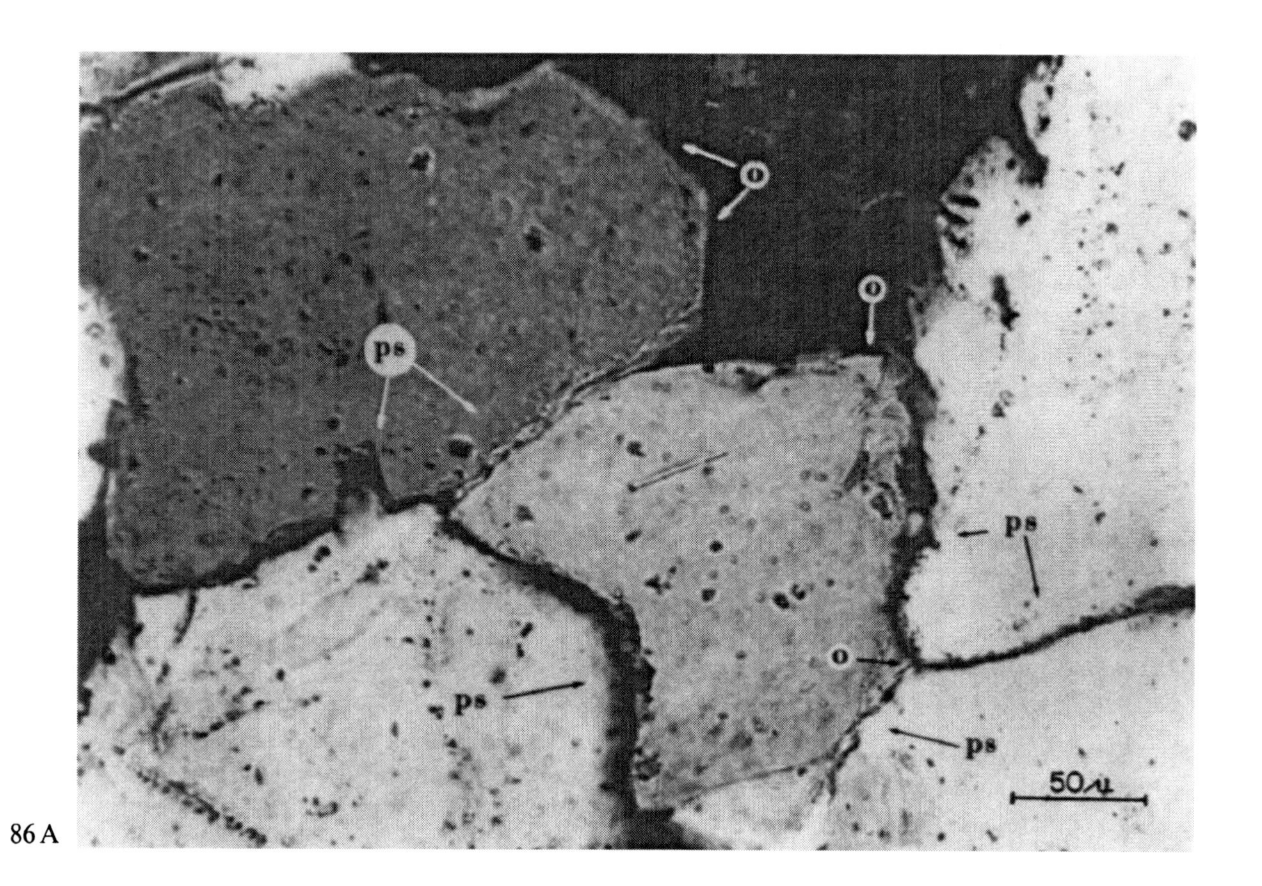


86 A

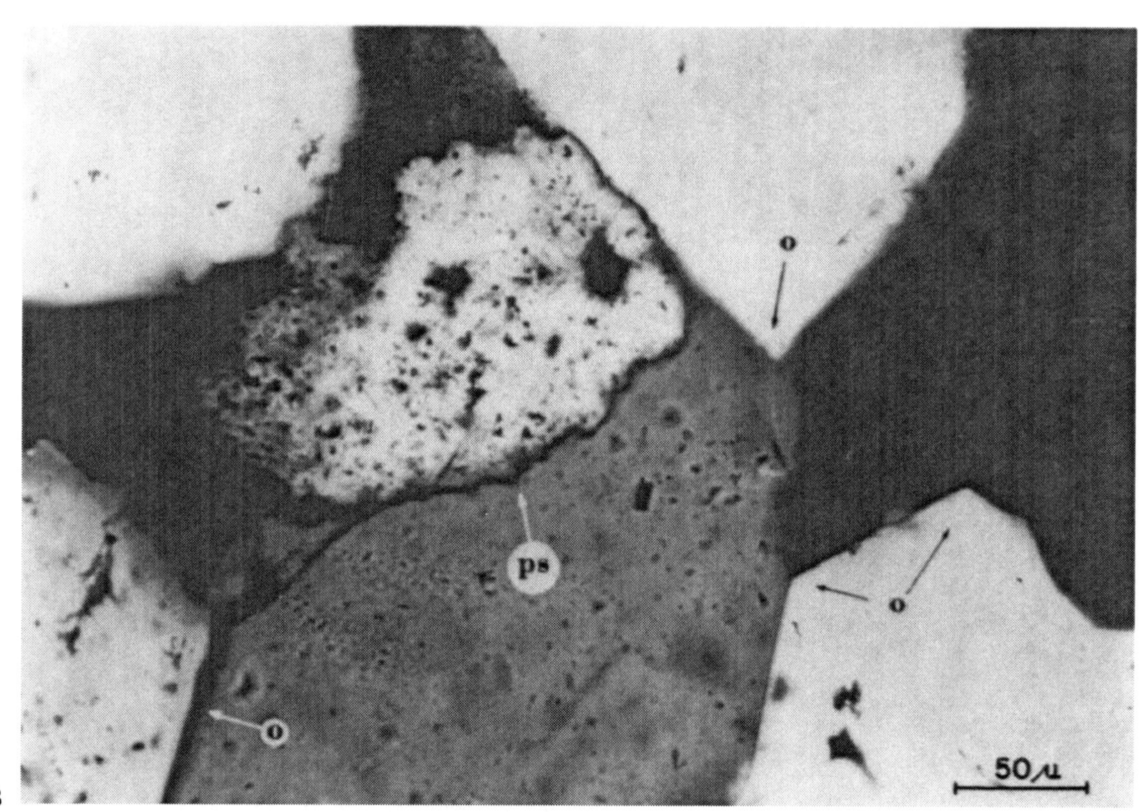


86 B

# 87. Pressure Dissolution of Coral and Associated Shale

T. O. Wright and E. K. Wright

Corals from the Mahantango Formation (Devonian) 6 km northwest of Stroudsburg, Pennsylvania, U.S.A., show obvious loss of outer walls and septae. Concentrations of opaque insoluble residues between corals indicate material loss from shale between corals.
Cleavage at this locality is not obvious in outcrop but is near vertical. The beds dip gently southwest. The thin sections are cut parallel to bedding and the trace of cleavage is indicated by *broken lines* on the photographs.

---

**A. Dissolution of corals.** Several examples of material loss from coral walls can be seen parallel to the general trace of cleavage. Calcite filling voids is not present outside these corals. The coral pair in the *lower right* are enlarged in **B**, the pair in the *upper left* in **C**. PPL

**B. Dissolution of coral pair without intervening shale.** Several millimetres of coral wall, septae, and early sparry calcite chamber filling have been removed from each corallite. Little insoluble residue is concentrated along the suture. Note that internal features of the corals and encrusting bryozoan are not distorted. PPL

**C. Dissolution of coral pair with intervening shale.** Several millimetres of coral wall, septae and chamber-filling calcite have been removed from this pair of corals. However, material has also been lost from the shale originally present between the corals as shown by the concentration of opaque insoluble residues and the bladed shapes of clasts in the matrix. A considerable (but unknown) amount of material has been removed from the shale by cleavage-related pressure solution. PPL

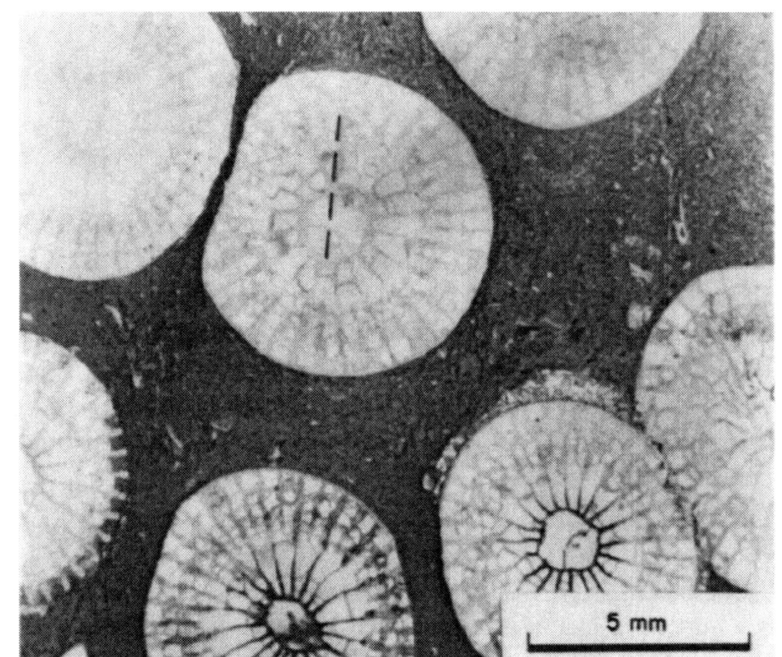


87 A

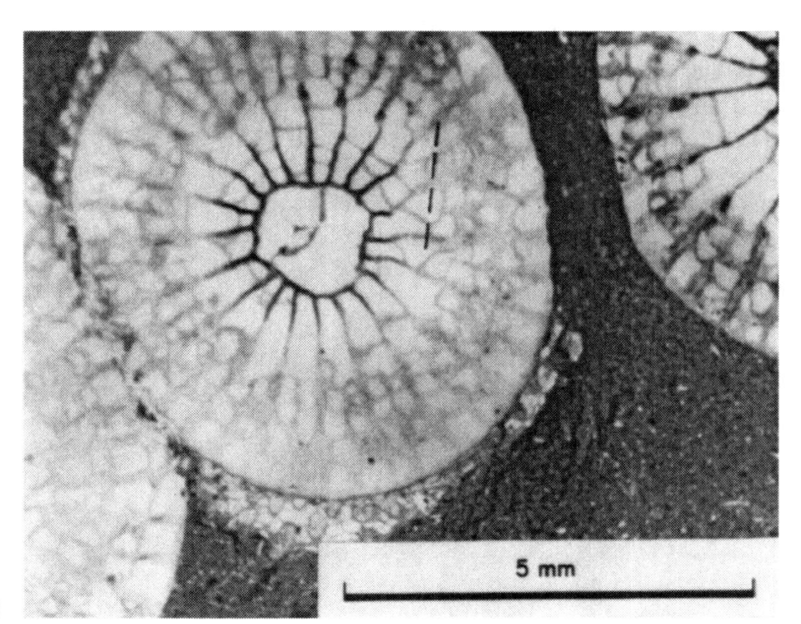


87 B

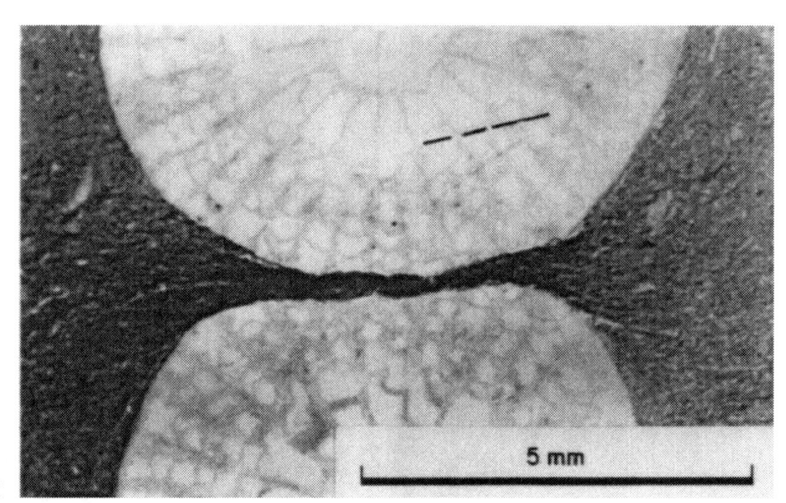


87 C

# 88. Dissolution of Carbonate to Form Cleavage

A. E. Oldershaw, J. H. Spang, and M. Z. Stout

All of the samples illustrated are from the same outcrop of the Mississippian Banff Formation at Pigeon Mountain in the Front Ranges of the Canadian Rocky Mountains. (See also Plates 95 and 102.) The cleavage planes are complex structures formed by clay-rich seams bounded by zones of dissolution pitting in the adjacent carbonate host rock. On a larger scale, the occurrence of truncated crinoids along cleavage planes also indicates that solution has taken place preferentially at these sites. Energy-dispersive and X-ray diffraction analysis indicates that illite is the most significant component of the clay seams. All photographs were taken on a Cambridge 600 S.E.M using gold-coated samples.

---

**A. Dissolution-pitted crinoid.** The photograph represents a portion of one crinoid grain that has been truncated along a narrow clay seam. The crinoid exhibits a relatively wide pitted zone (*between arrows*) that terminates along a sharp boundary (*white arrow*) approximately 50 $\mu$m away from the clay seam contact (*black arrow*). Similar pitted zones are observed in skeletal grains on both sides of clay seams in other examples. S.E.M. fracture surface.

**B. Lens-shaped pods of carbonate outlined by thin clay seams.** Well defined, complementary, lensoid pods (*L*) of carbonate host rock separated by thin seams of well oriented clay mineral platelets (*arrowed*). The clay minerals are aligned in the plane of the mesoscopic cleavage which would appear normal to the surface of the photograph.

S.E.M. fracture surface

**C. Clay seam with dissolution pits.** Clay seam approximately 12 $\mu$m wide (*arrowed*) illustrates the open fabric typical of carbonate dissolution. Lens-shaped pits (*P*) represent removal of host rock pods of the type shown in **B**. Subsequent collapse of pits may lead to a more clearly defined cleavage.

S.E.M. fracture surface

88 A

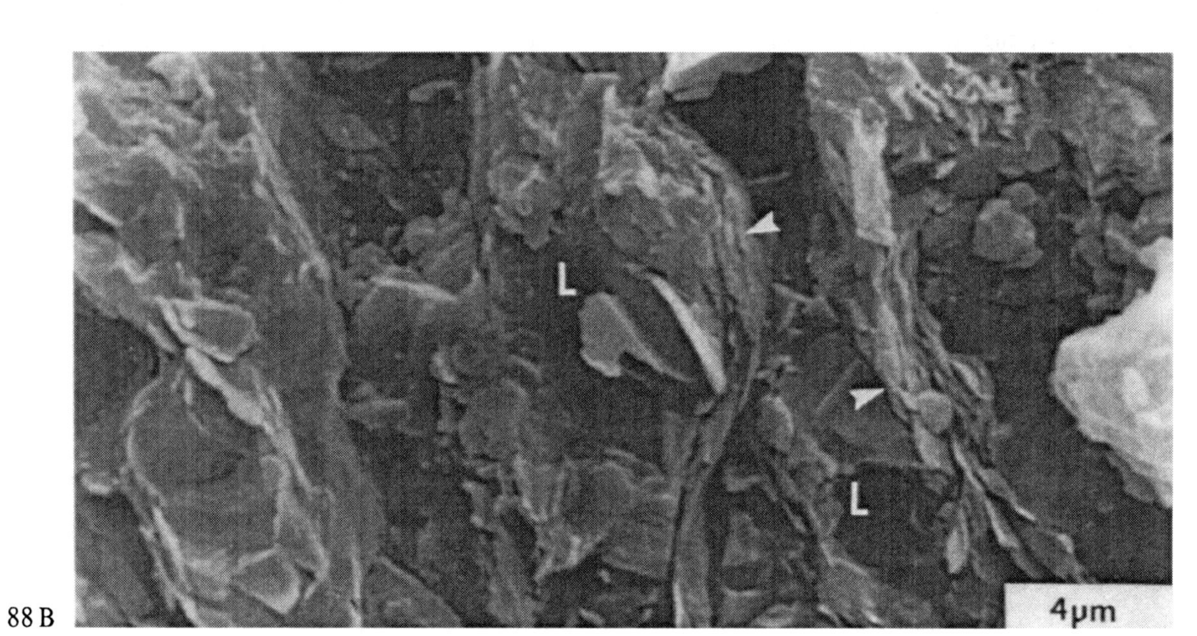

88 B

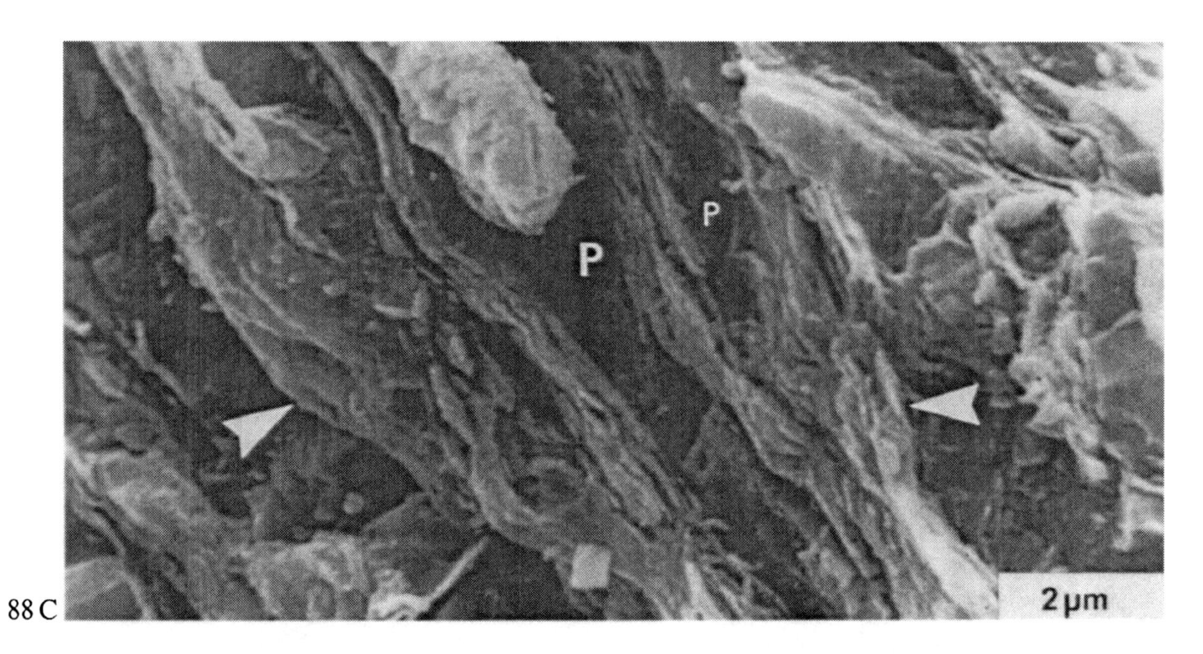

88 C

# 89. Stylolitic Cleavage and Strain

D. Cook

**A. Homogeneous and inhomogeneous strain**[1]. Middle Cambrian Chancellor Formation, from near Field, British Columbia in the western Main Ranges of the Canadian Rocky Mountains (Cook 1975, fig. 32, p. 57). Outcrop exhibiting contrasting structural styles in adjacent beds. Tectonic shortening has been accomplished by (1) essentially flexural folding of bed (*A*), (2) a combination of buckling and penetrative strain of beds at (*B*), and (3) homogeneous penetrative strain with no resultant folds at (*C*). The prominent cleavage is probably stylolitic and is parallel to a penetrative mineral alignment that can be seen in thin section (**B**). Deformation has been in conjunction with metamorphism to subgreenschist to lower greenschist grade. Deformation cannot be dated more accurately than Late Jurassic to Eocene (Price and Mountjoy 1970). Total strain has not been calculated for Middle Chancellor rocks but slates of the Upper Chancellor have been elongated parallel to cleavage by 25% to 70% and correspondingly shortened perpendicular to cleavage by 40% to 70% (Cook 1975). For scale, the bed at (*A*) is approximately 5 cm thick.

**B. Stylolitic cleavage**[2]. Detail of **A** in thin section (Cook 1975, fig. 34, p. 59). Photograph showing contrasting deformation styles in adjacent beds of limestone and argillaceous limestones. A limestone bed has been tectonically shortened by development of closely spaced stylolites. Equivalent shortening in adjacent more argillaceous beds has been by development of penetrative cleavage probably largely through recrystallization of clay minerals and pressure solution of quartz and carbonate. PPL

1 Geological Survey of Canada Photo No. GSC 123302

2 Geological Survey of Canada Photo No. GSC 199044

89 A

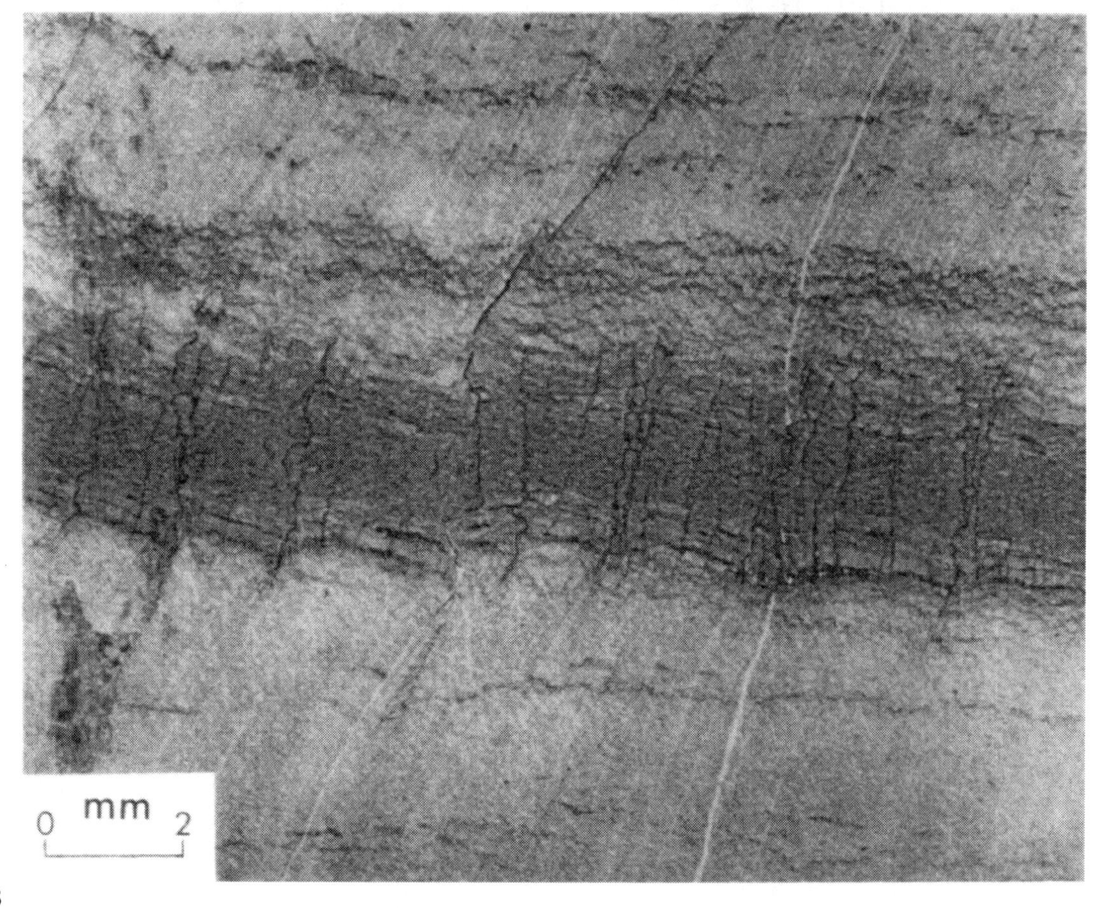


89 B

# 90. Cleavage in Siltstone from the Precambrian Rocky Cape Group, Tasmania, Australia

C. A. Boulter

The Rocky Cape Group includes two sequences of quartz arenite, each 1000 m thick. These control the regional fold characteristics. Axial surfaces are steep, hinge lines are gently inclined, fold wavelengths average 6 km, and interlimb angles are about 75° (Gee 1971). The metamorphic grade is subgreenschist and cleavages are only developed in the interlayered siltstone. **A** and **B** are siltstone cropping out at Rocky Cape. **C** is from the Cowrie siltstone, 2 km further west. A prominent feature of all the fabrics is alignment of detrital/diagenetic micas parallel or nearly parallel to bedding.

---

**A. Dissolution zones in quartz-rich layers.** These zones show essentially two orientations, one nearly parallel to the cleavage in the mudstone, the other markedly refracted. Both show offsets of the layer boundary and a higher concentration of "dark seams" within the mudstone. The seams at high angle to layering may have developed prior to the phase of pronounced fold amplification while interlimb angles were 20° to 40°. Any fabric produced in this phase maintains an axial plane relationship until fold amplification takes over, when it would be rotated away from the axial plane. Later flattening could bring about convergence of the early-formed fabric and the final axial plane fabric. PPL

**B. Detail of an anastomosing dissolution zone in siltstone.** Bedding is *horizontal* in this photograph whilst cleavage is nearly *vertical.* White micas are not involved in the dissolution process and hence often cause considerable deviation of the cleavage planes. The latter are concentrations of opaques and extremely fine micas. (In adjacent mudstone layers remnant pre-tectonic micas are in the minority and cleavage planes are defined by clearly resolvable micas). Locally the pre-tectonic fabric has been enhanced by soft-sediment slumping and the first tectonic cleavage may be recognized in some thin sections as a crenulation cleavage. PPL

**C. Large detrital micas subparallel to bedding.** Cleavage is defined solely by dark seams. The pre-tectonic micas vary from being equant to elongate parallel to bedding; few have long axes parallel to the cleavage. PPL

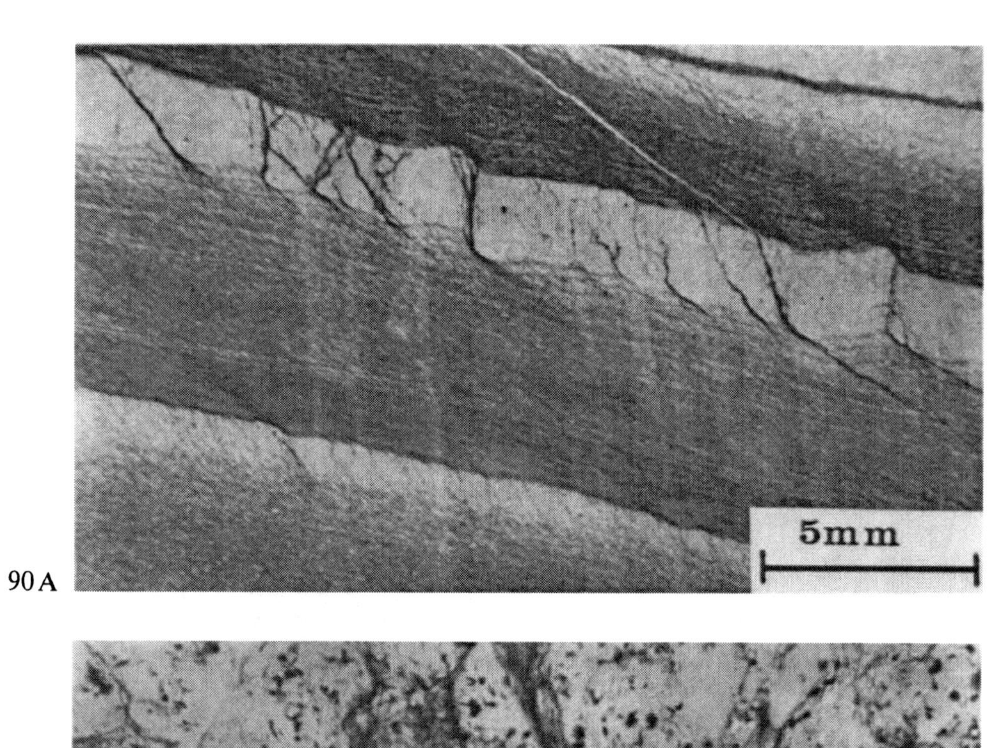


90 A

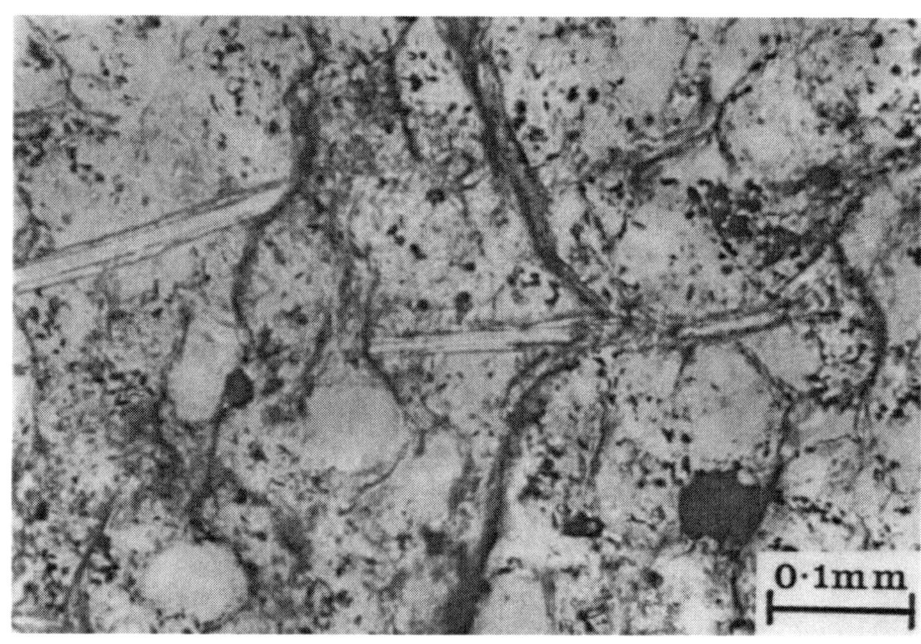


90 B

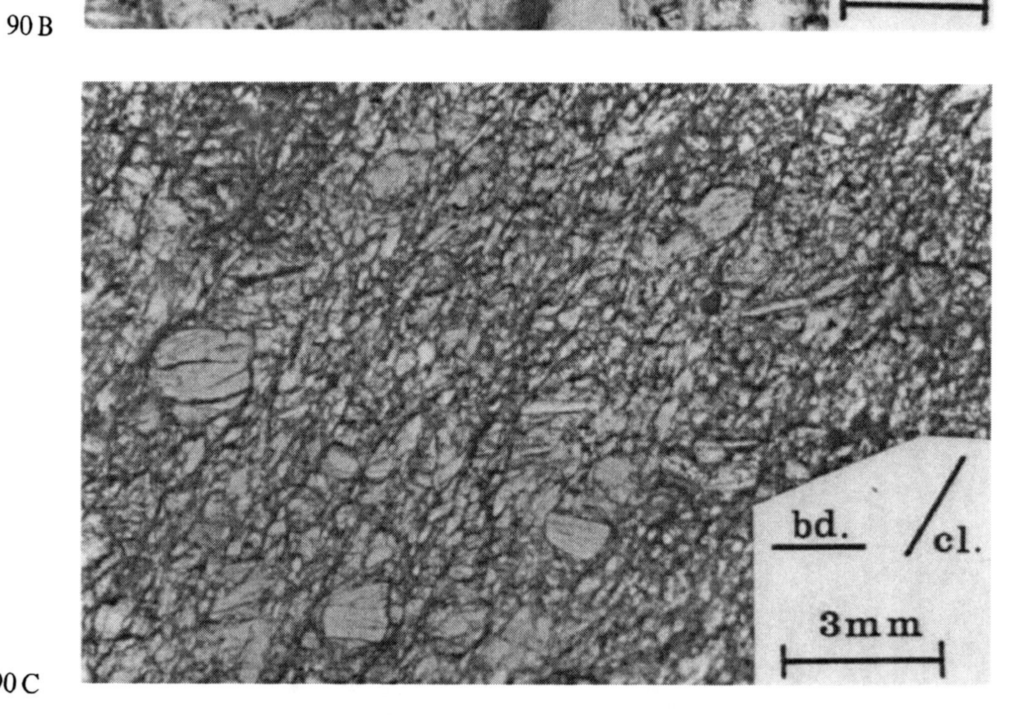


90 C

# 91. Characteristics of Solution Cleavage

S. A. Sansone

Solution cleavage, consisting of a seam of insoluble material, forms when calcite is dissolved along the cleavage surface. Fossils are commonly truncated along the cleavage surface providing direct evidence for calcite dissolution. Other evidence is provided where cleavage-seam thickness is a function of the concentration of insoluble material in the layers which are cut across by the cleavage. Cleavage seams are thicker in layers with higher concentrations of insoluble material.
Photographs are of etched limestone slabs cut normal to cleavage and bedding. The etching, done with 10% HCl for 30 min, provides information on relative concentration of insoluble material and calcite. Cleavage seams consist mainly of clay minerals with minor amounts of quartz, carbonaceous material and galena; thus they appear as black seams in positive relief.

---

**A. Thickening and thinning of solution cleavage.** Helderbergian Lower Thacher at Rosendale, New York, U.S.A. The light-colored bed in the center of the photograph has a negative relief, low concentration of insoluble material, and the segment of cleavage seam in this bed is thin. The layers above and below have a higher relief, higher concentration of insoluble material, and the cleavage seam is respectively thicker in these layers.

**B. Thickening and thinning of solution cleavage.** Helderbergian Lower Thacher at Rosendale, New York, U.S.A. The rock has two generations of cleavage. The first-generation cleavage is represented by a long, large seam that runs down the middle of the slab. The second-generation cleavage is represented by the smaller, more numerous seams. In the upper 3 cm of the rock, the layers have a concentration of insoluble material between 25% and 30%, whereas in the lower lighter layers the concentrations are 5% to 9%. The first-generation cleavage is thick in the upper layers, but thin in the lower layers; likewise for the second-generation cleavage. The average thickness for thc sccond cleavage in the upper layers is 0.09 mm, whereas in the lower layers the average thickness is 0.03 mm. Hence, cleavage seam thickness is a function of concentration of insoluble material in the layers across which the cleavage cuts

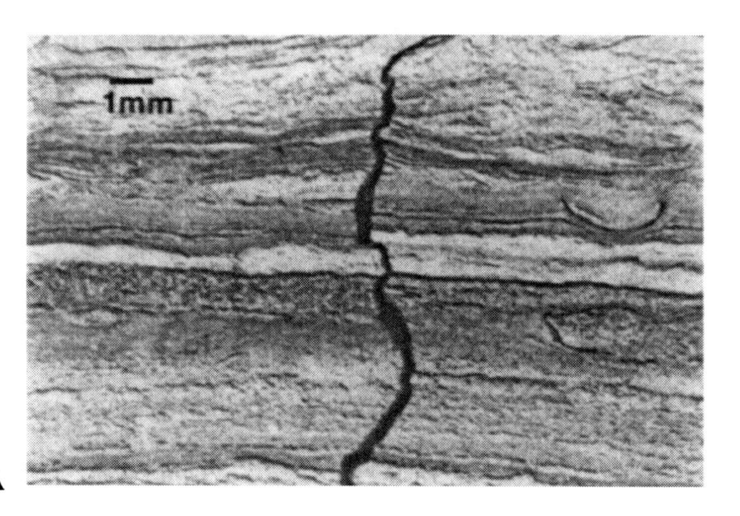


91A

91B

# 92. Spaced Cleavage of Pressure-Solution Origin

O. A. PFIFFNER

**A. Pressure solution seams.** Middle Eocene Bürgenschichten at Kistenpass (Infrahelvetic complex, eastern Switzerland).

The pressure solution seams are developed in calcareous glauconitic sandstones and limestones containing up to 50% by volume of large nummulites (*Assilina* and *Nummulites*). The fabric is of post-early Oligocene/pre-Miocene age (Calanda phase, MILNES and PFIFFNER 1977, PFIFFNER 1977). The Calanda phase deformation occurred during increasing burial and progressively more ductile behavior of the rocks. Metamorphic mineral assemblages indicate P-T conditions of around 3 kbar and 300 °C at the end of the Calanda phase (FREY 1978, PFIFFNER 1977). The sample was collected from the hinge zone of a Calanda phase syncline. The cleavage, defined by the pressure solution seams, is an axial plane cleavage to this fold. It can be traced up section into slates, where it becomes a slaty cleavage, and down section into pure limestones, where rock cleavage is controlled by the preferred shape orientation of the calcite crystallites.

The rock cleaves along a set of undulating parallel spaced dark pressure-solution seams (*arrows*). Beside pressure solution, which is responsible for the truncation of fossils, there were other deformation mechanisms active which allowed the rotation and in part the folding of these fossils (PFIFFNER 1977). *Reference line:* bedding trace. Weathered surface of hand specimen.

**B. Stylolites.** Upper Jurassic "Mergelband" near Walenstadt in the Lower Glarus nappe complex (eastern Switzerland).

The stylolites occur in pure limestone beds of around 20 cm thickness, which alternate with very slightly argillaceous limestone beds of about 5 cm thickness. The fabric is of post-early Oligocene/pre-Miocene age (Calanda phase, MILNES and PFIFFNER 1977, PFIFFNER 1977, 1981). Metamorphic mineral assemblages indicate P-T conditions of 2–2.5 kbar and 250 °C at the end of the Calanda phase (FREY 1978, PFIFFNER 1977). The sample was collected from the limb of a mesoscopic fold (wavelength approx. 2 m). The rock cleaves along a set of parallel spaced stylolites (*arrows*) oriented at variable high angles to bedding and parallel to the axial plane of the fold. Apart from pressure solution, internal deformation of the fold was accommodated by bedding-plane slip, as indicated by bedding-parallel slickensides. *Reference line:* bedding trace. Weathered surface of hand specimen.

92 A

92 B

# 93. Microfossil Indicators of Pressure Solution in Slate

D. W. DURNEY

**A. Crinoidal ossicles.**
*Location:* L. Lias marl, anticlinal core of Morcles Nappe, Leytron slate quarry, Valais, Switzerland (photograph after DURNEY 1972b)
*Metamorphism:* Stilpnomelane zone (FREY et al. 1974)
*Deformation:* $X=4.0$ and $Y=1.2$ in plane of cleavage ($S_1$) from 18 measurements on microboudinage and pyrite pressure shadows at this locality (DURNEY 1972a, Table 4).
*Fossil Deformation:* Exterior shapes of ossicles (circular cylinders, *grey* on photo) are modified by pressure-solution truncations (25%) parallel to cleavage, and overgrowths (~25%) of calcite in the form of synkinematic syntaxial pressure shadows (*white*). The combined pressure solution and overgrowth demonstrate partial deformation of the ossicles by a solution-transfer mechanism (BATHURST 1975, DURNEY 1976). Presence of *e*-twin lamellae further demonstrates internal deformation by a dislocation flow mechanism, although not sufficient to produce a visible distortion of the ossicles, even in the present favorable orientation (with c-axes subparallel to $S_1$) for twinning.
*Cleavage:* Seams of clay and carbonaceous matter (*black lines*) appear as thin and slightly wavy, discrete cleavage seams adjacent to truncations of ossicles and pass gradually into closely spaced ($< 50\ \mu m$) domainal slaty cleavage of the matrix. These two cleavage morphologies are thus genetically related to each other. Their association with the truncated parts of the microfossils suggests that they are insoluble residues of pressure solution origin. The matrix cleavage shows essentially similar relationships to small crinoid fragments in the matrix (as the discrete cleavages do to the larger ossicles in the photograph) although the matrix cleavage domains are somewhat more diffuse. PPL

**B. Globigerina tests.**
*Location:* U. Eocene Schistes à Globigerines, inverted limb of Morcles Nappe, Rionda, Valais, Switzerland (specimen 7a, DURNEY 1972a). For further supporting information see Plates 108, 109.
*Fossil Deformation:* Subspherical chambers of two microfossils show clear truncations (30%) by pressure solution against discrete cleavage seams (*black*), and overgrowths (25%) of fibrous, syntaxial calcite (pressure shadows: *white*). These are similar to those in **A** and indicate external deformation by solution transfer. No internal deformation or recrystallization of the chamber walls is visible.
*Cleavage:* The discrete cleavages associated with the truncations of the microfossil are here a smooth and relatively thicker variety than those in **A**. This may be due to the higher overall clay content and the lower bioclastic content of this rock. Also the scale of observation is smaller here than in **A**. The discrete cleavages pass into rather diffuse domainal slaty cleavage ($< 30\ \mu m$ spacing) consisting of discontinuous domains rich in highly oriented clay and carbonaceous matter separated by lighter colored zones of equigranular calcite and less well-oriented flaky minerals. PPL

93 A

1mm

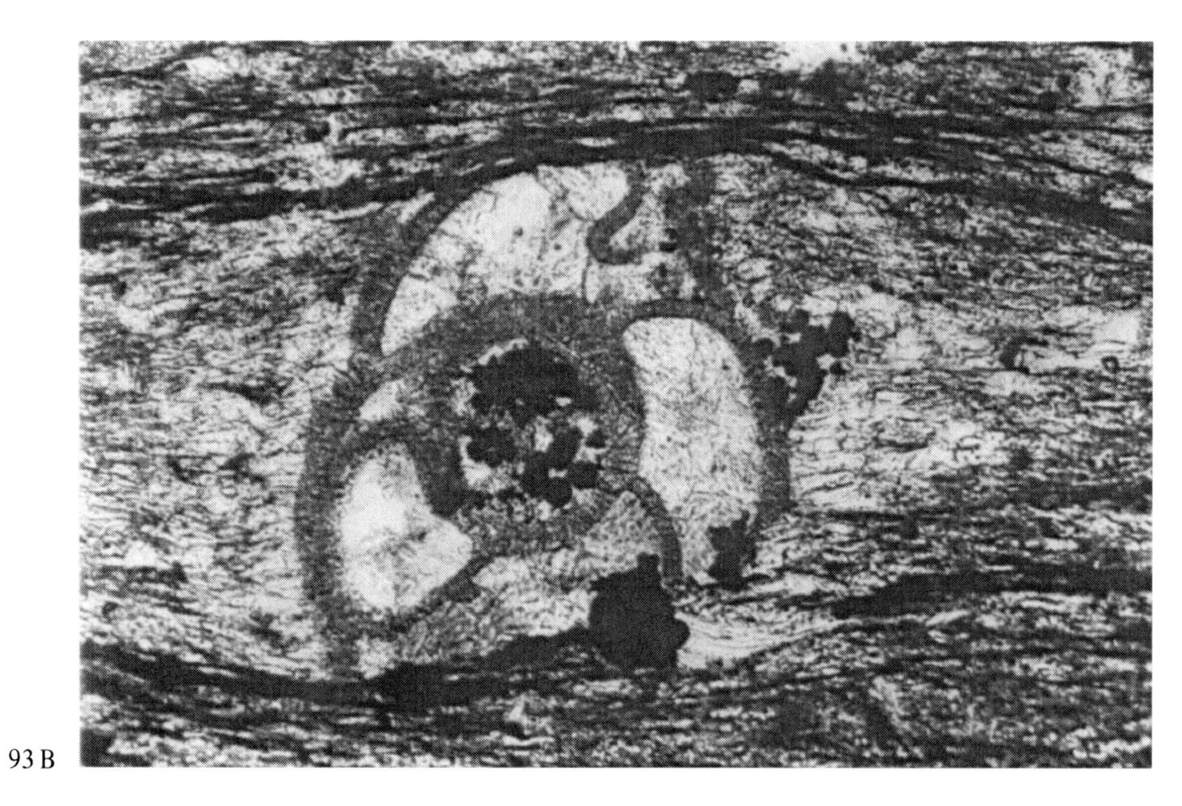
93 B

0·1mm

# 94. Finite Strain and Pressure-Solution Processes in a Deformed Limestone

D. B. SEYMOUR

The photographs show relationships between finite strain in limestones and microstructures indicative of pressure-solution processes operative during a single cleavage-forming event. All examples are from the Ordovician Gordon Limestone Subgroup, Mayberry area, northern Tasmania, Australia. For structural setting and comments on conditions of deformation see Plate 128.

---

**A. X-lineation defined by deformed ooids.** Oosparite, Grunter Hill, Mayberry area. Section is cut parallel to the planar fabric defined by shapes of deformed ooids, and is also parallel to the cleavage defined by the overall trend of tectonic stylolites (**B**). Assuming that the pretectonic three-dimensional fabric of the ooids was random, this section is the $XY$ plane of the finite-strain ellipsoid and the preferred orientation of ooid long axes on the section defines the $X$-direction, which pitches at close to 90° in the cleavage plane. The assumption of initial random fabric is supported by the close internal consistency of strain ratios measured on $XY$, $YZ$ and $XZ$ planes (using method of DUNNET 1969). Strain ratio on this section 1.80; 3-D finite strain gives +66% for $X$, –7% for $Y$, and –36% for $Z$, which is close to plane strain. Note that there are three different types of ooids present. One has a coarsely recrystallized core; measurements of such markers may underestimate the bulk strain of the rock. Another has a concentrically layered mantle surrounding a coarser-grained radially structured core which has a lower axial ratio than the overall grain. The third type is concentrically layered throughout and is probably the closest approach to a marker which reflects the strain. All three types of ooids have coatings which are finer-grained than the spar cement, raising the possibility of superplastic flow in the coatings (SCHMID et al. 1977) and consequent overestimation of the bulk strain by using them as markers. This possibility was not supported, however, by a check against the center-to-center method of RAMSAY (1967 p. 197) which should yield the true bulk strain. (See also Plate 206). PPL

**B. Tectonic stylolites in a deformed oolitic limestone.** Oosparite, Grunter Hill, Mayberry area. On the $YZ$ section of a deformed oosparite, the $Y$-direction defined by preferred orientation of ooid long axes is parallel to the overall trend of the trace of a tectonic stylolite, against which some of the ooids show truncation due to pressure dissolution. The wavy, anastomosing tectonic stylolites are spaced 0.5–2 cm apart and in 3-D display overall parallelism with each other and with the $XY$-plane of the finite-strain ellipsoid, as defined by the ooids. Strain ratio on this section, 1.45; 3-D finite strain gives +70% for $X$, –8% for $Y$, and –36% for $Z$ (close to plane strain). PPL

**C. Calcite-fiber vein.** Dolomicrite, Mayberry area. This vein follows a contact between fine-grained dolomite (*above*) and micrite. Crystallization took place upward from the lower side of the vein, and was synchronous with progressive folding of the dolomite-micrite interface, preservation of the orientations of *incremental* strains resulting in the concave-inward final geometry of the fibers. The upper ends of the fibers are subparallel to cleavage in the dolomite layer. In the hinge zone in the center of the photograph the non-rotational strain history resulted in the formation of straight fibers which probably parallel the $X$-direction of the *finite*-strain ellipsoid. This is an approximate $XZ$-section. PPL

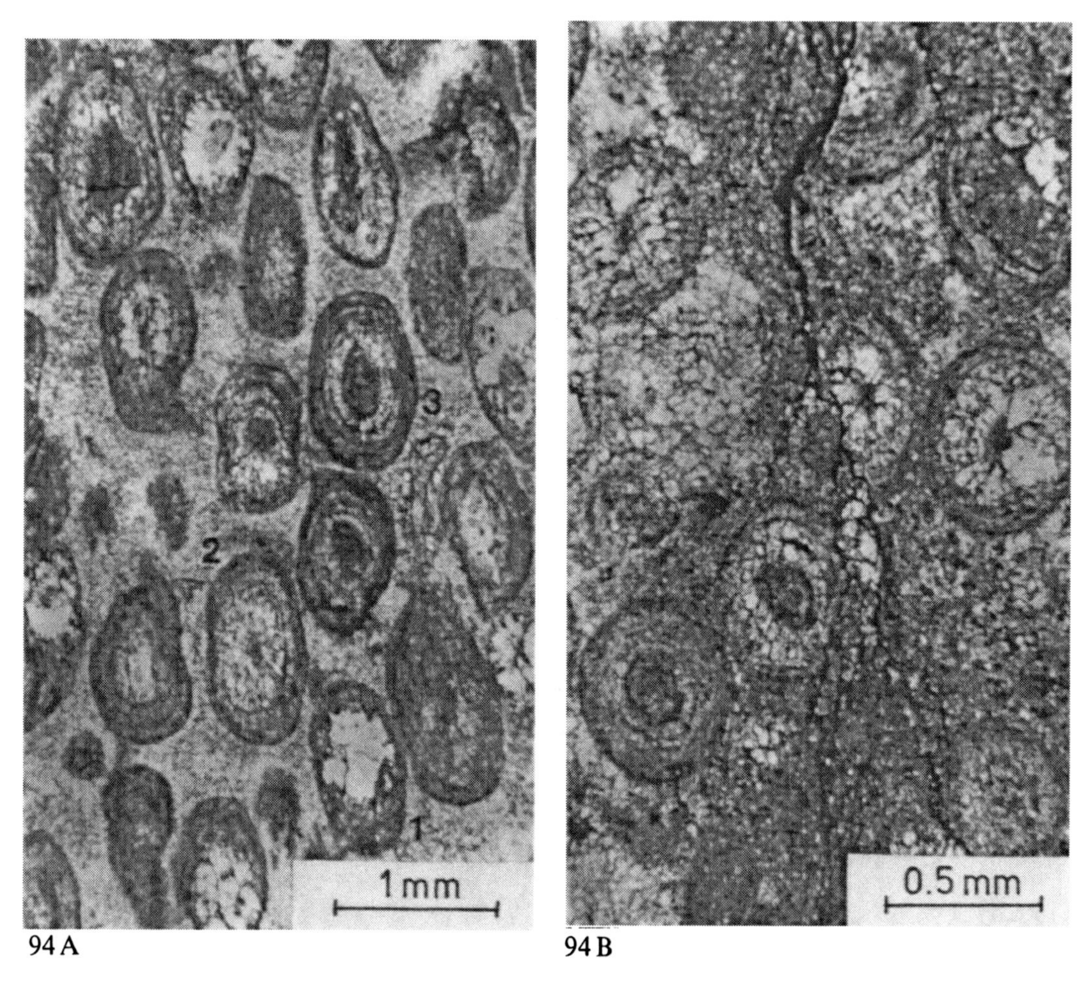


94 A 94 B

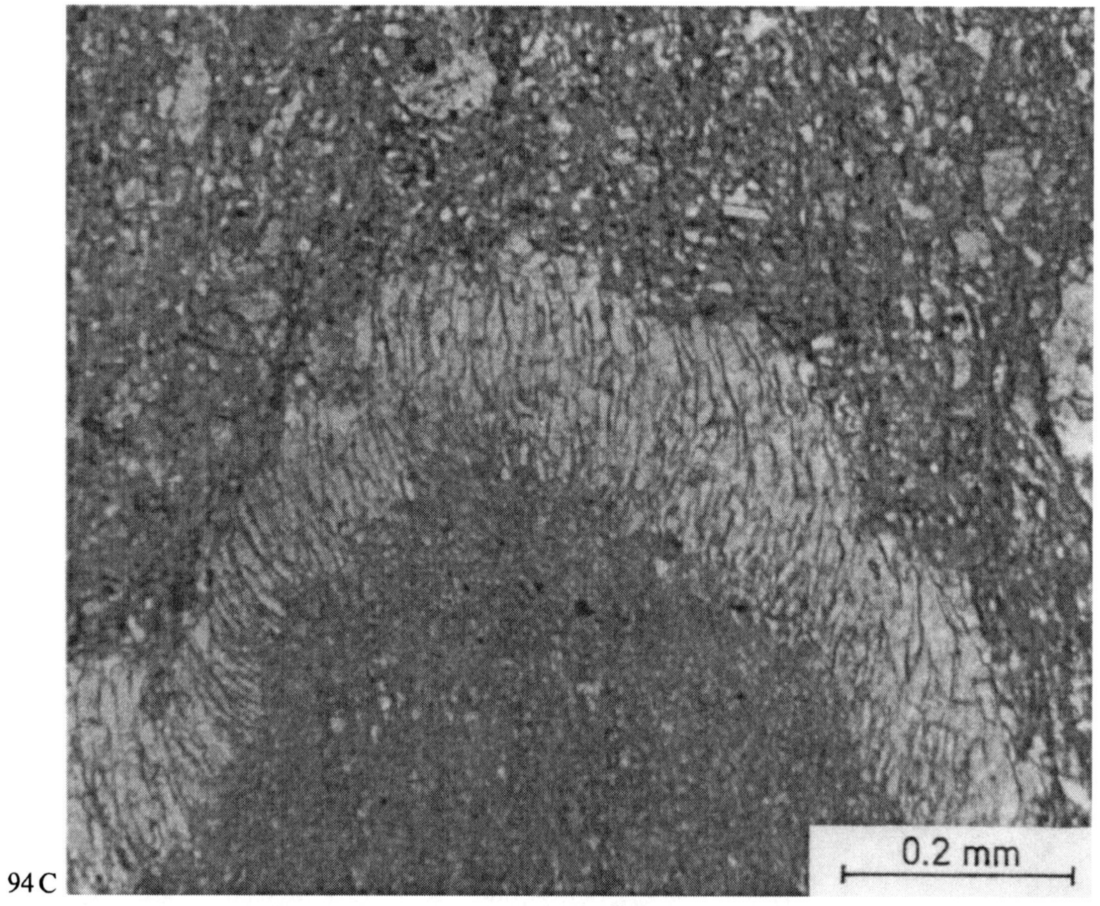


94 C

# 95. Cleavage in the Banff Formation, Canadian Rocky Mountains

J. H. Spang, A. E. Oldershaw, and M. Z. Stout

The cleavage is developed in the McConnel thrust plate at Pigeon Mountain in the Front Ranges and is systematically related to the transport direction of the thrust. (Also see Plates 88 and 102.) The nature of the cleavage varies from non-penetrative cleavage in argillaceous limestone to penetrative cleavage in the same orientation in shale. Numerical dynamic analysis of calcite twin-lamellae indicates that the cleavage lies in a plane of high resolved shear. $NDA_1$ (the maximum principal compression) makes an angle of 13° with bedding and 40° with the cleavage plane, and the twins seem to have formed in a single irrotational stress field. If the cleavage formed in the plane of flattening, the differential stress must have been very small and possibly less than the critical resolved shear stress necessary to cause *e*-twinning in calcite (Spang et al. 1976). The cleavage occurs within distinct zones and is defined by parallel or anastomosing clay seams oriented oblique to bedding. The spacing of the zones and the density of clay seams varies with the clay content of the rock (see *figure*).

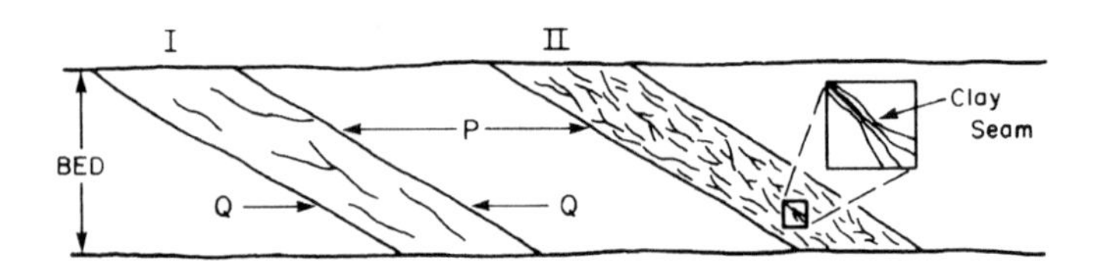


I Low clay zone : widely spaced seams

II Clay-rich zone : closely spaced seams

P Zone separation : centimeters to fractions of a millimeter

Q Zone width : dependent on clay content

---

**A. Cleavage-bedding relationships.** Outcrop photograph of cleavage in argillaceous limestone dipping southwesterly at 28°. The angle between cleavage and bedding is 27° based on repetitive measurements of cleavage (50 measurements) and bedding (25 measurements) orientation within one outcrop.

**B. Truncated crinoid.** The large crinoid has been truncated (*white arrows*) along two sides by dissolution of carbonate along clay seams. Away from the clay seams whole crinoids are found. This crinoid contains no evidence of distortion. Away from the truncated portions of this grain, there are pressure shadows (*black arrow*) of fibrous calcite (length of fibers = 0.5 mm or less) built out from the original grain. NX

**C. Anastomosing nature of clay seam.** Photograph shows the nature of one of the clay seams (*labelled*), which form the cleavage at this locality. In the clay seams, the clay platelets are oriented with their long dimensions parallel to the clay seams, whereas in the host rock they are randomly oriented. NX

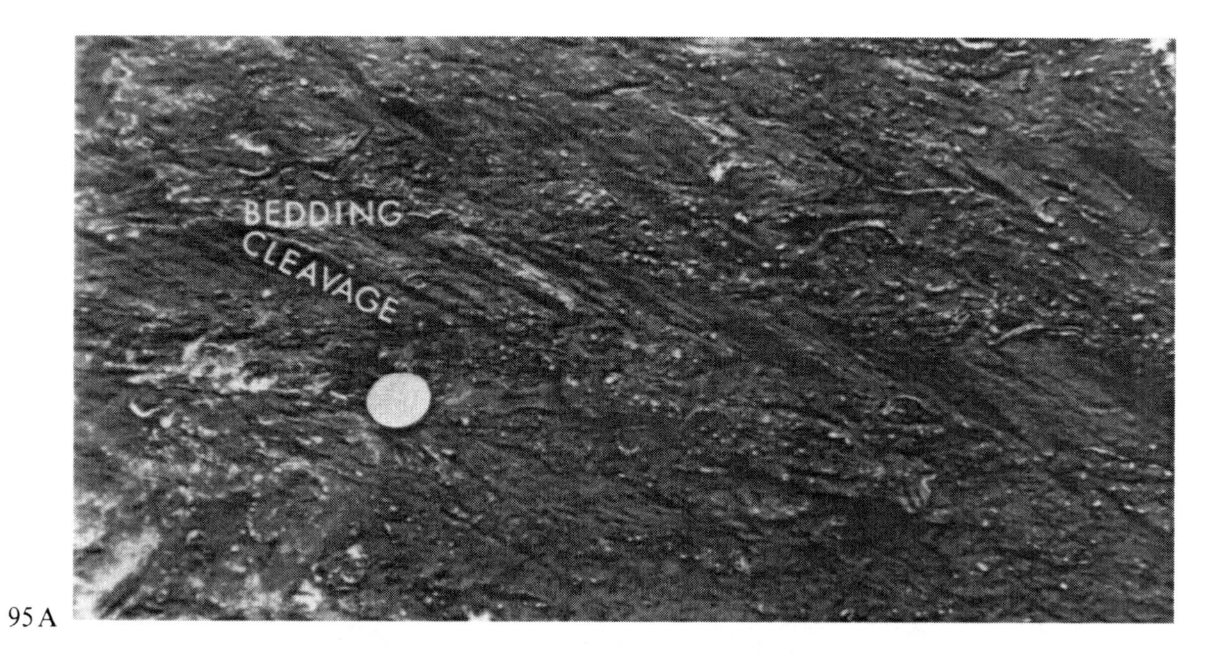


95 A

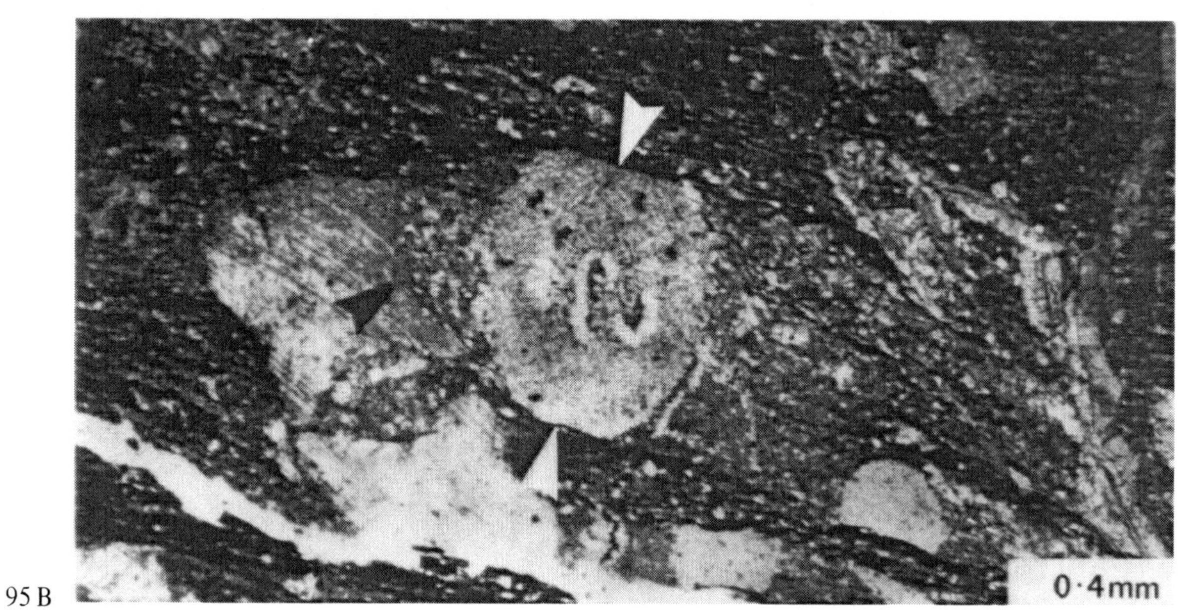


95 B

95 C

# 96. Fossil Indicators of Solution on Stylolitic Cleavage

G. J. BORRADAILE

Devonian slates, near Bad Berleburg, Rheinische Schiefergebirge, West Germany.
The cleavage in the following rocks is defined by spaced, stylolitic cleavage surfaces. The cleavage may have formed relatively late in the deformation history when the rocks had already accumulated significant strains (BORRADAILE 1977). Strains determined using geometry of large sand dikes were:

| $X$ | $Y$ | $Z$ |
|---|---|---|
| $\sim 1.48 \pm 0.07$ | $\sim 1.48 \pm 0.07$ | $\sim 0.44 \pm 0.01$ |

with $XY$ oblique to the stylolitic cleavage.

**A. Overview.** Spaced cleavage surfaces (*vertical*) cut through bedding (*left to right*). Carbonate organic detritus is aligned subparallel to bedding and truncated by sutured cleavage surfaces along which irresolvable dark residue is concentrated. The "stylolitic" form of the sutures prohibits any considerable slip and the truncation of the fossil fragments indicates that solution is responsible for the formation of the cleavage. PPL

**B. Detail (of a different specimen).** Stylolitic cleavage surfaces (*left to right*) show low amplitude/wavelength ratios of the order of 1 : 10 or less. They sharply truncate the ostracods. The surfaces anastomose, merging together where they pass around the ostracods. PPL

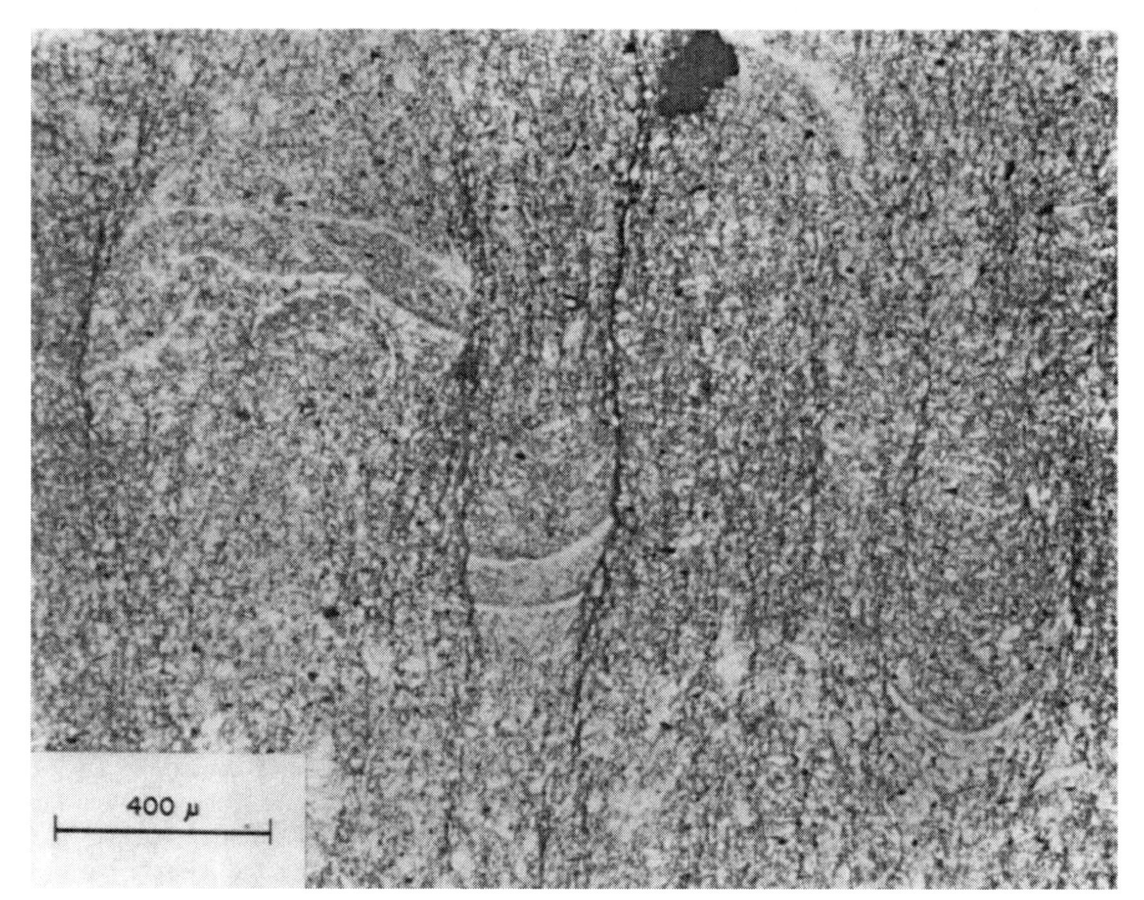


96 A

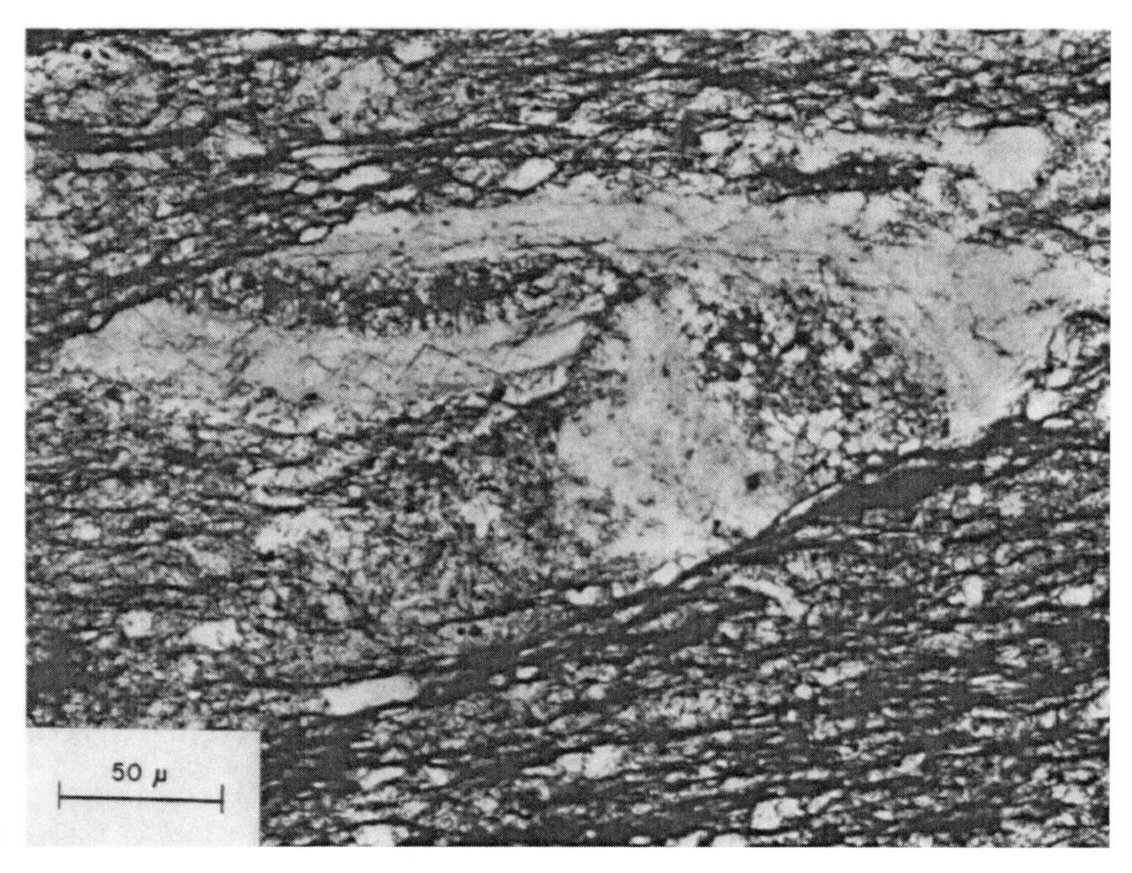


96 B

# 97. Spaced Cleavage: Pressure-Solution Stripes

A. BEACH

Spaced cleavage stripes are often well developed in sandstones weakly deformed at low metamorphic grade. They vary in thickness and spacing on a scale of a few millimetres and centimetres respectively. The term "fracture cleavage" should not be used because of the genetic implications – fractures often develop along stripes during uplift and weathering. A number of authors have attributed the stripes to a pressure solution origin (WILLIAMS 1972, HARRIS et al. 1976, BEACH and KING 1978, BEACH 1979). The stripe is defined by a zone of clay and mica minerals plus accessories, these minerals often being absent from the lithons of rock separating the stripes. The stripes do not merely record accumulations of inert minerals left behind after solution of quartz, feldspar, etc., but of new minerals produced by metamorphic processes localized in the sites of developing stripes (BEACH 1979). The stripes are usually best developed in immature stones such as sandstones and greywackes, where reactive minerals are most abundant.

---

**A. Variation of spacing and thickness of cleavage stripes.** Devonian alluvial sandstones and siltstones, Kenmare Bay, Co. Kerry, Eire. The photograph shows a cross-section through bedding from a quartz-rich sandstone (*Q*) at the base, upwards through a feldspathic and chloritic cross-laminated sandstone (*XL*) to a fine plane laminated sandstone and siltstone (*PL*). The stripes are thickest and most widely spaced in the cross-laminated sandstone. Spacing and thickness decrease downward to the quartzite where the stripes appear to die out. In thin section, the effects of pressure solution around individual quartz grain contacts are observed. This change is related to rock maturity: the thickest stripes show the greatest accumulation of new minerals formed from an immature sandstone. From the middle to the top of the photograph, the decrease in thickness of stripes is related mostly to a decrease in grain size from fine sandstone to fine siltstone. Stripes often die out into, and are continuous with, a slaty cleavage in a pelite.

**B. Spaced cleavage stripes.** Devonian cross-laminated sandstone in cross-section, Kenmare Bay, Co. Kerry, Eire. The mica-rich stripes have clearly defined margins and are separated by lithons of sandstone consisting of quartz, feldspar and chlorite. Much of the chlorite and feldspar has broken down in the stripe and some of the quartz has dissolved (BEACH and KING 1978). In the lithons, the cross-lamination often appears over-steepened, suggesting that a certain amount of ductile shortening occurred.

**C. Spaced cleavage stripes.** Trace of stripes on bedding plane of sandstone in Carboniferous flysch, Spekes Mill Mouth, north Devon, England. The stripes are discontinuous and slightly wavy in trace. In some examples the stripes may be arranged en echelon

---

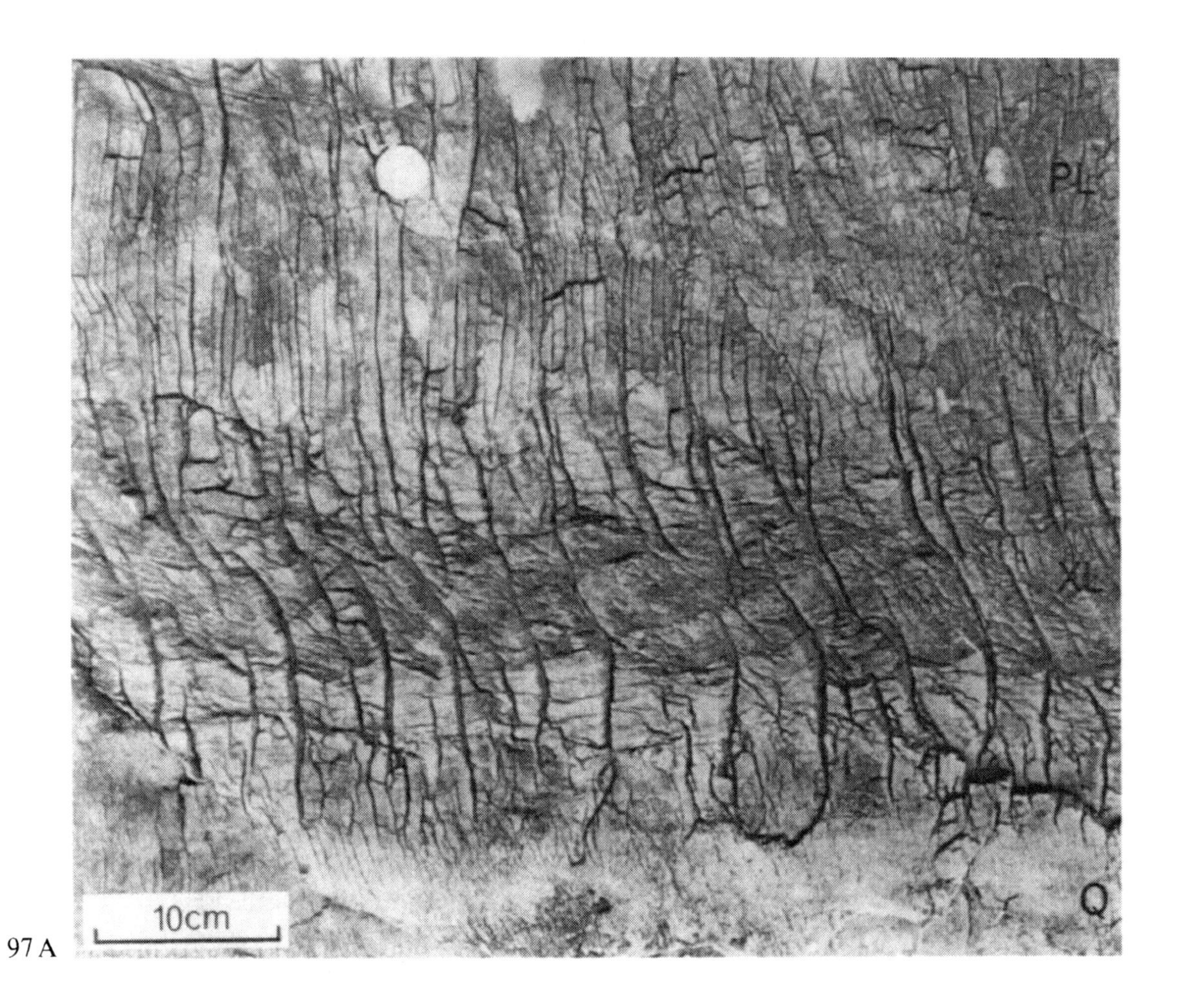


97 A

97 B

# 98. The Relation Between Spaced Cleavage Stripes and Veins

A. BEACH

Also see Plate 129

In many sandstones that show development of spaced cleavage stripes, it is often seen that veins (usually quartz with subsidiary carbonate) developed syntectonically and simultaneously with these stripes. A mechanism of solution transfer of material from the stripe (site of dissolution) to the vein (site of precipitation) has been invoked to explain this relation (BEACH 1974, 1977).

---

**A. Pressure solution stripes and en echelon veins.** Greywacke sandstone from Carboniferous flysch with recumbent chevron folding, Millook Haven, north Cornwall, England. En echelon vein systems represent a common and particular pattern of vein development (DURNEY and RAMSAY 1973, BEACH 1975). Individual veins record a maximum dilation at their midpoints and are often sigmoidal in shape. En echelon patterns are commonly interpreted to have developed in shear zones. In the photograph, the spaced cleavage stripes are better developed within the zone of en echelon veins than outside the zone. Chemical analyses of samples within and around such a zone (BEACH 1974) show a small but persistent decrease in silica content of the rock within the zone. The solution of quartz from the spaced cleavage stripes and its simultaneous precipitation in veins oriented approximately perpendicularly to the stripes helps to limit the increase in area of the plane shown during the development and dilation of en echelon veins.

**B. Pressure solution stripes cutting veins.** Greywacke sandstone from Carboniferous flysch with recumbent chevron folds, Crackington Haven, north Cornwall, England. If pressure solution stripes intersect pre-existing veins obliquely and dissolve them the veins appear offset across the stripes. Assuming that the displacement vector during solution is oriented perpendicularly to the stripe, the offset vein can be restored to its original continuity and a measure of the minimum shortening resulting from pressure solution in the stripes is obtained. The example shown, thus interpreted, records a shortening of 25%–30%. (See also Plate 117).

**C. Pressure solution stripes truncating veins.** Greywacke sandstone from Carboniferous flysch, Millook Haven, north Cornwall, England. Pressure solution stripes often form the boundaries to veins that are dilating as solution proceeds, and the veins then appear to be truncated by the stripe as shown here. The best analogy is with plate tectonics – the vein is the spreading ocean ridge, while each stripe is a transform fault. It is not possible to relate different vein segments either side of a stripe, as in **B**, to measure shortening across the stripes. The dilating veins may develop as single thick and rectangular veins, or the same amount of dilation may be achieved by the formation of many thin veins

---

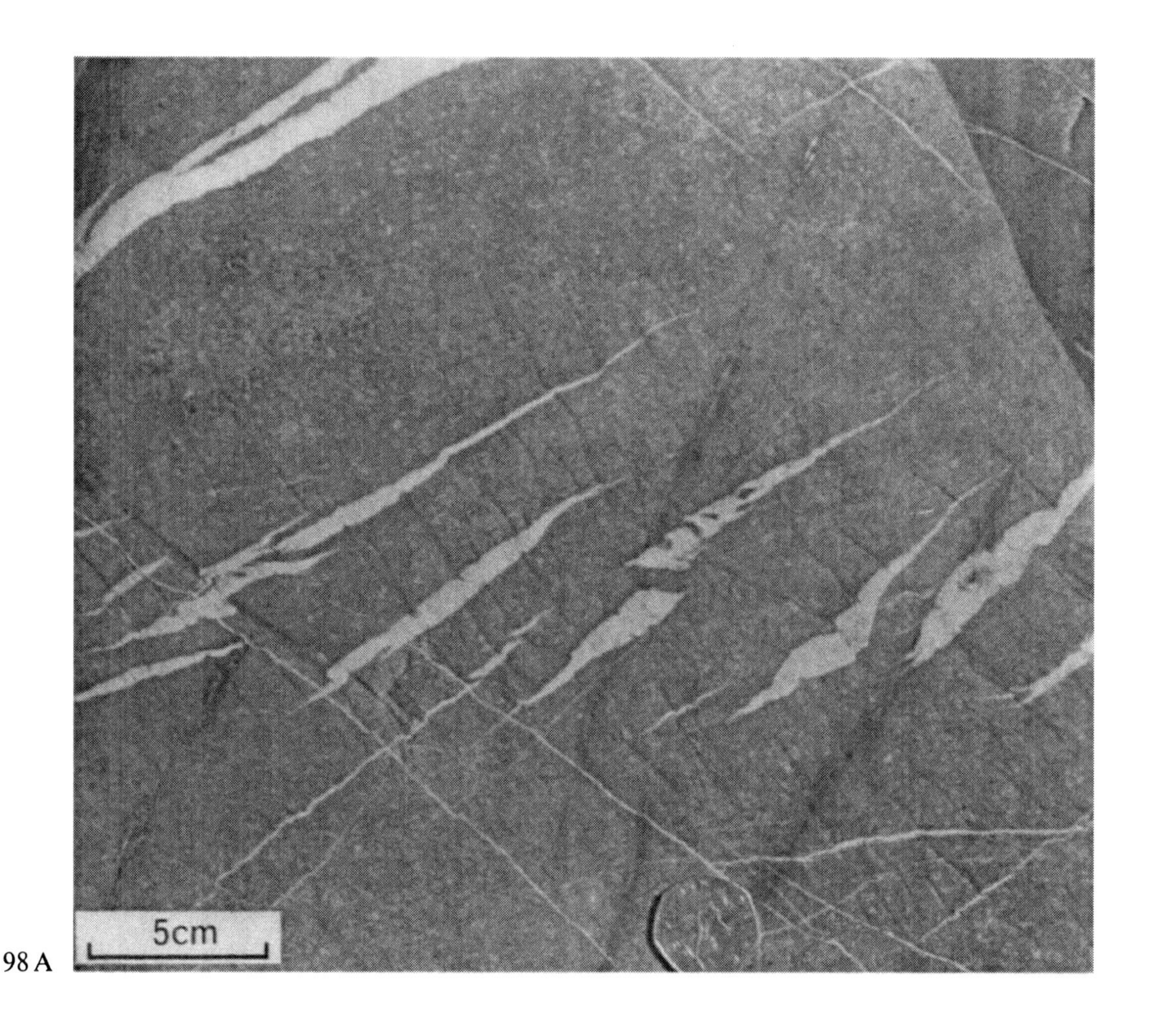

98 A

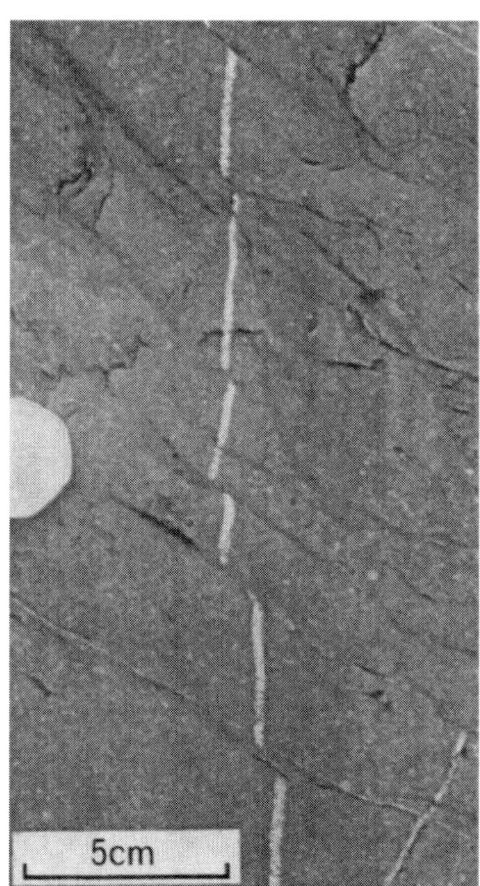

98 B

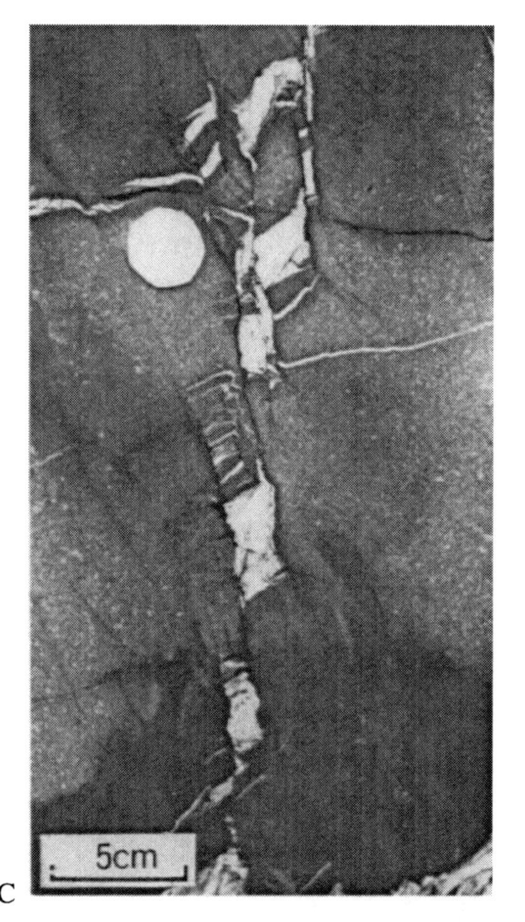

98 C

# Spaced Cleavage

## Section 9: Cleavage Defined by Anastomosing Seams

(Plates 99 to 103)

This small group of five plates draws attention to cleavage defined by seams which anastomose about grains or groups of grains. The contributors draw attention to various special aspects of the rocks with which they are concerned, but we have nevertheless grouped these plates together to emphasize this distinctive morphology.

---

**Plates**

# 99. Schistosity from a Shear Zone and its Relationship to a Fluid Phase

R. Ahmad

A retrograde schistosity is associated with amphibolite-facies shear zones that intersect lower-granulite facies rocks of the Willyama Complex at Broken Hill, Australia (Vernon and Ransom 1971). The retrograde mica-schist described here is taken from North Mine diamond drill core that penetrates the Globe-Vauxhall Shear Zone. A comparison of U, Li, and/or B distribution in the retrograde schists to fresh sillimanite gneiss suggests that retrogression and schistosity development involved a considerable redistribution of these elements, introduction of fluids rich in U, K, Ti, Li and/or B and an overall increase in U and Li and/or B contents in shear zones. The Lexan plastic print technique (Wilson 1977) is used to monitor the distribution of these elements in a thin section. The print records the fission tracks due to $U_{235}$ and sub-micron-sized etch pits due to light ions of Li and/or B as revealed by chemical etching of the plastic after irradiation.

---

**A. Microstructure of mica-schist.** A typical retrograde schistosity (oriented east-west), is dominated by concentration of muscovite (*A*) and strain-free quartz (*B*) and contains minor biotite and chlorite. Euhedral staurolite (*C*) and chlorite (*D*) together with rutile and ilmenite aggregates (*E*) overgrow the schistosity. Small zircon crystals (*F*) are ubiquitous. Biotite (*G*) is the relic high-grade mineral. This assemblage suggests an enrichment of K and Ti in shear zones. PPL

**B. Lithium and boron distribution in mica-schist.** In reflected light the Lexan print of the area corresponding to **A** shows varying etch-pit densities. This can be related to the Li and/or B content in the different minerals recognised in **A**. The mica-rich zones (*A*) are characterized by high pit densities and are highlighted as gray areas (*A*) with white streaks which are due to low Li and B in the quartz grains. These zones also have a uniform distribution of fission tracks. Quartz (*B*), staurolite (*C*) and ilmenite-rutile aggregates (*E*) are prominent, having low etch-pit densities which reflect a low Li and/or B concentration compared with muscovite zones (*A*). The grain boundaries (*H*) contain fine micaceous aggregates. In contrast, the high-grade assemblage lacks muscovite-rich areas and has a lower Li and B concentration suggesting an enrichment of Li, B, U and K during the development of retrograde schistosity. Reflected light

**C. Uranium distribution in mica-schist.** The distribution of U-bearing minerals and sites where U is held is revealed by fission tracks on the Lexan print corresponding to **A**. Four distinct patterns of fission tracks are apparent: (1) uniformly distributed tracks in areas of muscovite (*A*). (2) tracks associated with ilmenite-rutile aggregates (*E*). The rutile is characterized by a high concentration of tracks compared with ilmenite. (3) high track concentrations, in places forming fission-track stars, are associated with zircon (*F*). (4) The grain-boundary areas within staurolite (*H*) are outlined by tracks whereas the grain boundaries and fractures associated with other minerals, e.g., quartz (*B*) are devoid of tracks. This contrasts markedly with unretrogressed assemblages where the only U-bearing sources are zircon grains occurring as inclusions. It is therefore the muscovite and titanium oxide grains, and certain grain boundaries, that appear to have trapped U during retrograde metamorphism. PPL

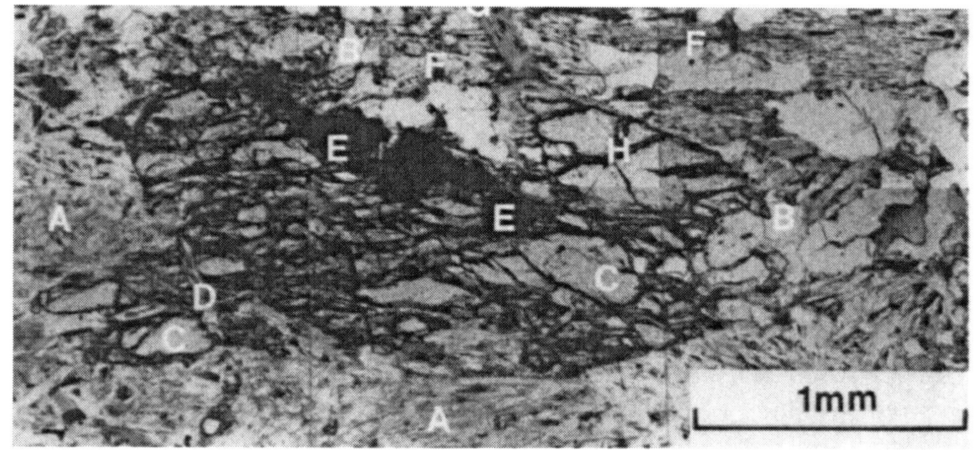

99 A

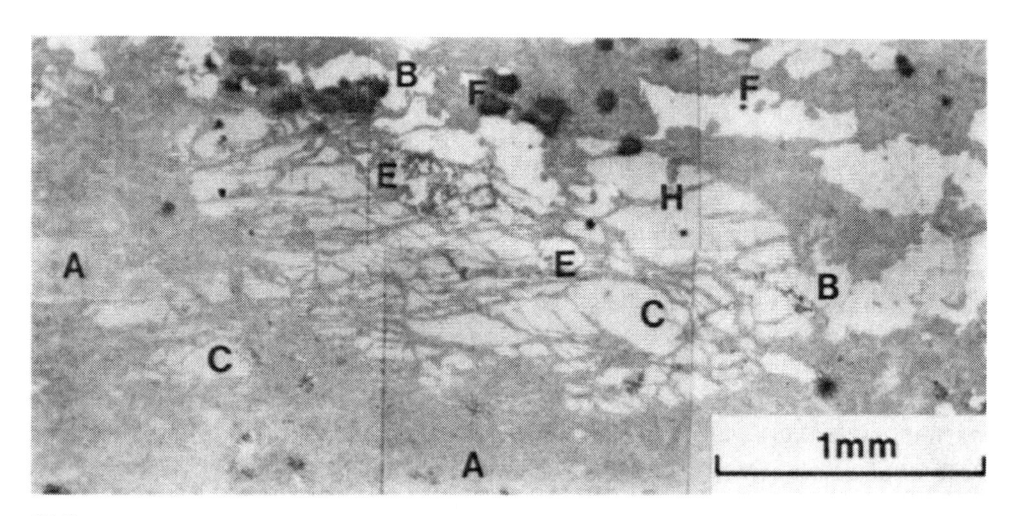

99 B

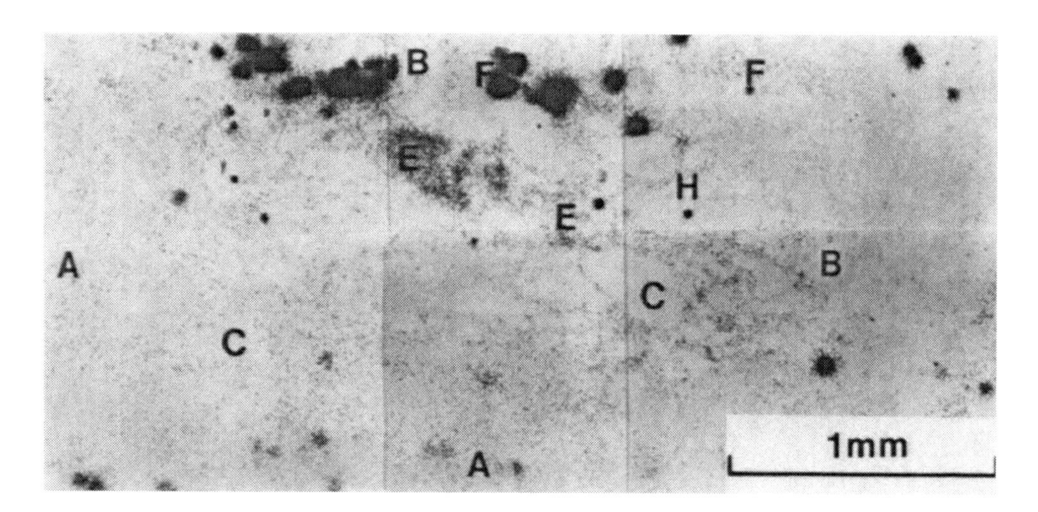

99 C

# 100. Cleavage Development in a Slate and its Relationship to a Fluid Phase

R. AHMAD and C. J. L. WILSON

Microstructures associated with cleavage in many low-grade rocks have been explained by chemically controlled and water-assisted diffusive processes (STEPHENS et al. 1979). It appears that the fluid phase is highly reactive and enriched in certain mobile elements. The specimen of slate described here was collected from an Ordovician slate sequence in the Brisbane Ranges, Victoria, Australia. The uranium distribution in these rocks has been studied using the fission-track method (as used by WILSON 1977). The U-distributions suggest that cleavage development is related to an extensive redistribution of material associated with a fluid phase. The cleavage domains are associated with U-bearing white mica and titanium oxide.

---

**A. Mineralogy and microstructure.** Localized compositional and strain distributions are responsible for the variation between cleavage domains (*A* and *B*). The cleavage is characterised by near-planar domains (trending north–south) within which there is differentiated crenulation cleavage, a preferred orientation of white mica and a differentiated layering. Marked cleavage development occurs along the short limbs and adjacent to hinge zones of the mesoscopic folds. This results in regularly spaced mica-rich domains (*A*) where bedding appears to be rotated and may sub-parallel domain boundaries. In domain *A*, fine white-mica blades (30–60 $\mu$m) are oriented parallel to the domain boundaries and either cross-cut or form anastomosing trains around chlorite and fine quartz aggregates (*C*). Differentiation into mica-rich zones versus quartz-rich zones occurs in the siltstone (*D*). Grains of rutile and leucoxene always coexist with the metamorphic white mica and are not associated with the apparently earlier chlorite. In domain *B*, planar bedding occurs at a high angle to a weakly developed cleavage in the pelite areas. Metamorphic white mica grains are associated with rutile needles. The mica (10–40 $\mu$m) and rutile grains are much finer and less abundant than those of domain *A*. The siltstones have little cleavage development and contain detrital zircon (*E*) and tourmaline. PPL

**B. Uranium distribution and nature of metamorphic fluid.** The fission-track distribution recorded on the Lexan plastic correlates with cleavage development. Domains of marked cleavage development and differentiated layering (*A*) are outlined as areas of high fission-track concentrations. Refraction of cleavage in siltstone layers is marked by a change in the orientation of linear fission-track zones (*D*) and their anastomosing pattern defines the differentiated layering (*C*). The fission tracks are associated with white mica, rutile and sporadic zircon grains. Pelite areas in domain *B* are characterized by a uniform fission-track distribution, except for occasional fission-track stars (as at *E* and *F*) that can be related to detrital zircons. The majority of fission tracks in domain *B* are associated with metamorphic mica and rutile grains and their lower concentration can be related to a weak cleavage development. The variation in fission-track distribution and concentration suggests an extensive redistribution of material and a high degree of fluid activity in domain *A* as compared to *B*. It appears that a metamorphic fluid phase was enriched in U and Ti complexes together with other elemental components (STEPHENS et al. 1979). PPL

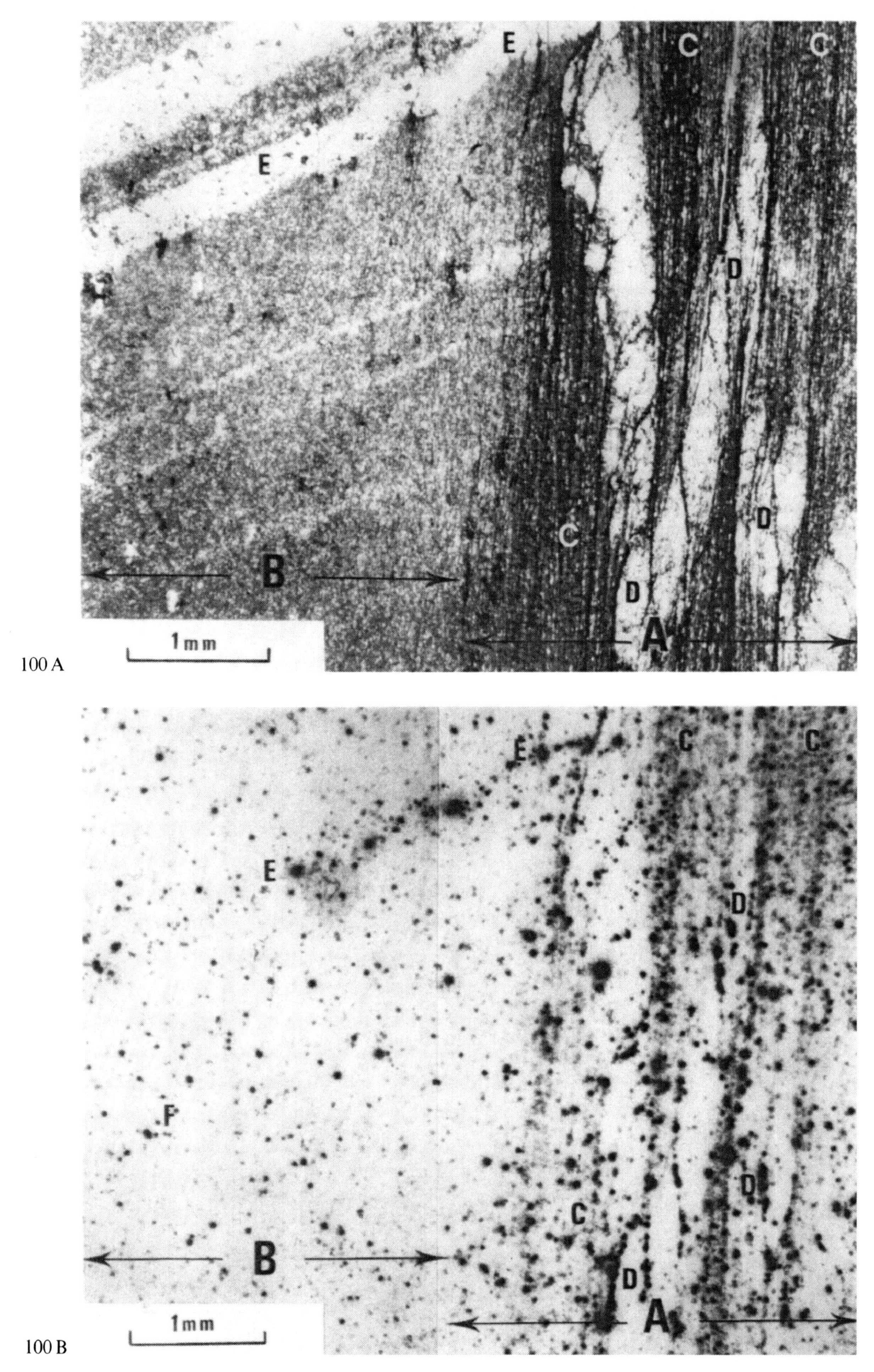

100 A

100 B

# 101. Microprobe Analysis of a Slaty Cleavage Lamella

R. J. Holcombe

The analysis shown is of a single cleavage lamella in chlorite grade slates of the Upper Precambrian Walden Creek Group in Ocoee Gorge, SE Tennessee, U.S.A. (see also Plates 39, 46).

---

**A. Electron microprobe scan.** The distributions of eight elements are shown in separate images of the same area. The intensity of the image is a function of the element concentration. A secondary electron surface image is also shown. The cleavage lamella is about 0.03 mm wide and runs horizontally across the center of the image area.
Na, Ca and Ti concentrations reflect the distribution of albite, carbonate and rutile respectively. The other elements shown occur in several minerals. The carbonate is ankeritic and quartz, muscovite and chlorite are also present. Other elements detected in minor amounts are Mn (follows the Fe distribution), Ba (follows the Ti distribution), Zr (zircon), P (apatite), S (sulfide) and C (graphite). The last four occur in irregularly scattered grains with no detectable spatial preference.
Relative to the microlithon area, the cleavage lamella is enriched in mica (K, Al) and rutile (Ti) and depleted in albite (Na) and carbonate (Ca, Mg, Fe). Analysis also shows that the lamella is depleted in quartz.

**B. Slaty cleavage microfabric** of the rock analyzed in **A**. The foliation is a smooth disjunctive spaced cleavage (Powell 1979) at about 70° to bedding. PPL

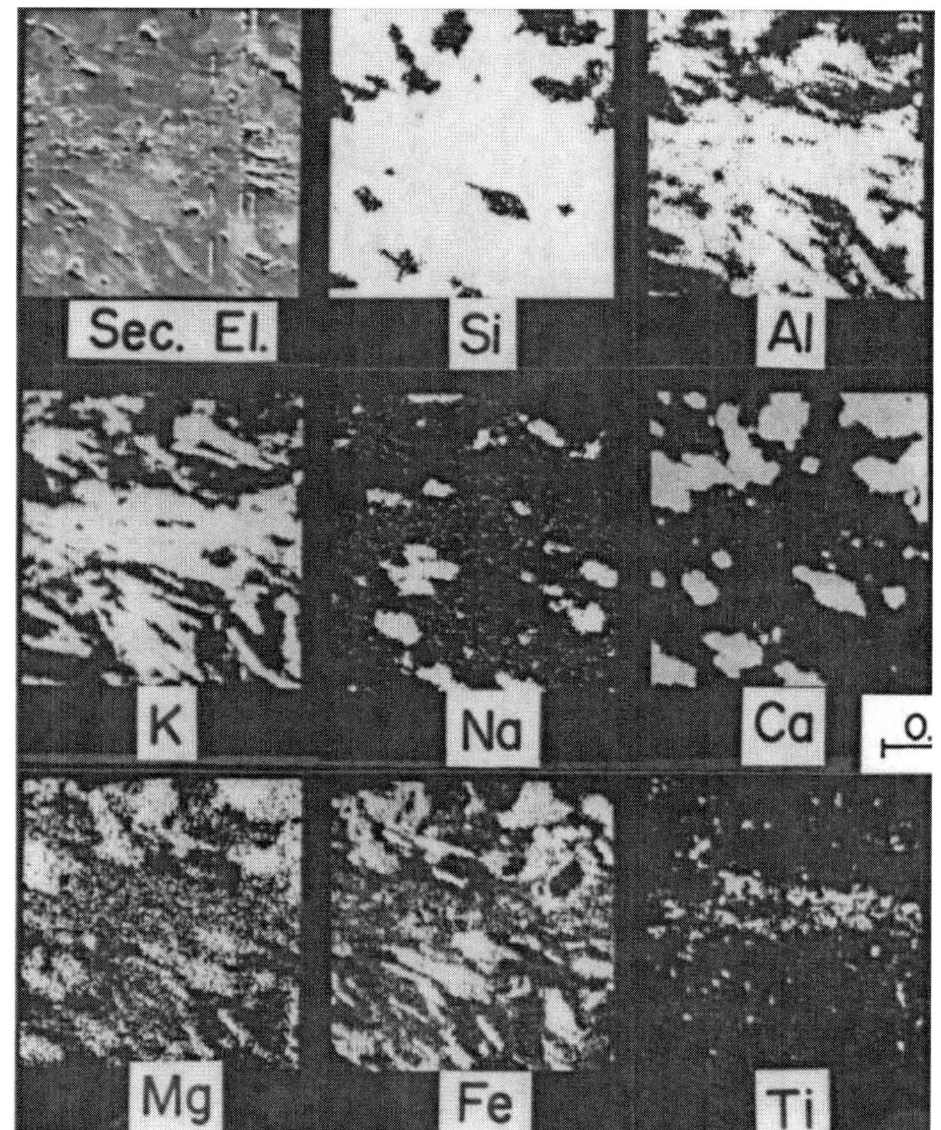

101 A

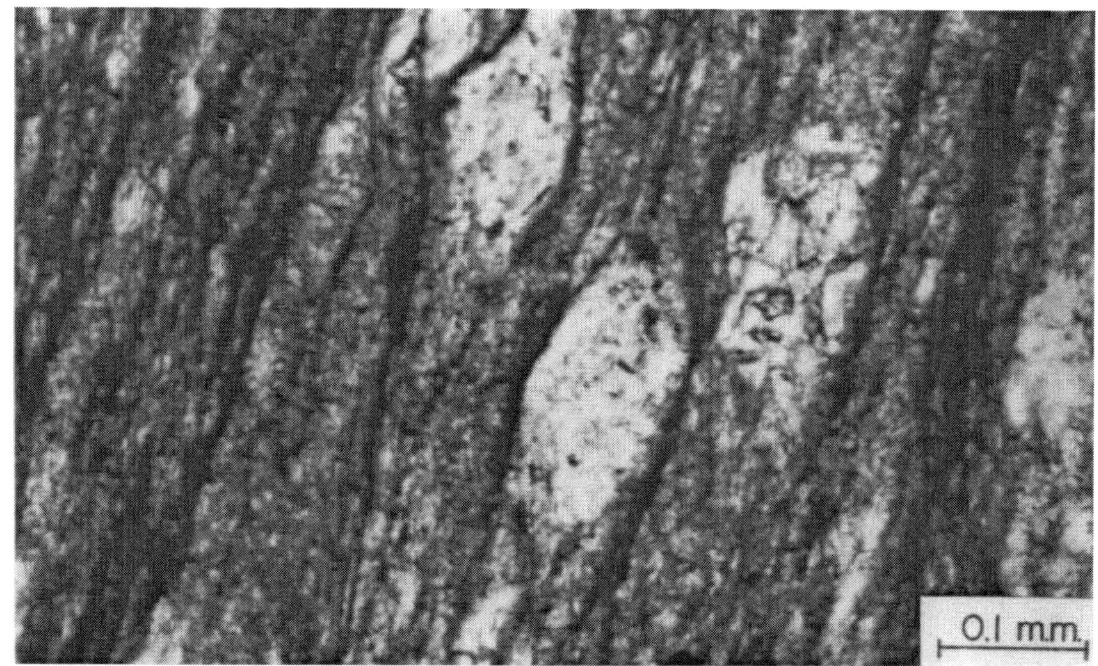


101 B

## 102. Chlorite/Muscovite Fabrics in Low Grade, Spaced Cleavage

A. E. OLDERSHAW, J. H. SPANG, M. Z. STOUT, and E. D. GHENT

All samples illustrated are from outcrops of calcareous, sandy and silty argillite of the Upper Proterozoic Miette Group, approximately 2 km NW of Lake Louise in the Rocky Mountains of Alberta, Canada. The photographs were produced on a Cambridge S-600 SEM equipped with a Kevex energy-dispersive unit as an aid to mineral identification; in each case the plane of cleavage is oriented normal to the photograph. On a submicroscopic scale the cleavage consists of well-oriented, subparallel to anastomosing zones of sodium-bearing muscovite which form the boundaries to pods, lenses, and occasionally irregular patches of unoriented iron-chlorite or chlorite and calcite.

Possible reactions which operated during formation of the cleavage are: illite → muscovite, illite + dolomite + pyrite → iron chlorite + calcite.

---

**A. General fabric.** General view of unoriented chlorite plates (*CL*) which form lensoid pods 20 $\mu$m to 60 $\mu$m across, separated by well-oriented muscovite which defines the cleavage (*arrowed*). Textural relationships are similar to those observed in cleavage from the Mississippian Banff Formation (Plates 88, 95) but the mineralogies are different.

S.E.M., gold coated fracture surface

**B. Muscovite/chlorite relationship.** Detail of textural characteristics of an individual chlorite pod. The chlorite is well crystallized but exhibits no preferred orientation. Muscovite (*M*) in the cleavage is well oriented and conforms to the lensoid shape of the pod. Framboidal pyrite (*P*) is common as inclusions within the cleavage.

S.E.M., gold coated fracture surface

**C. Muscovite/chlorite/calcite relationship.** Calcite (*CA*), with well-defined rhombohedral cleavage, and chlorite (*CL*) form a complex intergrowth fabric outlined by muscovite (*M*). The boundary between muscovite and calcite or chlorite is discordant but there are no obvious cross-cutting relationships.

S.E.M., gold coated fracture surface

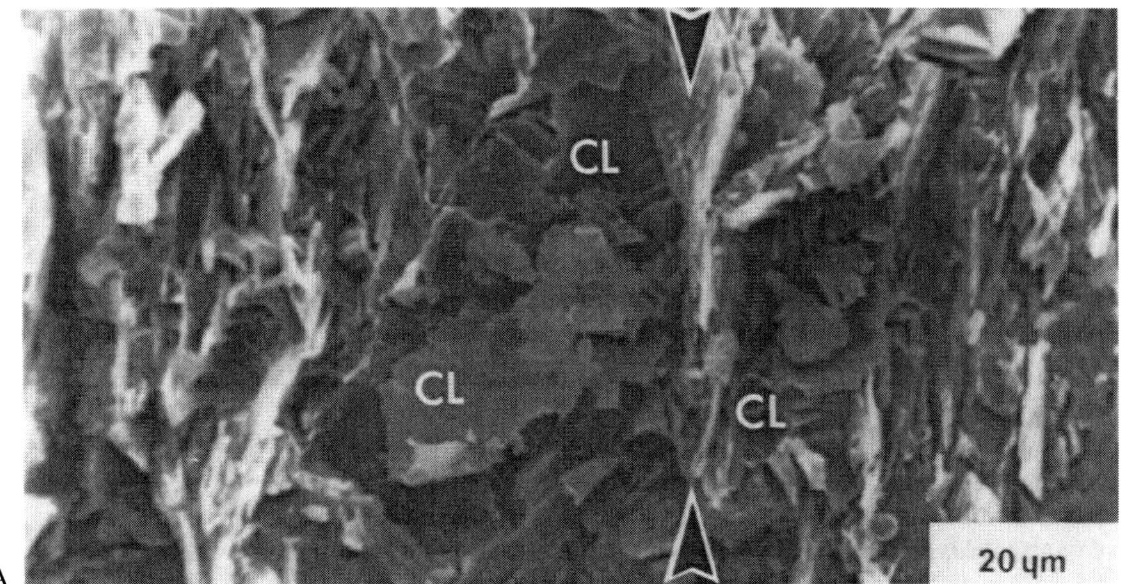

102 A

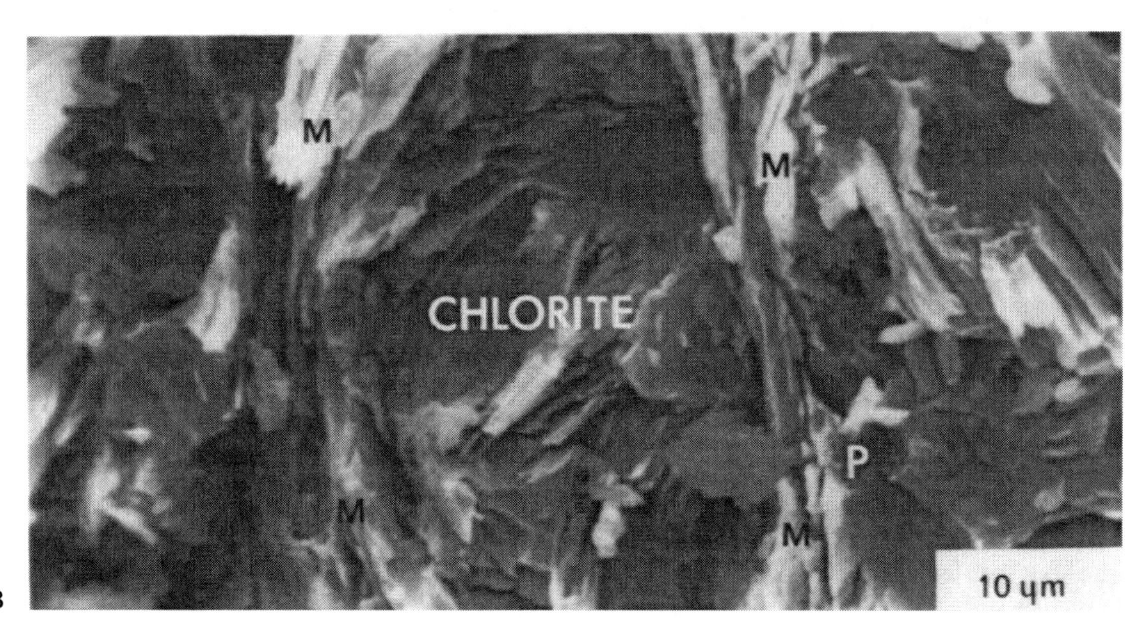

102 B

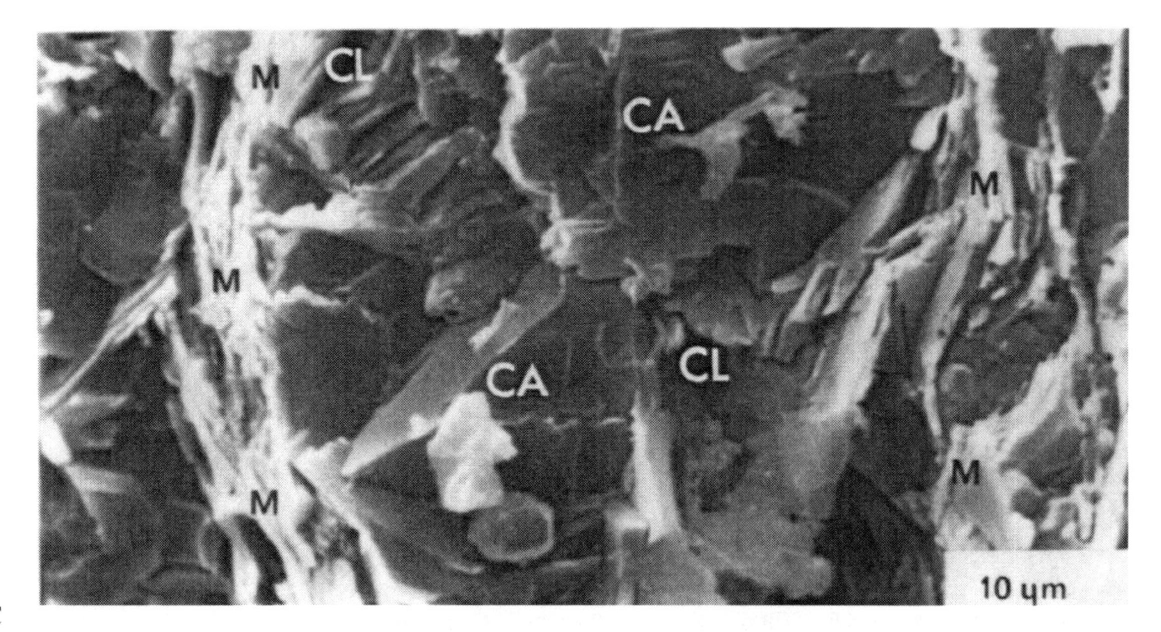

102 C

# 103. Significance of Mechanical Heterogeneities in the Formation of Cleavage

N. B. LEBEDEVA

Mechanical heterogeneity at the grain scale plays a very prominent role in cleavage formation. Especially important is the contrast in deformation of different constituents. Deformation leads to flattening along surfaces normal to compression; the most ductile grains are most deformed to bladed forms and help define cleavage planes.

---

**A. Calcareous siltstone with weak cleavage.** Flysch synclinorium, Caucasus. Section is oriented perpendicular to the hinge line of a fold. Large grains are of quartz, feldspar, and calcite. Between them a vertical system of rudimentary cleavage surfaces is defined by fine-grained micaceous bands. The surfaces pass around large grains and either completely attenuate or merge with other surfaces of the same kind. The preferred shape fabric and cleavage surfaces are due to the differing deformation of mica versus quartz and feldspar.

Width of view: 2 mm PPL

**B. Fine siltstone with weak cleavage.** Southern limb of Svanetia anticlinorium, Caucasus. Section is oriented perpendicular to hinge line of a fold. Large grains are quartz, feldspar and muscovite. Between them are seen vertical, elongated, undulatory lenses of iron oxides indicating possible former water channels. In this rock there was a drastic contrast of ductility between fluid-filled pores containing dissolved iron oxides and the mineral grains. In the course of rock flattening, water was exuded between the solid grains along the cleavage surfaces.

Width of view: 1.5 mm PPL

**C. Experiment confirming significance of viscosity contrast in formation of cleavage (drawing from photograph).** The sample was made up of isometric grains of a resin mixed in different proportions with engine oil (of viscosity between $10^5$ and $10^6$ poises), and dry resin grains (their viscosity was $10^9-10^{11}$ poises) that did not deform. The grains were poured to produce horizontal layers. Then horizontal compression reduced the length of the sample by 75%. The grains of dry resin did not undergo any significant deformation (*black in the figure*), but deformation of the less viscous grains ($10^5$ poises) was much greater than deformation of the whole sample: after flattening they became 10 times longer (*lines in the figure*). The grains with intermediate viscosity stretched out 5 to 7 times and occupy the *white background area* in the figure. Flattened grains which wedge in between undeformed ones and which pass around hard grains to merge with one another, form surfaces similar to the natural cleavage surfaces. (See also Plates 139, 225)

---

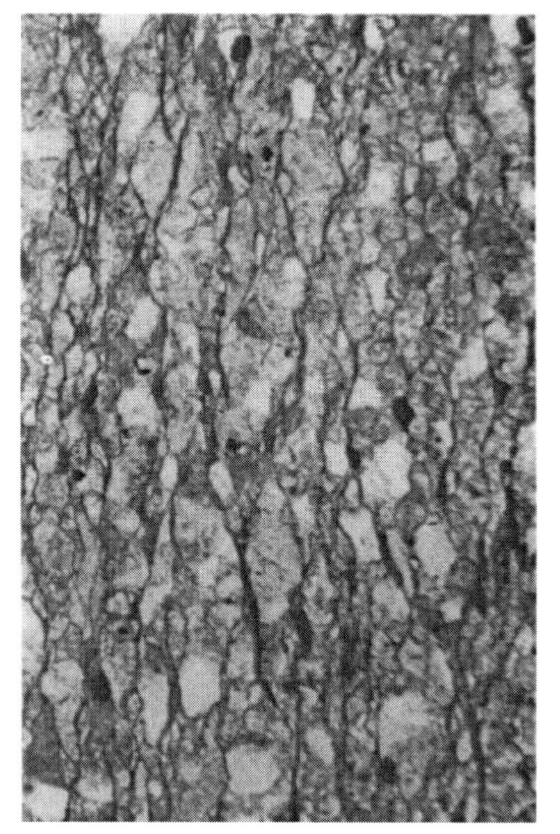

103 A

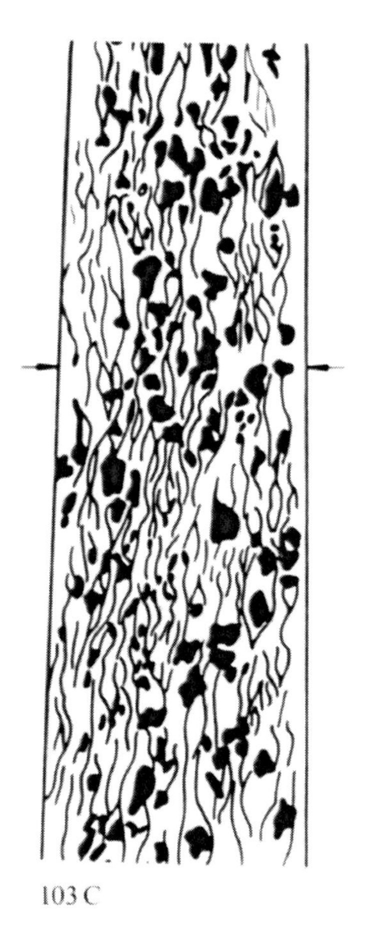

103 C

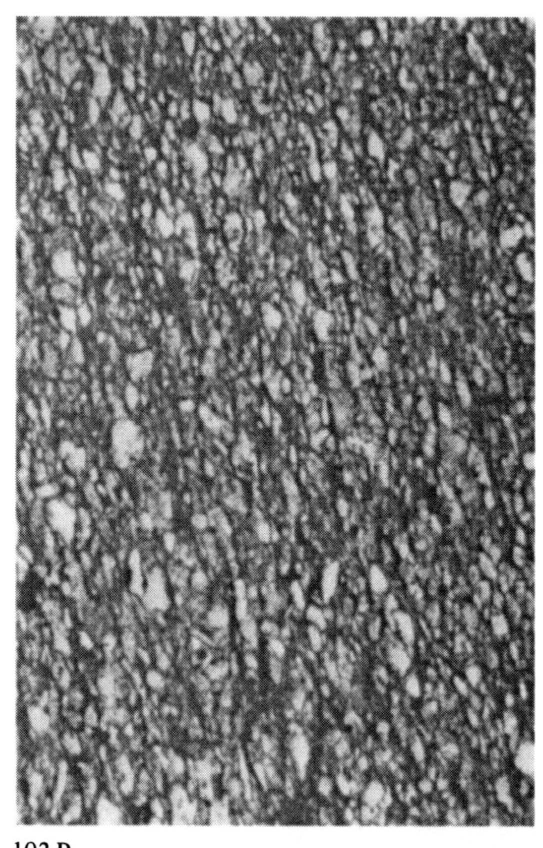

103 B

# Spaced Cleavage

## Section 10: Cleavage Defined by Planar Seams

(Plates 104 to 126)

Twenty-three plates are grouped to illustrate this type of cleavage. The cleavage type differs from those of the preceding two sections in that the seams are planar. Some contributions may at first appear to present examples of continuous cleavage but close examination reveals the discrete nature of cleavage-defining seams.

---

**Plates**

## 104. Closely Spaced Disjunctive Cleavage

G. J. BORRADAILE

This calcareous slate is from a Devonian formation in the Rheinische Schiefergebirge, West Germany (BORRADAILE 1977, locality 3). The rocks were deformed in a noncoaxial strain history as indicated by the "sand dike test" (see Plate 205). The rocks were penetratively deformed once and the cleavage is sufficiently closely spaced that the rock is worked as a roofing slate.

The photograph illustrates the regular, homogeneous anisotropic character of this disjunctive cleavage. In detail the cleavage surfaces are wavy and they truncate carbonate fossil fragments (see Plate 96) and quartz clasts. Magnetic anisotropy data for this rock cannot be correlated with strain because the magnetite grains ($< 10\mu$) along the cleavage surfaces are of metamorphic origin with their preferred dimensional orientation not directly induced by strain. The magnetite between cleavage surfaces records a depositional fabric and interferes with the magnetic fabric of the cleavage surfaces. The resultant hybrid, bulk magnetic fabric has maximum susceptibility parallel to the intersection lineation of cleavage and bedding (BORRADAILE and TARLING 1981). PPL

104

## 105. Chlorite Grains and Mica Films in Calcareous Slates

D. A. C. Gardner

Calcareous slates of the Upper Cambrian Upper Chancellor Formation near Golden, British Columbia in the Western Main Ranges of the Southern Canadian Rocky Mountains, contain elliptical chlorite grains varying in long dimension from 10 to 200 microns, within which the basal (001) cleavage is generally oriented at 35° to 90° to the statistical orientation of slaty cleavage in the matrix material. Slaty cleavage is statistically parallel with axial surfaces of observed folds and is defined by anastomosing films, a few microns wide, of interlocking crystals of white mica which abut abruptly against the chlorite grains. The chlorites represent remnants of an earlier bedding plane fabric similar to that reported by Aitken (1966) in nearby uncleaved, stratigraphically equivalent rocks. Textures in adjacent limestones are illustrated in Plate 118. (Also see Plate 198).

---

**A. Chlorite grains and mica films.** The variously sized, white, elliptical grains scattered throughout the photograph are chlorite grains whose long axes are subparallel with external slaty cleavage. PPL

**B. Kinked chlorite grain.** The chlorite grain (*C*) is folded into a pair of chevron-style microfolds, the axial surfaces of which are approximately subparallel with the external slaty cleavage. PPL

**C. Detail of chlorite grain.** The chlorite grain (*C*) is abruptly bounded by white mica (*M*) defining slaty cleavage. Quartz (*Q*), bounded by white mica, forms a pressure fringe at the end of the chlorite. SEM

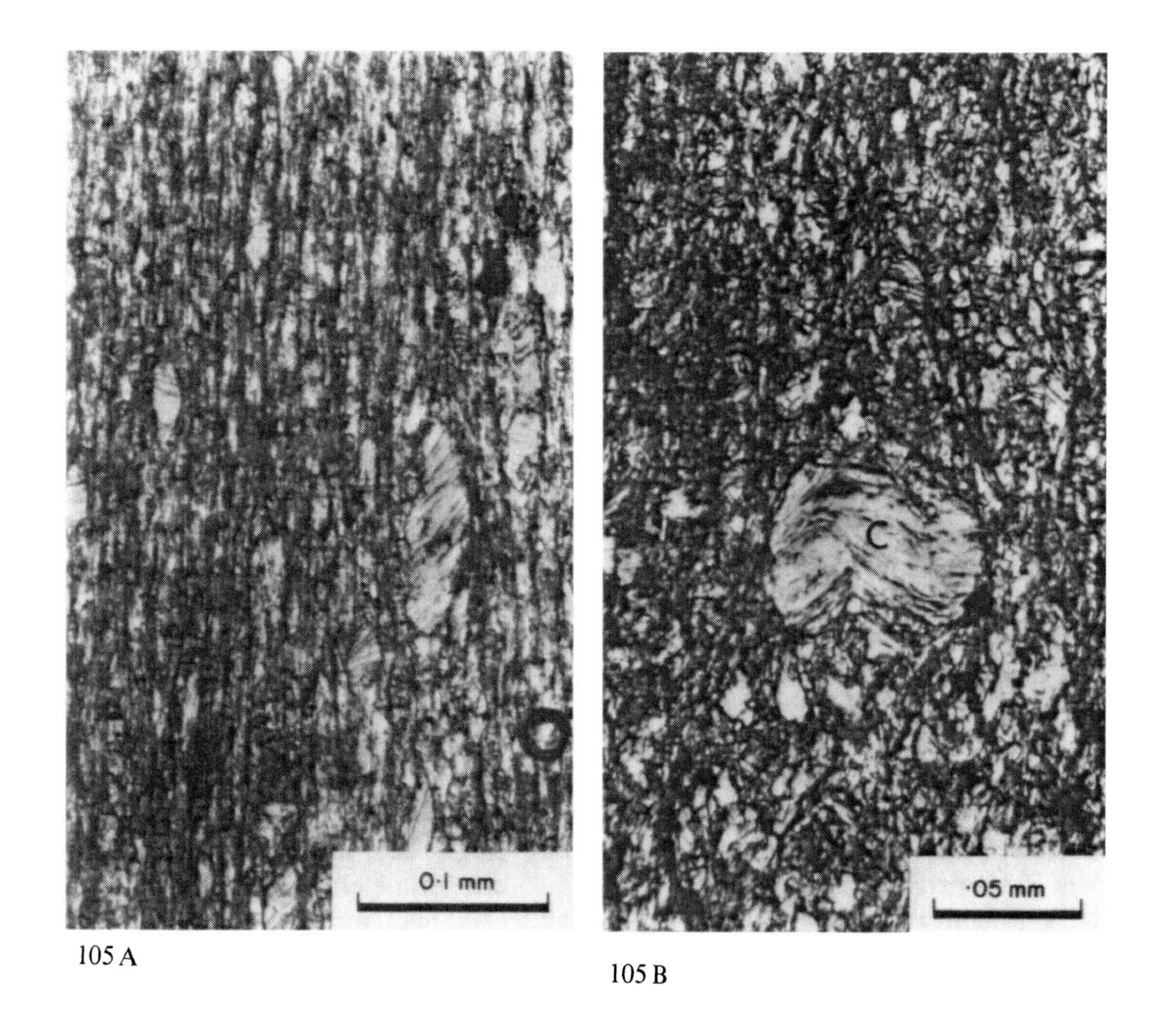


105 A

105 B

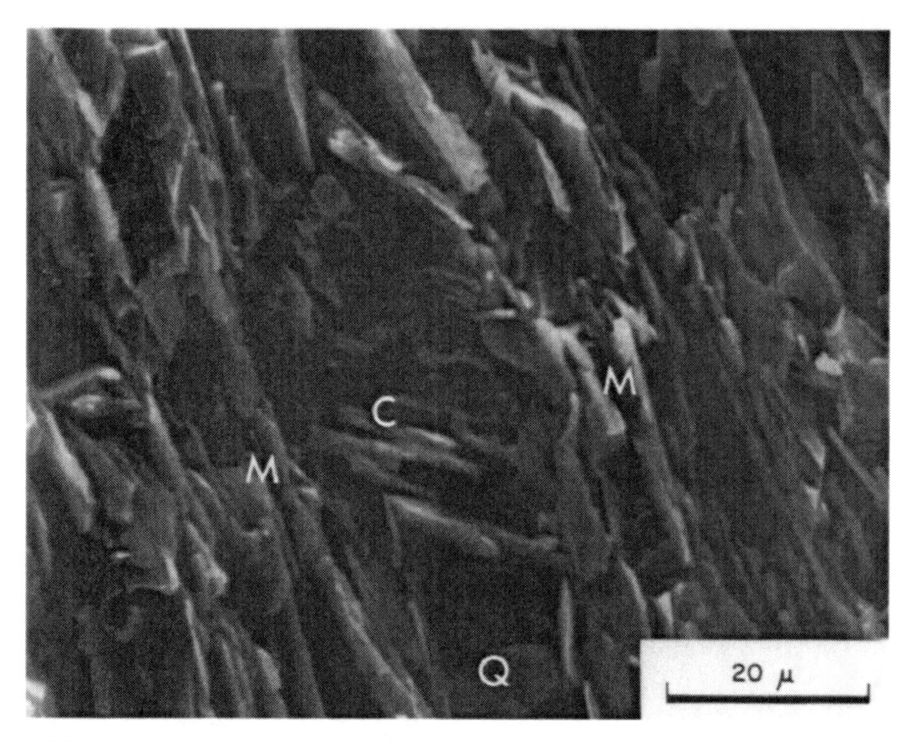


105 C

## 106. Conjugate Cleavage in Quartzose Sandstone (I)

C. McA. Powell

The conjugate disjunctive cleavage described in this and the following plate comes from the Mandagery Sandstone, part of the Late Devonian Hervey Group, in a roadcut about 1.5 km west of Murga, on the Orange-Eugowra road in central-western New South Wales. The Mandagery Sandstone is approximately 700 m thick, and consists of fine- to medium-grained white sandstone interbedded with thin red and green siltstone and shale lenses (Conolly 1965). In the roadside outcrop, thick sandstone beds (some exceeding 2 m) are interbedded with very thin (10 cm thick) siltstone layers, and form a cliff-like outcrop 20 m high. The beds dip gently west (20° to 30°) on the western limb of the open, meridionally trending Nangar Anticline, which has an axial surface inclined about 80° westward.

An anastomosing, near-vertical disjunctive cleavage, spaced at intervals of 2 to 5 cm with cleavage domains occupying about 10% of the rock, pervades the whole outcrop, and the cleavage traces have been accentuated in outcrop by weathering. The photograph (looking south at a near-vertical outcrop), is parallel to the fold profile. In the orthogonal horizontal plane parallel to the fold axis and perpendicular to the cleavage, the spaced cleavage domains anastomose to form elongate lenses. Between the spaced cleavage domains, a thin, planar, conjugate cleavage pattern is developed. The acute angle between the conjugate cleavages (mean $68.4° \pm 12.0°$, $n = 17$, in the fold-profile plane) is bisected by the trace of the thicker, spaced cleavage domains. The thin-section appearance, and possible origin of the cleavage, is discussed in the caption to the next plate.

Regionally, this cleavage is at the outer, western perimeter of cleavage developed in the Late Devonian quartz sandstones. Folds are tight, and locally overturned, in the eastern part of the Lachlan Fold Belt, and become gentle warps further west (Powell et al. 1977, 1980).

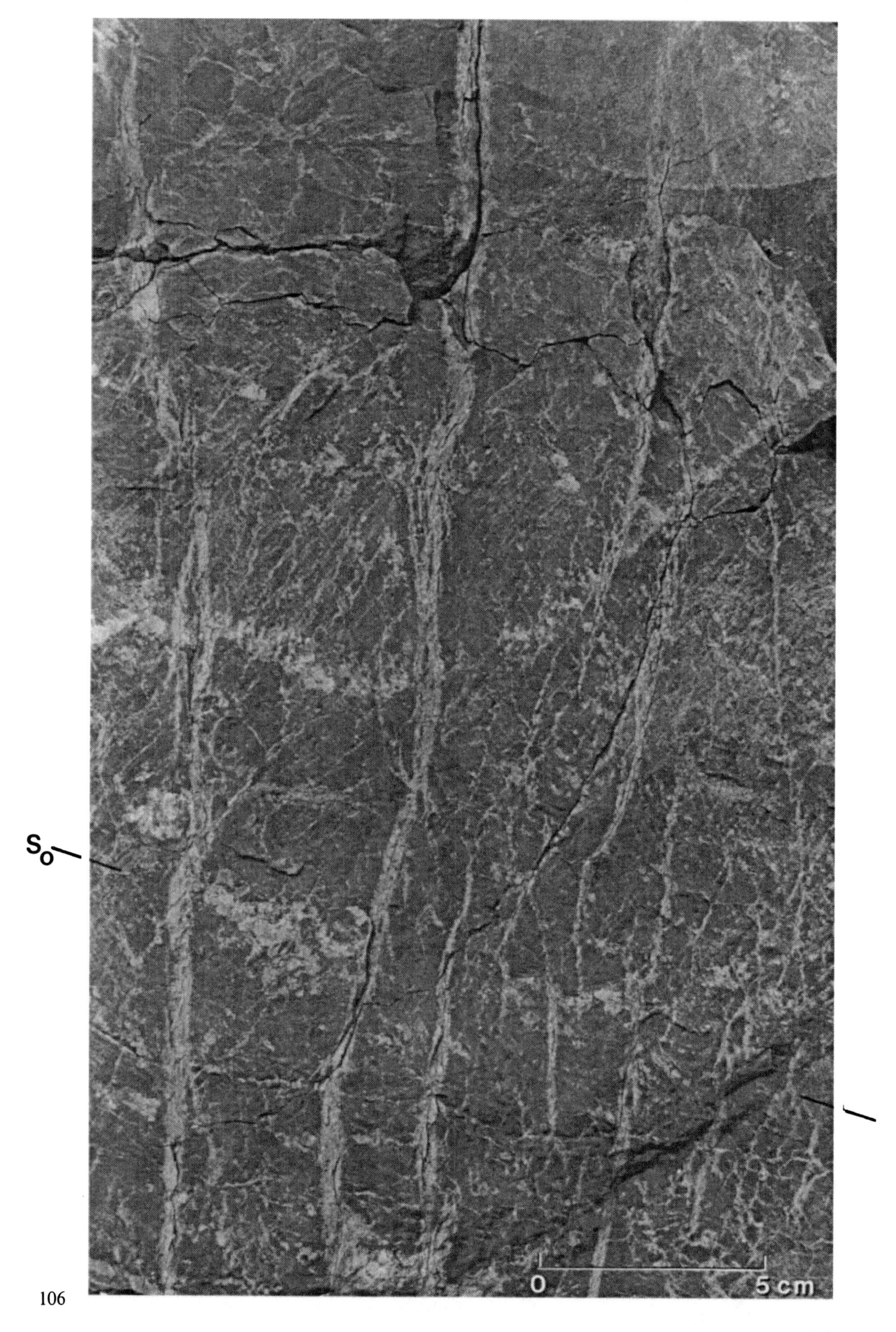


106

# 107. Conjugate Cleavage in Quartzose Sandstone (II)

C. McA. Powell

Late Devonian Hervey Group, Central-western N.S.W., Australia

The field occurrence and appearance of the cleavage is described for the preceding plate. The photographs have the same orientation as the outcrop photograph. In thin section, the rock fabric is seen to consist of:

1. Principal cleavage domains corresponding to the near-vertical, spaced cleavage in outcrop, and consisting of multiple micaceous folia in which any preserved detrital quartz grains are considerably smaller than grains in the microlithons.
2. Planar conjugate zones 0.01 to 0.05 mm wide, and commonly more than 2 cm long, marked by phyllosilicates. Measured angles between the conjugate directions range from 41° to 94°, and average around 68°, with the acute angle bisected by the trace of the principal cleavage domains.
3. Beard-like overgrowths and intergrowths of quartz and phyllosilicates, with fiber elongations either parallel to one of the two conjugate directions, or elongate at some intermediate orientation in the acute angle between.
4. Quartz grains with interpenetrating and interlocking grain boundaries, and commonly showing undulose extinction and deformation lamellae.

The origin of the rock fabric is unclear. The region has been subject to only one deformation (post-Late Devonian to pre-Late Carboniferous), and all indications in thin section are that the principal spaced cleavage and conjugate cleavage developed synchronously with the beard fabric. The principal cleavage domains appear to be zones of concentrated quartz-grain dissolution, accompanied by recrystallization and growth of phyllosilicates. The conjugate cleavages may have been initiated as shear surfaces in a deforming granular body (Means 1977), their orientation having been modified progressively by rotation during shortening perpendicular to the principal cleavage zones.

Assuming plane strain and an initial acute angle of 45° to 50° between the conjugate shears (corresponding to an angle of internal friction of 45° to 40°, respectively), ductile shortening of between 40% and 50% would be required.

No independent strain measurements have yet been made to test this estimate, nor has any estimate been made of how much strain is accomplished by displacement on the conjugate zones. Nonetheless, despite uncertainties about the mechanism by which the conjugate zones developed, they have acted as zones of phyllosilicate growth, with the long axes of individual grains either parallel (**A**) or slightly oblique to the zone (**B**). Where the (001) traces of the phyllosilicates lie oblique to the conjugate zones, they are oriented in a fashion concordant with growth by (001) extension during shear along the zone.

The mica beards are similar to those in other deformed psammites (e.g. Siamo Slate, Powell 1969, and Plates 124, 125), and can be inferred to have grown during extension of the rock in the plane of the principal cleavage. Curiously, in some places, mica beards are oriented parallel to one or other of the conjugate cleavage directions, and on some quartz grains there are different (001) orientations in the mica beard on different sides of the grains. It appears as though the conjugate cleavages, rather than being discrete, brittle, shear failures, may have been zones of dilatant shear with growth of phyllosilicate (001) parallel to the local extension direction.

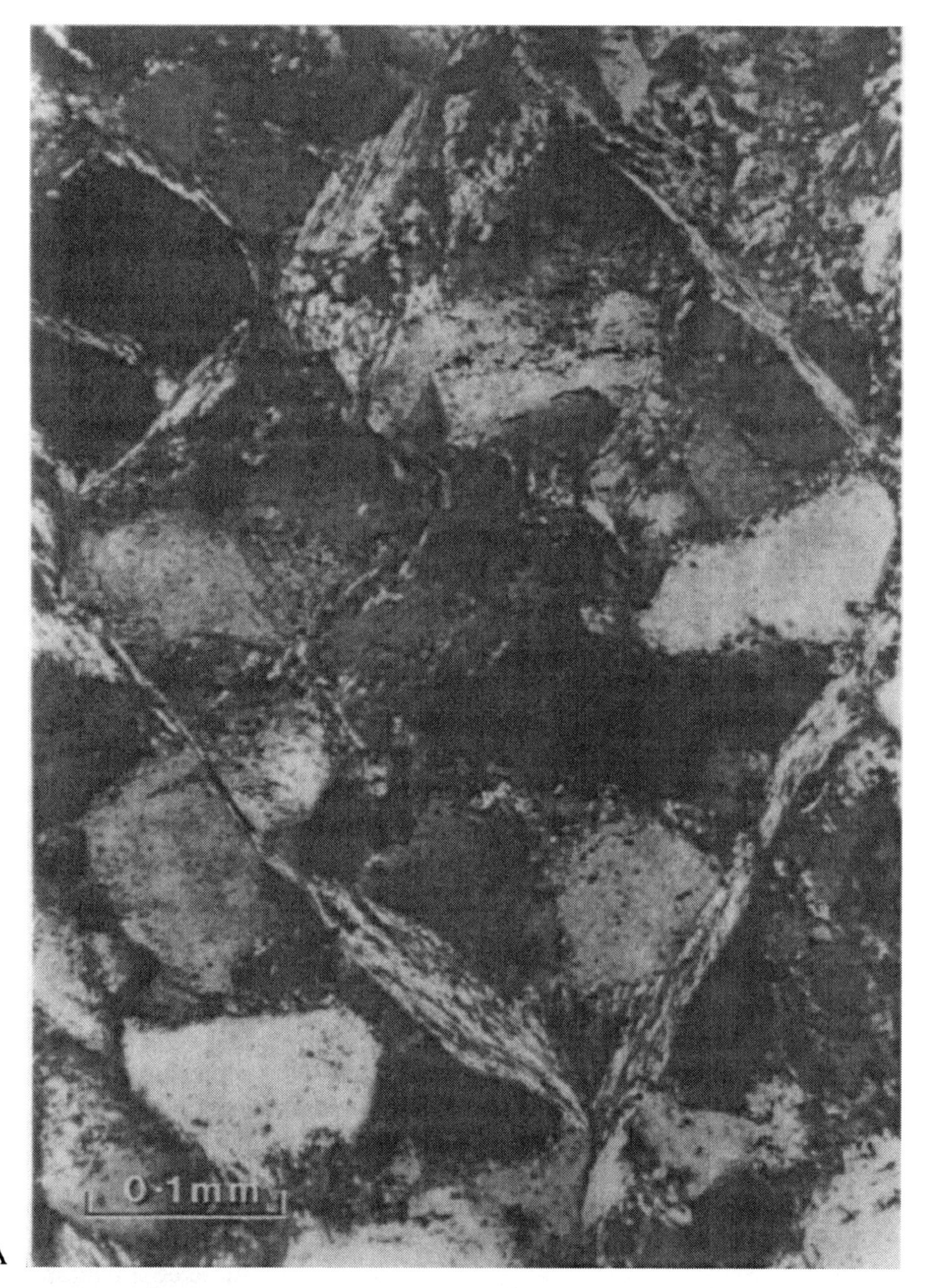

107 A

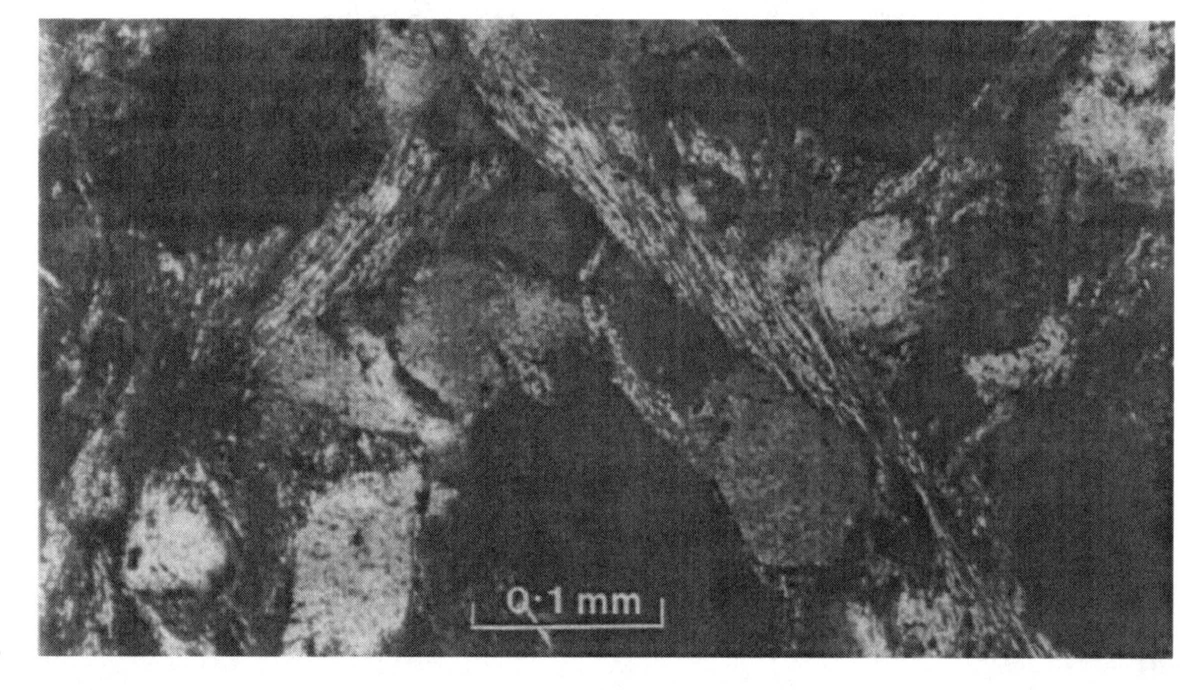

107 B

## 108. Slaty Cleavage in Flysch (I)

D. W. DURNEY

The thin section in this plate is sectioned normal to $F_1$ and details are shown in Plate 109. Here, the following are shown:

1. Two features indicating a post-lithification origin of the cleavage.
2. Five features pointing indirectly towards a pressure-solution origin of the cleavage.
3. One feature indicating little or no recrystallisation or growth of muscovite.
4. Two features indicating non-Marchian rotation of phyllosilicates into cleavage planes.

These points are identified in the captions as "(evidence 1 A)", "(evidence 1 B)" for the first and second lines of type (1) evidence, etc., in this caption page and the following one.

Location: parautochthonous Eocene Flysch beneath inverted limb of Morcles Nappe, Rionda, Valais, Switzerland (specimen 3c, DURNEY 1972a).

Period of Deformation: some time within Middle Oligocene to Middle Miocene (DURNEY 1972a, FREY et al. 1974).

Structural Setting: cleavage concordant with tight gently reclined $F_1$ minor folds and with major (nappe) structure.

Metamorphism: laumontite to prehnite-pumpellyite facies, anchimetamorphic (FREY et al. 1974). Illite from the area typically shows poor (1 Md.) crystallinity (DURNEY 1974, spec. 32b) and this appears to preclude the possibility of new growth of the mineral as a significant factor in the cleavage development.

---

**General description of microstructures, listed in time sequence of formation.**

$s_0$ Bedding – laminated and unlaminated poorly sorted, fine, micaceous silts grading stratigraphically upwards (toward bottom of photo) into clays.

*c* Ribbon-structured calcite layer – composed of large tabular layer-parallel crystals; interpreted as a bedding-slip vein; the rock must have been in a lithified state when it was produced (evidence 1 A).

*m* Tectonic microstylolites – fine dark wavy lines now approximately 45° to cleavage; folded by $F_1$ and partly cutting calcite layer; these also indicate previously indurated rock (DUNNINGTON 1954) (evidence 1 B).

*f* Folding – gentle buckling of calcite layer was probably initiated during *m*. However, the main harmonic folding in the surrounding slate is $F_1$; this dies out rapidly on *left* but more gradually on *right* where bedding is more prominent.

$s_1$ Slaty cleavage; *d* differentiation; *tr* transposition; *p* pressure-solution truncation of calcite layer – all axial planar to (i.e., coeval with) $F_1$ folding. The dark differentiation bands, due to depletion of calcite and quartz, are developed preferentially along $F_1$ fold limbs causing incipient transposition of compositional layering. Slaty cleavage fabric is strongest in these dark bands (evidence 2A) especially near the pressure-solution truncated portions of the buckled calcite layer (evidence 2B). Stubby, fibrous antitaxial calcite veinlets (*v*) (DURNEY and RAMSAY 1973) are associated with this pressure solution (*p*).

*h* Recent hairline cracks formed by natural weathering processes. These show the actual cleavage (i.e., splitting) orientation in the rock. PPL

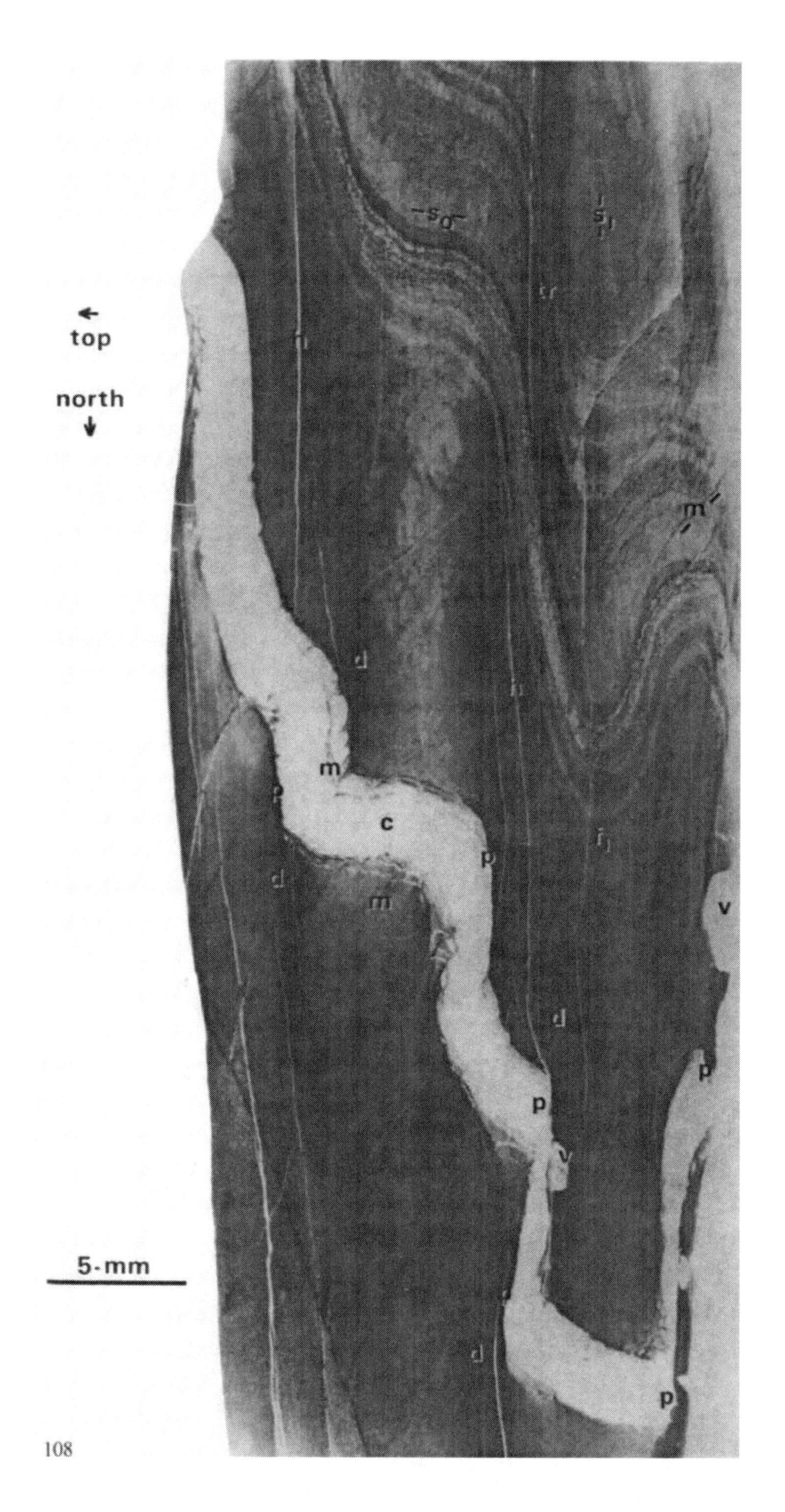
top
north
5-mm
m
c
p
m
v
p
p
v
p
d
d
d
d
108

# 109. Slaty Cleavage in Flysch (II)

D. W. Durney

Detail of specimen in Plate 108. The sections shown are normal to the $F_1$ fold hinges.

**A. Stylolites passing into slaty cleavage.** Small, conical stylolites ($m_1$) on the inner arc of an $F_1$calcite-layer fold pass continuously into thick, flat, residual clay seams ($r_1$), thence into normal slaty cleavage fabric ($s_1$). The lateral continuity between them suggests a common or related origin, i.e., pressure solution, because stylolites are known to form in this way (evidence 2 C). The contrasting morphologies of the three features are perhaps due to differences in clay content of the rock in which they occur. PPL

**B. Deformed microstylolite.** An early microstylolite ($m$) cuts obliquely across bedding ($s_0$) and is transected by slaty cleavage ($s_1$). The microstylolite shows a tendency to adjust to the changing strain orientation by becoming subdivided into en echelon segments. The tips of the segments curve toward the slaty cleavage direction suggesting a continuation of pressure solution up to at least the initial stages of cleavage development (evidence 2 D). PPL

**C. Slaty cleavage independent of mica fabric.** Micaceous siltstone layer showing coarse, domainal slaty cleavage ($s_1$) at 35° to bedding ($s_0$) and detrital muscovites (*white flakes* in photo) at 6° to bedding (see Fig. **1 a**). The muscovite fabric can be explained as an initially bedding-parallel fabric which has suffered passive (March model) rotation of the flakes during bulk strain. This interpretation is supported by a lack of obvious systematic alteration of shapes and sizes of the muscovites (evidence 3 A). The orientation of the clay minerals in the cleavage cannot be explained in the same way (by the March model) because of the large divergence from the muscovite orientation (evidence 4 A). PPL

**D. Inhomogeneous rotation of mica.** Same layer as for **C**, here at the hinge of an $F_1$ fold. The detrital muscovite fabric (Fig. **1 b**) is bimodal with one weak mode subparallel to bedding ($s_0$) and a stronger, cleft one 15° to cleavage ($s_1$). Bimodality is not a prediction of the March model (evidence **4 B**) but may reflect a greater rotation, and therefore greater deformation, of the medium between the cleavage-mode flakes due to heterogeneous negative dilation across the cleavage, i.e., pressure solution (evidence 2 E). PPL

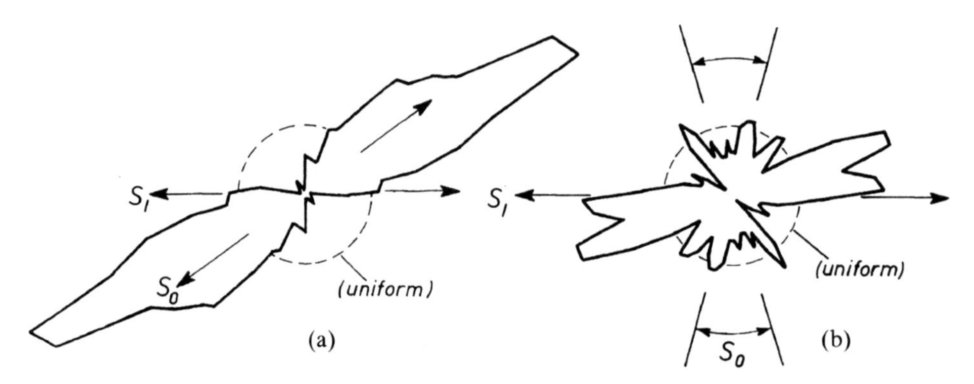


**Fig. 1.** Orientation density of detrital muscovite on a polar plot. Frequency per 10° at 5° intervals referred to uniform; $s_0$ bedding; $s_1$ cleavage; **a** from Plate **C** (N = 156), **b** from Plate **D** (N = 216)

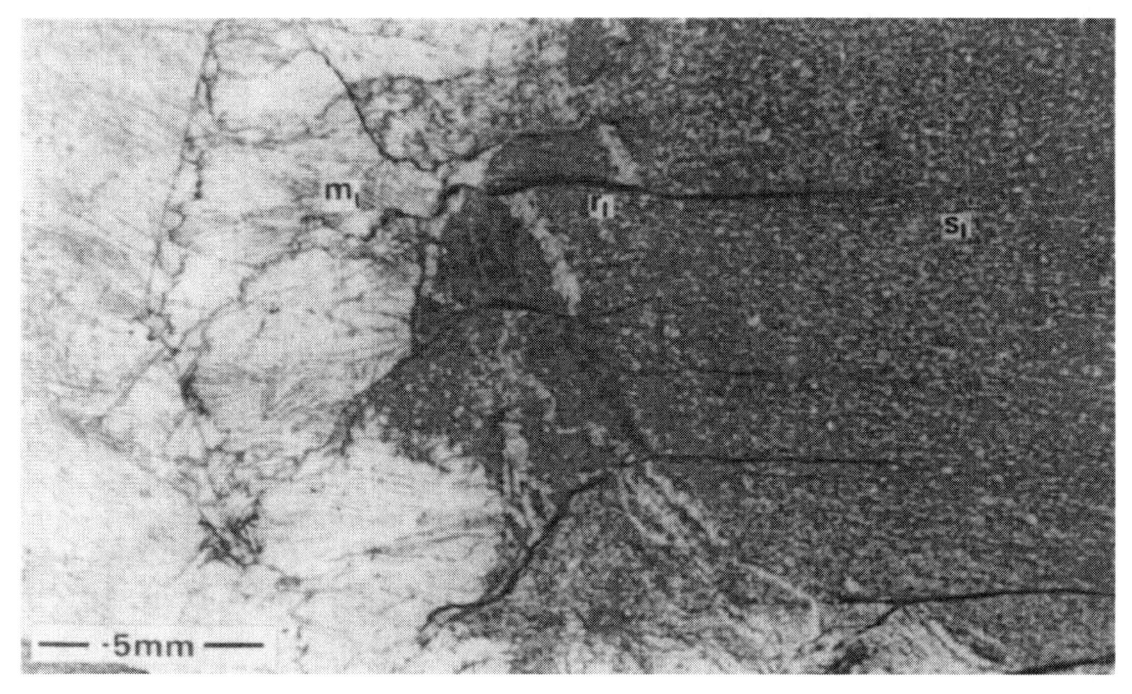


109 A

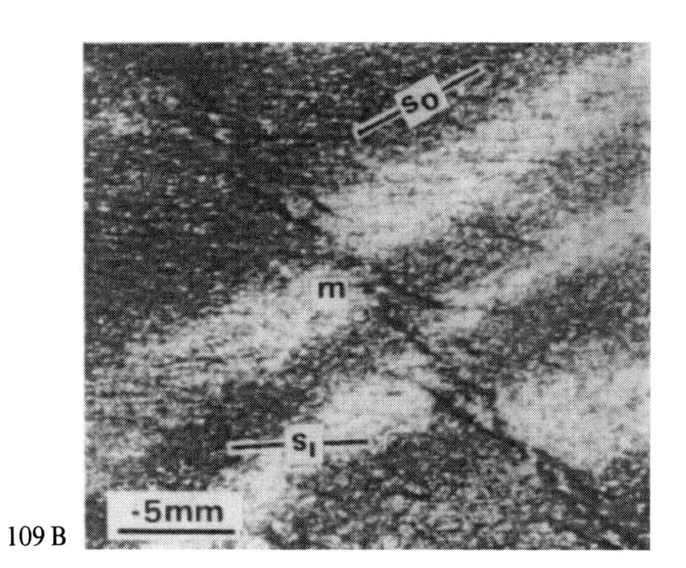


109 B

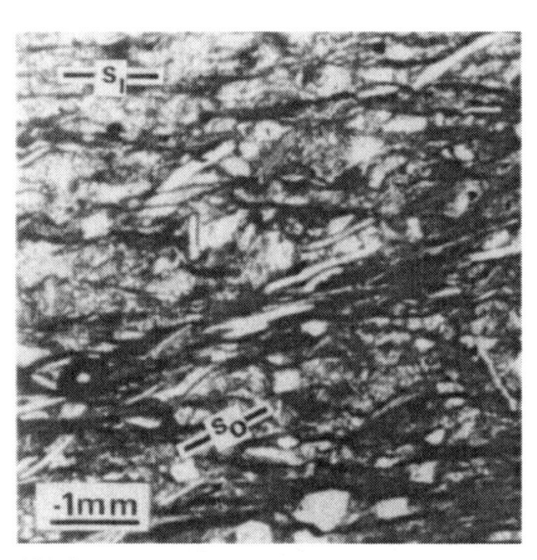


109 C

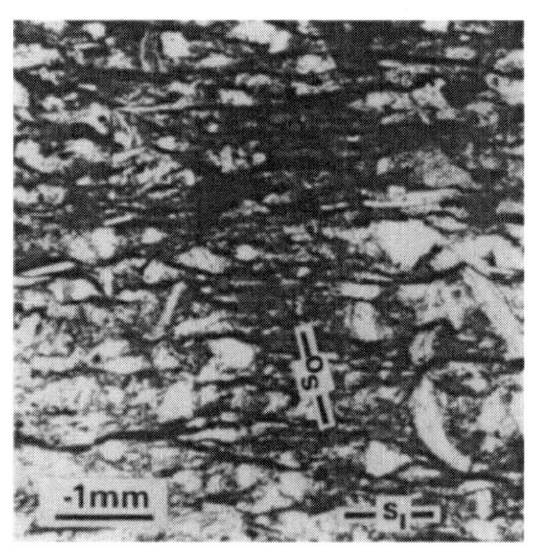


109 D

# 110. Seams Defining Cleavage

P. R. Williams

These photographs are of a rock from Bathurst Harbour in SW Tasmania, Australia (see also Plate 140). It has been subjected to four deformation events. The dominant cleavage from top to bottom of the photographs is the $S_3$ cleavage, which is the dominant cleavage throughout the area.

---

**A. Seams and folds.** Cleavage seams ($S_3$) developed within interbedded siltstone and mudstone. Probable load casts developed at the base of the siltstone layer, and compound cleavage seams formed on the same wavelength as the load casts. Deformation of the thin siltstone bed followed the same wavelength. In places the cleavage seams truncate this thin bed, but in others pressure solution has resulted in the concentration of mica and opaque grains parallel to the fold limbs. The latter process forms sigmoidal seams where the concentration zones become continuous. Note the early development of seams parallel to bedding. Seams in the clean siltstone are rarer and more widely spaced. PPL

**B. Morphology of seams.** Seams have developed along the long limbs of asymmetric microfolds. In detail the seams represent concentrations of opaque minerals and mica (phengite) in zones spatially related to the folds. Quartz grains are truncated by the zones. The seams therefore define a crenulation cleavage. PPL

**C. Overprinting of seams.** The vertical cleavage direction is $S_3$. The horizontal seam direction is the direction parallel to bedding in **A**; the seams are contorted into small folds of irregular form. The form is controlled by the $S_3$ direction, the folds being either cuspate and bisected by $S_3$ or step-like, with the step risers parallel to $S_3$. Clearly the horizontal seams are affected by the $S_3$ seam formation process. They are also truncated by $S_3$ and are therefore earlier. PPL

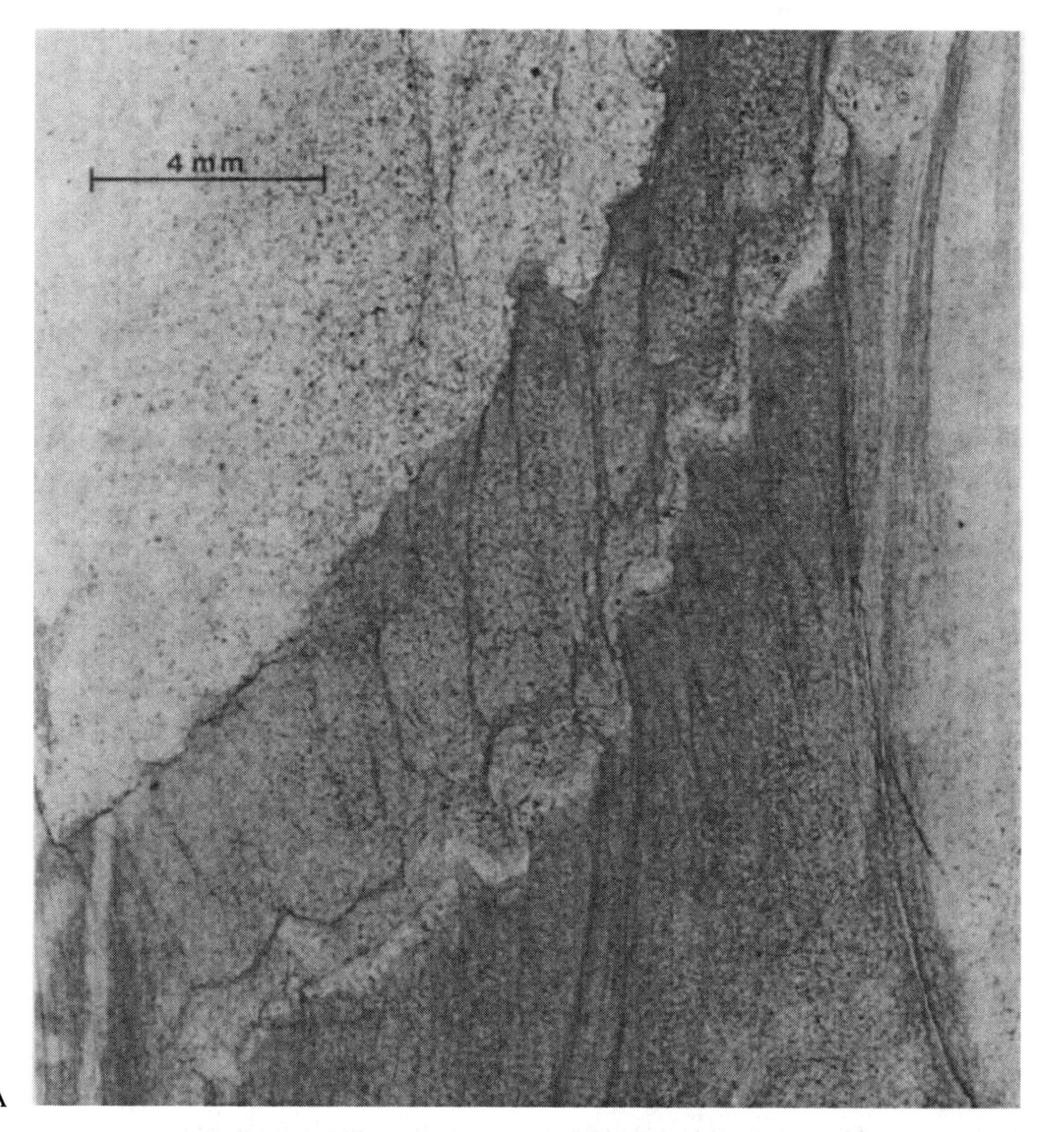

110 A

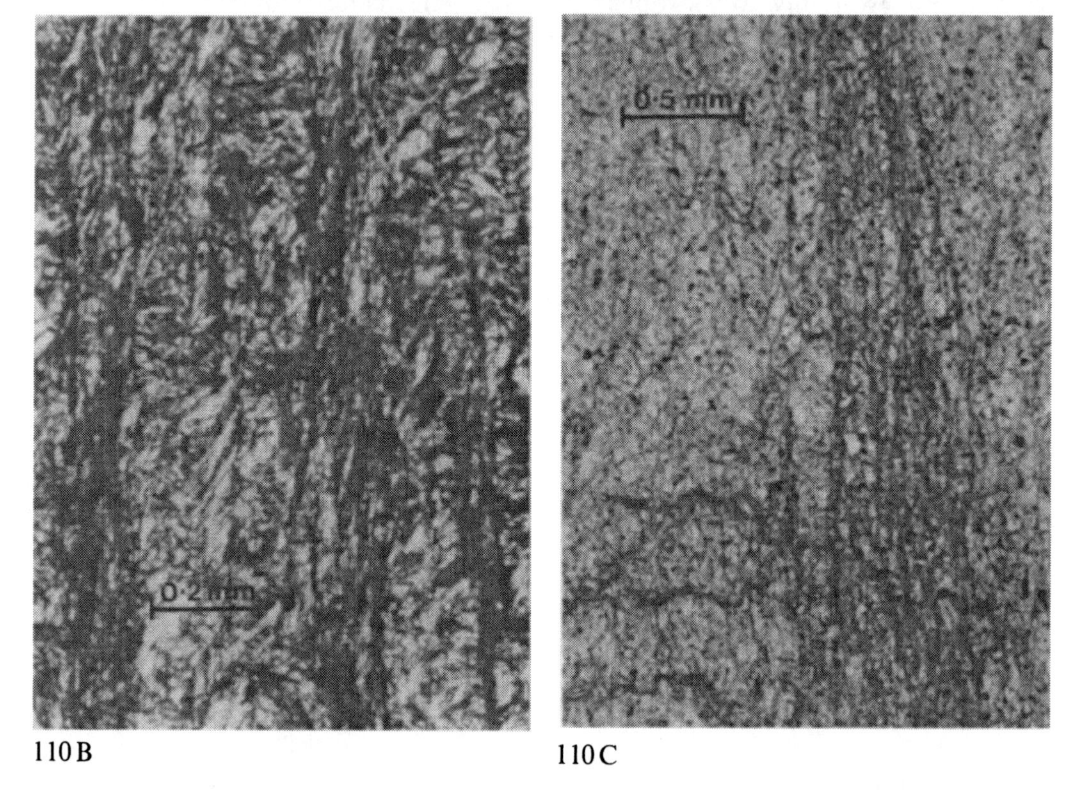

110 B

110 C

# 111. Cleavage in a Kinked Phyllite

B. BAYLY

Ordovician Walloomsac Formation, outcrop on Taconic Parkway near Chatham, N.Y., U.S.A. (**A**) crossed nicols (**C**) same field, plane light (**B**) and (**D**) detail of two cleavage planes.

The rock is a silver green phyllite, highly homogeneous with only the faintest layering. Its texture is shown in detail in Plate 9B. Here attention is given to the kinking, on planes that run straight and parallel for distances at least five times greater than the photographs show – a very regular rock.

There are several points of contrast between this rock and the many other examples of crenulation. The profile is angular rather than rounded; there is very little sign of differentiation, and very little sign of cleavage developing by rearrangement of flakes in positions more parallel to the axial plane. There are few cleavage planes, as shown in **B** and **D**, not at the hinge but very close alongside it. Despite the differences, I suggest that we are seeing here the same effects that give crenulation cleavage; the different appearance arises only from the very high anisotropy of the rock, as follows:

Assume the rock is highly anisotropic, with easy slip parallel to the flakes. This causes the highly angular profiles and narrow hinge zones. We still expect a contrast in the stress field, with smaller vertical stress in the hinges than on the limbs, as in normal crenulation, and a tendency for solutes to migrate. But in this rock, the amount of potential void space in the hinge zones is so small that dissolution and migration, the "robbing of the limbs to feed the hinges", only has to occur over a very small volume. Only the part of the limb next to the hinge is affected; the cleavage seen is a normal differentiated cleavage except for dissolution being localized very close to the hinge.

Points arising from this suggestion are:

i. The asymmetry: cleavage develops only on one side of a hinge. I attribute this to the maximum compression being not quite perpendicular to the axial planes.

ii. Can we find intermediate cases? If the above suggestion is correct, intermediate cases could exist as in diagram (b) below.

Scanning this Atlas suggests that the intermediate state, if ever developed, is easily masked by normal irregular departures from ideal profiles.

Notes: 1. The idea that kinking involves brittle behavior receives no support from this example. The hinges, though narrow, are perfectly continuous arcs.

2. The same geometry, with a differentiated cleavage close to but not at the hinge, was described by Dewey from southern Ireland (1965, pp 485–6 and Fig. 25C) under the designation of "pelitic strain band".

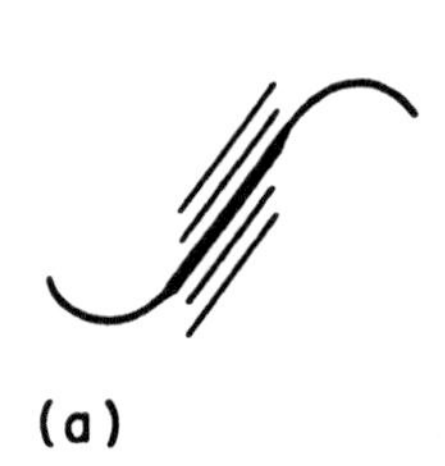

(a)

Ideal crenulation; concentration of residues is max. at center of limb as in Plate 24

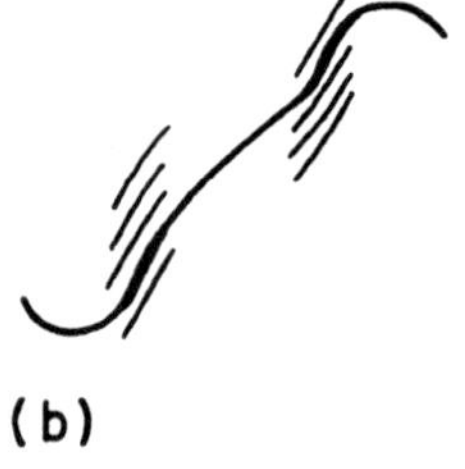

(b)

A possible intermediate, with residues concentrated all down limb but more so at the ends, close to the hinge zones

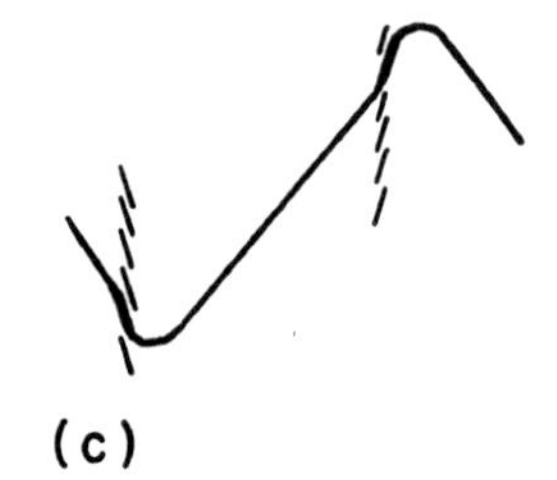

(c)

Kink cleavage: concentration of residues is max. just at end of limb. This Plate

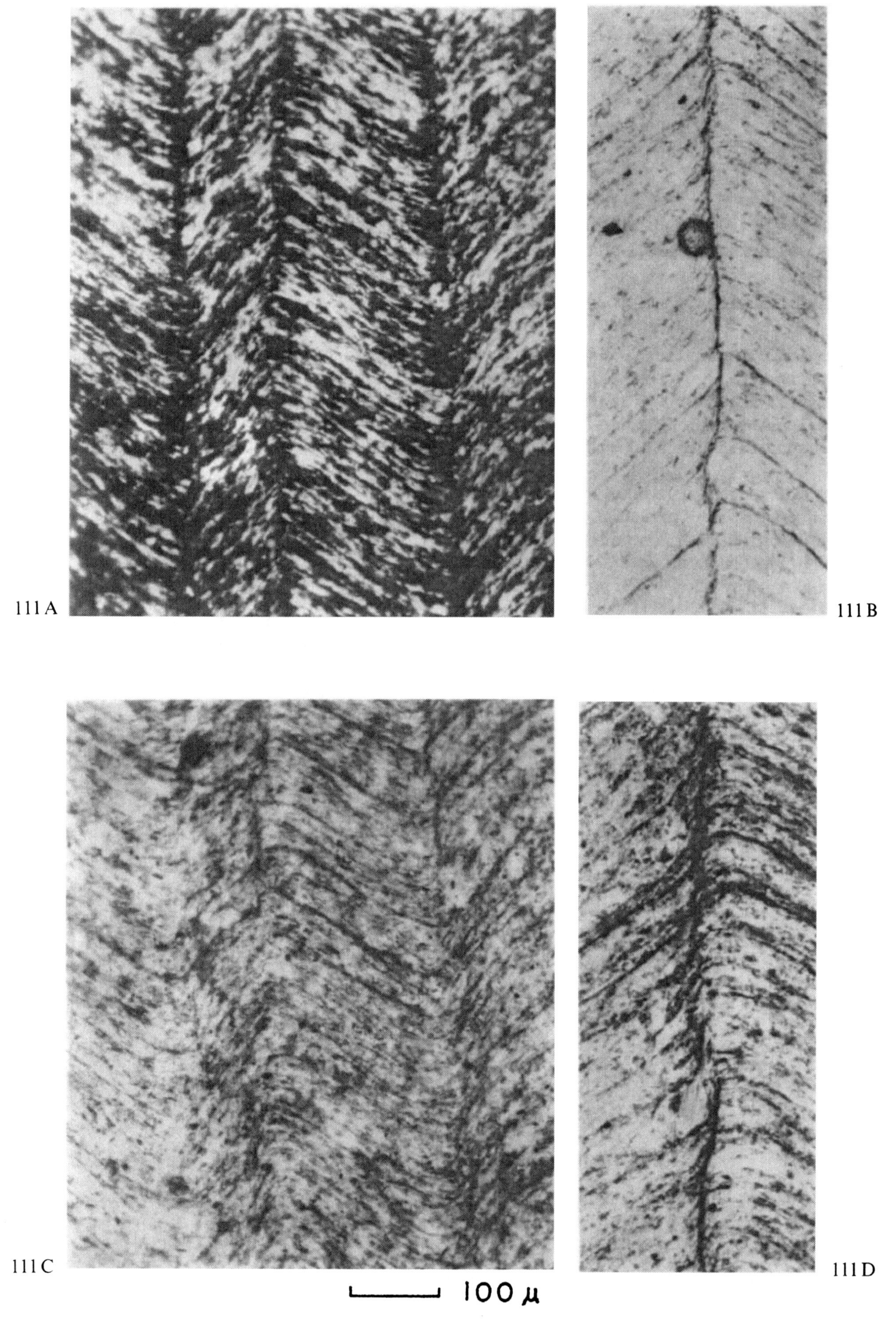
111A
111B
111C
111D
100 μ

# 112. Poly-deformed Slate with Buckled Quartz Veins

D. R. Gray

Silurian Chesleigh Formation, Sofala, N.S.W., Australia. (Specimen by courtesy of M. J. Hordern).

The slate is from a poly-deformed zone within 2 km immediately west of the Wiagdon Fault Zone (Powell et al. 1976). Four crenulation cleavages with associated folds post-date the regional slaty cleavage. A muscovite-albite-quartz-opaques assemblage indicates lower greenschist facies metamorphism.

---

**A. Mesoscopic structural relationships.** Three S surfaces are present:
$S_0$: bedding (*oblique*) defined by thin (up to 1 cm) siltstone laminae within the pelite. Load structures and truncated small-scale cross lamination indicate facing.
$S_1$: slaty cleavage (*oblique*) defined by a dimensional preferred orientation of muscovite, minor chlorite, elongate quartz and finely disseminated opaques.
$S_4$: discrete crenulation cleavage (*horizontal*) defined by thin dark wispy traces which are approximately axial planar to microfolds in the buckled quartz veins. PPL

**B. Buckled quartz vein** (segment enlarged from **A**). The vein shows thinning and truncation along the discrete crenulation cleavages ($S_4$). PPL

**C. Thinned and truncated vein segment** (enlargement from **B**). Note the irregular thinning and the sutured boundaries of thinned vein segments adjacent discrete cleavages ($S_4$). The origin of these features is discussed in captions to Plate 113. PPL

**D. Microstructure of discrete cleavage** ($S_4$). The cleavage (horizontal) consists of a zone of densely packed muscovite flakes which show strong parallelism to the cleavage boundary (see also Gray 1979, Plate 2).
Opaques and clots of carbonaceous material are interspersed with the mica. Microprobe analyses of these cleavages are given by Gray (1977b, Table 4). S.E.M.

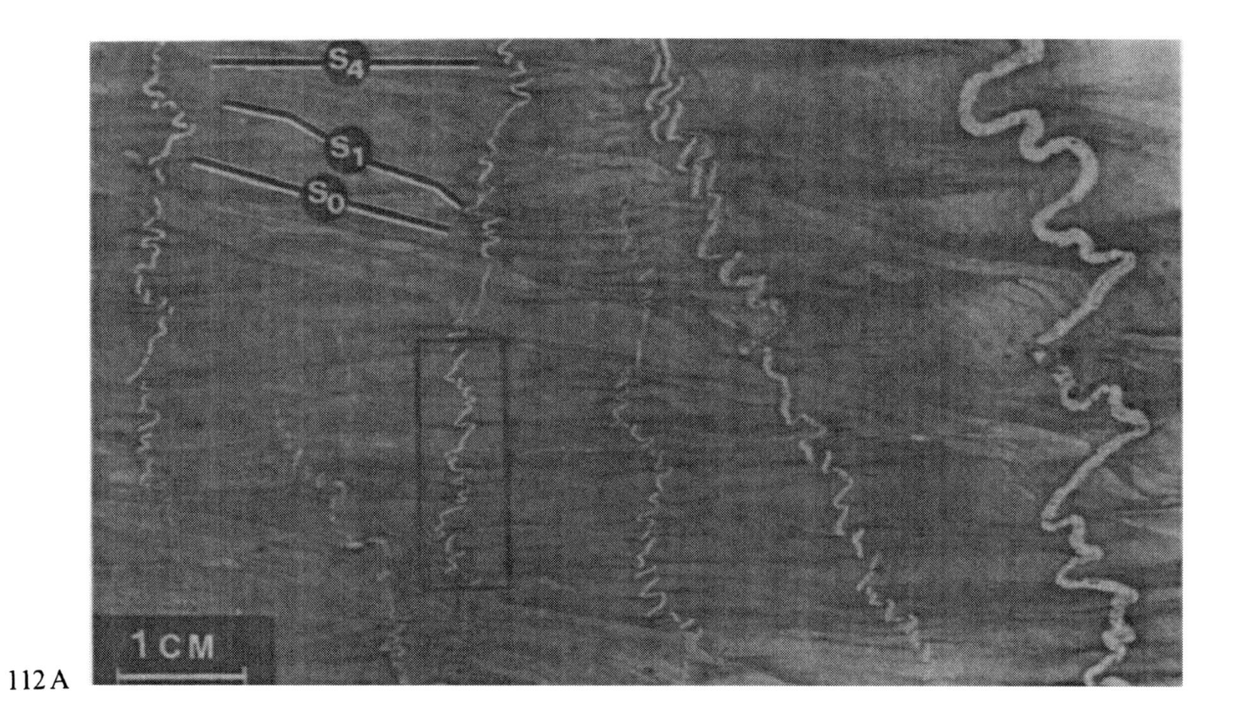

112 A

112 B

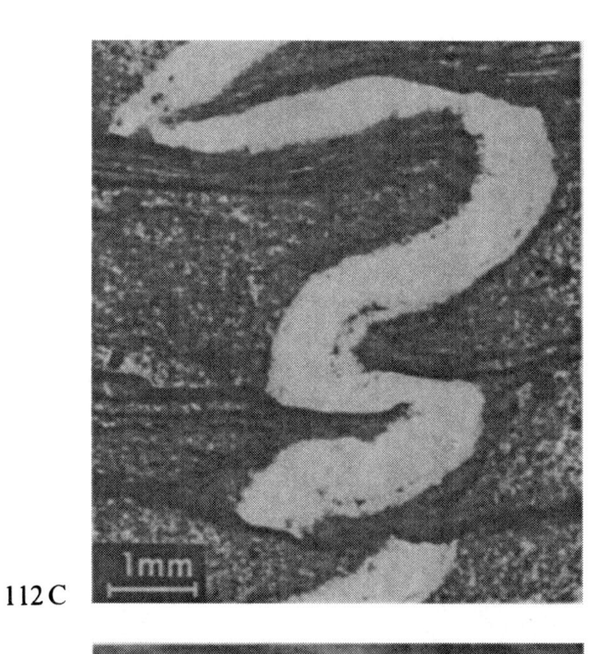

112 C

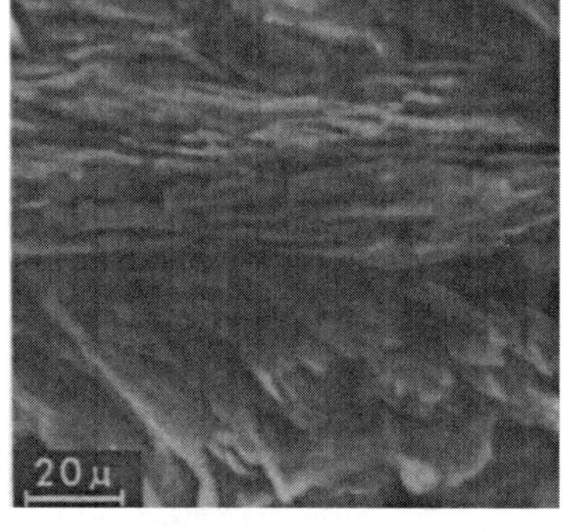

112 D

# 113. Dissolution Features: Discrete Cleavages

D. R. GRAY

These photographs illustrate dissolution features associated with the discrete cleavages in the poly-deformed slate of the previous plate. Further discussion is given in GRAY (1979, p. 113–116).

**A. Partially dissolved buckled vein.** Thinning and truncation of the vein are due to partial dissolution of limbs bounded by discrete cleavages. Reconstruction (see GRAY 1979) shows that this portion has undergone a total shortening ($\varepsilon_T$) of 60%, due to a *buckle shortening* ($\varepsilon_B$) of 52% accompanied by a *solution shortening* ($\varepsilon_S$) of 16.7%. PPL

**B. Interacting microbuckle-fold limbs.** Where the limbs are in contact the veins display irregular thinning and sutured interlocking boundaries. The geometry of the buckles is strictly "parallel" even up to their points of truncation. This feature is incompatible with stretching due to plastic deformation. These veins have undergone an approximate *solution shortening* ($\varepsilon_S$) of 22%. PPL

**C. Dissolved fold hinge.** The sutured, irregular boundary of the truncated vein shows no evidence of brittle deformation. Furthermore the truncation of the muscovite inclusion trail in the vein, and the lack of changes in grain shape and subgrain structure along the thinned vein-segment, preclude stretching due to plastic deformation. Note the accumulation of opaques and iron oxides along the adjacent cleavage. These features together suggest cleavage development is due to a dissolution mechanism. PPL

The diagrammatic sketch below is based on observations of the sample pictured above. It illustrates the progressive development of apparent offsets in a buckled vein due to pressure solution along discrete crenulation cleavages

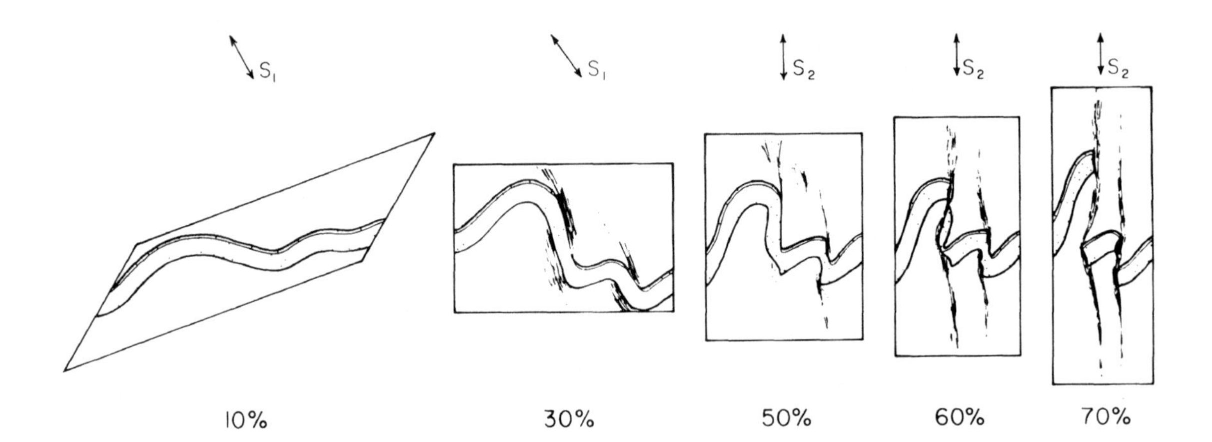


A and the sketch are reproduced from GRAY (1979) with permission of the American Journal of Science

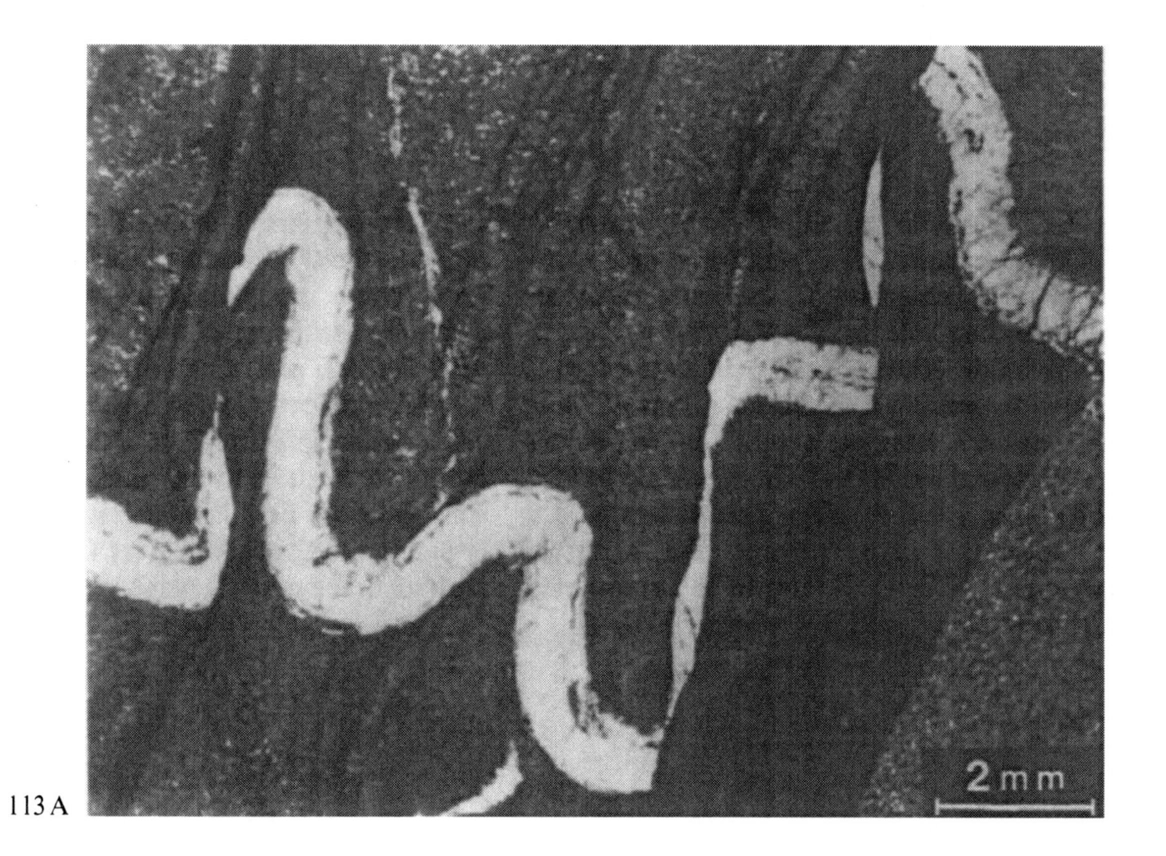

113 A

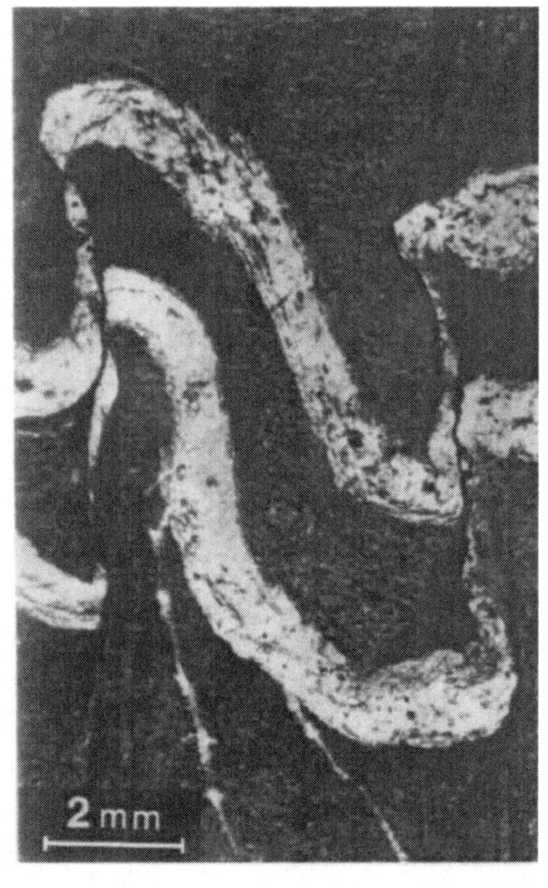

113 B

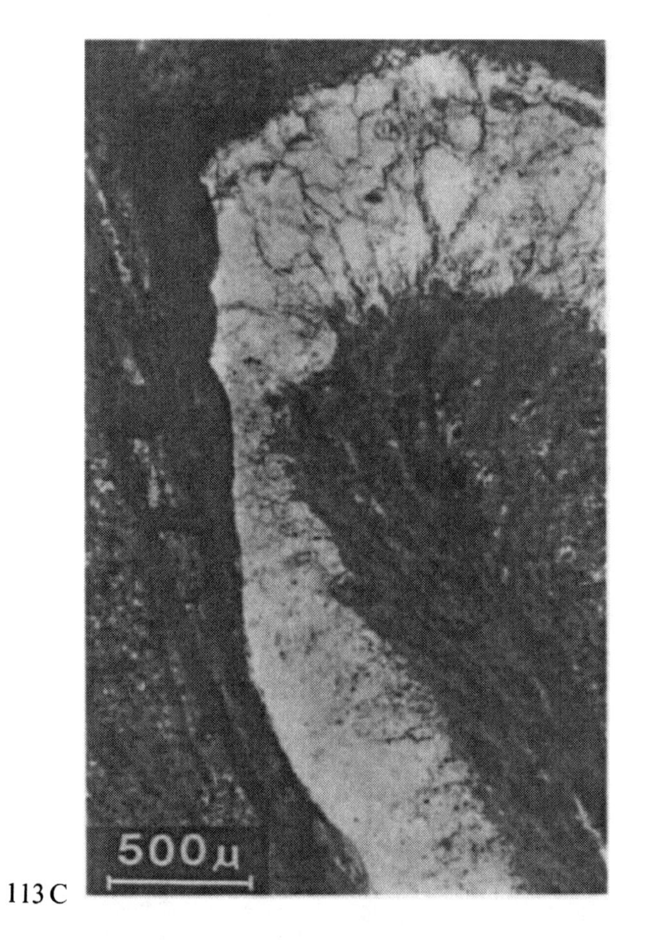

113 C

## 114. Domainal Slaty Cleavage

P. F. Williams

Folded Paleozoic turbidites from Bermagui, N.S.W., Australia. This pelite has been subjected to greenschist facies metamorphism below biotite grade.

The foliation (*from bottom right to top left*) is defined by weakly anastomosing films of white mica, chlorite, heavy minerals, and iron oxides that enclose domains containing mainly quartz, white mica, and chlorite. This cleavage is a first-generation structure and the preferred orientation of layer silicates in the quartz-rich domains is parallel to bedding and is an inherited, though modified sedimentary structure. There are numerous "chlorite stacks" generally consisting of a white mica core (possibly detrital) plus a chlorite overgrowth. Note that layer silicates in the cleavage films have a very strong preferred orientation and commonly are inclined to the length of the film. Locally there are bent layer silicates that are partly parallel to the sedimentary fabric and partly parallel to the slaty cleavage. NX

114

## 115. Spaced Cleavage in Silty Shale

R. P. NICKELSEN

**A. Detrital phyllosilicates and quartz oriented in cleavage.** Pennsylvanian Llewellyn Formation, Anthracite Coal Region, Bear Valley Strip Mine, near Shamokin, Pennsylvania, U.S.A. (NICKELSEN 1979). Anastomosing cleavage planes ($S_1$) are paralleled by phyllosilicates (*p*). The long dimension of quartz grains is either parallel to bedding as at (*b*) or parallel to cleavage as at (*c*). Orientation of quartz and phyllosilicates parallel to $S_1$ is thought to be due to pressure solution. Photographs below are from same formation and locality. NX

**B. Phyllosilicates in different domains oriented parallel to either bedding or cleavage.** Spaced cleavage ($S_1$) is expressed as clay-carbon partings (NICKELSEN 1972), that contain parallel phyllosilicates at (*c*). Detrital phyllosilicates in domains between clay-carbon partings are parallel to bedding as at (*b*). Clay-carbon partings owe their origin to pressure solution and residual concentration of insoluble matter. During this process, phyllosilicates became parallel to $S_1$ by loss of support and rotation. NX

**C. Local cleavage variation with rock type.** Clay-carbon partings in *upper left*, phyllosilicate-rich part, converge into more widely-spaced, thicker partings in the quartz-rich lamina. Bedding-parallel phyllosilicates between partings at (*t*) are slightly crenulated. All specimens on this plate were collected east of 2.5 CAI (Conodont Alteration Index), indicating a temperature range 110°–140 °C (EPSTEIN et al. 1976). Percentage of fixed carbon in associated anthracite coal is 92% (DAMBERGER 1974), suggesting a temperature of approximately 160 °C. NX

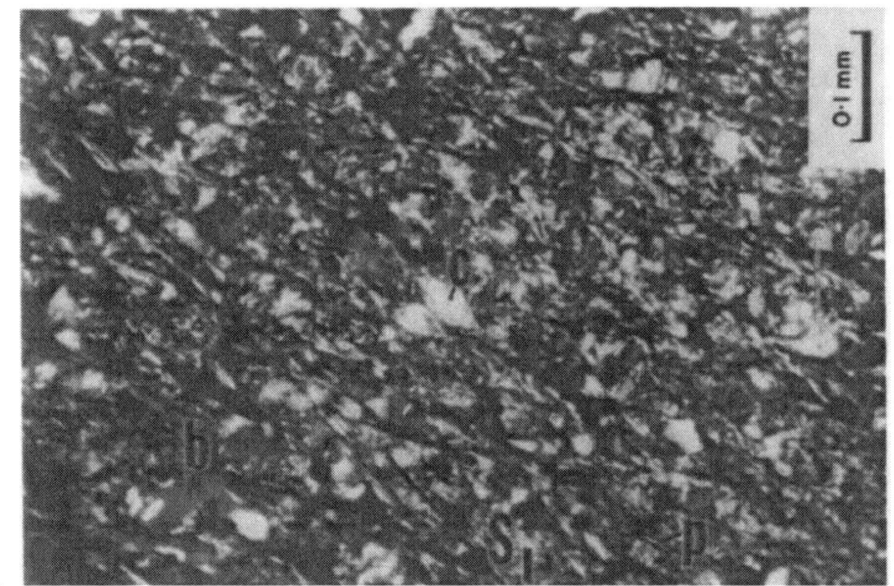

115 A

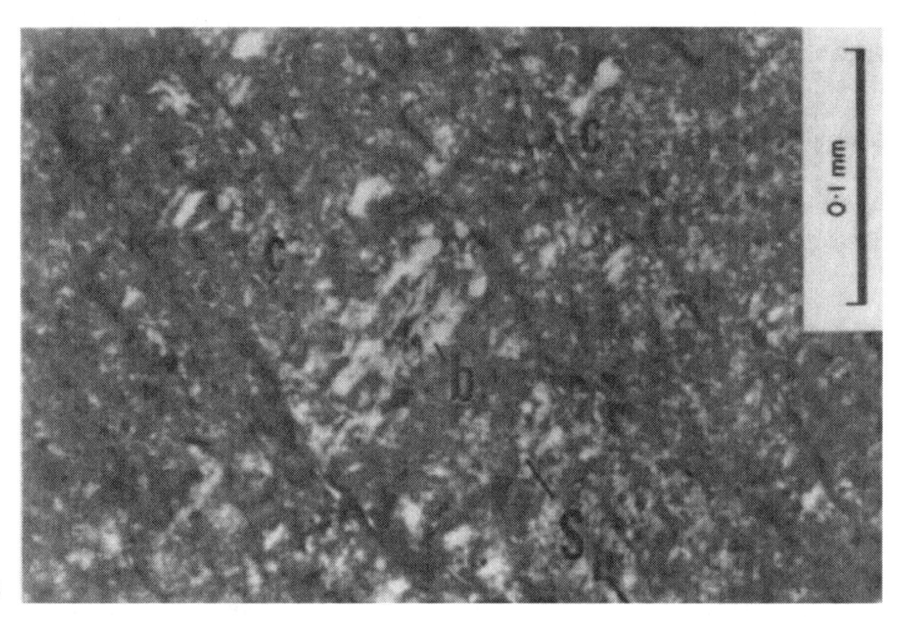

115 B

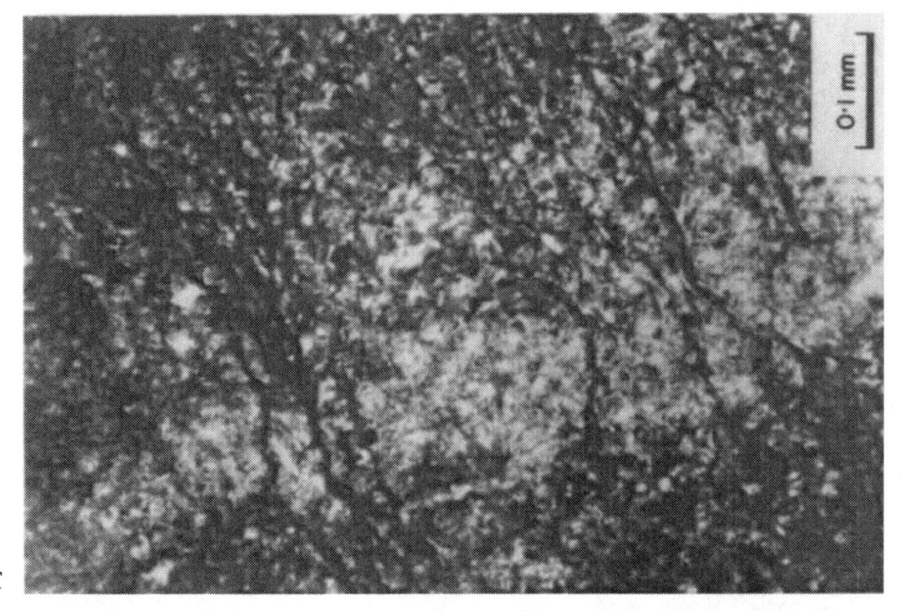

115 C

## 116. Spaced Cleavage from Kootenay Arc Rocks, Washington, U.S.A. (I)

A. J. WATKINSON

The examples are of axial-plane cleavages from multiply folded Cambrian rocks in the Kootenay Arc, NE Washington, U.S.A. (See also Plate 119.) The stratigraphic sequence is Lower Cambrian quartzite (Gypsy) followed by Maitlen Phyllite and Middle Cambrian limestone (Metaline). Metamorphism reached a peak in the greenschist facies during the development of the first folds' cleavage, $S_1$. The first folds are tight, overturned to recumbent and plunge gently to the southwest or northeast. The folds have a pervasive axial plane cleavage, $S_1$, which has a different expression as it crosses the different lithologies. In the Gypsy quartzite and in the quartzitic phyllite units in the Maitlen, the cleavage has a field appearance of a discrete, spaced cleavage.

---

**A. Spaced cleavage in a quartzite bed.** Maitlen phyllite. In this instance, the bedding interface, $S_0$, between the quartzite and the phyllite remains planar. The spacing between the cleavage surfaces is generally about 5 mm, but is not strikingly regular.

**B. Cleavage associated with a cusp interface.** Gypsy quartzite. This negative print emphasizes the discrete cleavage associated with cusp interfaces between phyllite (*light*) and quartzite (*dark*). The light cleavage traces emphasize the concentration of micaceous material along the cleavage surface. The spacing is related to the cusping at the quartzite/phyllite interface. Notice that the spaced cleavage dies out away from the cusps.

**C. Disrupted phyllite/quartzite interface.** Gypsy quartzite. The origin of the flame-like interface is not clear, but may be due to strongly flattened cusps accentuated by pressure solution

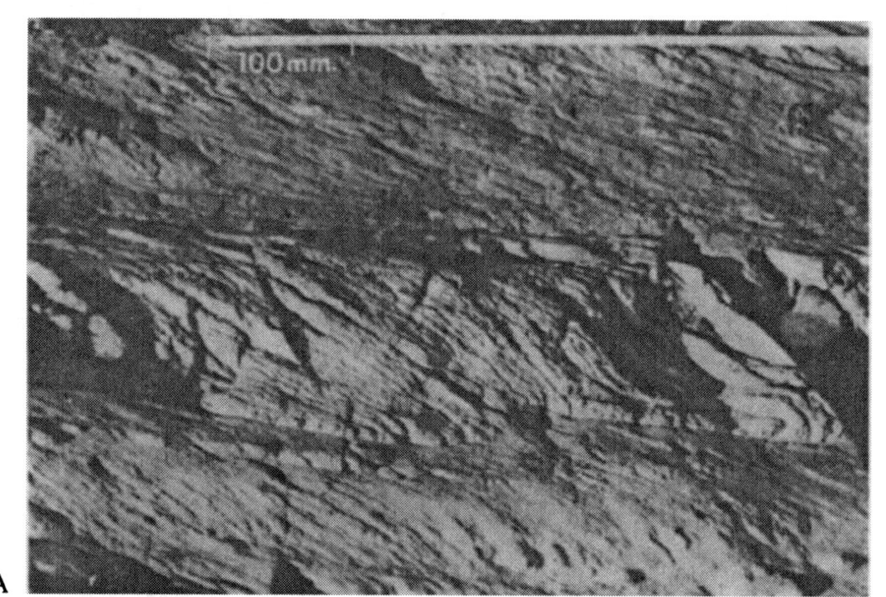

116 A

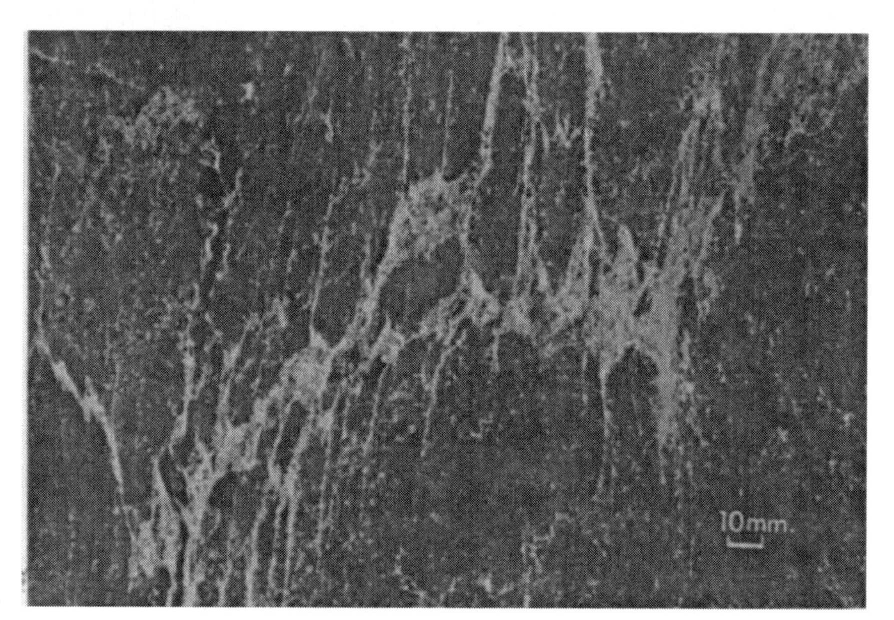

116 B

116 C

# 117. Anastomosing, Discrete Seams: Solution Cleavage?

G. J. Borradaile

Devonian silty shales between Kautenbach and Fronhofen near Traben-Trarbach, Mosel Valley, West Germany.

The photograph shows the typical character of the first cleavage in silty layers of the Mosel Slate Belt, where the second cleavage of this region is absent.

Discrete cleavage surfaces anastomose as they are deflected around grains or groups of grains which appear less affected by cleavage. Each cleavage surface is rich in micaceous or opaque minerals and is depleted in quartz. At higher magnification cleavage surfaces may be likened to stylolites of subdued amplitude. The cleavage surfaces terminate diffusely.

A common feature of rocks with supposed solution cleavage is the production of "dip-slip", step-like displacements of pre-cleavage layers (Plessmann 1964). This is illustrated in (**a**) and (**b**) below. If this displacement is due to solution removal, it is possible to determine the amount of solution removal ($\varkappa$, in arbitrary units) from the angle between the displaced layers ($\Phi$) and the displacement ($d$, in arbitrary units) using the graph.

Where only one set of layers shows such a displacement the possibility of shear parallel to cleavage may not be excluded. In this example, however, the sedimentary layers have two different orientations because the silts are cross-bedded. Both bedding and cross-beds are displaced across the cleavage surfaces so that solution appears to be a likely component in the formation of this cleavage.

Using the graph, the amount of supposed solution removal amounts to between 10% and 15% of the present width of the specimen.

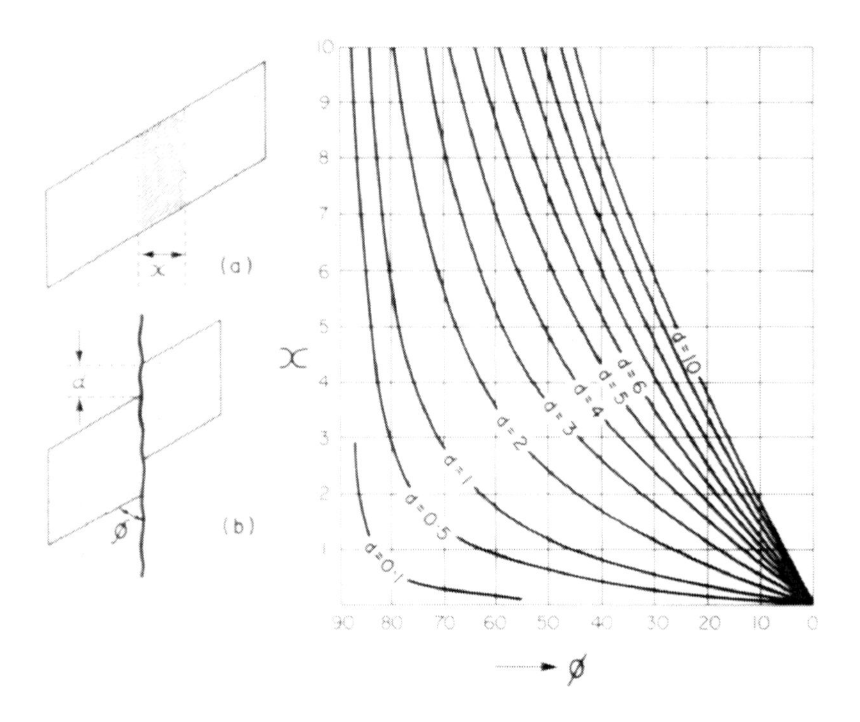

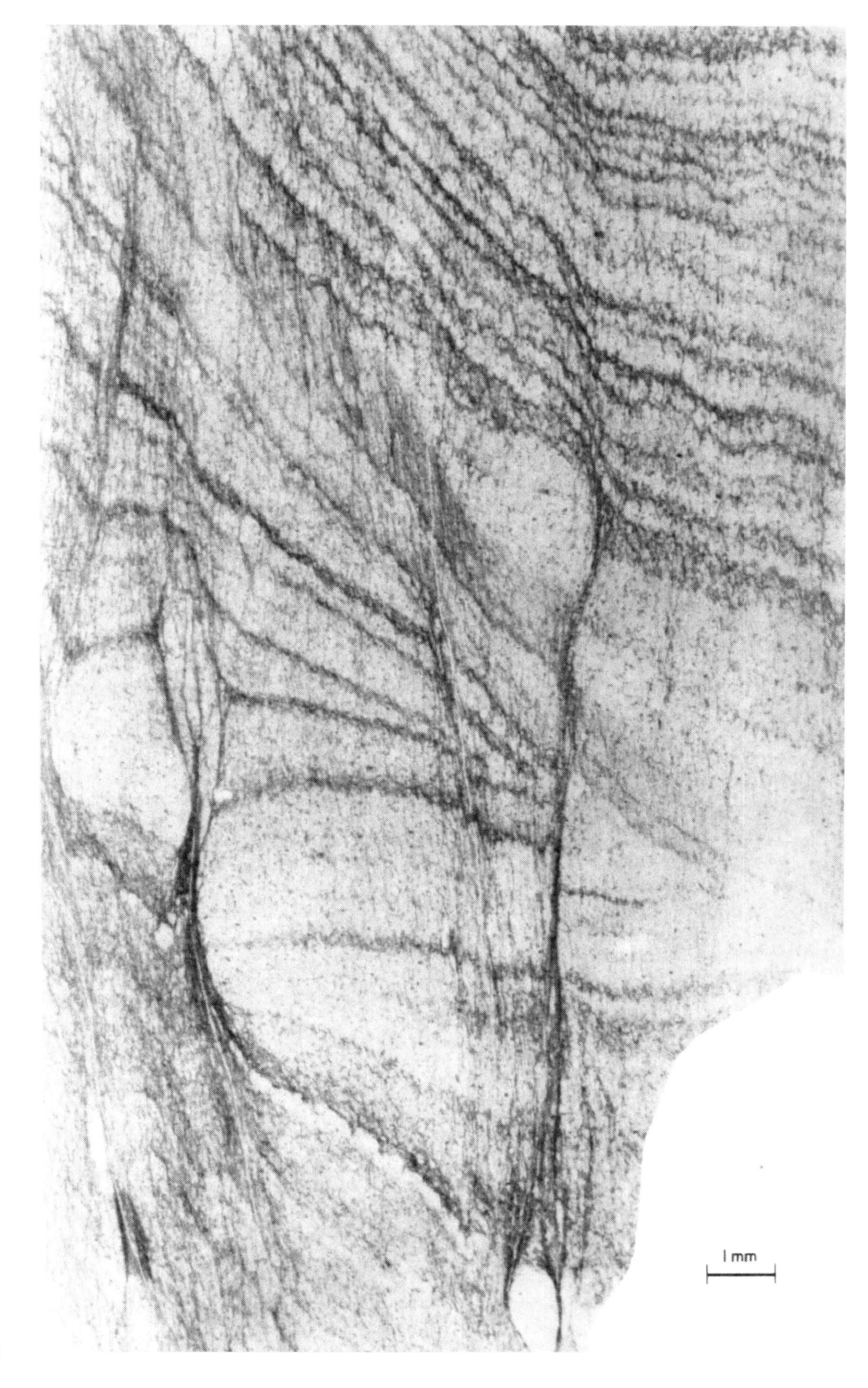


117

## 118. Spaced Solution Cleavage in Argillaceous Limestones

D. A. C. Gardner

Limestone laminae interbedded with calcareous slates of the Upper Cambrian Upper Chancellor Formation near Golden, British Columbia in the Western Main Ranges of the Southern Canadian Rocky Mountains, commonly exhibit a spaced cleavage which is formed by thin zones, up to 40 microns wide and spaced about 1 cm apart, composed of strongly aligned white mica, opaque minerals, insoluble residues and remnant carbonate grains, all of which suggest a pressure solution origin. Sedimentary laminae are commonly offset across cleavage zones where cleavage cuts obliquely across bedding, suggesting net shortening of layering has taken place through pressure solution within cleavage zones. Penetrative slaty cleavage in adjacent slates is illustrated in Plate 105.

---

**A. Spaced solution cleavage.** The scattered cleavage planes (*A*) which cut the thin limestone laminae are funnelled together into two larger cleavage zones (*B*, *C*) in the thicker limestone lamination. Note that zone *B* continues through this lamination while zone *C* disappears. PPL

**B. Enlargement of cleavage zone *B*, above.** This enlargement illustrates the texture of the cleavage zone, in particular the strong alignment of mica, and the progressive change in grain size and shape, from the cleavage zone to the adjacent limestone. PPL

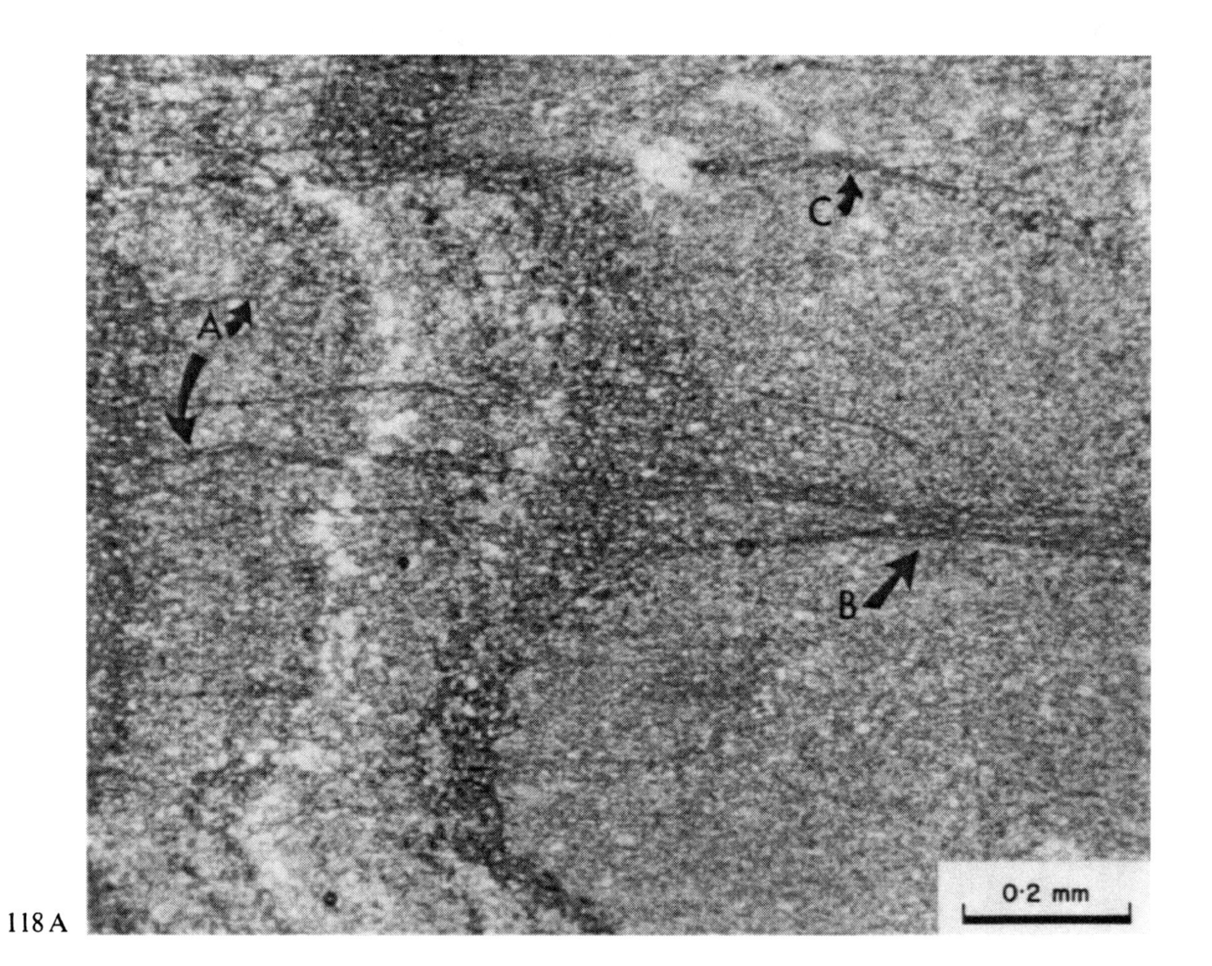

118 A

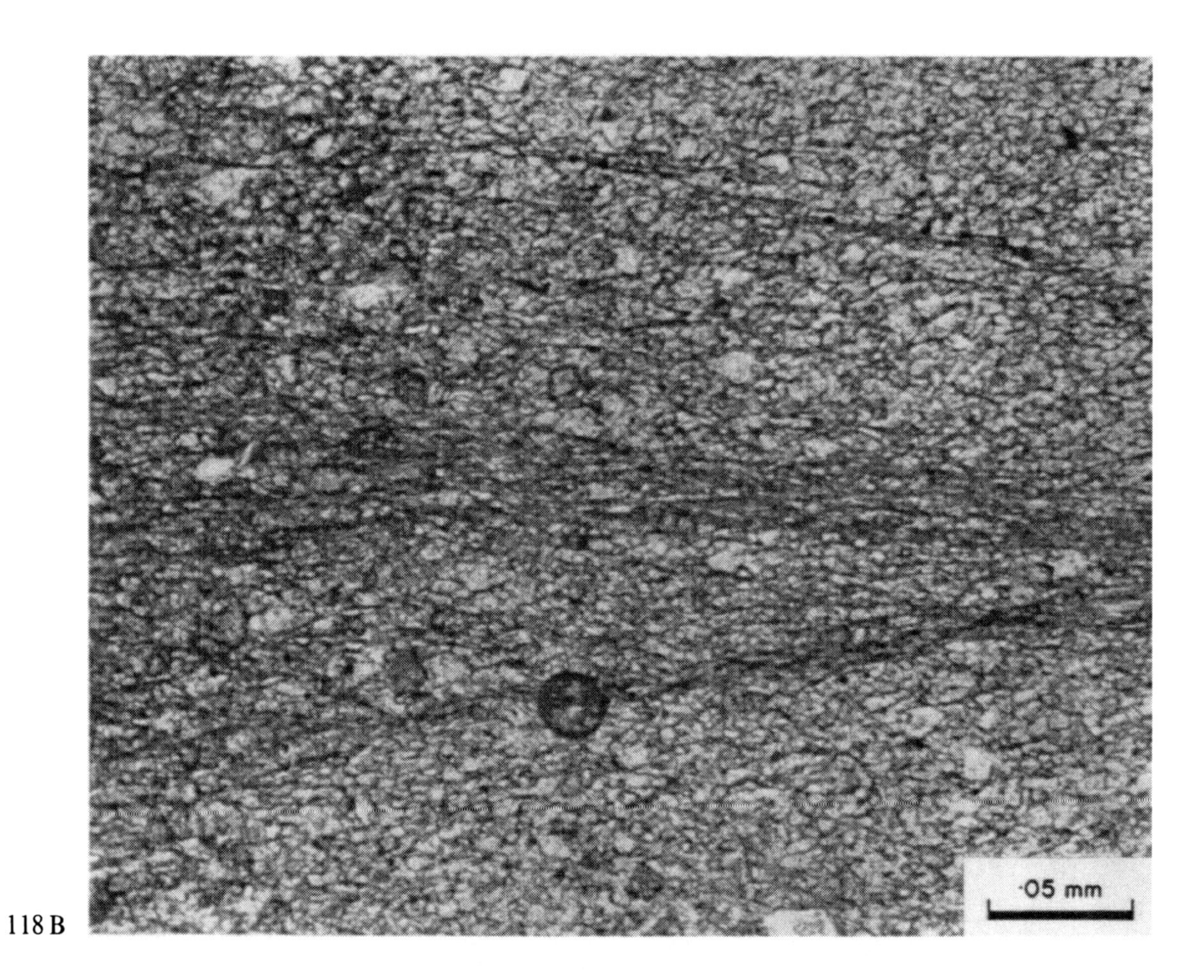

118 B

# 119. Spaced Cleavage from Kootenay Arc Rocks, Washington U.S.A. (II)

A. J. WATKINSON

Also see Plate 116

**A. Axial-plane cleavage in quartzitic and phyllitic units.** Gypsy quartzite. In this instance the bedding interface, $S_0$, between the phyllite and the quartzite is not disrupted by the cleavage, $S_1$. The cleavage just refracts slightly across the boundary. In the phyllite, the cleavage is penetrative and closely spaced. In the quartzite the expression of the cleavage changes to that of microlithons with an internal fabric, bounded by pelitic intercalations of strongly oriented chlorite and muscovite. The cleavage is spaced at about 5 mm to 1 cm, with no obvious regularity. Scale: marked at 100 mm.

**B. Discrete nature of the cleavage.** The cleavage traces have a greater concentration of very fine-grained chlorite and muscovite than the quartzitic microlithons. Negative print

**C. Quartzite microlithon domain.** From the quartzite in **A**. The fabric within the microlithons between the spaced cleavage consists of mica folia, elongate quartz grains, and opaque augen. The folia and grain elongation are generally parallel to the thicker intercalations that constitute the spaced cleavage. PPL

**D. Photomicrograph of the cleavage separating the microlithons.** Opaques and zircons are concentrated near and along the cleavage plane. The quartzite matrix framework appears to have collapsed in the vicinity of the cleavage. The texture appears compatible with that predicted for the preferential removal of quartz by pressure solution along a discrete zone leaving residue opaques and other "insolubles". PPL

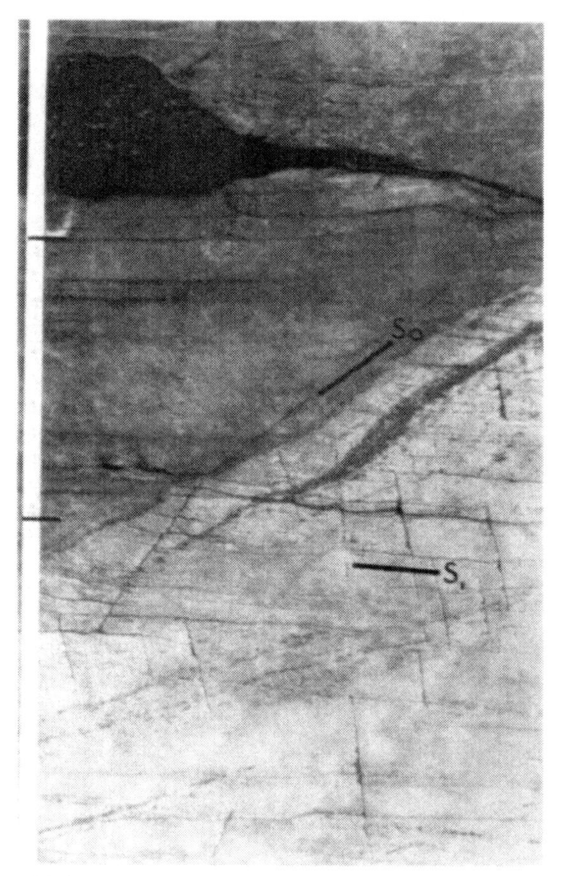

119 A

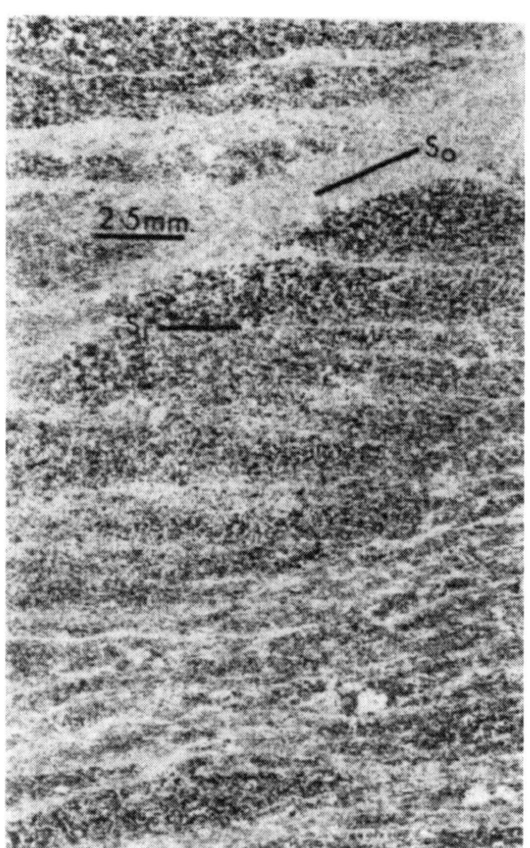

119 B

119 C

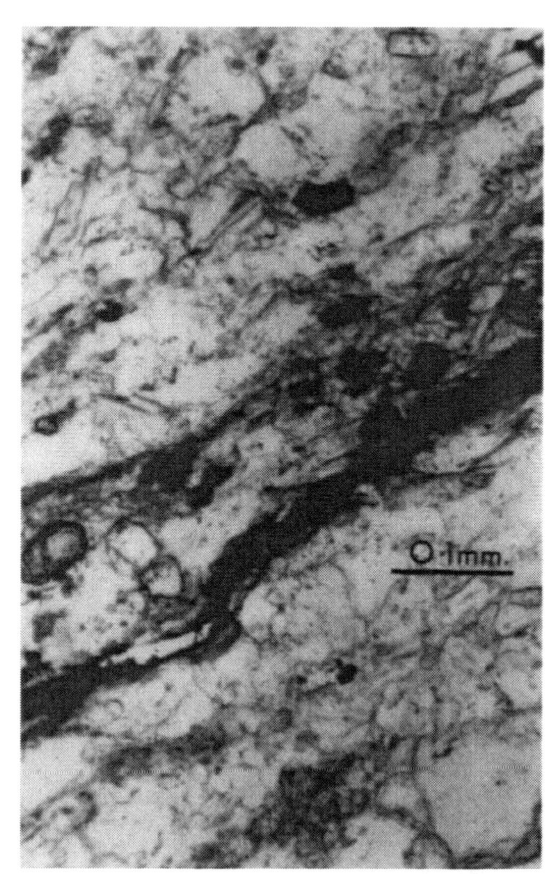

119 D

## 120. Schistosity Passing from Graywacke into Slate

J. G. DENNIS

Eocambrian Pinnacle Formation near Franklin, Vermont, U.S.A. Schistosity passes from coarse granular texture to slaty texture. In the slate layer, the schistosity may be described as "slaty cleavage", in the graywacke as "rough cleavage", after the German "Rauhschieferung" (BORN 1929).

---

**A. Outcrop.** Slaty layer near bottom of picture shows gentle dip to right (east). Scale: 25 c coin

**B. Thin section.** *Black* and *dark gray* grains are quartz and feldspar, *white* grains are opaque minerals. In the slaty lithology, schistosity is defined by sericite-chlorite fabric. This fabric penetrates the coarser lithology, where schistosity is defined by dimensional orientation of both micaceous and granular minerals near the transition, but predominantly by micaceous strands surrounding granular minerals away from the contact. However, even here weak dimensional orientation of granular minerals is evident. $S_1$ in both lithologies is continuous cleavage (CHIDESTER 1962, DENNIS 1967). Negative print

120 A

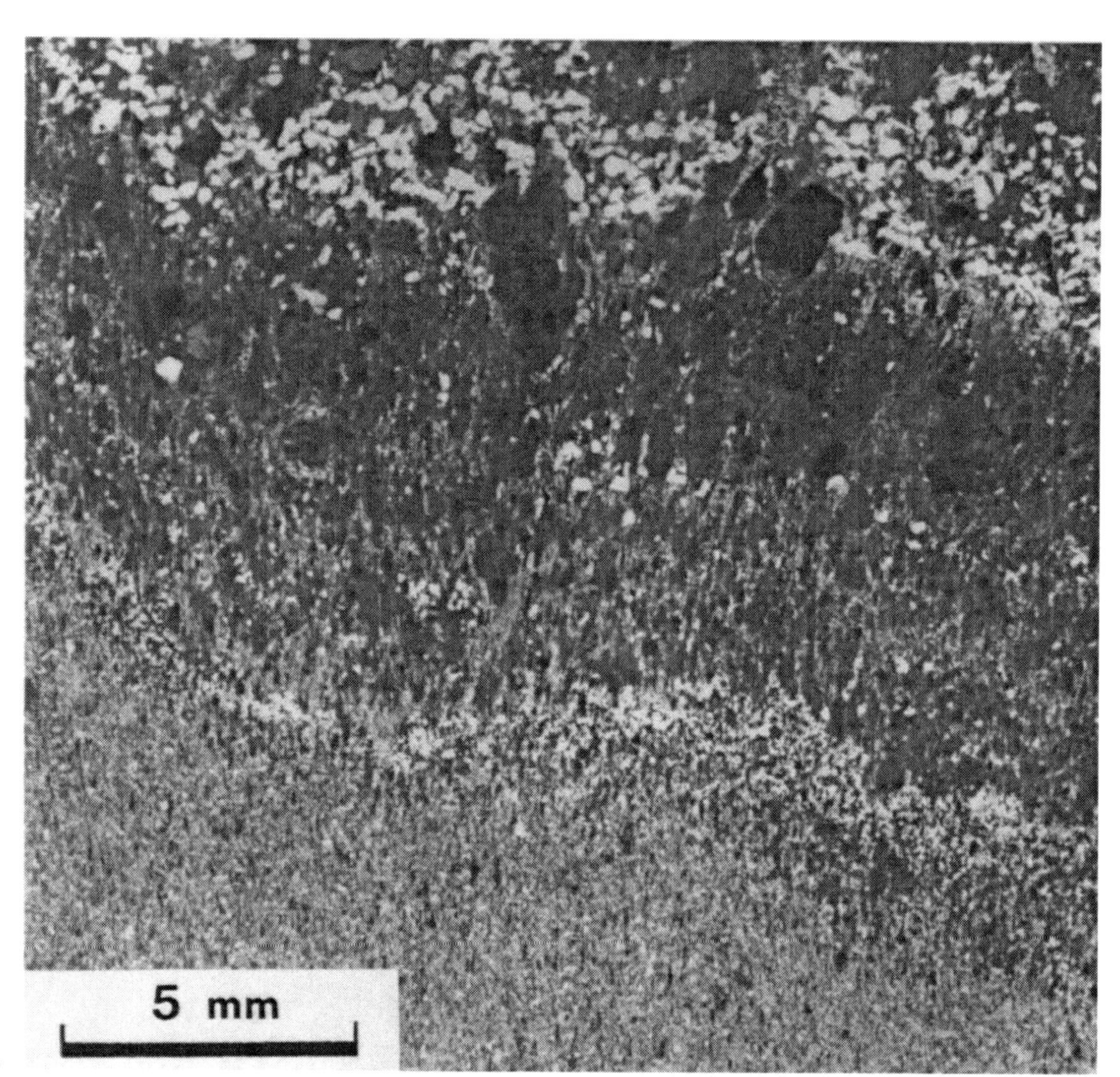

120 B

## 121. Dissolution Around Resistant Units

B. M. Bayly

Whitehall Slate, Ordovician Indian River formation from quarry north of Truthville, near Granville, N.Y., U.S.A.

The resistant grains are now chert, possibly replacing former euhedral crystals, e.g., of sulfate. Where a resistant grain has a high concentration of flakes at its margin, dissolution is suspected with formation of a residue. Such films with abnormal abundance of phyllosilicates tend to link up and contribute to the cleavage of the rock. They also permit slip: the flat surfaces on the chert grains probably result from a combination of dissolution and slip, but it is impossible to say which effect was the more important; compare with Plates 122 and 123.

In this instance, localized dissolution at the chert grains is believed not to be the main source of the rock's cleavage. Even away from the chert grains, there is a reasonably good preferred orientation of phyllosilicate flakes, shown in Plate 9C. Surfaces along which the rock has parted are parallel to the foliation but do not seem to run preferentially through the concentrations of flakes at chert-grain margins.

NX

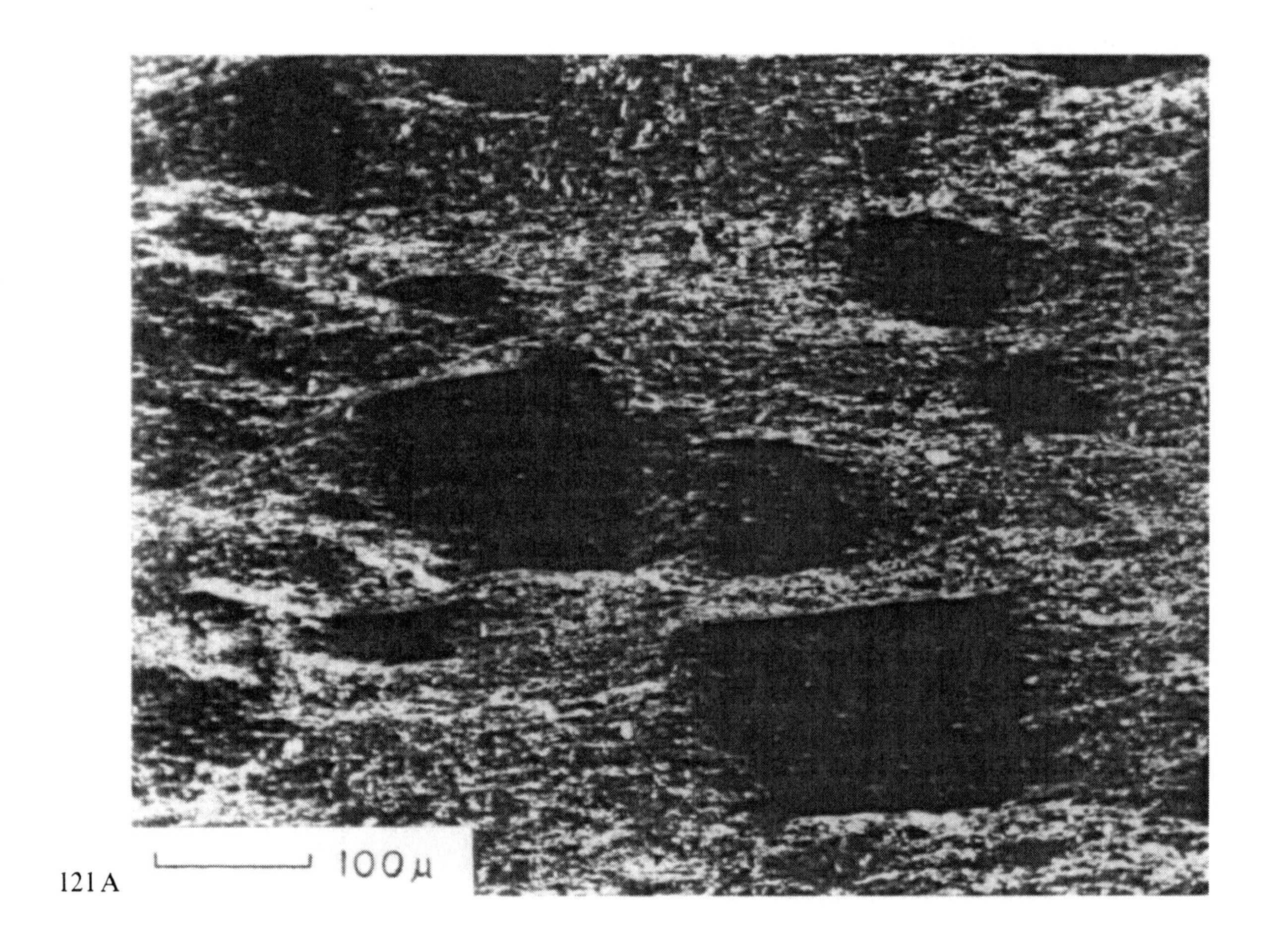


121 A

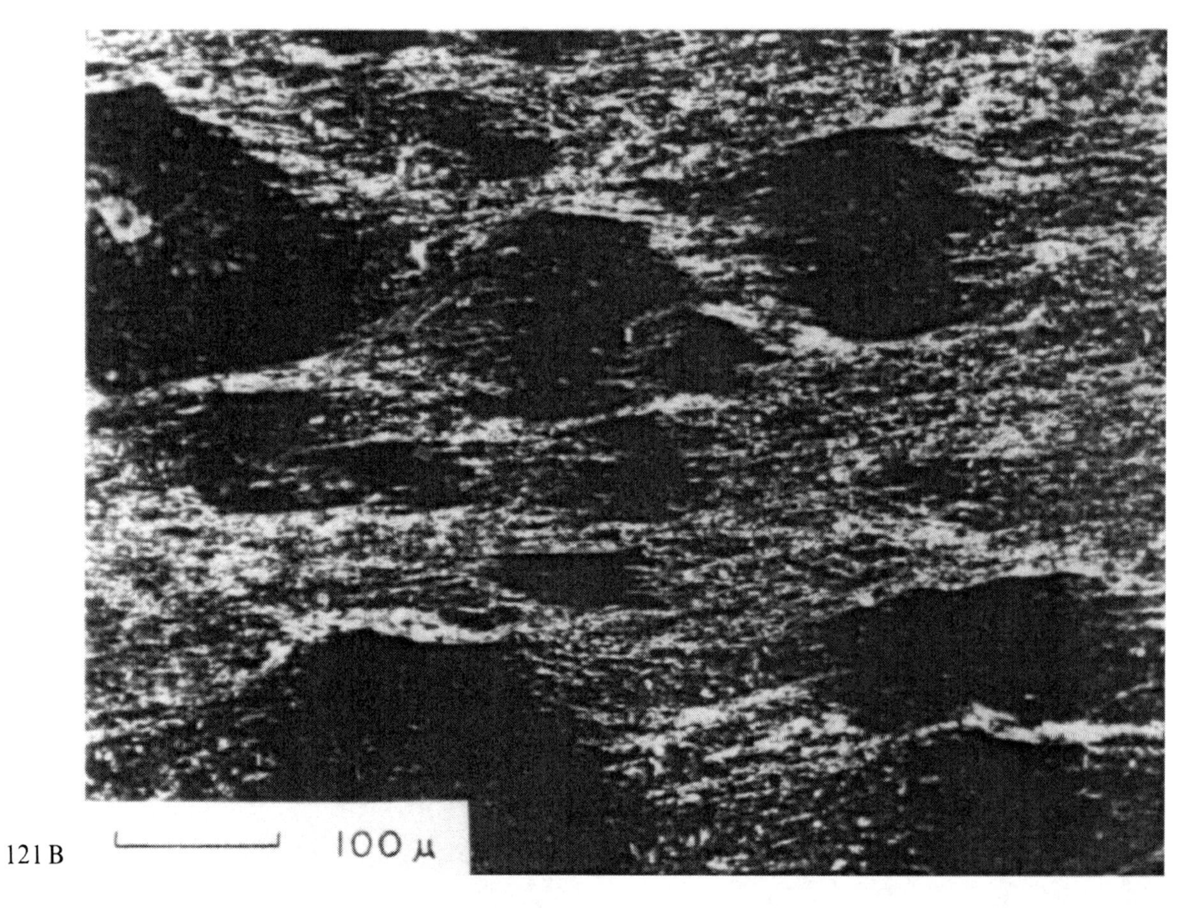


121 B

## 122. Flat-sided Garnet Crystals

B. M. BAYLY

Lower member of the Cobble Mountain formation at Cobble Mountain Reservoir, Blandford, Mass., U.S.A. (STANLEY 1975).

The quartz-muscovite schist contains many normal garnet crystals, but it has more instances than can be thought accidental of crystals with flat sides, as shown. The possibility that they grew in that shape is rejected and we look in particular at two possibilities, *shearing* and *dissolution*. The flat surfaces are in contact with sheaves or layers of mica concentration, but that fact fits well under either hypothesis. Comparison can be made with Plates 36, 121 and 123.

It is suspected that we see here a coarsened crenulation fabric. If continued dissolution of quartz were to convert a rock like that in Plate 42 to one like that in Plate 44, garnets such as those shown would occupy the full width of one microlithon; then it is possible that, with a fluid of sufficiently aggressive chemistry, garnet shapes could be modified in the same way that chert grains, ooids and phyllosilicates are truncated. Slip parallel to the mica-rich zones could also have played a part. Grain growth after the differentiation process would be postulated to account for the present irregular texture in the microlithons, where only faint suggestions of former crenulation can be found. One has to imagine a rock like that in Plate 44 being coarsened; the result could resemble the texture seen here. PPL

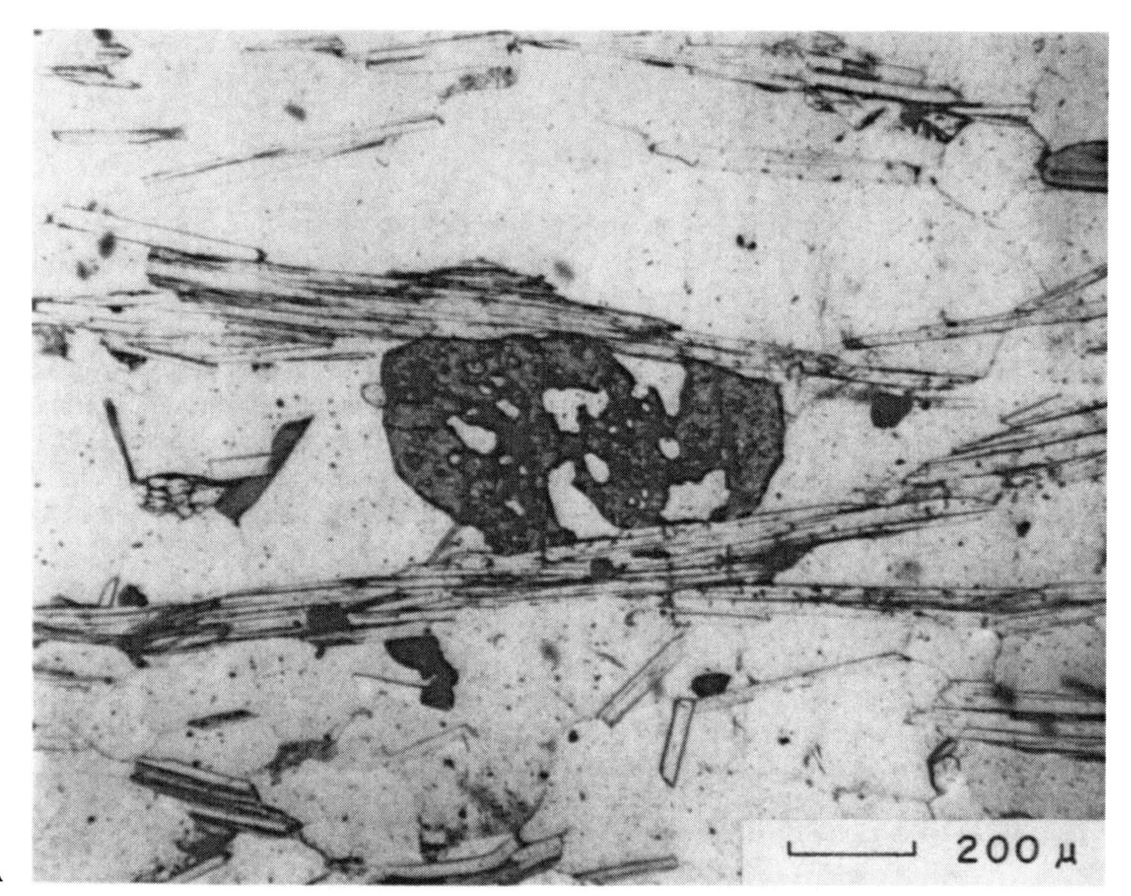


122 A

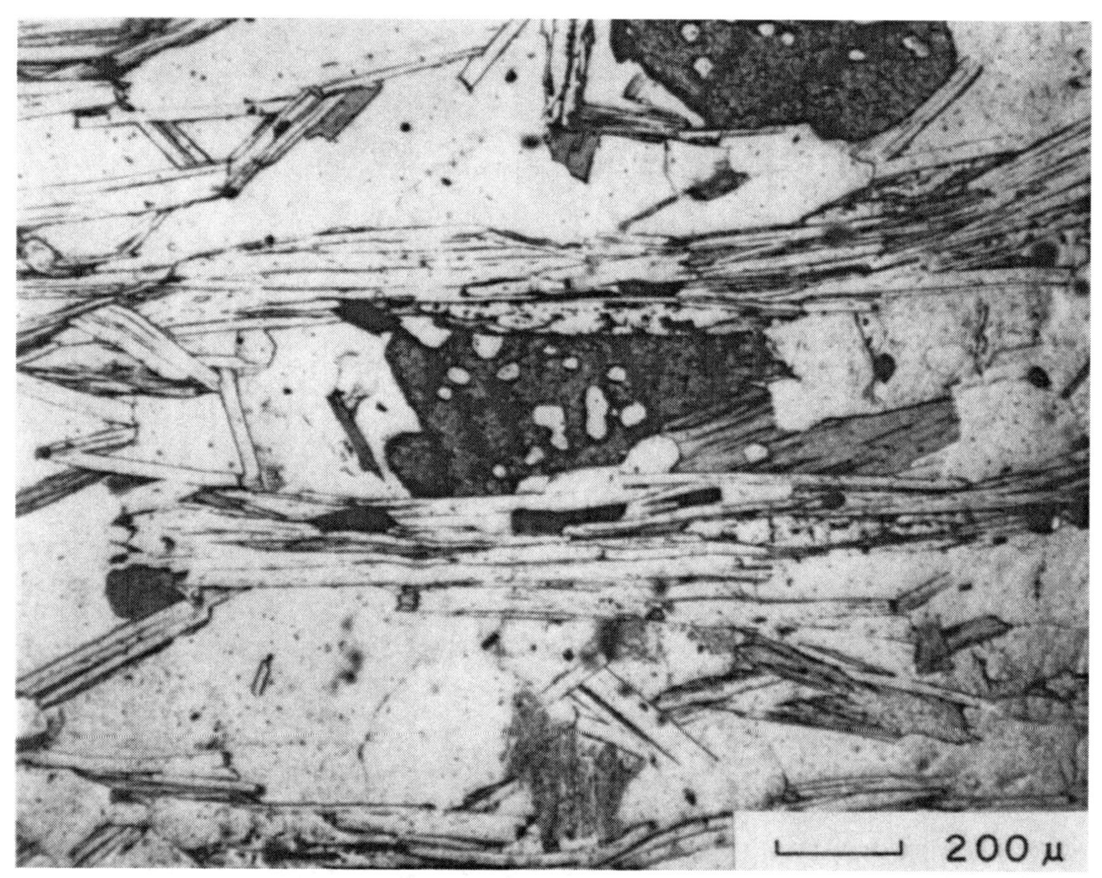


122 B

# 123. Schistosity-Parallel Shear Displacements in Medium-Grade Tectonites

W. J. Gregg

The specimen illustrated is from the Moretown member of the Mississquoi Formation in southeast Ludlow township, Vermont, U.S.A. The rocks are epidote-amphibolite grade tectonites containing distinct secondary layering with no traces of bedding preserved. The schistosity contains no features indicative of mylonitization or cataclastic deformation, and is indistinguishable in this regard from any other schists in the Ludlow area. However, the presence of large porphyroblasts of garnet and biotite allows the identification of shear displacements parallel to layering. Studies have shown that these porphyroblasts exhibited idioblastic postkinematic textures in the schistosity prior to the imposition of shearing (Gregg 1978). Shear displacements are conclusively demonstrated by reconstruction of grain segments to their initial shapes as shown in the sketch below.

**A.** Garnet porphyroblasts are sliced into tabular sections by shear displacements along the schistosity. (See sketch of reconstruction.) The surfaces of translation are marked by quartz-rich layers. Many small dark porphyroblasts of biotite are also visible in the schistosity. These grains have also suffered shear displacements, in this case along (001), in the same shear sense indicated by the garnet grains.

**B. and C.** Polished surface photographs of cross-biotites from the same specimen shown in **A**. The typical "slipped card deck" appearance of the biotite porphyroblasts is illustrated. Cross-biotites can be reconstructed to their original idioblastic shapes by translations along (001), whereby early internal fabric is brought into registration.

**D.** The upper surface of the specimen in **A** shows the idioblastic texture of the garnet slices in the plane of schistosity. A weak lineation defined by the edge-on views of the cross-biotites is also shown. Cross-biotites in these rocks are typically parallel to an earlier secondary layering and thus form the noted intersection lineation

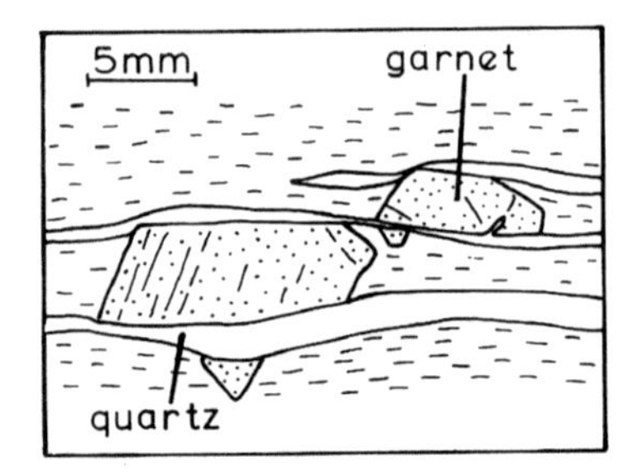


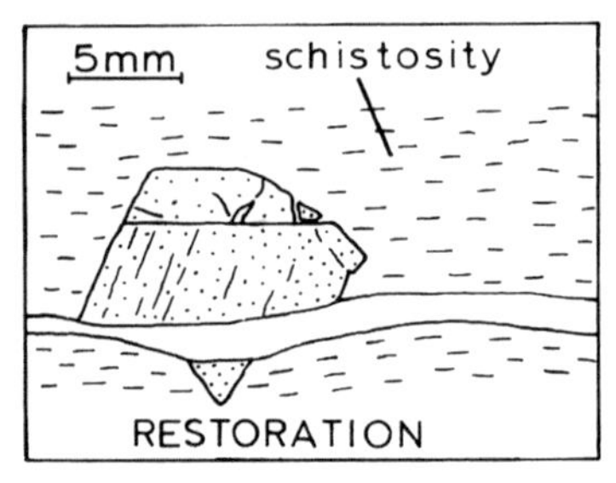

123 A

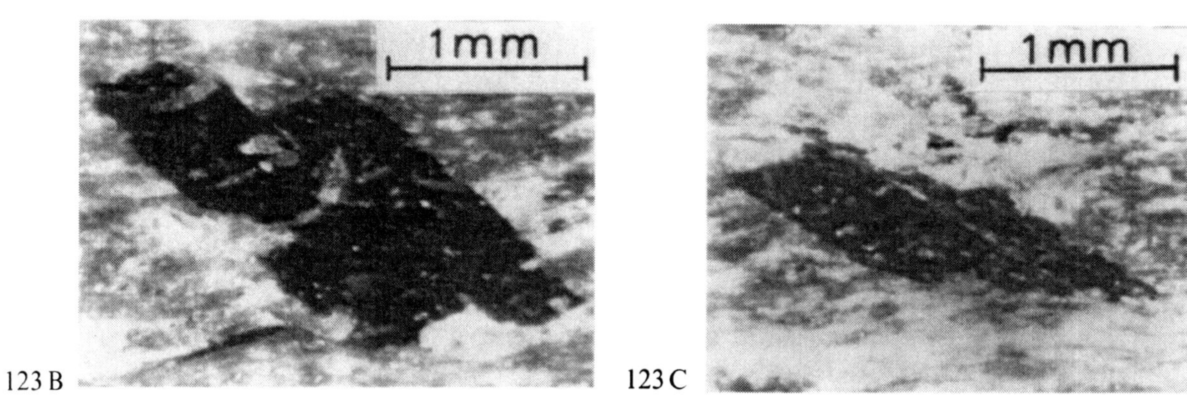

123 B
123 C

123 D

# 124. Overgrowths and Mica Beards on Rounded Quartz Grains Enclosed by Cleavage Folia

C. McA. POWELL

Early Proterozoic Siamo Slate, Northern Michigan, U.S.A. (POWELL 1969, 1970).
The upright slaty cleavage is approximately parallel to axial planes of gently plunging, close to tight, latitudinal folds that were subsequently overprinted by a regional thermal metamorphism (JAMES 1955).
The photographs show the microstructure of the pervasive slaty cleavage, this example being from the north side of highway US41 just west of Negaunee (Co-ords. 450520E, 5150780N on the 1000 m Universal Transverse Mercator Projection, Zone 16). In this fine sandy rock, the fabric is composed of three elements, viz:
1. Detrital quartz grains–commonly rounded,
2. Mica beards and quartz overgrowths–generally intergrown, and
3. Anastomosing, dark cleavage folia.
The rounded outlines of the detrital quartz grains (see also POWELL 1970, Plates 1 A, 3, 4, and 5) are marked by iron oxide dust rings, fluid inclusions and terminations of very small "sericite" crystals. The mica beards and quartz overgrowths (see also POWELL 1969, figs. 3 and 5) are closely intergrown, the beard mica being a fine, colorless "sericite". Mica flakes in these beards are strongly oriented parallel to the mean cleavage direction, and are intergrown with quartz that is optically continuous with the rounded quartz grains (POWELL 1970, Plates 3, 4, and 5). Individual dark cleavage folia, commonly 0.01 mm wide, can be followed along the cleavage trace for centimeters. Layer silicates in the cleavage folia are strongly pleochroic chlorite (POWELL 1969), and there are concentrations of iron oxide and other unidentified opaque minerals in the folia.
The cleavage folia were originally thought to have been formed as clay or mica films deposited by the flow of high-pressure muddy pore fluid through the rock (POWELL 1969, 1972a, b, 1973), but this interpretation has recently been rejected (GEISER 1975, BEUTNER 1980). No matter how these anastomosing cleavage folia were initiated, they now show abundant evidence of recrystallization (i.e., growth of new layer silicates and quartz), and removal of quartz by dissolution (POWELL 1970, Plates 2 F, 4 C, and 4 A to H).

---

**A. Cleavage folia enclosing rounded quartz grains.** The anastomosing cleavage folia produce an augen texture where they wrap around the rounded quartz grains, mica beards and quartz overgrowths. The size of the augen varies according to the size of the largest quartz grain enclosed. The newly recrystallized "sericite" flakes in the quartz-mica beard intergrowth do not penetrate the rounded (detrital) quartz grain boundary. PPL

**B. Rounded quartz grains partially truncated by cleavage folia.** The dark cleavage folia in this example are relatively planar, and several quartz grains abut on to them. The curvature of the large rounded grain in the middle of the photograph changes sharply at the contact with the cleavage folium at the base of the grain, suggesting that some of the grain has been removed by dissolution. Other planar quartz-cleavage folia boundaries can be seen in the photograph. Quartz dissolved at these boundaries may have been redeposited as quartz overgrowths intergrown with the mica beards. PPL

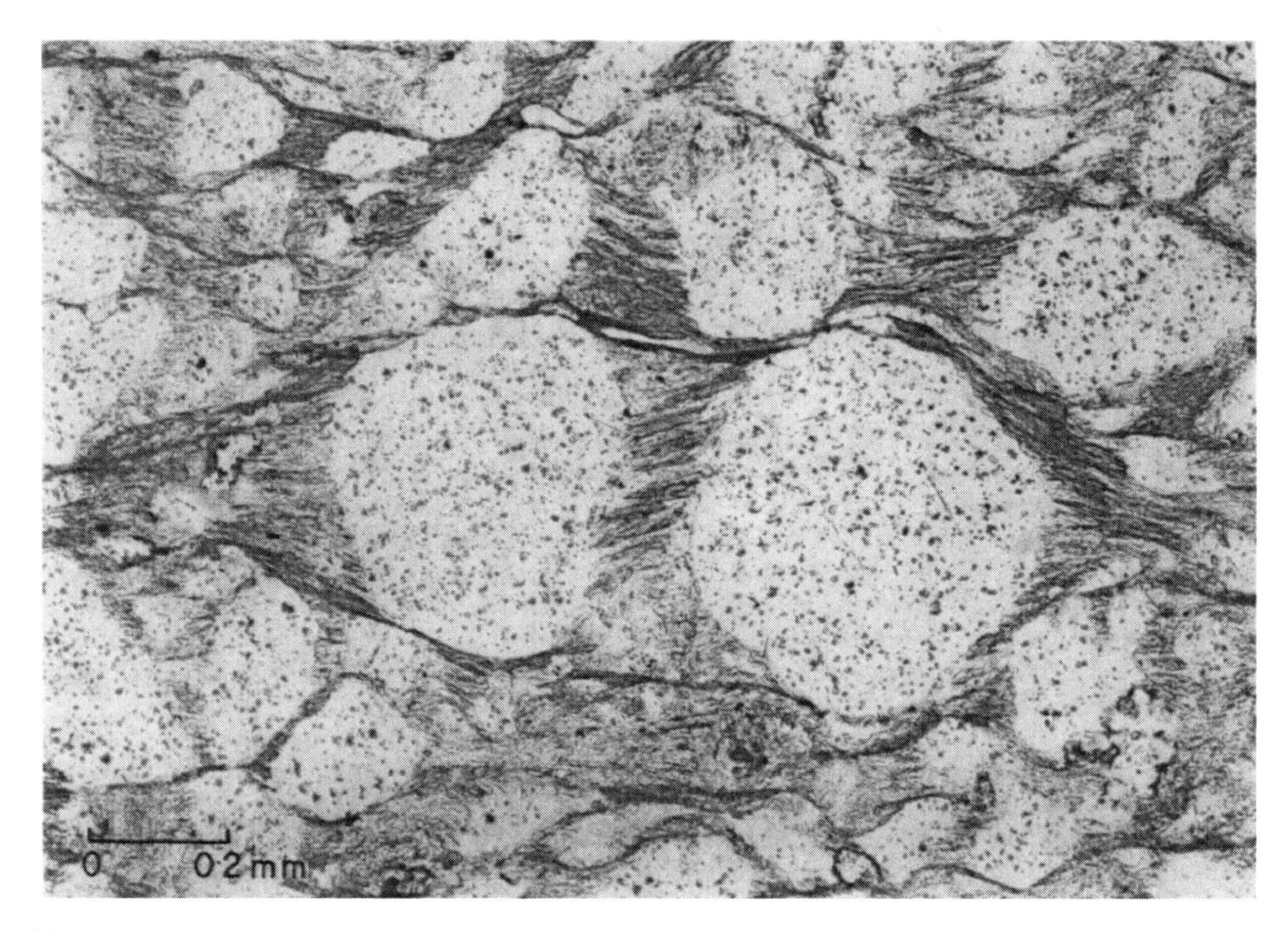


124 A

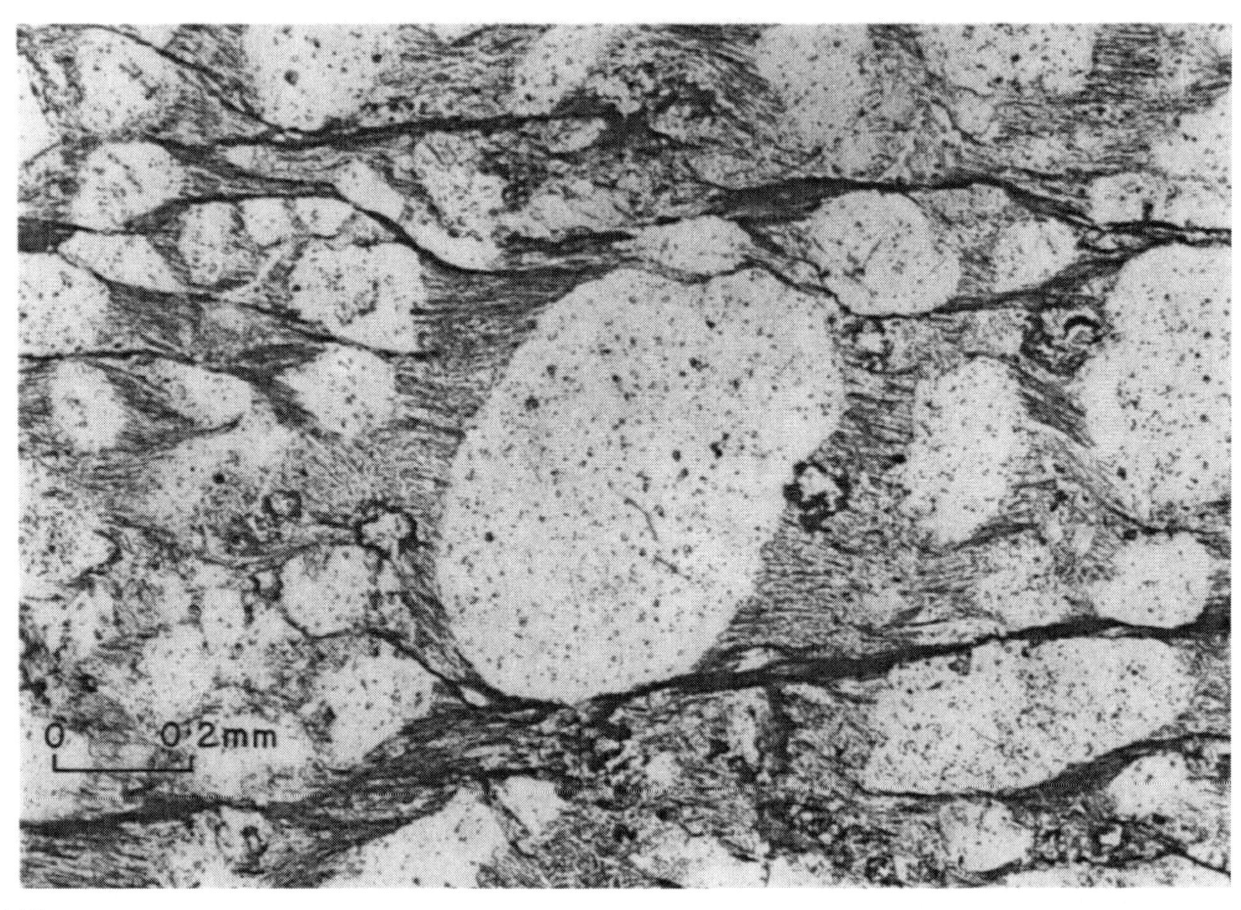


124 B

# 125. Reduction of Clastic Grain Size within Cleavage Zones

C. McA. Powell

The photographs in this plate are from the same location as the preceding plate.

**A. Broad view of anastomosing cleavage folia, and their effects on clastic grains.** Although the spacing of cleavage folia is generally determined by the size of the largest clastic grains between adjacent folia, some cleavage zones consist of two or more very closely spaced cleavage folia. In thin section, these cleavage zones are noticeably darker, and can be traced further than the normal cleavage folia. One such cleavage zone runs horizontally through the middle of this field of view. Close examination shows many small elongate quartz grains enveloped by the folia within the cleavage zone. PPL

**B. and C. Small lensoid grains enveloped by folia within a cleavage zone.** These photographs are enlargements of two areas of the cleavage zone indicated in **A**. The lensoid quartz grains are considerably smaller than the average quartz grain showing rounded outlines, and **A** shows that similar small quartz grains elsewhere in the rock are invariably enveloped by dark cleavage folia. The thin-sectional area of these lensoid quartz grains in the cleavage zones is less than the area of the quartz grains with rounded outlines elsewhere in the rock, but the length of the small grains is approximately the same as the diameter of the more equidimensional quartz grains. These small elongate grains may thus have formed from originally larger grains (viz. grains of the same size as quartz grains with rounded outlines elsewhere in the rock) by dissolution of the sides in contact with the cleavage folia. In this way, the rock may have shortened considerably perpendicular to the cleavage folia, and elongated in other directions by precipitation of the dissolved quartz as overgrowths in the interfolial domains (cf. Williams 1972, Durney 1972b, Gray 1976, and this Vol., for similar processes in crenulated rocks).

The minimum shortening strain associated with the cleavage zones may be estimated by comparing the spacing of the centers of the elongated grains with the spacing of the centers of the quartz grains with rounded outlines between the cleavage zones. This estimate is a minimum because (1) some grains may be completely removed by dissolution in the cleavage zones, and (2) the domains between cleavage zones most likely have suffered a similar, if smaller strain. PPL

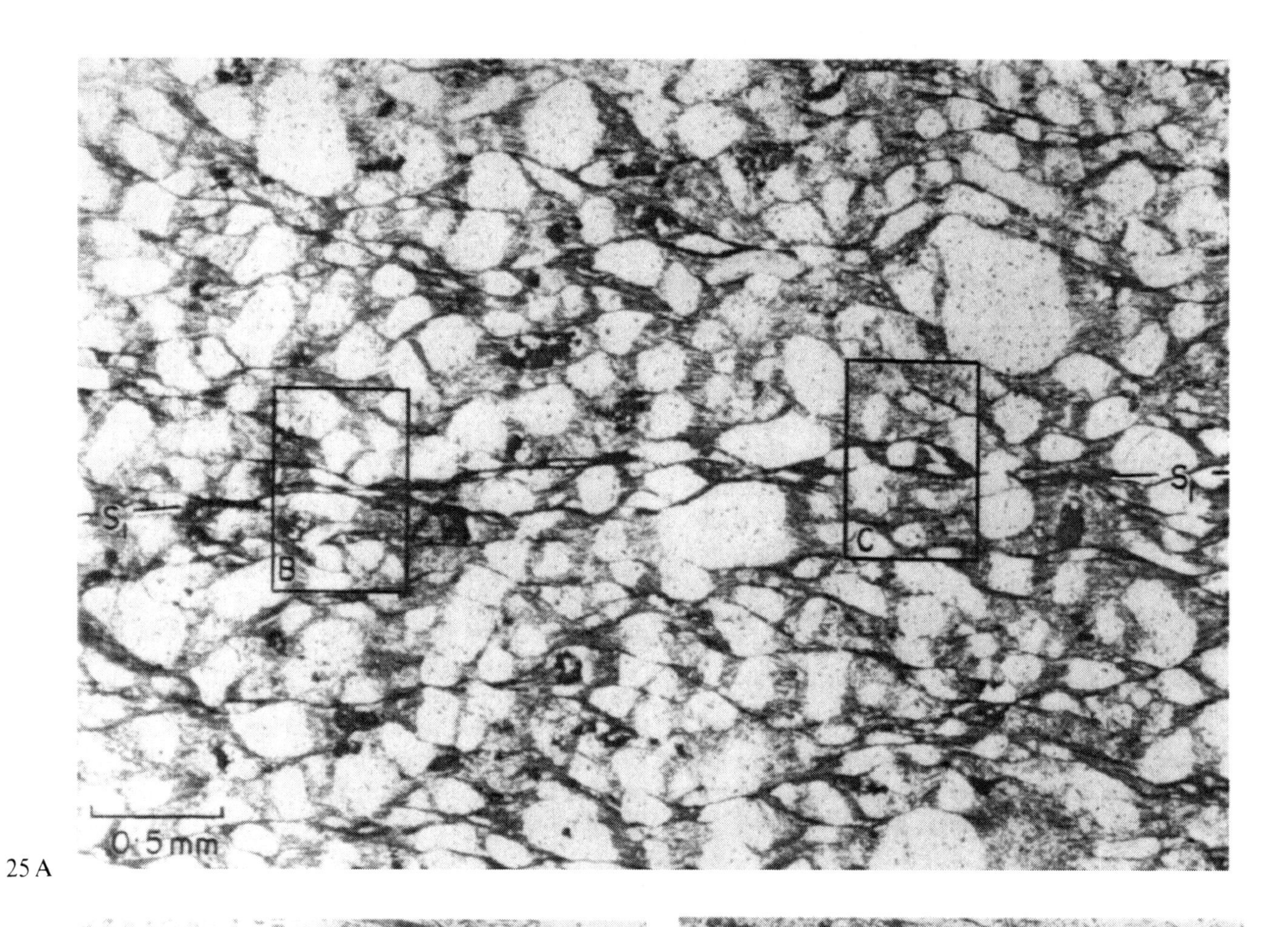


25 A

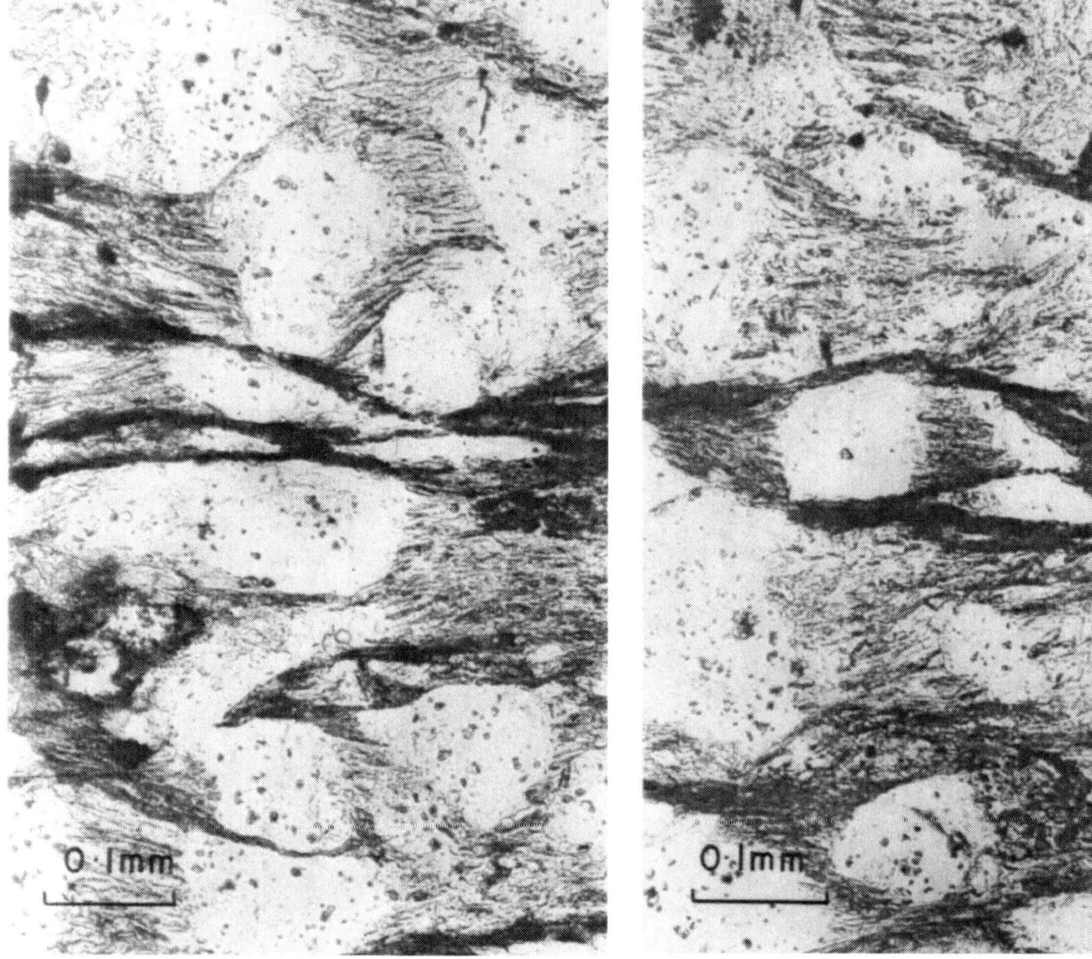


125 B

125 C

## 126. Fiber Development in Deformed Hydrothermally Altered Acid Volcanic Rock

S. F. Cox and M. A. Etheridge

Cambrian Mt. Read Volcanics near Queenstown, Tasmania, Australia. See also Plates 180 and 181.

---

**A. Fiber development in intragranular extension site.** Fibrous quartz and muscovite in a pulled-apart intragranular microfracture in a quartz phenocryst have fiber long-axes subparallel to the total extension direction. Muscovite fibers have (001) parallel to grain long-axis. The majority of quartz fibers are not in optical continuity with the host phenocryst and have abundant cross-cutting fluid-inclusion-rich planes at a high angle to the fiber long-axis. These are interpreted to be healed microfracture sites in which fiber growth has taken place. NX

**B. Fiber development in transgranular extension sites.** Fibrous layer silicates and quartz have grown in opened transgranular microfractures. Irregularities in microfracture wall geometry indicate that fiber long-axes are subparallel to the total extension direction. Partly crossed polars

**C. Fiber development in intergranular extension site.** Oriented mica fibers have grown on the end of a quartz phenocryst forming a mica "beard". The dark inclusion-rich line marks the original surface of the phenocryst. This has been overgrown by quartz in optical continuity with the host grain. Mica fibers closest to the original grain boundary have (001) and long-axes subparallel to the host grain surface, but micas further out have long-axes and (001) subparallel to other foliation-defining micas in the adjacent matrix. NX

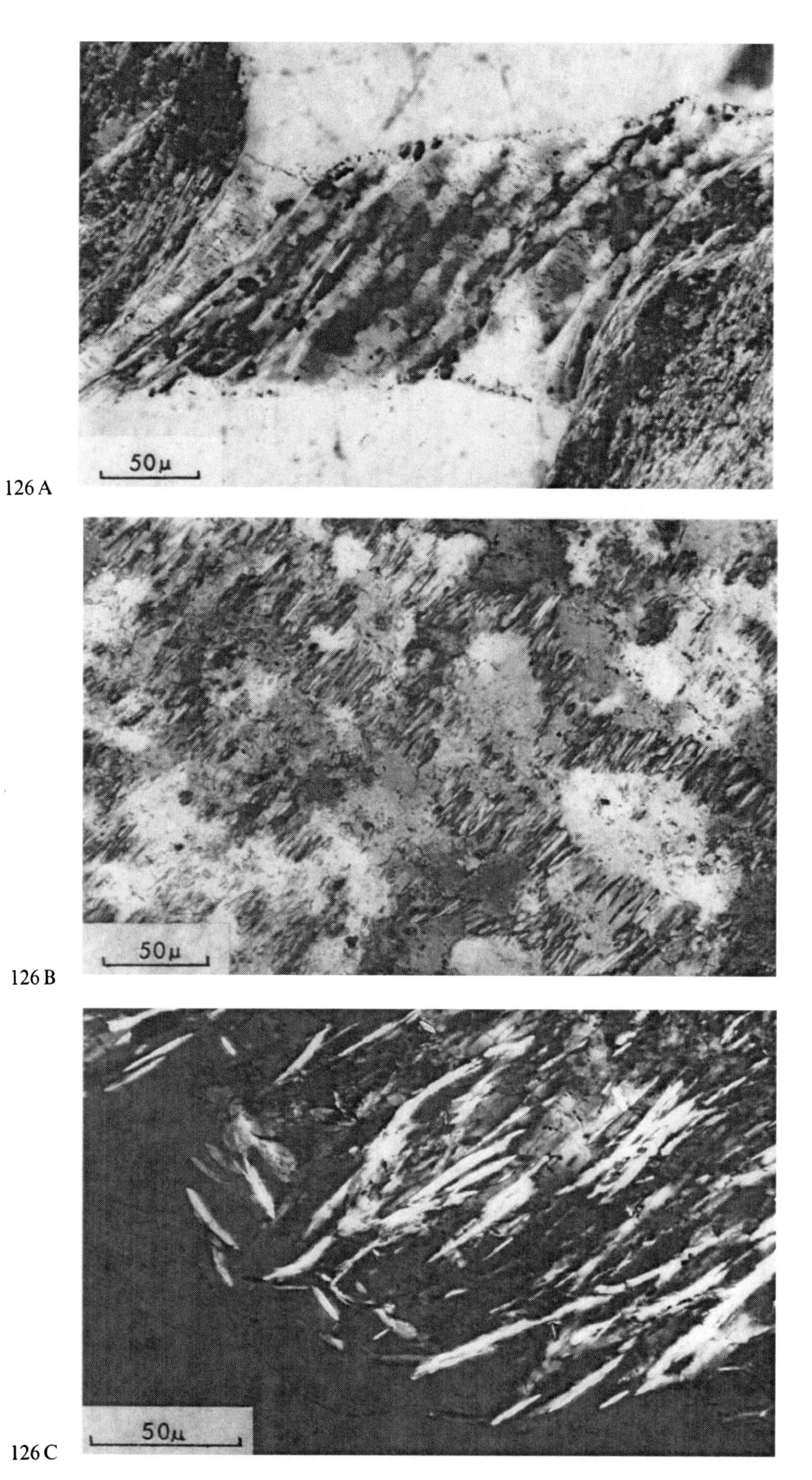

126 A

126 B

126 C

# Spaced Cleavage

## Section 11: Cleavage Defined by Wispy Seams

(Plates 127 to 131)

Five contributions group together limestones, sandstones, and slates in which the cleavage is partly, or solely, defined by wispy seams. Some solution-removal process is postulated in each case to account for the wispy seams.

---

### Plates

127 Spaced cleavage in argillaceous limestone

128 Pressure dissolution as a cleavage-forming process in quartz sandstone and limestone

129 Solution cleavage in shear zones

130 Pressure dissolution decoupling of thin pre-cleavage calcite vein

131 Continuous fine cleavage with wispy seams

# 127. Spaced Cleavage in Argillaceous Limestone

R. P. NICKELSEN

**A. Cleavage zones and uncleaved pure limestone nodules.** Upper Silurian-Lower Devonian Keyser Limestone, Meckley Quarry, Manadata, Pennsylvania, U.S.A. (NICKELSEN 1972). Cleavage zones, composed of groups of clay-carbon partings, occur in argillaceous beds or between pure limestone nodules. Clay-carbon partings are accumulations of insoluble residue that truncate partially dissolved fossils. The specimen is a polished, etched slab cut perpendicular to bedding ($S_0$) and cleavage ($S_1$). Photographs below are from the same rock and in similar orientation.

**B. Clay-carbon partings and calcite overgrowths.** Dark residual accumulations of clay and carbon mark spaced planes of pressure solution. Calcite overgrowths on the large crinozoa plate are oriented in the plane of the cleavage, perpendicular to the bedding-cleavage intersection. Twin lamellae cutting the overgrowths prove some dislocation after diffusion. The temperature of formation is indicated by Conodont Alteration Index (CAI) of 4.5, 190°–300 °C (EPSTEIN et al. 1976) and percentage of fixed carbon 92% = 160 °C in adjacent anthracite coal (DAMBERGER 1974). NX

**C. Cleavage zone.** Pressure solution of crinozoa plates and residual accumulation of clay and carbon. Crinozoa plate at *f* is 50% dissolved. NX

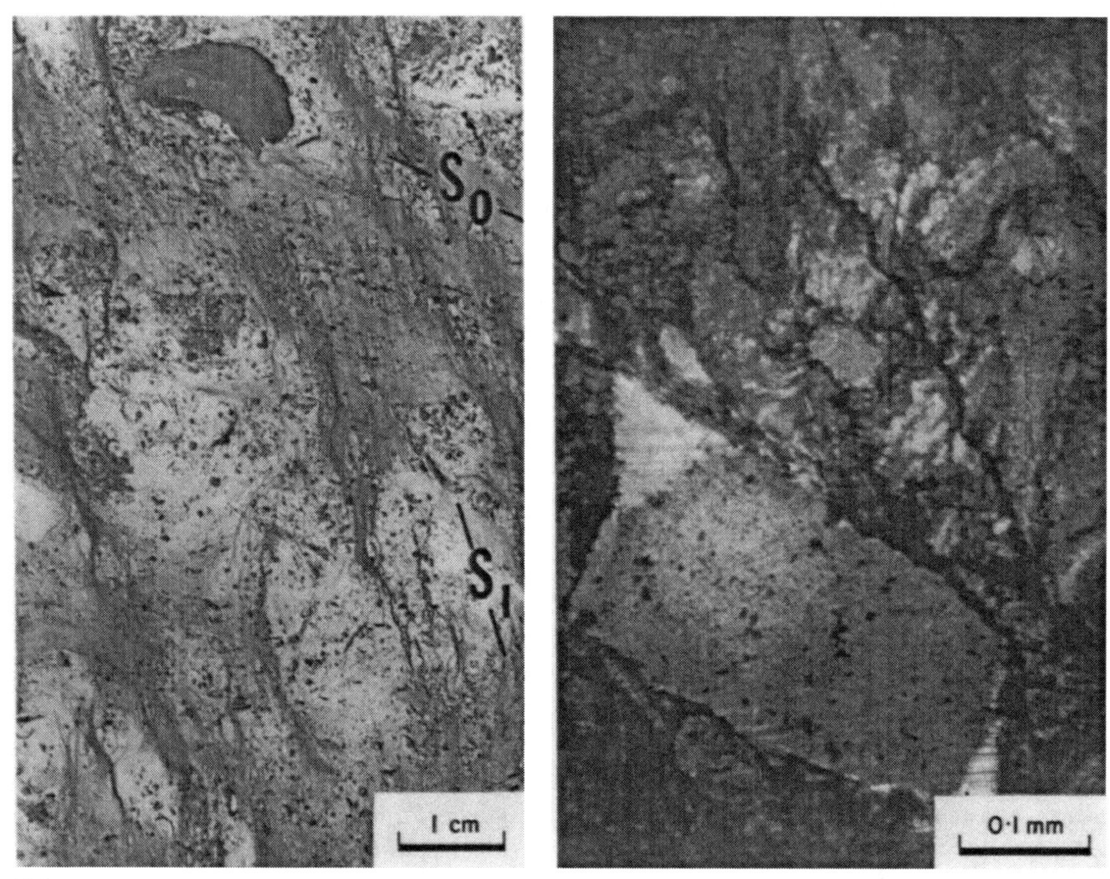

127 A 127 B

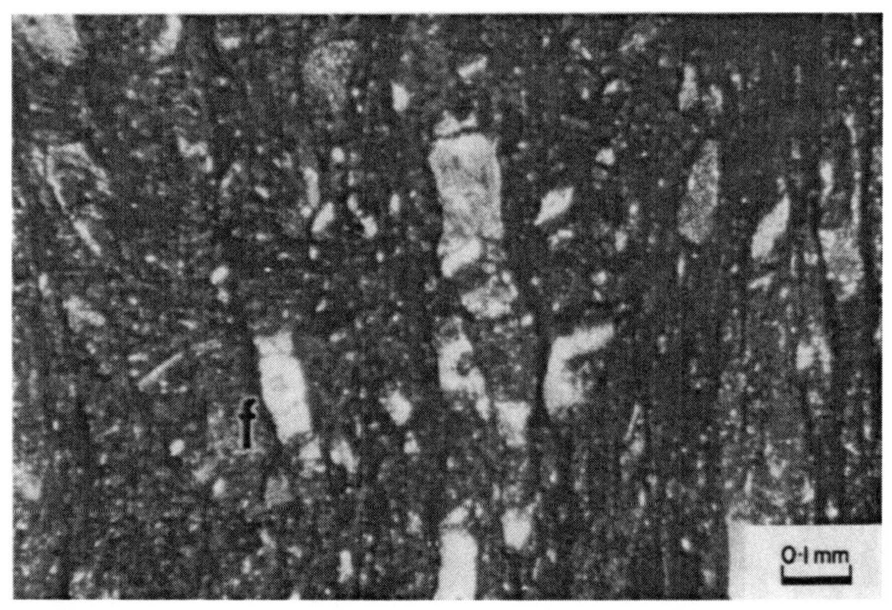

127 C

# 128. Pressure Dissolution as a Cleavage-Forming Process in Quartz Sandstone and Limestone

D. B. SEYMOUR

The photographs show examples of pressure dissolution as a cleavage-forming process during deformation of Lower Ordovician quartz sandstones and overlying limestones in central northern Tasmania, Australia. Upright, horizontal to gently plunging folding on several trends, and development of steeply dipping cleavage in these rocks is attributed to the Middle Devonian Tabberabberan Orogeny. From stratigraphic considerations, material at the base of the Gordon Limestone Sub-Group was buried to a depth of approximately 1.5 km at the time of deformation. Results of conodont geothermometry suggest that temperatures of > 300 °C may have been attained at this time (BURRETT 1978).

---

**A. Detail of cleavage zone in sandstone.** From a correlative of Lower Ordovician Moina Sandstone, Black Bluff area. Pressure dissolution of quartz has played an important role in the formation of discrete, sharply bounded, silica-poor cleavage domains in an argillaceous quartz arenite. In detail the domains consist of bunches of semi-continuous dark seams which anastomose on a small scale around truncated quartz grains. This section is approximately parallel to bedding, and a cross-section of a bedding-perpendicular worm burrow occupies the central part of the photograph. Where the cleavage zone crosses the burrow it is widened and enhanced. Note that the pre-cleavage matrix : grain ratio was probably higher within the burrow. The mesoscopic appearance of these features is that of a "fracture" cleavage, the wavy anastomosing domains being spaced 0.5–2 cm apart and oriented at a high angle to bedding. Microfabric of the material outside the domains is practically unmodified. PPL

**B. Tectonic stylolites.** Ordovician Gordon Limestone Sub-Group, Mayberry, northern Tasmania, Australia. A pair of tectonic stylolites have modified the cleavage-parallel edges of a desiccation vug filled with coarse calcite in a dismicrite. The overall trend of the stylolites is parallel to a pervasive grain-elongation fabric in the extremely fine grained micrite matrix (grainsize 2–4 $\mu$). The fineness of the matrix raises the possibility of superplastic flow as an important deformation mechanism (e.g., SCHMID et al. 1977). One of the large calcite grains within the vug has deformed by formation of twin lamellae (*dipping right to left* in photo) and kink bands (*dipping left to right* and displacing the twin lamellae). PPL

**C. Dissolution of fossil fragments.** Ordovician Gordon Limestone Sub-Group, Mayberry, northern Tasmania. In a deformed biomicrite, inhomogeneous strain between fossil fragments and fine micrite matrix has been expressed in concentration of pressure dissolution (producing insoluble-residue seams) at their cleavage-parallel edges. Pervasive grain-elongation fabric in the matrix tends to "wrap" around the fragments. PPL

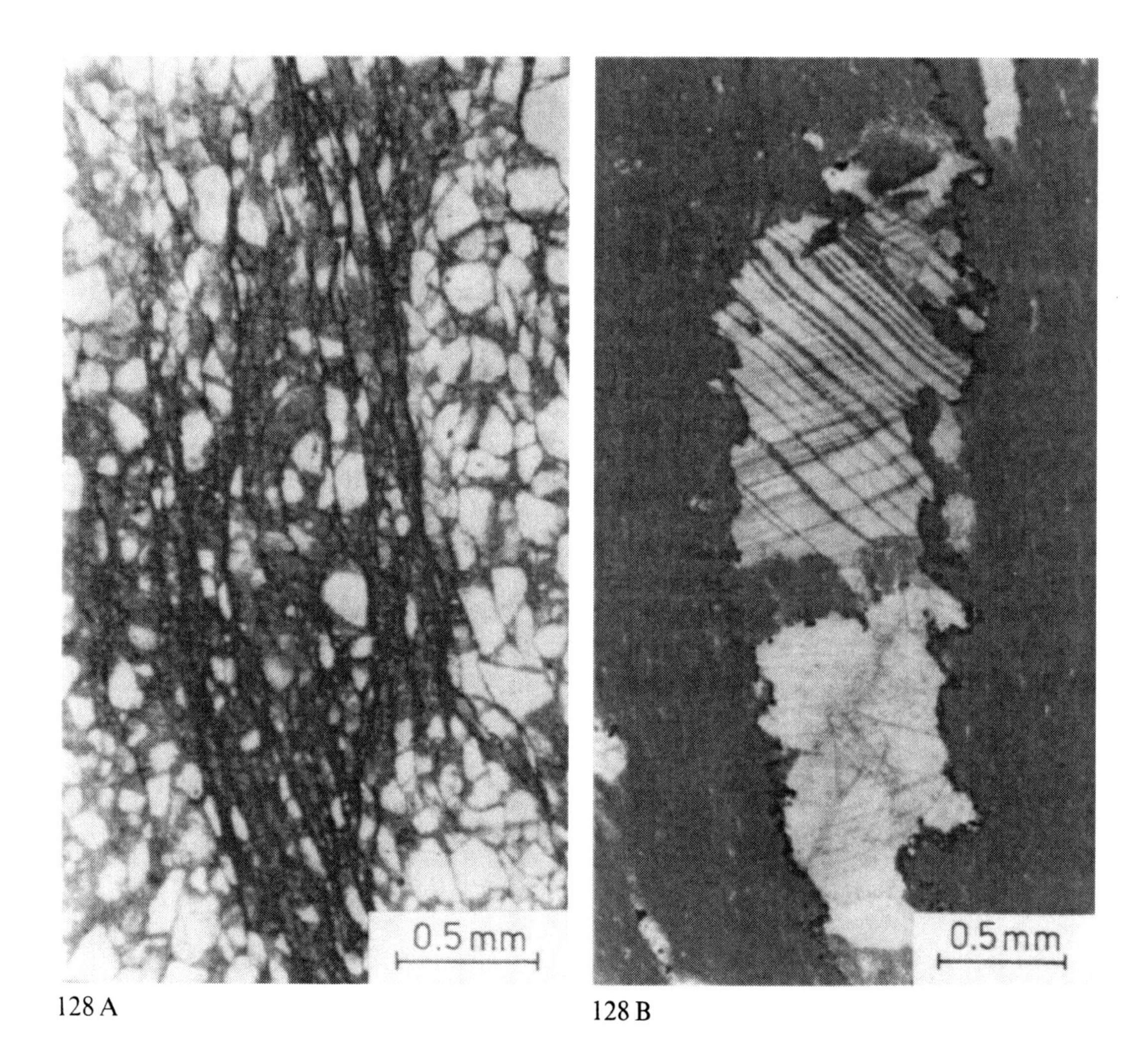


128 A

128 B

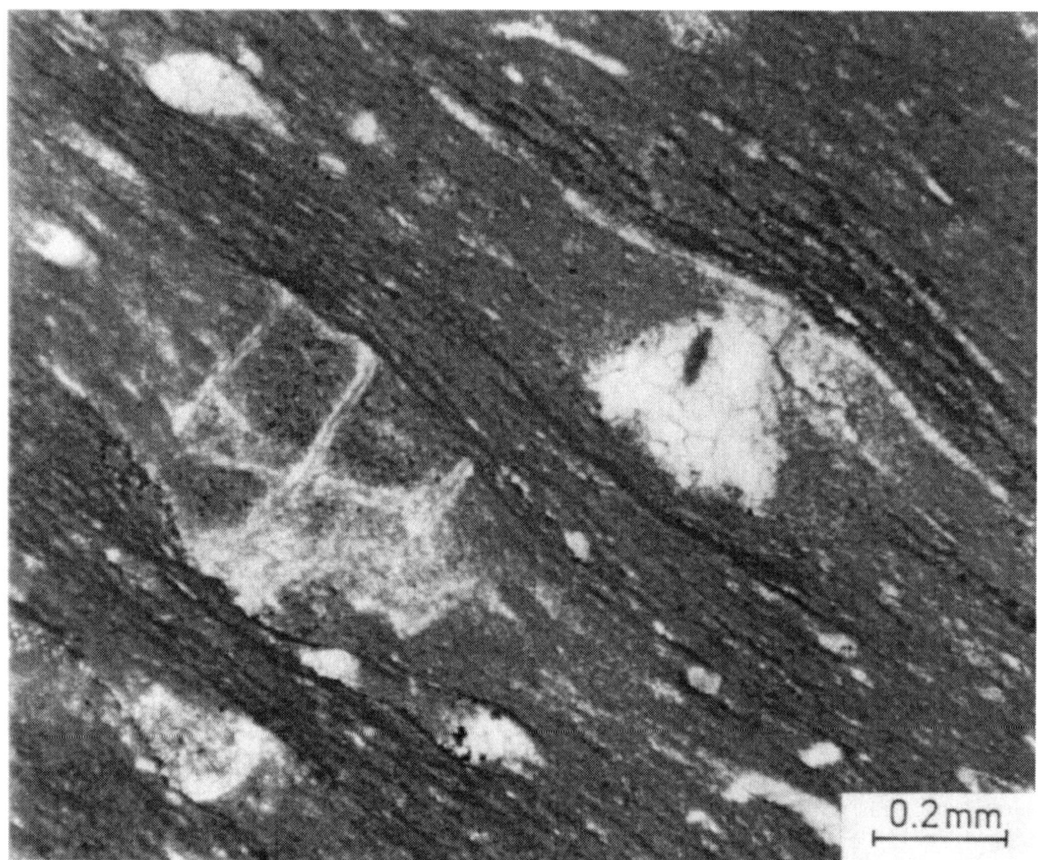


128 C

## 129. Solution Cleavage in Shear Zones

H. R. Burger

Also see Plates 97 and 98

**A. Solution cleavage in shear zone.** Old Red Sandstone, Gateholm, Pembrokeshire, Wales, U.K. (Beach 1975). Shear zones developed at this locality are often conjugate and normally exhibit a coarse solution cleavage. Each shear zone contains an array of en echelon veins (*white*) which are cut, and portions of which have been removed (as at *Q*), by a solution cleavage. Scale: cm

**B. Solution cleavage in shear zone.** Close view of a portion of area illustrated in **A**. Solution cleavage is oriented from top to bottom of photograph. Note that segments of veins terminate abruptly against the cleavage (as at *R*). Also note that there is clear evidence of grain elongation (in the rock matrix) with the long axes of grains oriented parallel to the cleavage.

**C. Pelitic accumulations.** Negative print of thin section from area illustrated in **A**. Veins are dark. Solution cleavage is oriented approximately normal to trend of veins and trends from upper left to lower right of photograph. Cleavage is marked by light wisps that mark accumulations of pelitic material. Note that larger veins are disrupted by the cleavage (as at *S*) and that some small, thin veins terminate against the cleavage (as at *T*)

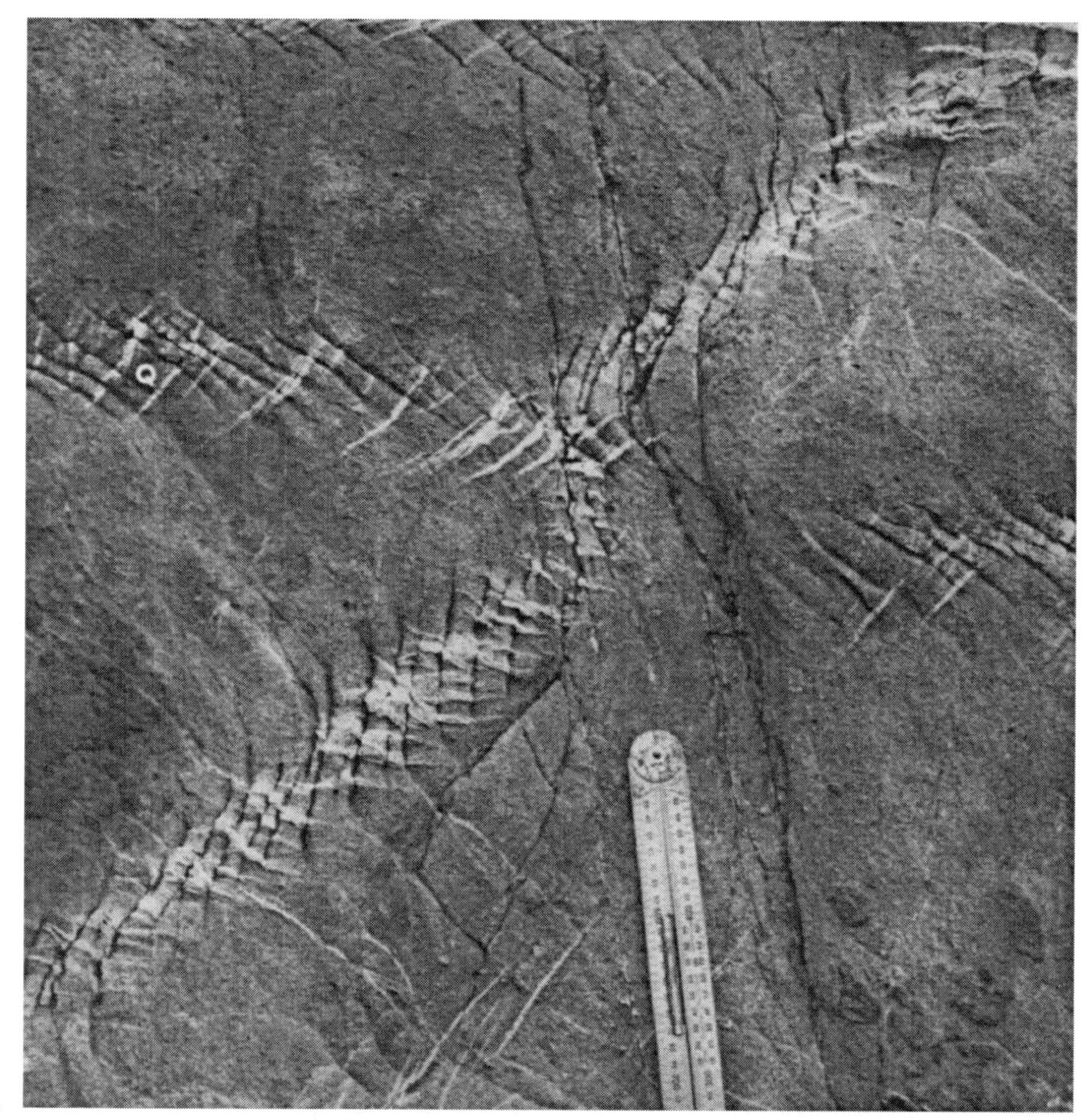


129 A

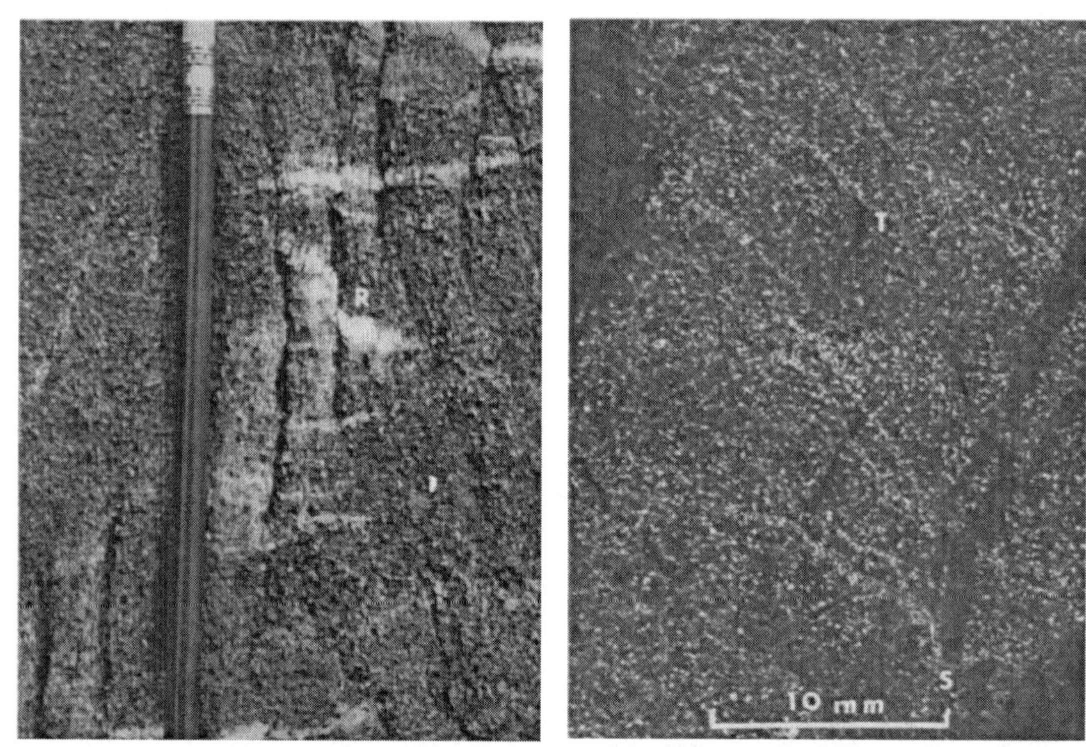


129 B

129 C

## 130. Pressure Dissolution Decoupling of Thin Pre-Cleavage Calcite Vein

G. GILL and T. O. WRIGHT

Thin pre-cleavage calcite veins in limy shale of the lower part of the Martinsburg Formation at Camp Hill, Pennsylvania, U.S.A., have been folded and offset by pressure dissolution.

**A. Decoupled pre-cleavage vein.** Bedding in this photograph is oriented *upper right to lower left*. Cleavage is nearly horizontal. The pre-cleavage calcite vein in the *upper right* appears to be offset by small faults. Note, however, the concentrations of opaque insoluble residues along the "faults" that separate the vein segments. The pre-cleavage vein on the *left* joins this offset vein near the *top* of the photograph. Its plane is inclined to the plane of the slide and it is therefore indistinct; however, it is obviously folded, not faulted. PPL

**B. Enlargement of decoupled pre-cleavage vein.** Bedding can be faintly seen (upper right to lower left) and cleavage is horizontal. Note the maximum concentrations of insolubles occur above and below each vertical vein segment with lesser amounts between offsets and outside of the offset. Also note the tendency of the vein segment's ends to curve and thin toward the adjacent vein segment. In two instances a small calcite fragment is present along the offset between vein segments. PPL

*Comment:* We think these features are the result of pressure dissolution of the surrounding matrix. We envisage the following steps, illustrated in the line diagrams (1) through (5).

1. Initial shortening of surrounding matrix and buttressing by the relatively insoluble vein.
2. Folding of vein during increased shortening.
3. Thinning of the limbs oriented parallel to cleavage.
4. Decoupling of vein segments by nearly complete loss of material in vein segments oriented parallel to cleavage.
5. Concentrated pressure dissolution at each end of the decoupled segments. Since the matrix dissolves faster than the vein, the ends of the vein segments pass each other and produce the "S" shape of the opaque insoluble residues. Detached segments of the horizontal limb remain in the offset

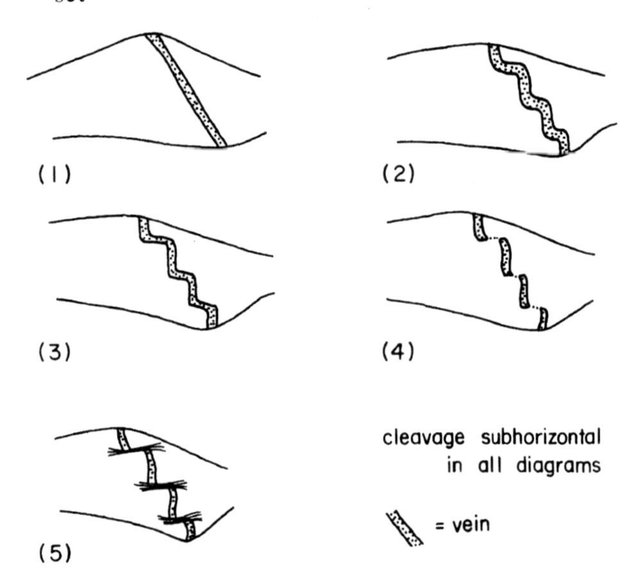

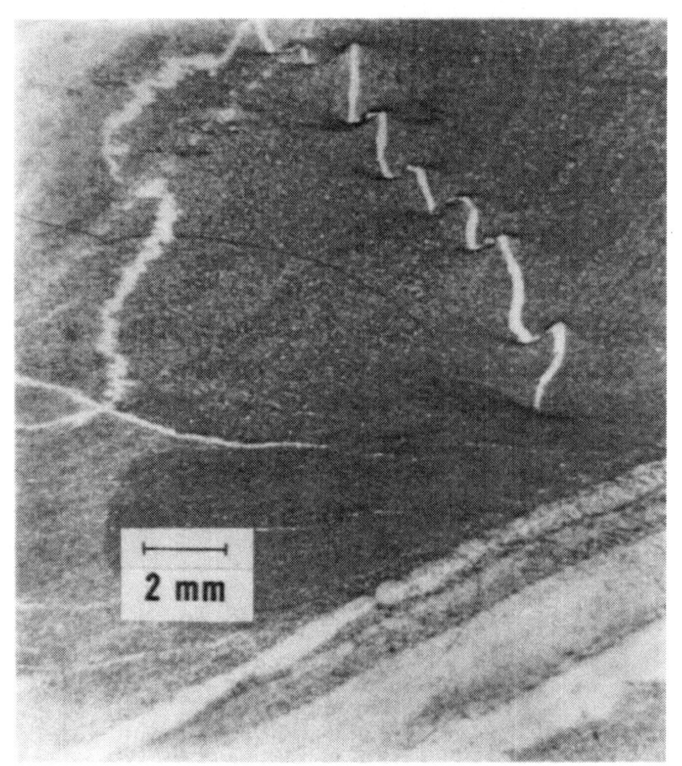

130 A

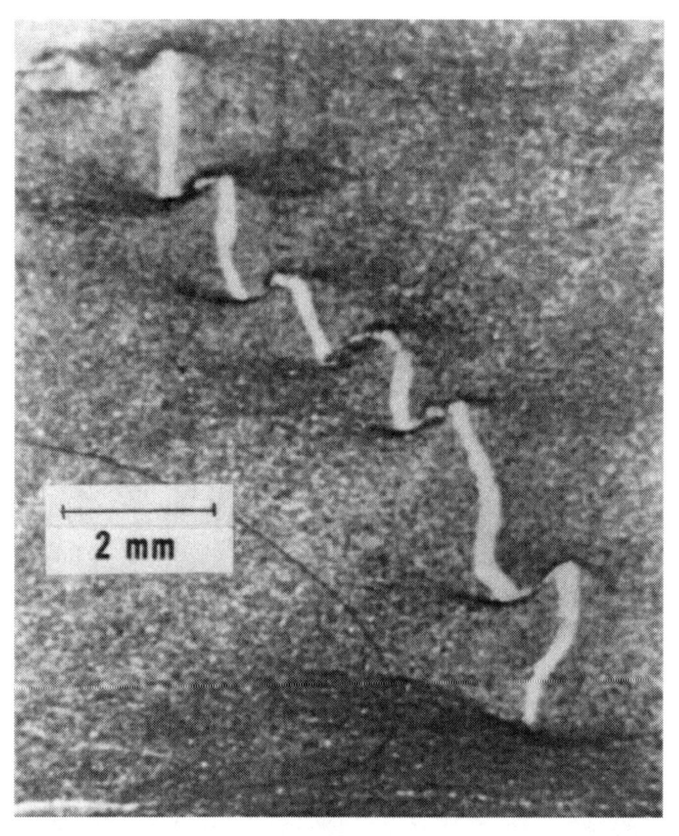

130 B

# 131. Continuous Fine Cleavage with Wispy Seams

G. J. BORRADAILE

Devonian siltstones and argillaceous rocks are cleaved and deformed at this locality near Nieder-Salwey in the Rheinische Schiefergebirge, West Germany. Two-dimensional strain analyses of brachiopods (LANGHEINRICH 1967) and three-dimensional strain analysis of worm tubes such as those seen in this photograph (*W*) give the following bulk strains assuming constant volume strain (BORRADAILE and TARLING 1981, outcrop II):

| *X* | *Y* | *Z* |
|---|---|---|
| 1.373 | 1.195 | 0.609 |

The strain history here is approximately coaxial so that the cleavage trace is approximately an *XY* trajectory. In the plane of this photograph this is confirmed to some degree by the fact that the cleavage trace lies in the acute angle between the originally orthogonal bedding and worm tubes. (See the "sand dike test" of BORRADAILE 1977, 1979a and Plate 205 in this Vol.).

Note that the cleavage is enhanced by dark, wispy seams adjacent to worm tubes (*a, b*). These locations might be interpreted as zones of locally higher strains, abutting the less strained silt of the worm tubes. The impoverishment in quartz at these points causes the darkness and it may be caused by the diffusive mass transfer through fluid films ("pressure solution") or in the solid state. PPL

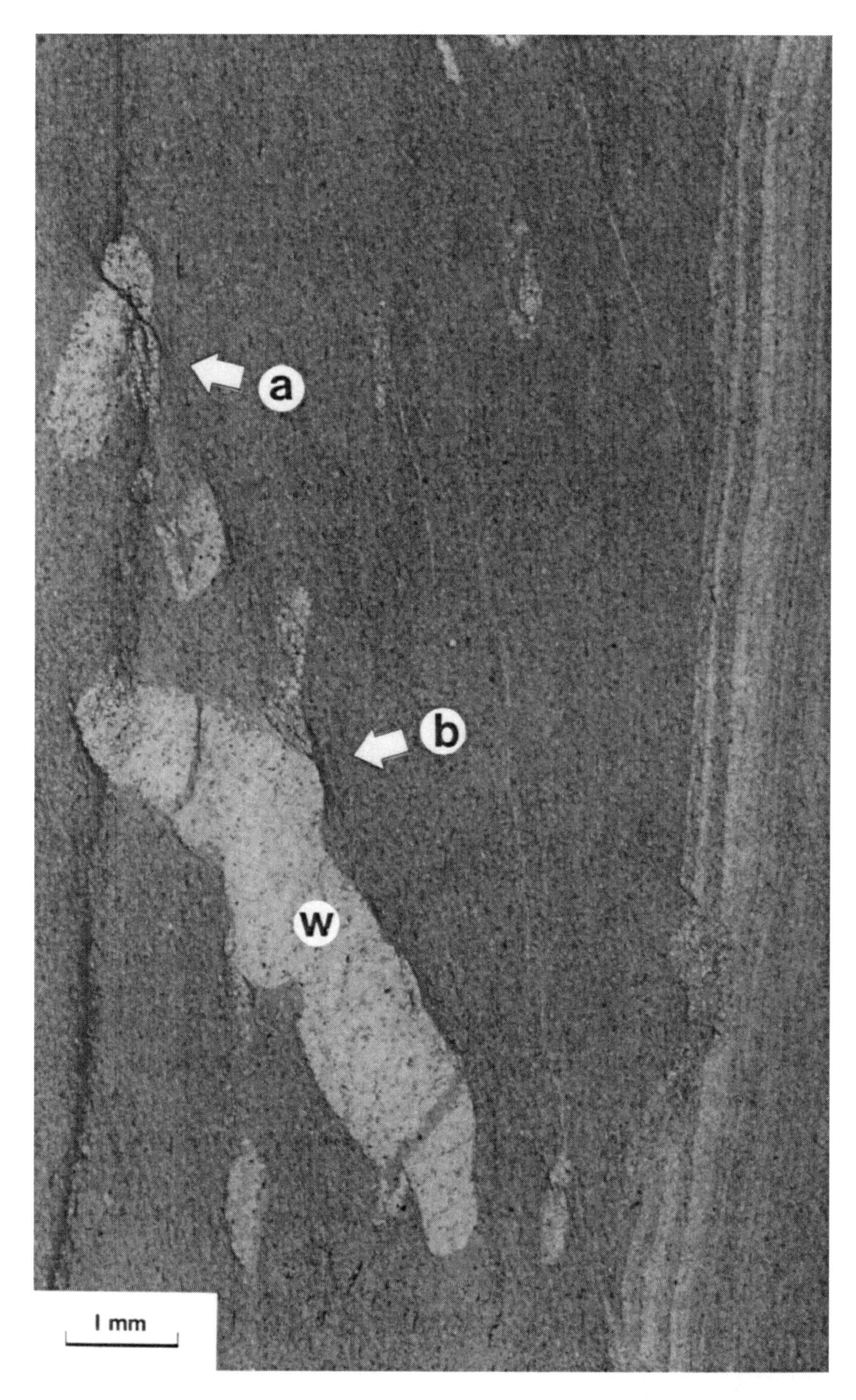


131

# Spaced Cleavage

## Section 12: Cleavage Defined by Flame-like Seams

(Plates 132 to 140)

This group of nine contributions illustrates cleavages completely or partly defined by the physical disaggregation of clastic layers. In the literature such features have sometimes been referred to as sand dikes or tectonic sand dikes. However, their flame-like appearance and field relations are usually different from sedimentary sand dikes and other pre-cleavage sand intrusions which are illustrated in Section 16 (Plates 201 to 205). In these examples clasts appear to have been moved (in a direction subparallel to cleavage) from clastic horizons where their absence may be noted. The disaggregation may be related to an early tectonic or diagenetic event in some cases. Where it is mentioned in connection with transposition, however, the disaggregation is probably a tectonic particulate flow (Borradaile 1981).

---

**Plates**

# 132. Liquefaction Structures in Slate

C. McA. Powell

Ludlovian Bannisdale Slates, near Backbarrow, Lake District, U.K. (Powell 1972a). These liquefaction structures (photographed from large thin sections) are from an outcrop on A590, ½ mile north of Backbarrow (Grid Ref: SD357854). Bedding is horizontal, and a spectacular vertical banding has been produced by multiple intrusions of clastic material. The largest dike, approximately 5 cm wide (Powell 1972a, Plate 5a) has intruded downward, but other dikes (Powell 1972a, Plate 6a) have intruded upward. Mixing of clastic material occurs on all scales, and commonly clastic dikes, clearly distinguishable from the intruded matrix in one place, can be traced upward or downward to a region where the intrusive boundaries are more diffuse, indicating that there has been mixing or homogenization of sedimentary material.

The rock is thoroughly pervaded by an excellent flaggy cleavage, which rings to the blow of a hammer, and which is marked by films of phyllosilicates that gleam in the sunlight. In many places, the rock has a "fissured" appearance (Powell 1972a, Plates 5b and 6d), and cleaves along the margins of the intruded dikes. Powell (1972a) argued that the parallelism of these clastic dikes and cleavage demonstrated initiation of cleavage in these rocks by liquefaction during tectonic dewatering. Subsequently, modification and enhancement of the initial fabric has occurred by pressure solution, neo- and re-crystallization, and mechanical rotation during continued strain. Geiser (1975) and, more recently, Beutner (1980) have discounted this view elsewhere. In this example, there is no doubt that the intrusive clastic dikes have imparted a strong compositional and textural layering to the rocks, and that any subsequent cleavage-forming processes have inherited the orientation of these early liquefaction structures.

---

**A. Downward-intruding clastic dike.** The variable color density and width of the dike reflects different mixing proportions of silty (*clear*) and muddy (*dark*) material during its downward intrusion. The internal fabric of the dike reflects a pipe-like funnelling effect during intrusion. A thin dikelet of fine silt at the top of the dike (enlarged in **B**) is derived from the immediately overlying source bed. In the middle part of the dike (Powell 1972a, Plate 6b), muddy material has been intruded through a silty bed, and a mixed rock has resulted further down. PPL

**B. Thin siltstone dikelet.** Enlargement of the thin siltstone dikelet at the top of **A**. A dark zone in the source siltstone marks the place from which the clastic material has been derived. The dark pelitic folia in the underlying "mixed rock" of the large clastic dike are parallel to the thin siltstone dikelet. PPL

**C. Mixed rock of the clastic dike.** This photograph from the upper part of the clastic dike in **A** shows variation in color density, which reflects variation in the mixing proportions of silt and mud. Dark bands along the margins show the intimate association of pelitic folia with the liquefaction structures. PPL

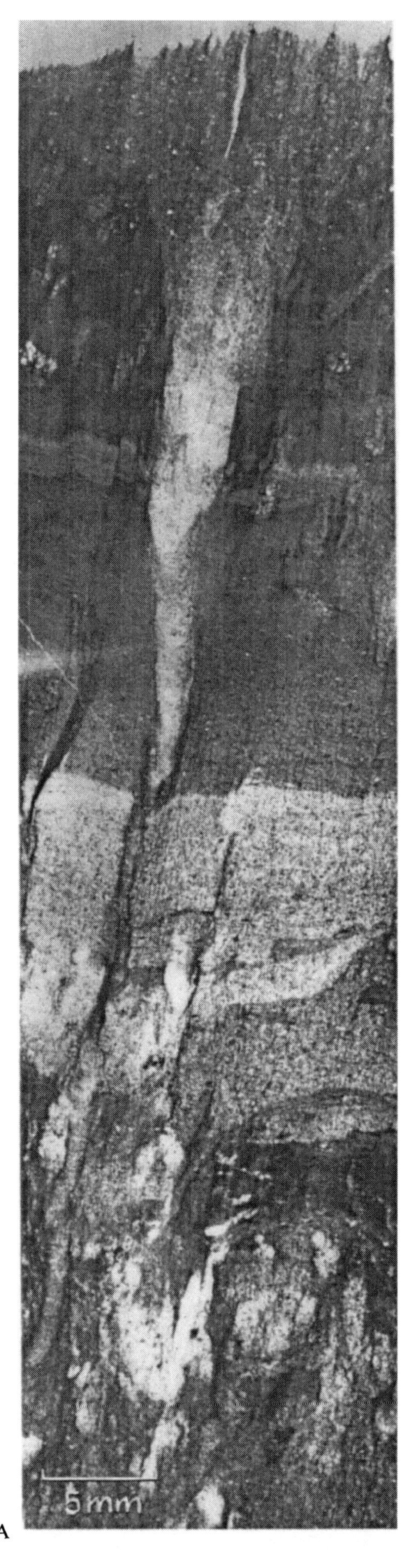


132 A

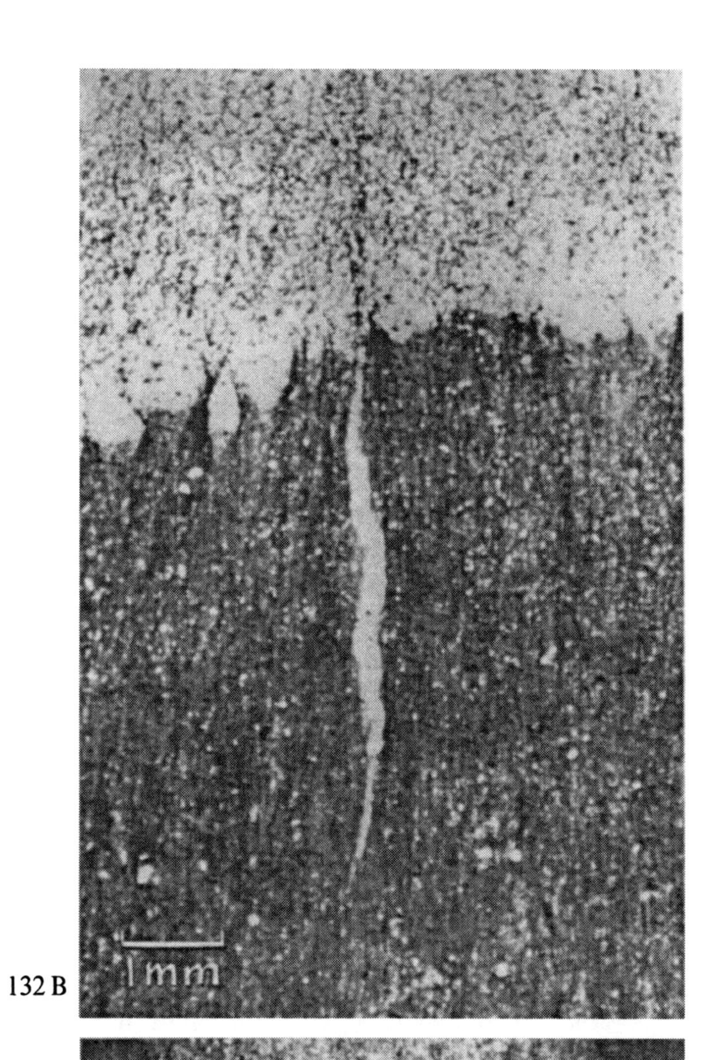


132 B

132 C

## 133. Cleavage Folia Transecting Silty Beds

C. McA. Powell

Ludlovian Bannisdale Slates, near Backbarrow, Lake District, England (Powell 1972a). These two photographs are from large thin sections from the same outcrop as the liquefaction structures in the preceding plate. The photographs are oriented with the vertical cleavage in outcrop across the width of the page, and the stratigraphic top to the left. Subhorizontal silty beds in the outcrop run at a high angle to cleavage. The cleavage folia tend to occur in zonal concentrations (cleavage zones) separating domains in which there are fewer, more widely spaced cleavage folia. Cleavage zones, 1 to 2 cm wide, appear as dark bands in outcrop, and individual zones may be traced for tens of centimeters. These cleavage zones, coupled with thin siltstone dikelets, give the rock a striped or "fissured" appearance (Powell 1972a, Plates 5a, 5b, and 6d).

The features displayed by the cleavage zones probably result from a combination of several cleavage-forming mechanisms. The thin siltstone dikes were intruded during liquefaction, probably at an early stage of the local strain history, and it is conceivable that some of the pelitic material now concentrated in the cleavage zones was also emplaced at this time. Precursors of the fine anastomosing folia within cleavage zones may have formed during flow of muddy pore fluid through the partially coherent, clastic framework. Alternatively, individual cleavage folia may have formed during deformation as residual concentrations where more soluble components (mainly quartz) were preferentially dissolved and removed from the rock. No matter how the cleavage folia were initiated, there is clear evidence (best seen in ultra-thin thin sections) that the dark material in many of the cleavage folia has recrystallized into new phyllosilicates dimensionally and crystallographically elongate parallel to the cleavage folia (see Siamo Slate, Plates 124 and 125). The Bannisdale Slates, deposited in the Ludlovian, were deformed during the Caledonian orogeny (Simpson 1968).

---

**A. Disruption of silty layers by cleavage folia.** The rounded blobs of silty material resembling load structures are bordered by concentrations of cleavage folia. These cleavage zones can be followed along the trace of the cleavage through several silty beds, and are commonly coupled with intrusive siltstone dikelets (see Powell 1972a, Plate 6a, b, c and d). Note the numerous, small, upward projecting siltstone dikelets in the *right hand part* of the photograph, and the markedly elongate pelitic zone parallel to the general cleavage. PPL

**B. Small fold with axial-surface cleavage folia.** The planar cleavage folia cut across several silty layers with little refraction, but do not penetrate other layers. PPL

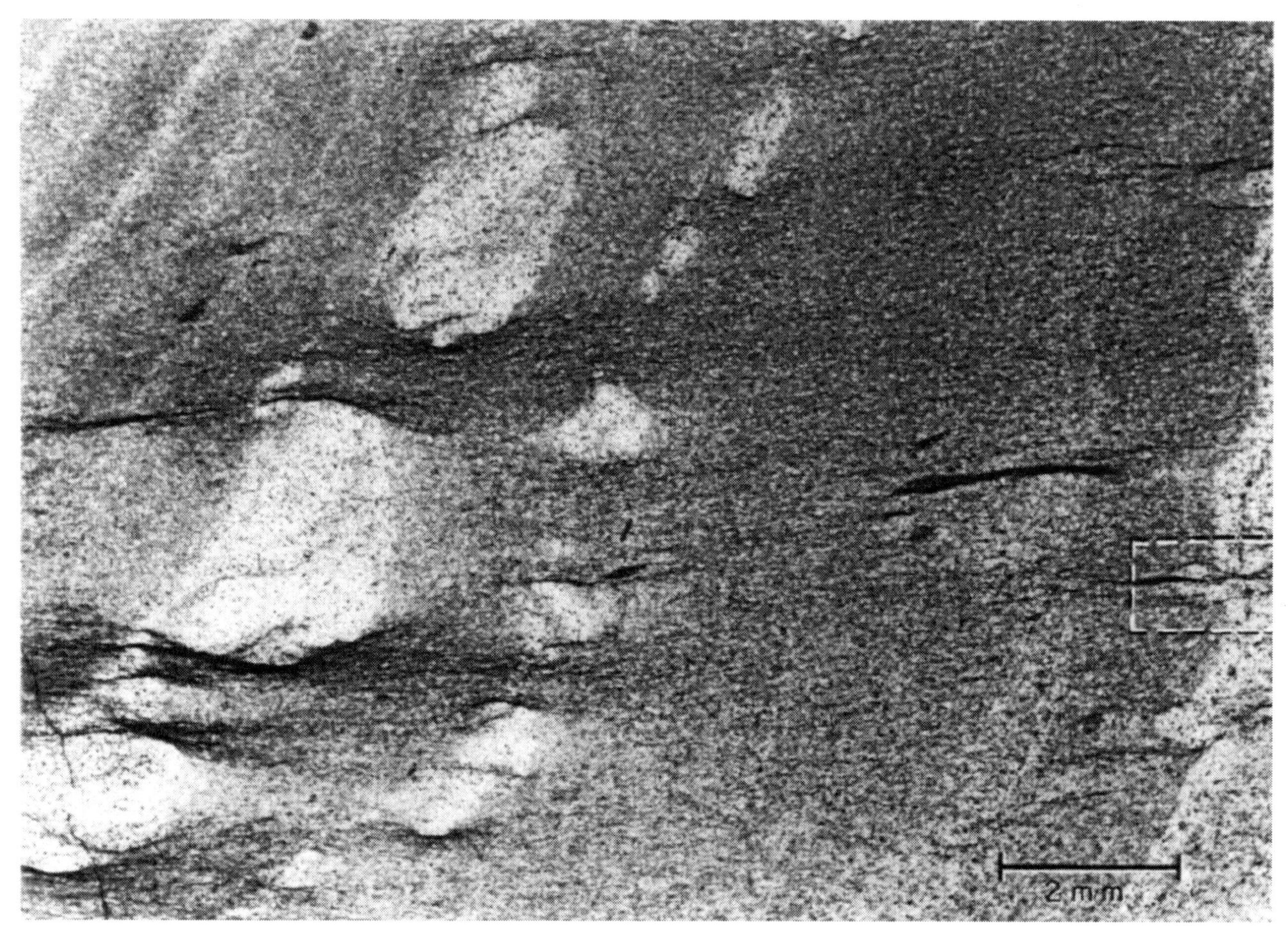

133 A

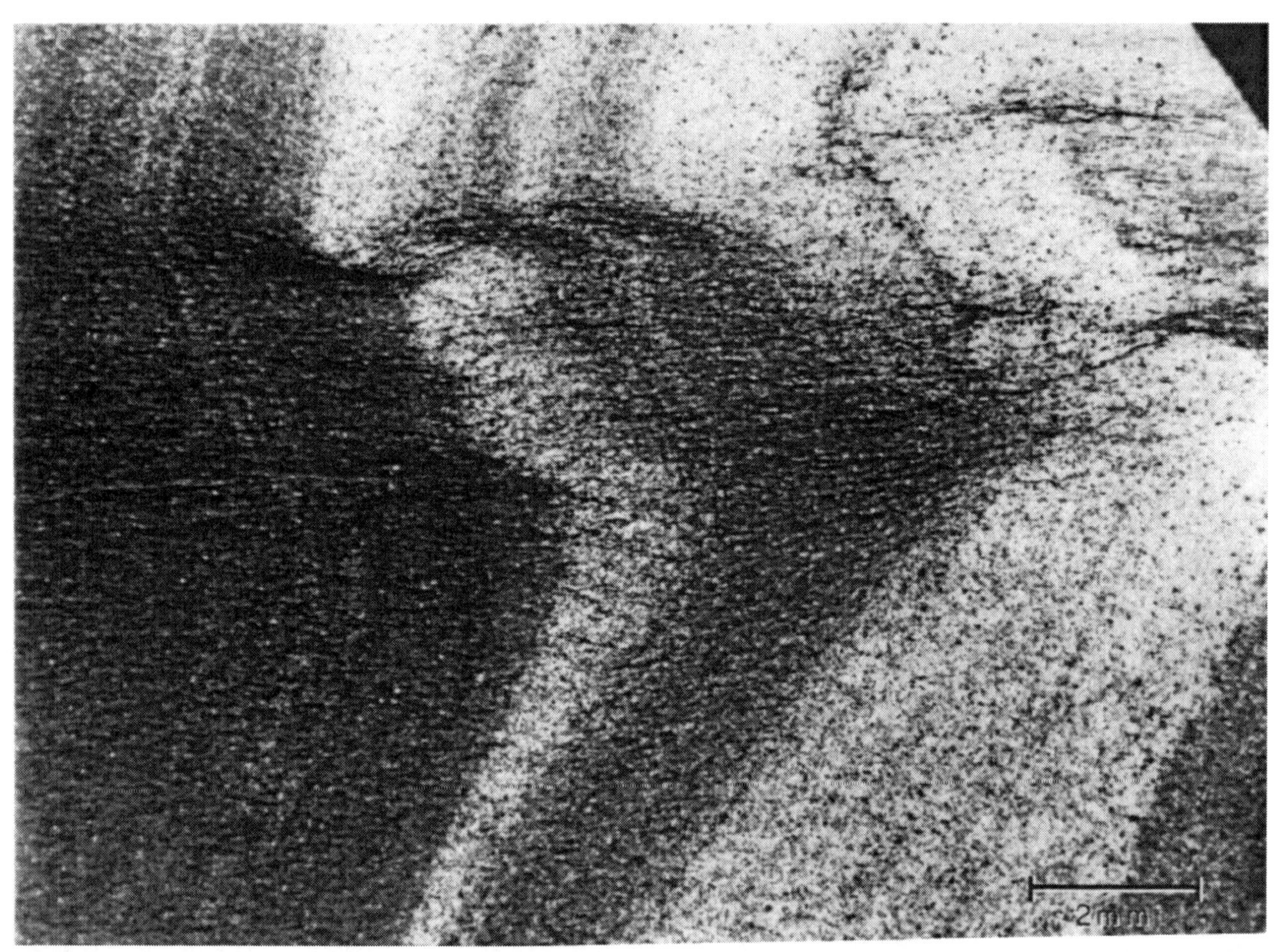

133 B

# 134. Disaggregation Structures in Slate

C. McA. Powell

These photographs from large thin sections show three varieties of disaggregation structure in slate. In each example, there appears to have been either complete or partial loss of cohesion of the clastic framework of the rock at some stage during its history, and rock components have been redistributed to form structures parallel to what is now a slaty cleavage.

---

**A. Cleavage zone passing from a coarse sandy bed into slate.** (Ordovician Slate from Dinorwic Quarry, North Wales; U.K. Grid Ref: 596608.) The coarse sandy layer is the upper part of a graded bed (green in weathered outcrop) interbedded with massive purple slate. The coarse sandy bed (composed dominantly of quartz) cleaves along pelitic zones spaced 1 to 2 cm apart. The orientation of the cleavage refracts continuously from an angle of 80° between bedding and cleavage at the base of the graded layer, to 30° in the slate. This photograph shows the upper part of one of these pelitic cleavage zones. Bedding runs obliquely at an angle of 55° to the length of the photograph, and in hand specimen is a sharp boundary. Clastic quartz grains are concentrated in dispersed groups where the cleavage zone passes out of the sandy bed. No such clastic grains occur at the base of the sandy bed, or away from the cleavage zones.

How did these clastic quartz grains become distributed in the slate? Their similar size and appearance to the clastic grains in the coarse sandy bed suggests that they were derived from that bed, and their distribution adjacent to, and within the cleavage zone suggests that a common process may have embedded the clastic grains in the slaty matrix and formed, or initiated, the cleavage zone. Is it possible that these grains were jetted out of the coarse sandy bed under high water pressure during deformation? PPL

**B. A siltstone dikelet, pelitic zone and cleavage folia – all parallel.** (Ludlovian Bannisdale Slates, near Backbarrow, Lake District, England [Powell 1972a]; Grid Ref: 357852.) In this photograph (an enlargement of the *dashed enclosure* in the *right hand edge* of **A** on the preceding page), bedding ($S_0$) is approximately horizontal and cleavage vertical. An elongate pelitic zone (approx. 0.1 mm wide) occurs up the middle of a small siltstone dikelet (approx. 0.5 mm wide – shown clearest on the preceding page). Cleavage folia (approx. 0.01 mm wide, and spaced 0.1 to 0.2 mm apart) pervade the rock, and are parallel to the siltstone dikelet and pelitic zone. In a number of places in the interfolial domains, subhorizontal detrital micas (*arrowed*) are parallel to bedding.

The siltstone dikelet and pelitic zone are probably the result of liquefaction early in the strain history of the rock. The pelitic folia may have been initiated as clay or mica films deposited by the flow of pore fluid through the clastic framework, or alternatively may have formed later as dissolution residues. If the pelitic folia are largely of this residue origin, the orientation of the dissolution surfaces may have been inherited from the liquefaction structures. PPL

**C. Downward-intruding siltstone dikelet with associated cleavage zone and folia.** (Same location as **B**). The dark cleavage zone marks the part of the silty source bed from which the siltstone dikelet has been derived. The cleavage zone ($<$ 0.1 mm wide in the silty bed) broadens into a zone of closely spaced folia as it is traced downward. As in **B**, the cleavage folia, however formed, have probably inherited an orientation from early liquefaction structures. PPL

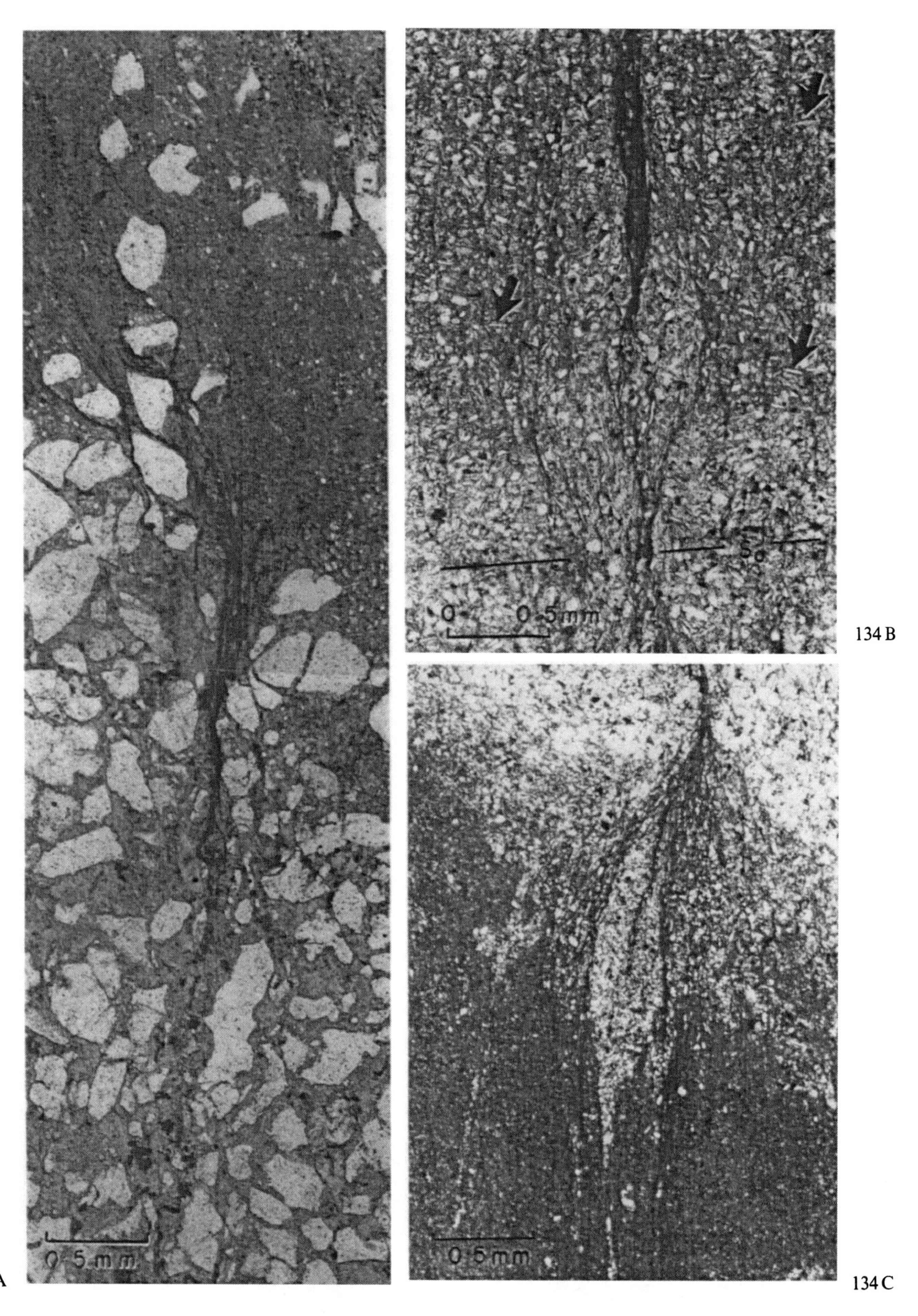


34 A

134 B

134 C

# 135. Disaggregation and First Cleavage

G. J. Borradaile

Devonian slates and sandstones, near Kautenbach, Traben-Trarbach, Mosel Valley, W. Germany.

The slates show only a first cleavage in this outcrop. Some very tight minor folds are present but the nearly vertical beds young to the right here, being situated on the right hand flank of an antiform.

The mean cleavage attitude (*S*) is refracted somewhat in sandy beds such as *A* and *B*. In outcrop the cleavage appears continuous or penetrative in the slates but it is associated with a disruption of bedding surfaces of the sand beds. "Flame structures" (*C*) and sand lenses (*D*) occur on both sides of the sand beds. It seems unlikely that these are deformed sedimentation features and it is suggested that these result from a hydromechanical disaggregation of beds during cleavage formation.

Scale: 10-Pfennig coin

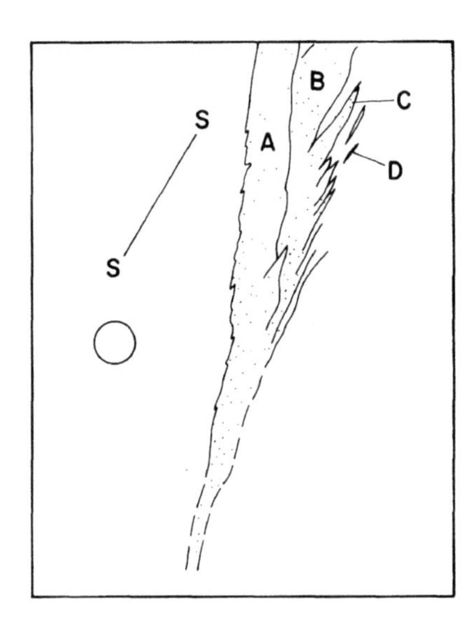

135

# 136. Intrafolial Folds Formed by Disaggregation

M. M. Kehlenbeck

Highway 527 (Armstrong highway), north of Thunder Bay, Ontario, Canada.
Basic metavolcanic rocks of amphibolite facies from the Wabigoon subprovince of the Superior province of the Canadian Shield develop intrafolial (rootless or limbless) folds. The folds affect layers and long lenses and on the flanks of the folds the layers become thinned to as little as 2% of their thickness on fold closures. This thinning is produced by disaggregating the layer to a thickness of just a few individual grains.

136 A

136 B

## 137. Solution Cleavage and Bedding Transposition

H. R. Burger

**A. Solution cleavage and bedding transposition.** Devonian slates at Tor Cross, South Devon, U.K. (Cosgrove 1976). Apparent solution cleavage in synform developed in slates. Light-colored silt layers are cut by a penetrative cleavage. The cleavage is defined by seams rich in pelitic material. Hammer length is 0.4 m.

**B. Bedding transposition.** Enlarged view of a portion of scene illustrated in **A**. Note projection of silt layers into darker gray pelitic layers suggesting initial stages of bedding transposition.

**C. Thin-section of A.** Bedding trends *from upper left to lower right* and is traversed at a high angle by cleavage. Darker cleavage folia are richer in pelitic material than surrounding matrix. Note disruption of light-colored silt layer by cleavage that suggests possible removal of material by solution. At this locality numerous examples of apparent offset of silt layers by cleavage are more satisfactorily explained by assuming removal of material by solution than by slip on the cleavage. In general, the offset segments do not fit together when the apparent slip on the cleavage is removed. PPL

137 A

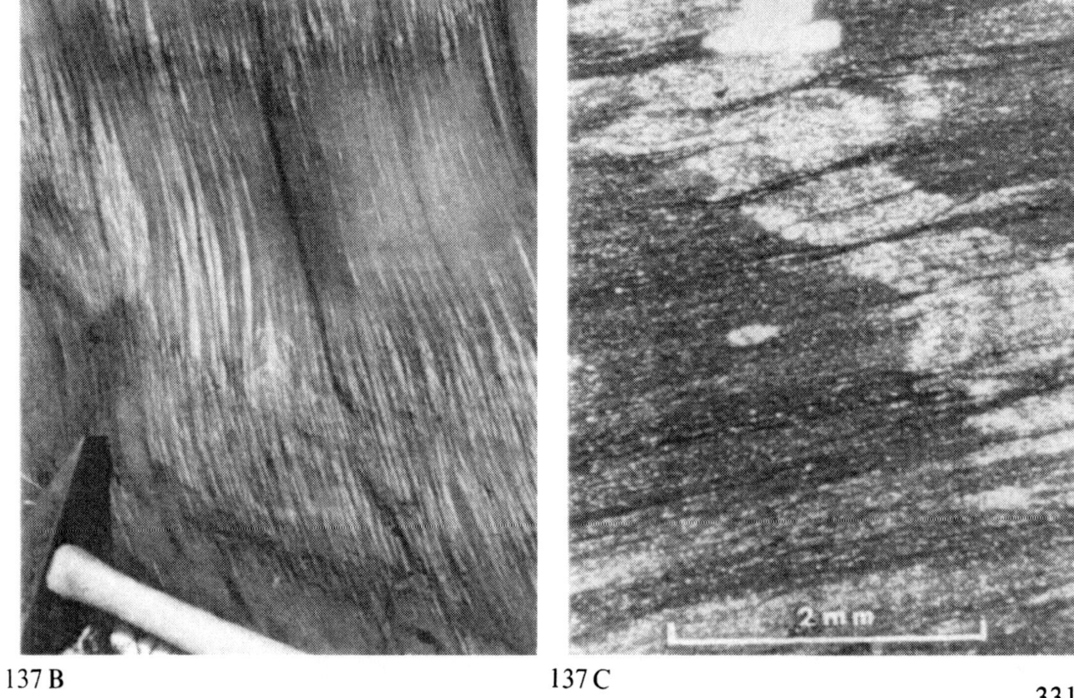


137 B

137 C

## 138. Cleavage Structures in Schists

D. G. BISHOP

The Otago Schists of southern New Zealand are derived from, and flanked by, uncleaved rocks of greywacke type. The greywacke-schist transition can be subdivided into a series of textural zones representing progressive stages in the intensity of cleavage development. Structures such as those illustrated opposite are not uncommon in textural zone IIA (BISHOP 1972a), in which cleavage first appears in psammitic rocks.

---

**A. Deformed and disrupted bedding.** Dansey Pass area, Otago, New Zealand (BISHOP 1972b, 1974). Thinly bedded sandstone-siltstone cut by a spaced (? solution) cleavage. The sandy layers are disrupted, either by the physical introduction of argillaceous material (water expulsion) or by the chemical removal of silica (pressure solution) along the cleavage. Note the concentration of discrete cleavage planes in the argillaceous zones between the segments of the pale-colored, more siliceous, sandy beds. The lobate forms may represent deformed load casts.

The rocks are of probable Triassic age, deformed and metamorphosed during the early Cretaceous. Estimated maximum PT conditions for this locality are around 280 °C and 4 kbar: the metamorphic climax, however, postdated the completion of folding.

**B. Transposition of bedding.** Dansey Pass area, Otago, New Zealand (BISHOP 1972b). The trace of bedding is approximately parallel to the hammer handle, but the sandy layers are separated into lenses, elongate parallel to the cleavage. The separation of the bed into lenses appears to result from the removal of silica by pressure solution along the cleavage planes. This outcrop is on the limb of the same fold as the outcrop **A** above

138 A

138 B

# 139. Slip-surfaces of Microlithons

N. B. Lebedeva

In rocks possessing microlithons of different scale and thickness, their relative motion may be distinguished. Motion is concentrated along spaced macroscopic cleavage surfaces where displacement may reach a few centimeters. Thinner lithons, down to microlaminae, may be distinguished inside thick ones; they are separated by surfaces, along which displacements are dispersed in character and which are smaller, proportional to the thinness of the laminae. The laminae or microlithons represent individual structural elements each of which undergoes flattening in the course of subsequent deformation. This process leads to relative gliding of microlithons that creates some characteristic features of the structure of cleaved rocks.

---

**A. Synclinal fold of mudstone with strong cleavage.** Flysch synclinorium, Caucasus. Different directions and amplitudes of relative motion of microlithons along cleavage surfaces produce jagged interpenetration of adjacent layers.
Width of reproduced part of fold is 2 m.

**B. Experimental confirmation of differential motion of microlithons.** Originally horizontal layers were made up in the sample with grains composed of varying mixtures of resin with engine oil. Due to subsequent compression the grains partly clung together, forming aggregates of different size and structure. Some individual grains were preserved as were the air-filled pores between them. Viscosity of the grains was $10^5$ poises. The sample was compressed along horizontal axis by 75% and layers were crumpled into folds. The grains and grain-aggregates of different structure due to the inclusions of air were flattened and underwent stretching by 5 to 15 times (*black strokes of different thickness* on the figure). That led to an oriented structure and to formation of surfaces along which relative motions of laminae occurred. Surfaces with motions up to 10 mm are relatively sparse: such motions lead to acute bending of layers.
The macrolaminae are divided by similar surfaces into large numbers of microlaminae with relative motions of fractions of millimeters. Different directions and amplitudes of displacements create jagged contacts between layers. Disharmonic "folds" formed around a major fold may have a similar origin; they result not from bending of strata but from relative displacements of microlithons along cleavage surfaces

---

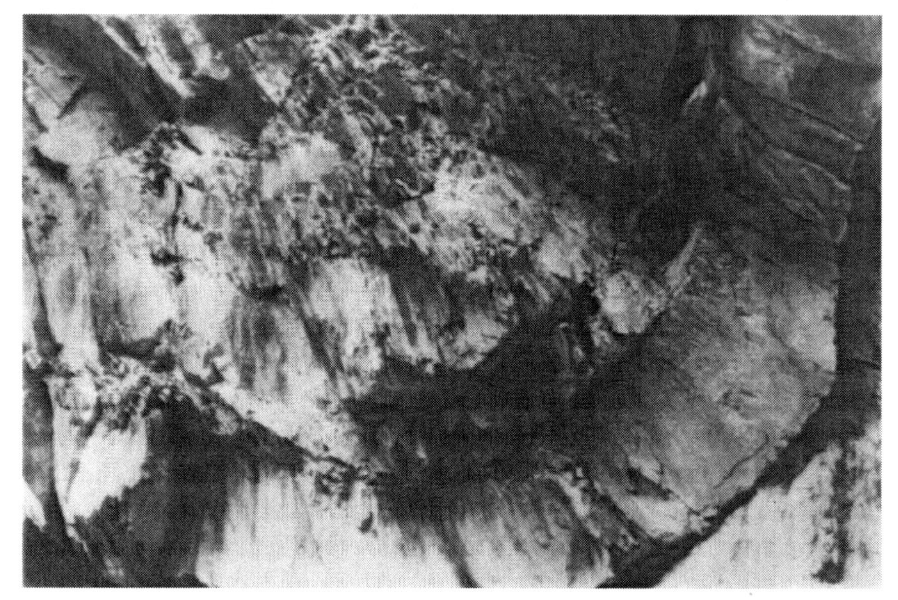

139 A

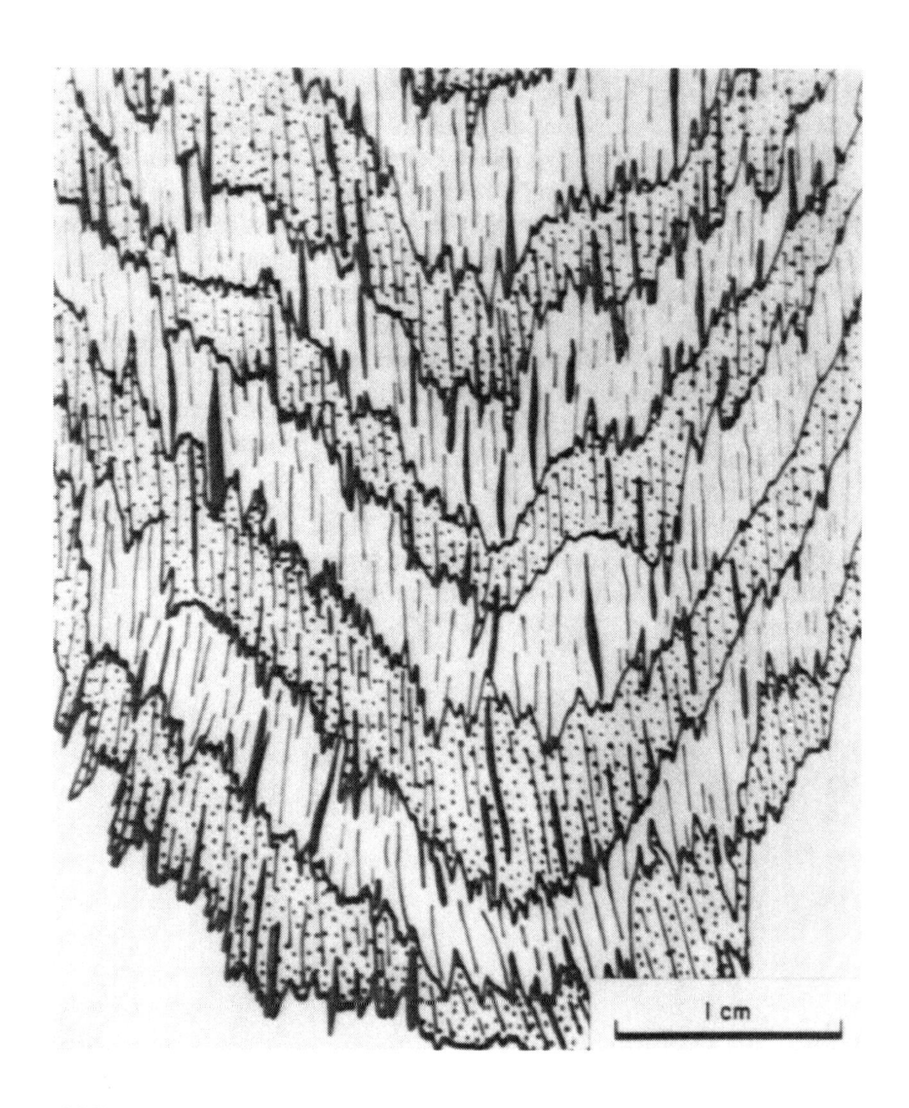


139 B

## 140. Lithological Banding Parallel to Cleavage

P. R. WILLIAMS

Photographs are of rocks from Bathurst Harbour, South-Western Tasmania, Australia, which belong to a group of relatively unmetamorphosed, probably Late Proterozoic, turbidites and resedimented conglomerate (see also Plate 110). The sequence has been deformed during four tectonic events which produced folds and cleavage. The last three of these events produced cleavage throughout the area.

A dominant lithological banding formed during the third deformation event in thinly interbedded mudstone and fine-grained sandstone and siltstone. In medium-grained sandstone beds, mudstone bands transecting bedding are more widely spaced than in the thinly bedded rocks and are restricted to fold core regions. The structures are most probably the result of metamorphic differentiation, the apparent offset of beds along cleavage seams being due to removal of silica from the limbs of microfolds. Grains of quartz, quartzite, and muscovite terminate abruptly against the zones, suggesting mobility of both mica and quartz. Seams composed of mica and opaque minerals formed in all three structural events. Mudstone flames of sedimentary origin also occur. Later structural events enhanced the original sedimentary structures, and provided zones of high strain along which cleavage seams developed.

---

**A. Lithological banding.** Lithological banding parallel to the cleavage is defined by dark seams. The zones are somewhat anastomosing around diamond shaped and lenticular lozenges of coarse siltstone. The isolated lozenges (e.g., *center left*) containing bedding remnants show that there has been some movement along faults parallel to the cleavage direction. Where the seams cross the silty layer (*bottom center–top right*) the envelope of which represents bedding, they do so in the limbs of small-scale folds. The apparent offset is due to the removal of silica from the limbs of these folds. The spacing between seams in the vicinity of the siltstone beds is controlled by the wavelength of the microfolds. This spacing is dominant for a considerable distance from the siltstone beds, and additional zones in the muddier rock are subsidiary.

**B. Cleavage overprinting a seam.** Seams define the $S_3$ cleavage and have been deformed by a later event which produced a crenulation cleavage at an angle of about 25° to the seams. The seam direction is vertical, and the crenulate nature of the cross-cutting cleavage is apparent. The section is from the rock shown in **A.** PPL

**C. Seams parallel to $S_4$.** Detail of the cross-cutting crenulation shown in **B**. Mica flakes terminate against zones enriched in opaque minerals and new mica. Some quartz grains also appear truncated by the cleavage. The termination of mica, folded to form the crenulation, suggests truncation either by solution or by mechanical processes. PPL

140 A

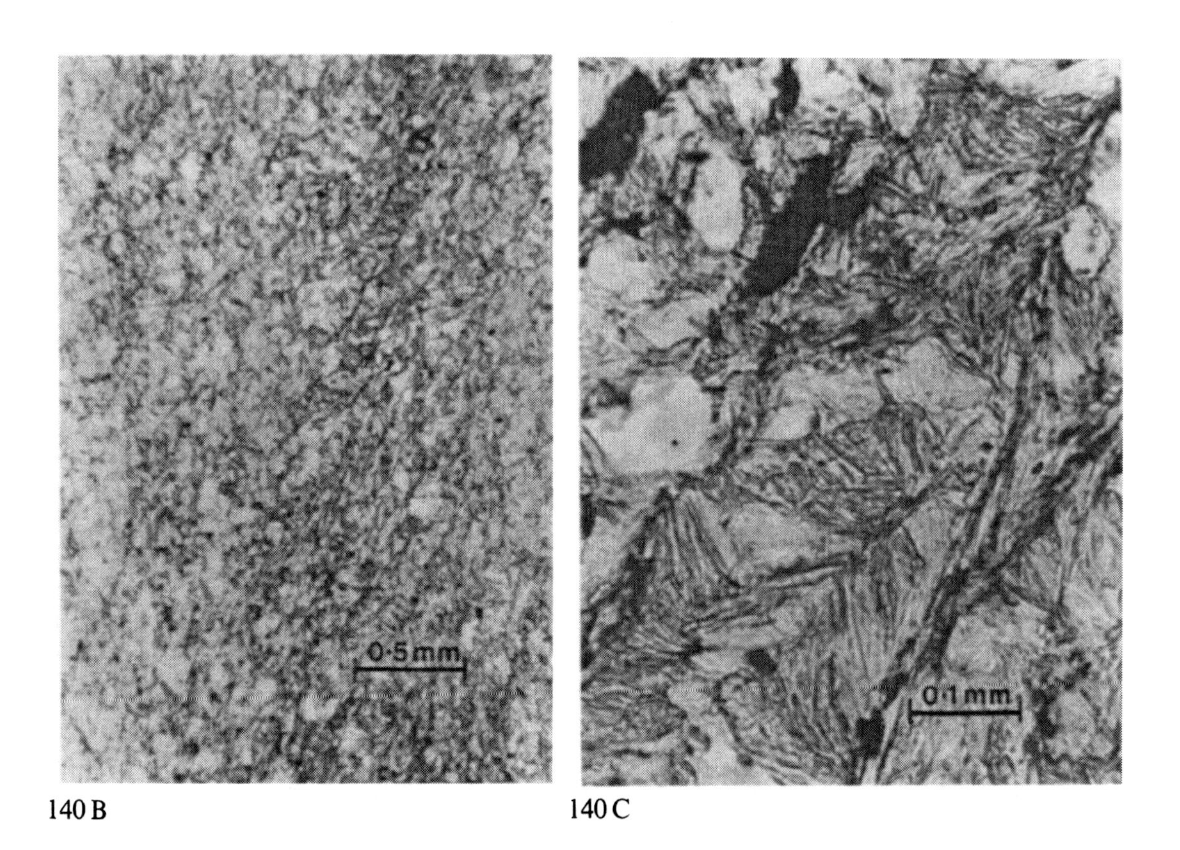


140 B

140 C

# Spaced Cleavage

## Section 13: Cleavage Defined by Seams Differentiated Without Dissolution or Disaggregation Aspects

(Plates 141 to 147)

The following seven contributions are mostly from high grade, crystalline rocks and cover aspects of differentiation associated with mylonites, shear zones, migmatites, and schists.

---

**Plates**

# 141. Foliation in Quartz Mylonite

C. J. L. Wilson

The photographs are taken perpendicular to the schistosity and lineation in two specimens collected from the Mount Isa Fault Zone, Queensland, Australia (Wilson 1975). The grade of metamorphism associated with the deformation is lower greenschist facies.

---

**A. Quartz mylonite.** A preferred dimensional orientation of quartz defines the schistosity in the quartz mylonite. Elongate and deformed quartz grains occur as distinct ribbons (Wilson 1975) surrounded by a mantle of fine recrystallized grains. These deformed quartz grains are small relics of older detrital grains. They are characterized by undulose extinction and a subgrain development (White and Wilson 1978). NX

**B. Quartz-mica mylonite.** There is a microstructural transition from the quartz mylonite in **A** to the fine-grained phyllitic mylonite illustrated here; this is a function of mica content. The extent of recrystallization in the quartz ribbons appears to decrease as the mica content increases. Similarly the grain size of the recrystallized quartz in the mica-free areas is larger than in the micaceous areas (0.06 mm versus 0.02 mm). The mica occurs as fine lenticular aggregates that define a lineation; where abundant mica is present (as at *M*), the foliation is planar. In quartz-rich regions the mica aggregates are deflected around the ribbon quartz resulting in an anastomosing fabric. NX

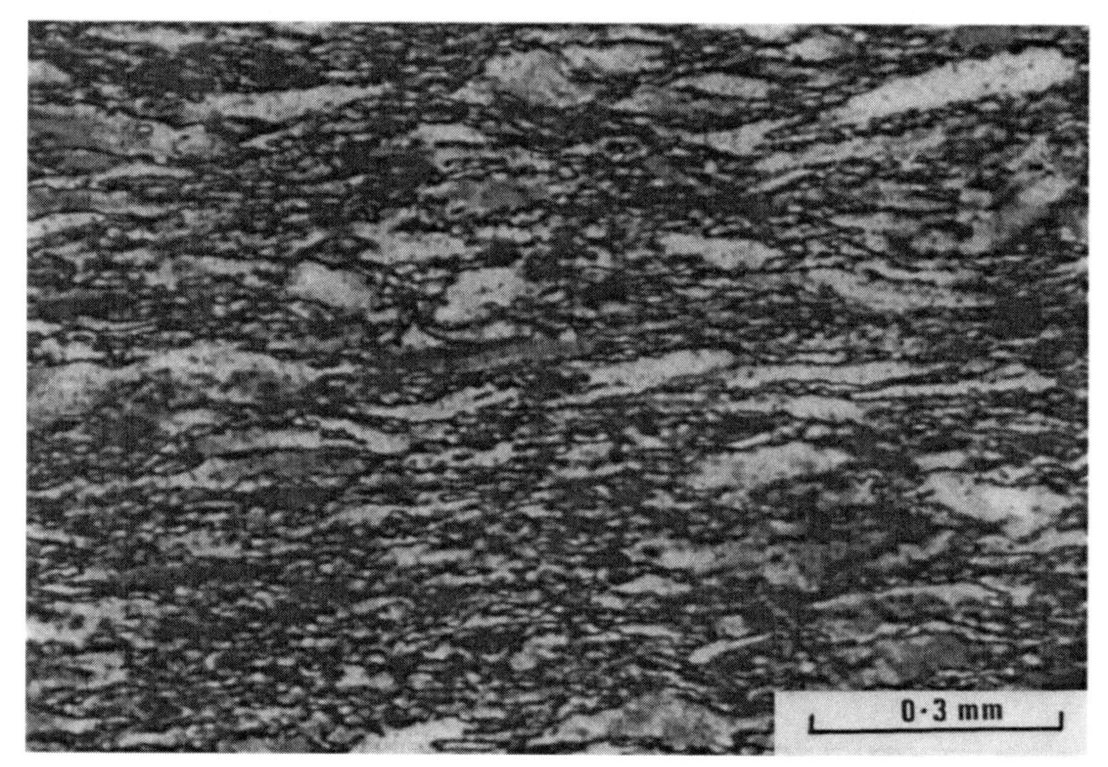


141 A

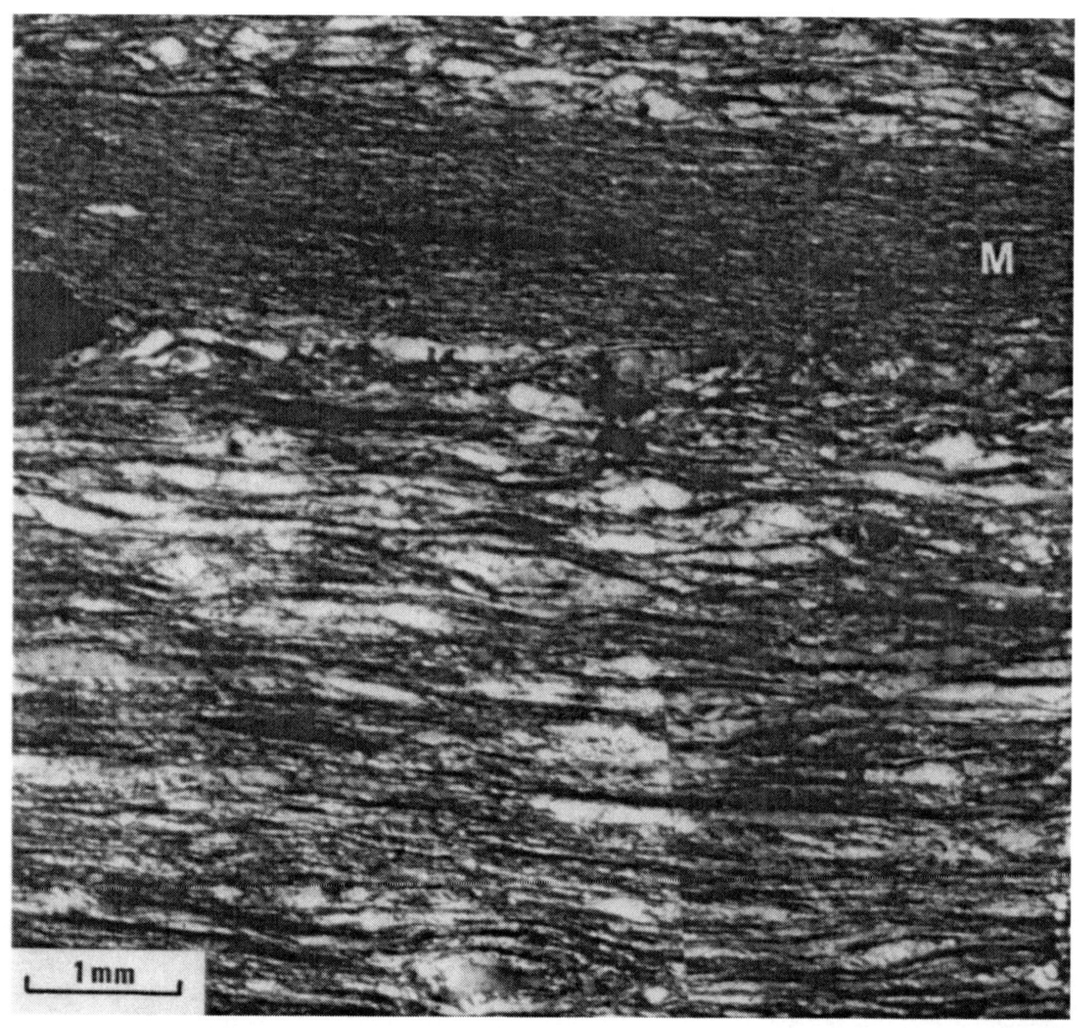


141 B

# 142. Shear Zones and Leucosome Bands in Migmatite

G. J. BORRADAILE and M. C. KENNEDY

Archean migmatite near Dexter Lake, Golding Township near Thunder Bay, N. Ontario, Canada. These rocks occur in the northern part of the Archean Quetico gneiss belt in which migmatites are well developed in biotite-schists. A general association has been recorded between shear zones and the development of an episode of formation of bands of more coarse-grained leucosome. Elsewhere this structure has been called diktyonitic by MEHNERT (1971).

---

**A.** The earlier leucosome banding is displaced by ductile, dextral-slip shear zones. In some cases, as here, a more coarse-grained leucosome developed along the center of the shear zone. (The coarse-grained leucosome is indicated by *coarser stipple* in the sketch.)

**B.** Usually the second phase of leucosome banding is less distinct, as in this photograph. However, the "net-veined" appearance of the migmatite is more regular on closer inspection. The undulations of the earlier leucosome constitute shear zones and along some of these, newer leucosome is discordantly developed

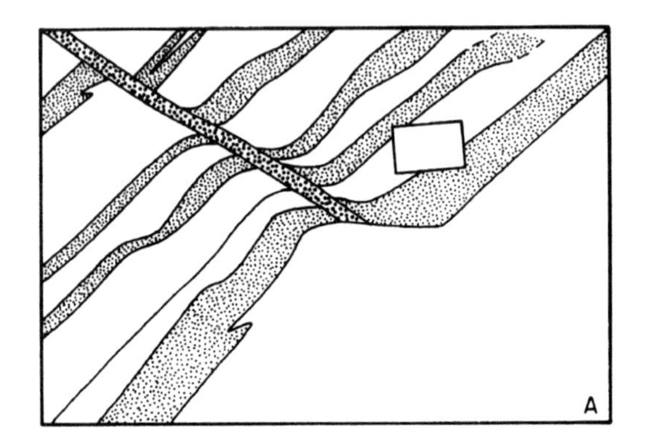


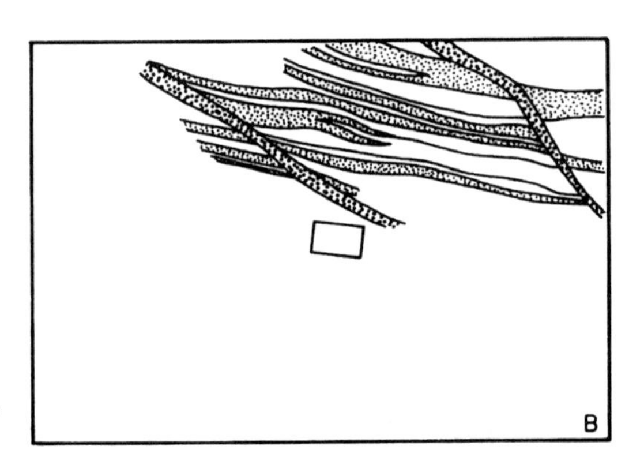

142 A

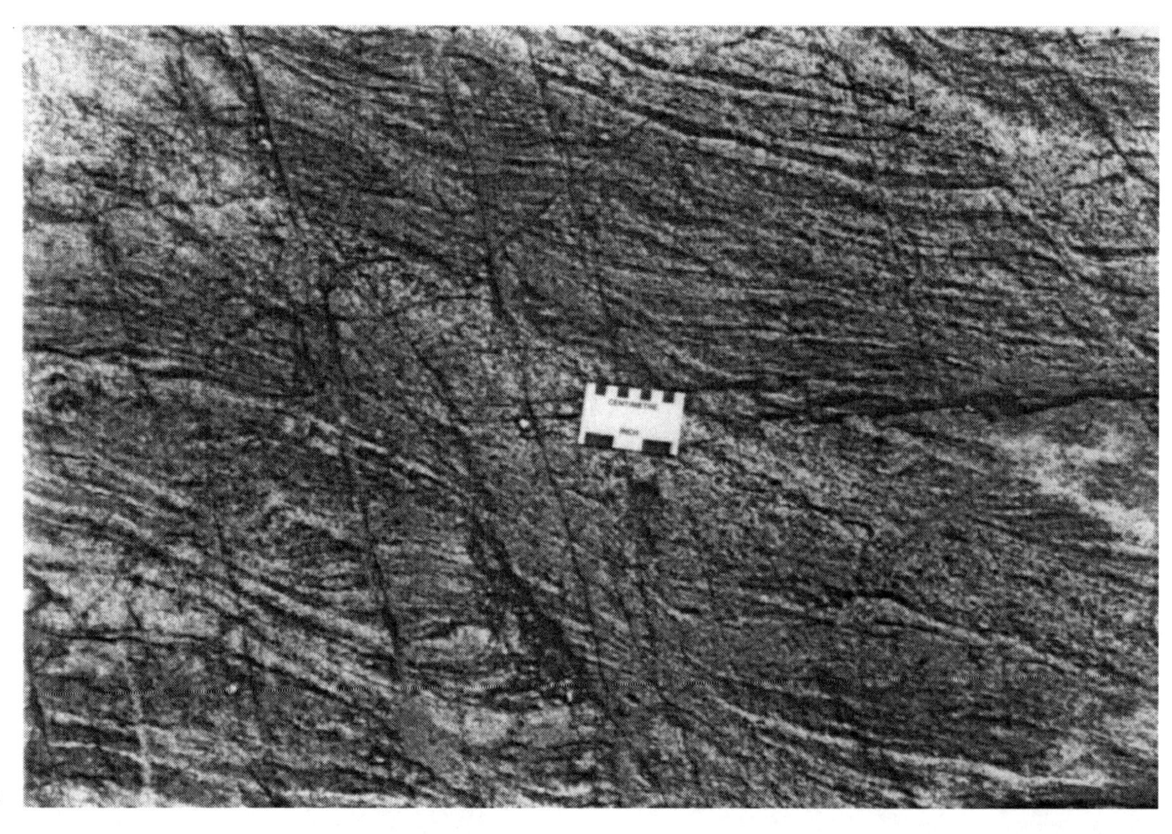

142 B

# 143. Structures of Migmatites (I)

F. W. BREAKS and W. D. BOND

Archean interlayered metawacke-metapelite sequence, Northern Supracrustal Domain, English River Subprovince at Ontario Hydro damsite, Ear Falls, northwestern Ontario, Canada.

**A. Combination of protometatexite and metatexite stages of migmatization.** Early stages of metasedimentary migmatite evolution are marked by incipient partial melting (possibly accompanied by metamorphic differentiation). Protometatexite (BREAKS et al. 1978) refers to a migmatite stage characterized by small quantities of irregular, podiform to lensoid cordierite-bearing leucosome segregations which are specifically confined to metapelitic layers. Such segregations have not undergone amalgamation into foliation-controlled, continuous leucosome layers as is the case of the metatexite stage distinguished by the stromatic structure of MEHNERT (1971 p. 18). Nevertheless, a layered distribution is usually evident since the protometatexite leucosome exhibits restriction to metapelitic bulk compositions which are generally intercalated with non-leucosome-bearing metawacke units. There is also often a tendency for leucosome masses in the metapelite to segregate within or proximal to low pressure zones between metawacke boudins (*left of knife*).

Scale: knife length = 8.3 cm

**B. Metatexite stage of migmatization in Archean metasediments.** This migmatitic stage is characterized by a conspicuous banded appearance manifested by low volumes of in situ anatectic segregation of leucosome, controlled by foliation and bedding anisotropies of host metapelitic and metawacke paleosome. In this example the stromatic structure, one of the prime identifying criteria of the metatexitic migmatitic stage, is relatively obvious, being affected by only relatively mild extensional deformation in the form of pinch and swell structures. Note especially the more prolific development of leucosome in metamorphically coarsened rocks of more favourable bulk composition, i.e., almandine-sillimanite-cordierite-K-feldspar-biotite-quartz-plagioclase metapelite (especially at *right side near knife*). In contrast, note paucity of leucosome in wide bands of pale grey, fine-grained, almandine-biotite-quartz-plagioclase metawacke, particularly *left of center.*

Scale: knife length = 8.3 cm

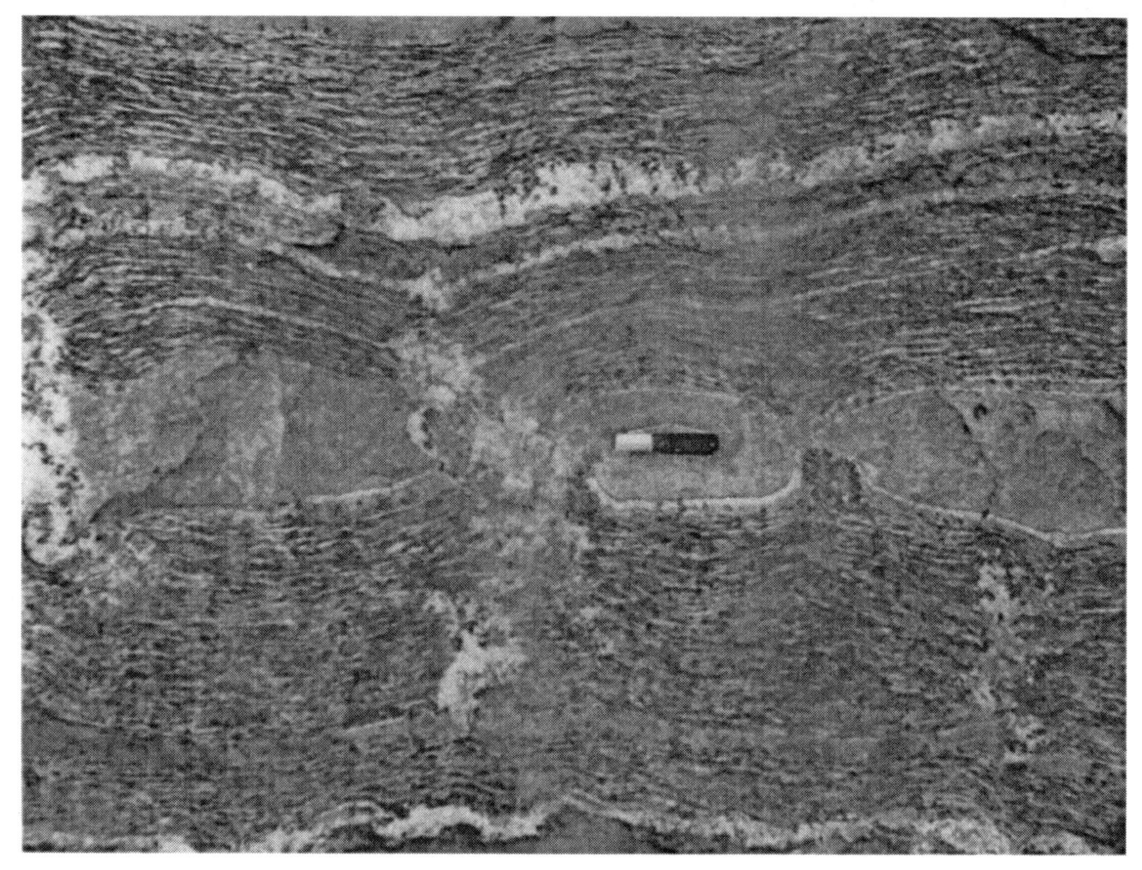

143 A

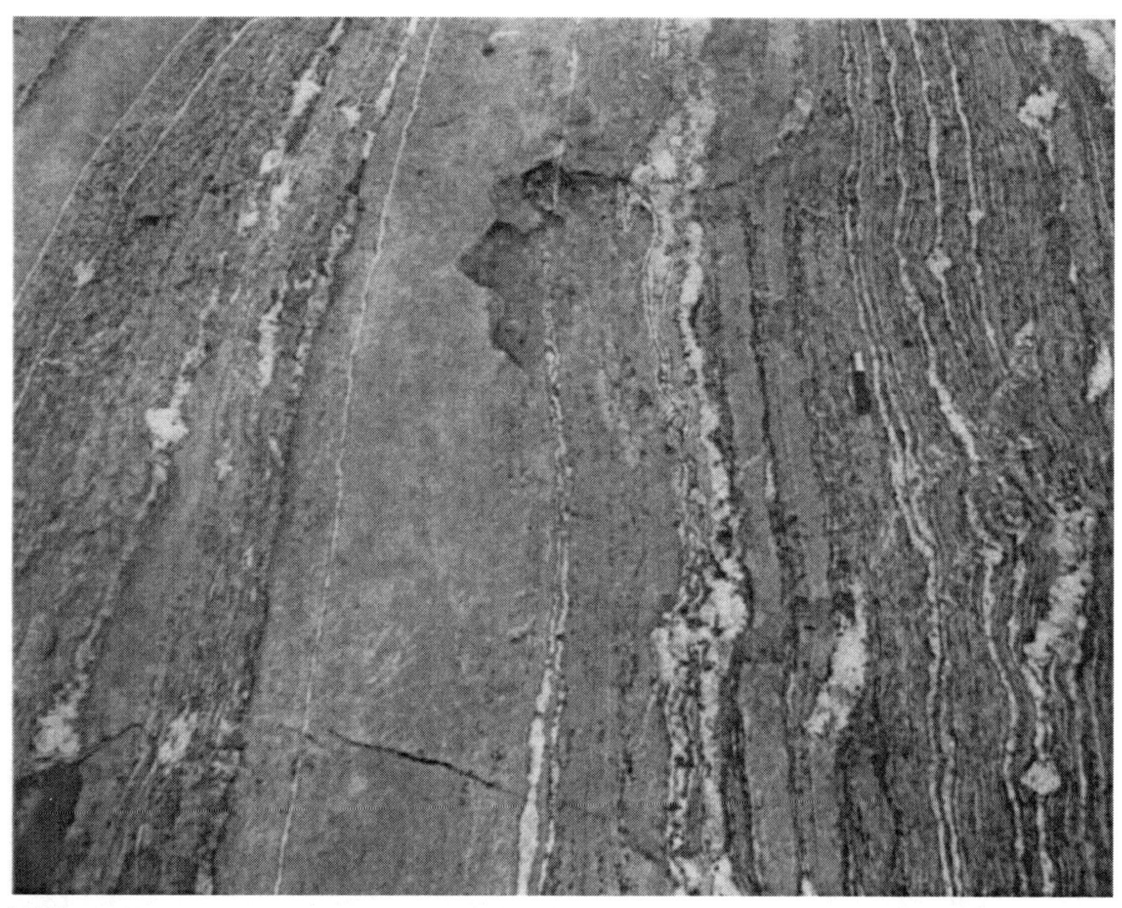

143 B

# 144. Structures of Migmatites (II)

F. W. Breaks and W. D. Bond

Archean metasediments of Northern Supracrustal Domain, English River Subprovince at Ontario Hydro damsite, Ear Falls (**A**) and at Pakwash Lake (**B**), northwestern Ontario, Canada.

---

**A.** Disharmonic folding of apatite-biotite syenite leucosome and host metawacke paleosome. Note exceptional development of thick, biotite-rich melanosomes along both the paleosome/leucosome interfaces. Quartz-impoverished leucosomes, such as this example, are quite rare in migmatized metasediments of the English River Subprovince and usually are associated with low-pressure granulite metamorphic zones. This particular occurrence is in the high grade almandine-sillimanite-cordierite-K-feldspar zone. Scale: knife length = 8.3 cm

**B.** This metatexitic metasedimentary migmatite exhibits a well-developed stromatic structure comprised mainly of arteritic leucosome which has been exposed to two periods of deformation. Early boudinage was followed by open, disharmonic folding (see especially area *immediately right of coin*). Metamorphic assemblage of metapelitic paleosome: almandine-sillimanite-cordierite-K-feldspar-biotite-quartz-plagioclase. Scale: diameter of coin = 2.3 cm

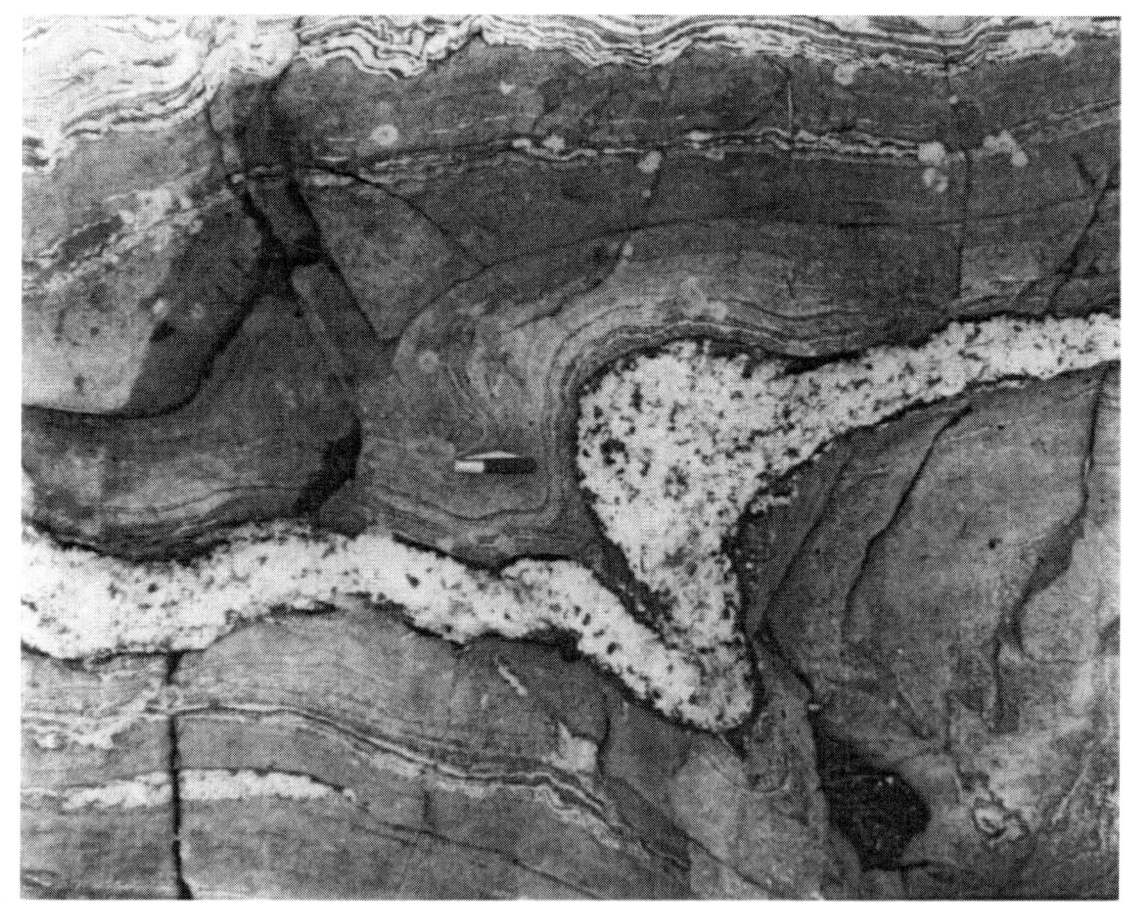

144 A

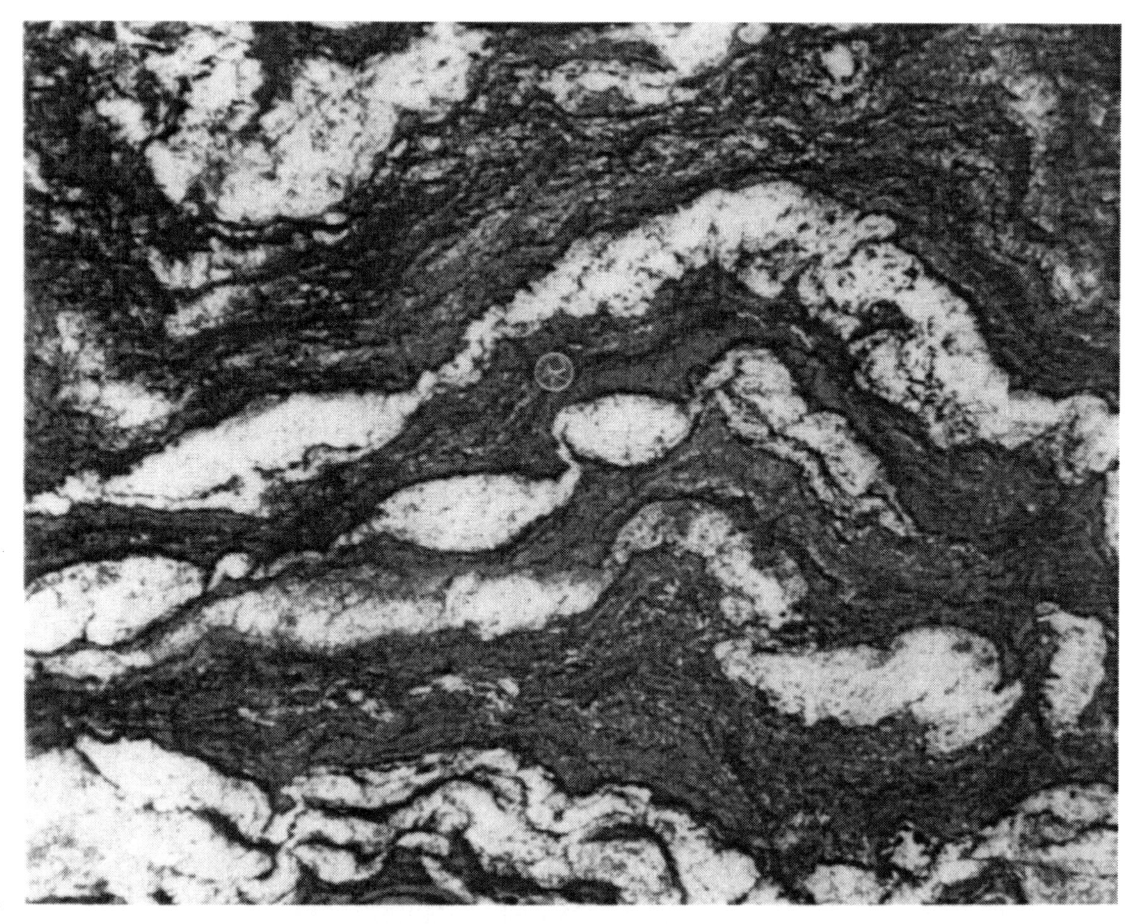

144 B

## 145. Quartz Blades in a Gneiss

S. FINLEY and B. M. BAYLY

Precambrian granitic gneiss of Adirondack type, from road-cut on Route 22, 2 miles north of Fort Ann, N.Y., U.S.A.

The plate shows an acetate peel from a polished and etched rock surface: quartz shows *dark* and feldspar *paler*. Other constituents are in very small amounts. In sections cut perpendicular to the section shown, the average length of the quartz blades is about two-thirds of the length seen here; that is, the blades are broad, more like leaves.

Elsewhere in the Adirondacks, quartz rods, pencils, and blades are abundant, some reaching 1 m in length; no tendency to be localized in zones of more intense deformation has been recognized. On the other hand, the rocks shown have affinities with ribbon mylonites, Plates 146, 151. The tendency for quartz to form blades or ribbons seems to be more a characteristic of the mineral than of any particular tectonic setting.

145

# 146. Quartz Microstructures Associated with the Himalayan Main Central Thrust

J.-L. BOUCHEZ

See also Plate 149

The Himalayan Main Central Thrust is a huge, low-angle, intracontinental shear zone (LE FORT 1975, MATTAUER 1975), studied in Central Nepal by PÉCHER (1978).
Five domains of quartz microstructures have been identified in the deformed quartz-rich rocks (BOUCHEZ and PÉCHER 1976) extending through a section several kilometers thick above and below the thrust "plane". The thrust is considered to be the base of the high grade rocks of the Tibetan plate.
Type 1 is an exaggerated grain-growth microstructure extending above the thrust plane; extensive post-tectonic recrystallization, due to persistence of high temperatures after deformation, has erased the deformation microstructures. Types 2, 3 and 4 are illustrated in Plates **A, B,** and **C**. Type 5 is a preserved sedimentary microstructure at depths greater than 3 km below the thrust, and developed at a low (chlorite) metamorphic grade.
Samples come from the Madi and Seti Khola sections (Central Nepal). The schistosity is normal to each photograph and oriented horizontally; the lineation is also horizontal in the plane of the photograph. Scales: 1 mm. Polarizer at 45° to analyzer.

---

**A. Ribbon microstructure.** Found near the thrust where the rocks have probably undergone maximum strain at a high metamorphic grade. The quartz ribbons, extending parallel to the schistosity, are made of one to several layers of subrectangular elongate grains. The inter-ribbon material consists mainly of fine-grained feldspars. Quartz is considered to be the more ductile phase segregating, perhaps with the help of diffusion, during the deformation (BOULLIER and BOUCHEZ 1978). Large grain size in the ribbons is attributed to grain growth, commonly observed in pure high temperature quartzites. The strong lattice preferred orientation induced by large strain in the dislocation creep domain of quartz, shows a [0001] unique girdle, perpendicular to the $<11\bar{2}0>$ main concentration (BOUCHEZ 1978). The angular discrepancy between this concentration and the prominent lineation is attributed to a rotational shear regime, the sense of which correlates with the sense of thrusting. Numerous visible prismatic subboundaries indicate the normal to the shear direction, giving here a dextral sense of shear.

**B. Elongate mosaic microstructure.** Developed down to 2 km beneath the thrust in highly strained rocks of high to medium metamorphic grade. These microstructures tend to become equant when impurities such as phyllites are lacking. Grain growth is limited since included impurities are scarce. Strong lattice orientations are observed; unequally populated [0001] girdles reflect the sense of shear.

**C. Relict porphyroclastic microstructure.** Developed more than 2 km beneath the thrust at medium to low metamorphic grade. The unrecrystallized quartz porphyroclastic cores are mantled by new grains. Elongate mosaic bands form progressively. Some feldspar porphyroclasts are also noticeable

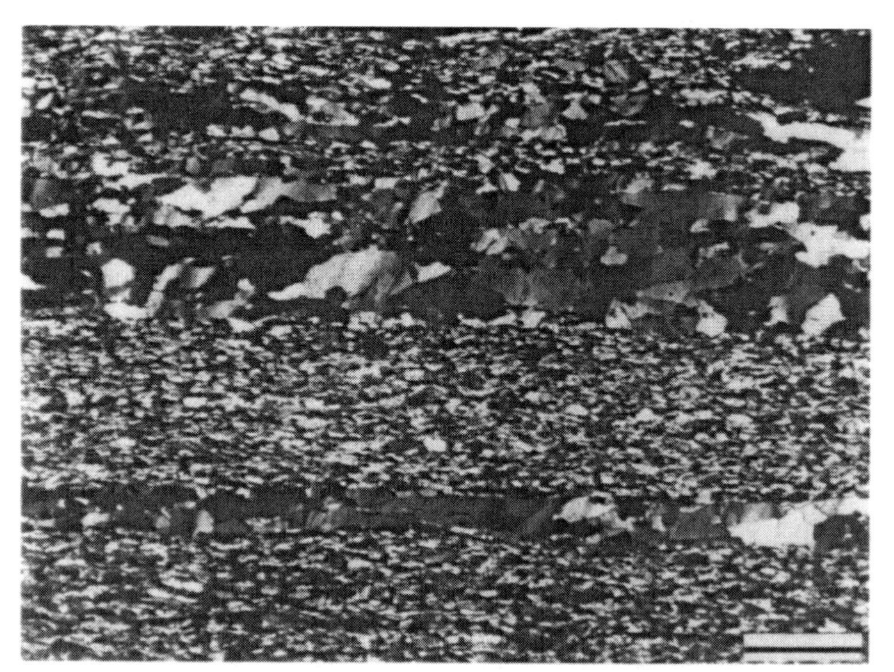
146 A

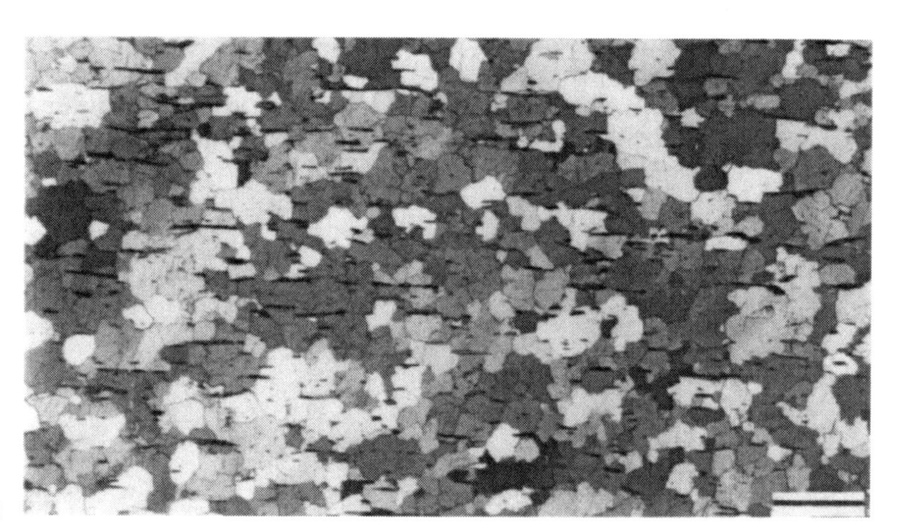
146 B

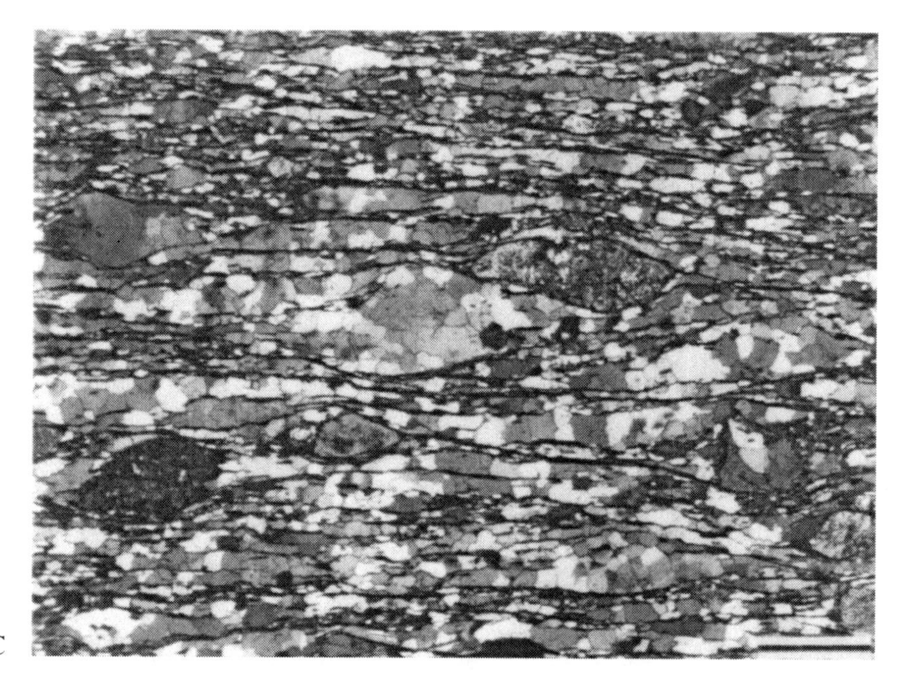
146 C

## 147. Cleavage Emphasized by Veining

J. G. DENNIS

Lower Devonian St. Francis Group, Caledonia County, Vermont, U.S.A. Schistosity ($S_2$) associated with Strafford-Willoughby arch (DENNIS 1956). $S_1$ is parallel to bedding in both outcrops. All schistosity is Acadian age.

---

**A. Gilman formation.** (Wheelock township.) Bedding trace going from top left to bottom right of picture separates staurolite schist (*bottom of picture*) from quartz schist (*top of picture*). Schistosity traces cross the bedding trace at a steep angle as seen in the picture. The staurolite schist has a sericite schistosity, with minor quartz, biotite and muscovite. Late biotite, muscovite, garnet and staurolite porphyroblasts are at best weakly oriented. Although the staurolites outline the schistosity as well as the bedding surface (erosion-resistant, crust-like appearance), they are not dimensionally oriented along these fabric planes; they post-date the schistosity, $S_2$. Metamorphic differentiation here imparts a composition foliation, but not a schistosity. The quartz schist consists chiefly of quartz, albite, and minor white mica. Its schistosity is emphasized by quartz veining which also post-dates $S_2$. Scale: half-dollar coin

**B. Waits river formation.** (Near Lyndonville.) Siliceous marble, schistosity defined by dimensionally oriented calcite, quartz, and phlogopite. Bedding revealed by quartz rich layers crossing picture *from top to bottom*. Schistosity emphasized by prominent veins of calcite and quartz. The veining is associated with post-$S_2$ metamorphism

147 A

147 B

## Section 14: Cleavage Defined by Differentiation on Grounds of Texture or Geometry but not Composition

(Plates 148 to 187)

This large group of forty contributions can be identified by differentiation into parts of differing texture or geometry. In most cases it is clear that the composition of the rocks is unchanged. Kinking, shock structures, cataclastic fabrics, pseudotachylite veining and some mylonitic schistosities are included in this section. The range of materials represented is also very wide: ores (158 to 161), peridotites (162, 163), salt glaciers (164), glacial ice (167), water (168, 169), experimentally deformed rock (165, 166, 175, 183–186), volcanic rocks (180, 181), meteorites (171, 172), as well as more conventional schists and slates.

---

### Plates

181 Foliation development in deformed hydrothermally altered acid volcanic rocks (II)

182 Kinking in a rotating stress field

183 Experimentally deformed clay (I): cleavage-like microstructures

184 Experimentally deformed clay (II): cleavage-like microstructures

185 Experimentally deformed clay (III): crenulations

186 Experimentally deformed clay (IV): other microstructures

187 Primary crenulation cleavage in clay shale

# 148. Lenticular Schistosities in Quartz-phengite Schists

C. J. L. Wilson

A variety of deformation events in a quartzite sequence at Zermatt (Pennine Alps, Switzerland) have developed a complex schistosity, characterised by either individual elongate phengitic micas or phengite aggregates (Wilson 1979). Superimposed on this schistosity are localized micro-shear zones. The shear zones are either subparallel or are oriented at an angle up to 45° to the pre-existing schistosity. They are responsible for the disruption of the phengite layers and the development of a strongly lenticular microstructure. This microstructure is similar to the "button" or "fish scale" structure described by Roper (1972).

---

**A. Boudinaged phengite micas.** Deformation of the phengites is often accompanied by boudinage. Individual grains are subdivided into separate elongate discs or lenses that are sometimes connected by narrow tapering necks. The ends of the neck are generally rotated, with the sense of rotation of the wider micro-shear zone. Phengite lenses have a high density of optically visible (001) cleavage planes. Thin misoriented elongate segments occur within individual lenses; they are bounded by closely spaced (001) cleavage planes and have a feathery appearance at their termination, which is generally within a grain and can be seen only at very high magnifications. These phengites are, in places, enclosed by a narrow phengite envelope (*A*) characterized by closely spaced (001) cleavage planes. In other phengite lenses (001) cleavage segments terminate along a grain boundary as a series of steps (*B*). In both cases there is a strong suggestion that there is slip on the (001) cleavage planes with internal disruption of some cleavage segments as the phengite grains are locally elongated during deformation. NX

**B. Lenticular phengite aggregates.** Phengitic aggregates intersected by micro-shear zones commonly have oblate ellipsoid shapes. Ends of individual aggregates either terminate as a narrow point (*C* and *D*) or as a series of lenticular grains (*E*). Individual lenses are themselves composed of disc-shaped phengite grains (*F*) with small displacement zones truncating the phengite lenses before disappearing into quartz-phengite grain boundaries. The enclosing quartz grains have a high degree of undulose extinction and highly serrated boundaries. NX

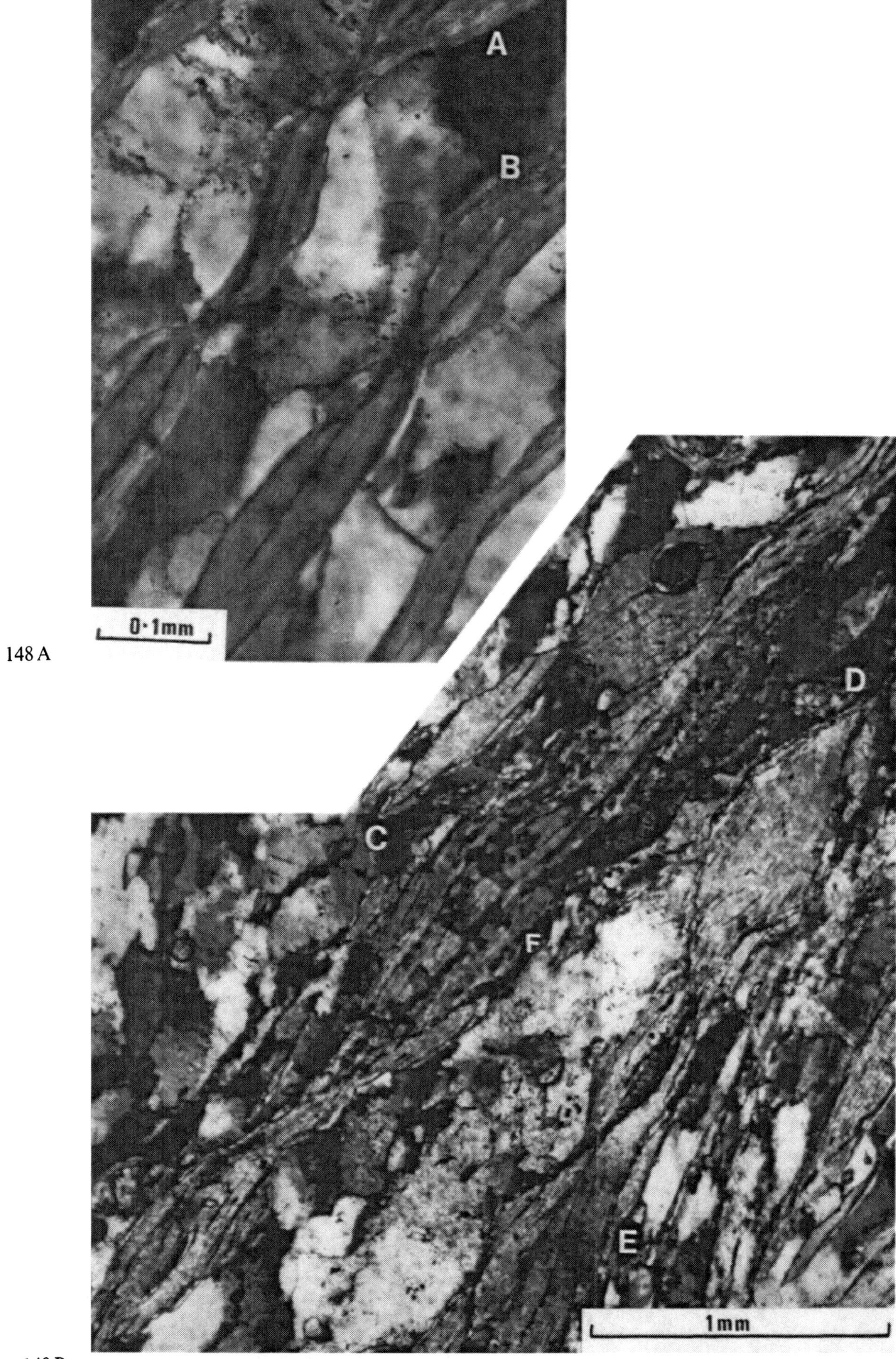


148 A

148 B

# 149. Microstructures in Low Temperature Porphyroclastic Quartzites

J.-L. BOUCHEZ

See also Plate 146

A sedimentary sequence enveloping an orthogneiss core exhibits a progressive increase in finite strain at a low grade metamorphic environment (up to biotite) in a narrow Hercynian shear zone crossing Angers, Massif Armoricain, France (BOUCHEZ and BLAISE 1976). The sandstone-quartzite formations have been studied (BOUCHEZ 1977) in terms of their microstructures and the plastic behavior of the progressively deformed and unrecrystallized original quartz grains, or porphyroclasts. These grains are embedded in a fine-grained quartzose matrix, the fractional volume of which increases with strain by dynamic recrystallization of the porphyroclasts. BOUCHEZ (1978) concludes that <a> slip directions predominate. The usual basal (0001) slip plane (common-type porphyroclasts) is considered to be complemented, at higher temperature, by slip along the prismatic planes, {1010} (ribbon grains). The [0001] axes preferred orientation patterns within the underlying orthogneiss suggest that prismatic slip dominated.

The average flattening plane of the grains (schistosity) is normal to each photograph and oriented E–W; the average elongation direction (lineation) trends E–W in the planes of photographs.

Scale: bars = 1 mm. **A** and **C**: polarizer at 45° to analyzer; **B**: NX

---

**A. Porphyroclastic microstructure.** Here illustrated in a highly deformed sample. Average porphyroclast size: $2.4 \times 0.95 \times 0.40$ mm. Matrix is over 50%. Numerous porphyroclasts display closely spaced, NE–SW trending, prismatic subboundaries, pointing to (0001) slip plane traces which make an angle with the grain elongation direction. That NW–SE direction fits with the normal direction to the strongest girdle of [0001] axes. It is equated with the major flow plane within the rock, thus indicating a dextral rotational component of shear. Post-kinematic cracks or bubble trails develop perpendicular to the shear direction.

**B. "Augen" grains.** These represent porphyroclasts that are unsuited for slip, as described in experimentally deformed peridotites (NICOLAS et al. 1973) and quartzites (TULLIS et al. 1973). Especially remarkable are the *globular* grains (*right*) with their (0001) plane nearly perpendicular to the lineation direction, and more precisely, often perpendicular to the shear direction. Notice the typical undulose kink-bands and the subbasal, bright, widely spaced, and flexuose deformation lamellae. The *tabular* grains (*left*) are subsequent to elongate in shape, with their (0001) plane parallel to the schistosity plane. They have typical closely spaced prismatic subgrain structure and no deformation lamellae.

**C. Ribbon grains.** These (*R*) only appear in the highly deformed, higher temperature zones, close to the orthogneiss core. Their [0001] axis is always perpendicular to the lineation direction within the schistosity. Their exceptional elongation ratios could be explained by prismatic kinking in long bands parallel to their lengths. Faint and closely spaced deformation lamellae are frequent. This ribbon (*R*) is wrapping around a globular grain (*X*). *C*: common-type porphyroclasts

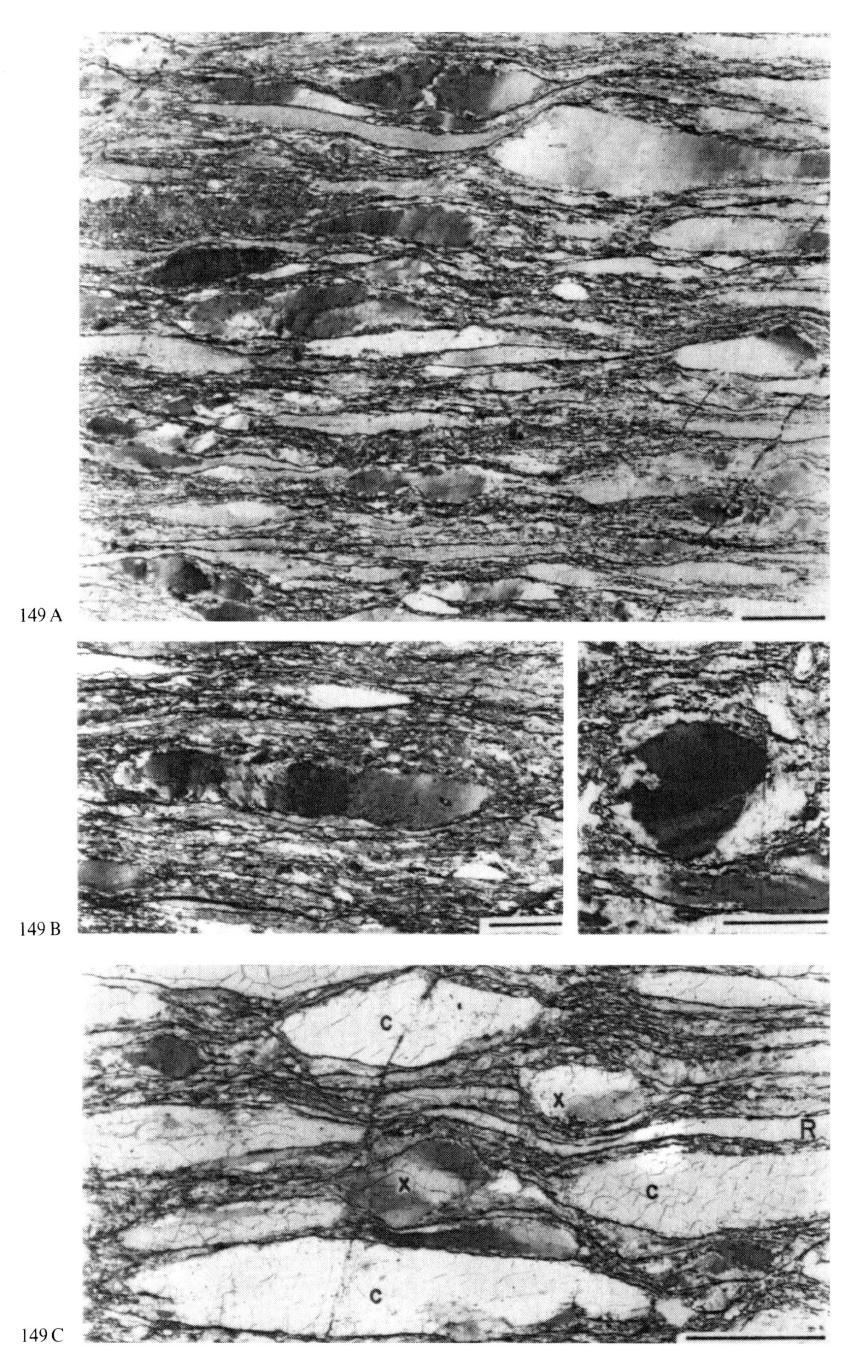
149 A
149 B
C
X
R
X
C
C
149 C

## 150. Schistosity of Meta-arkoses and Diminution of Grain Size

G. J. Borradaile

Bowmore Sandstone, Late Precambrian, Islay, Scotland (Borradaile 1979 b).
The Bowmore Sandstones form a major sedimentary unit of feldspathic sandstones and arkoses metamorphosed to lower greenschist facies. The first schistosity is feebly developed in most places. However, schistosity is better developed within a kilometer of the gently dipping thrust contact with the overlying nappe fold of Dalradian metasediments. The schistosity is illustrated in the photographs.

---

**A. Peripheral subgrain development in quartz: "core and mantle" structure.** Weak cleavage (*left* to *right*) is defined by the preferred dimensional orientation of quartz and feldspar clasts. These grains show undulose extinction. Recovery has removed distortions of the quartz lattices near the rims of the clasts so that the "cores" of the grains are now "mantled" by subgrains.
*Note:* sample preparation caused the cracks in the larger quartz grains. NX

**B. Advanced subgrain development in quartz.** This specimen is from a locality in a similar structural setting to the one illustrated in **A**. Schistosity is, however, more readily visible; it dips gently to the *left* in the photograph. It is defined by the now elongate shapes of the strained cores of originally more-or-less equant clasts left behind after the development of subgrains at preferred sites around the clasts. Thus the cores of the clasts have sutured outlines where they meet subgrains. The subgrains are also slightly elongate, subparallel to schistosity. NX

150 A

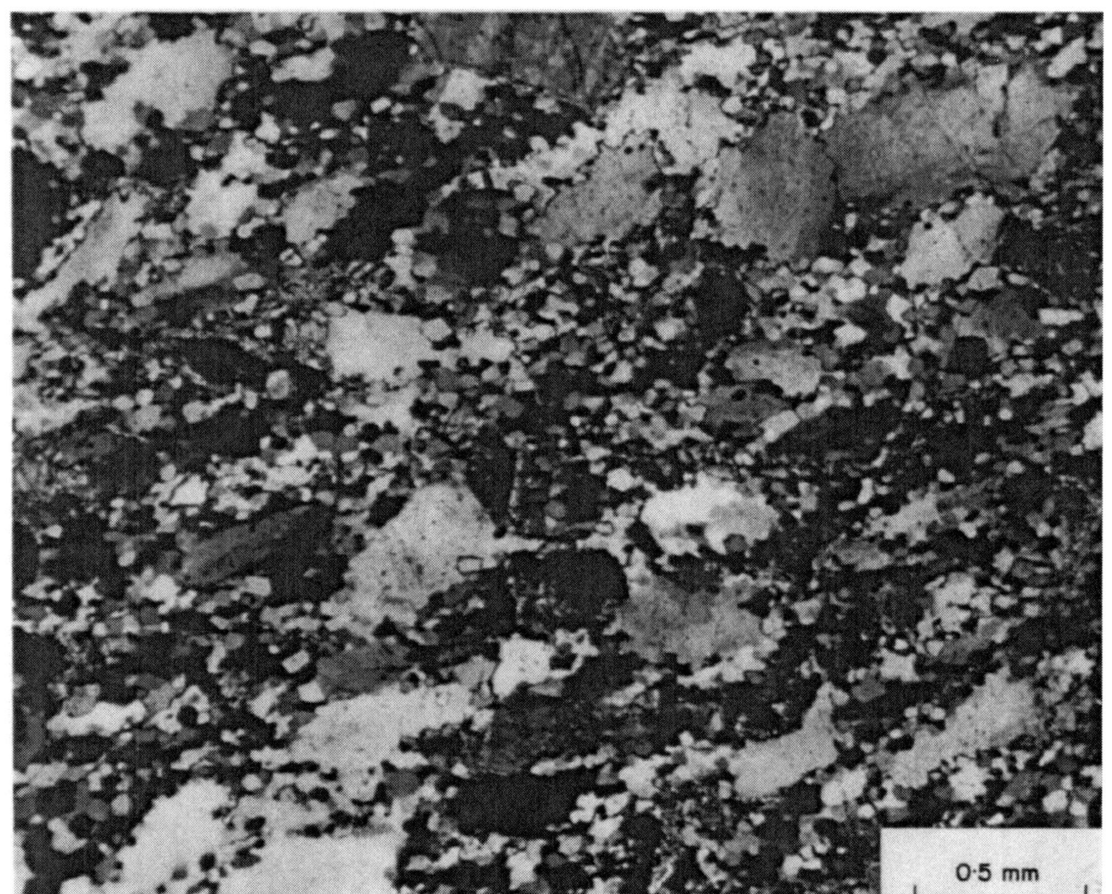

150 B

## 151. Mylonitic Schistosity and Quartz Blades or Ribbons

G. J. Borradaile

The specimens are from the Dalradian Jura Quartzite of Islay, western Scotland along the Loch Skerrols Thrust (Borradaile 1979a, b). Away from the thrust the Jura quartzite is little deformed and even grain-sized with only a subtle grain alignment defining schistosity. Within a few hundred meters of the thrust the quartz clasts show marked undulose extinction and appear to float in a sea of subgrains which developed from the margins of the original clasts. At or near the thrust contact with an older, Proterozoic formation, the clasts are severely deformed plastically to produce disc-shaped quartz grains ("quartz blades", see Plate 145; analogous to "quartz ribbons" see Plate 146). The quartz blades define a mylonitic schistosity at the thrust contact.

---

**A. Incipient mylonitization.** Near the thrust contact quartz forms peripheral subgrains and shows undulose extinction. NX

**B. Quartz blades define mylonitic schistosity.** At the thrust contact quartz blades with undulose extinction define the mylonitic schistosity. Note also the matrix of quartz subgrains. NX

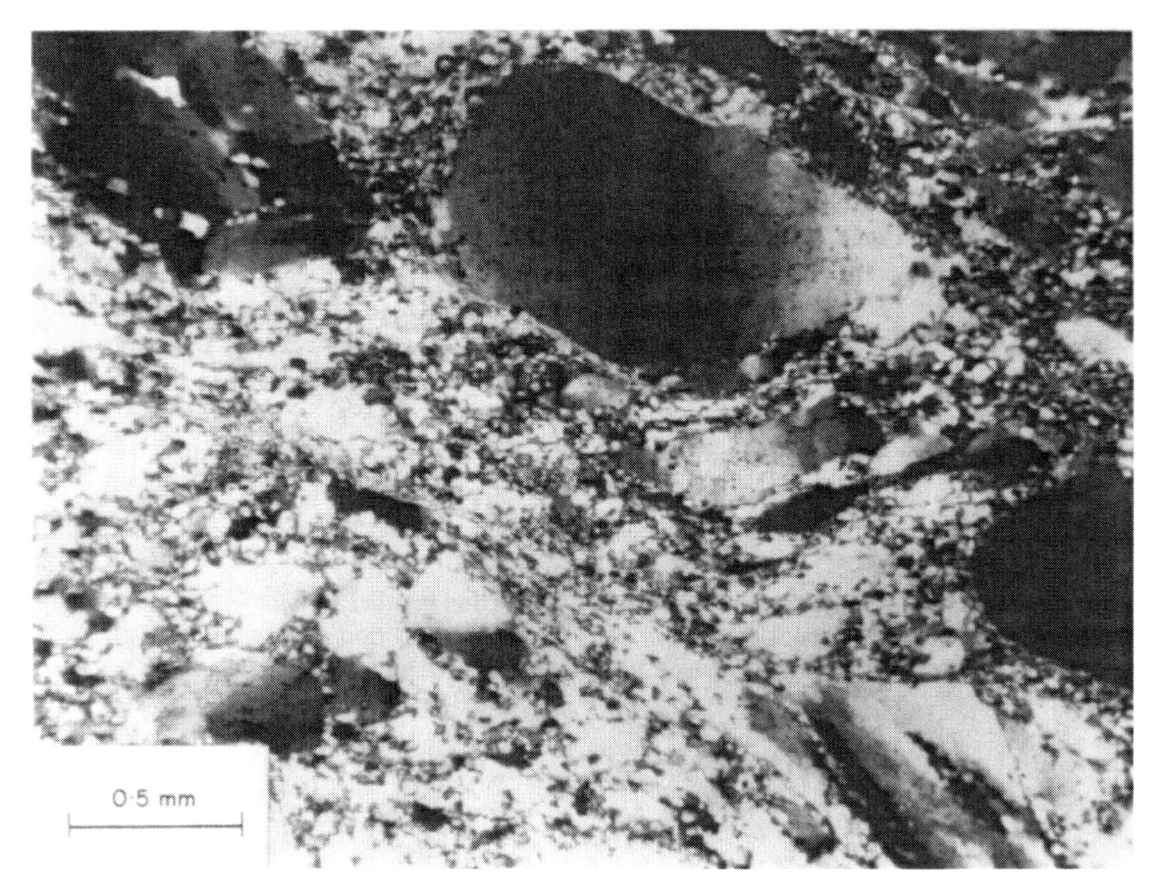

151 A

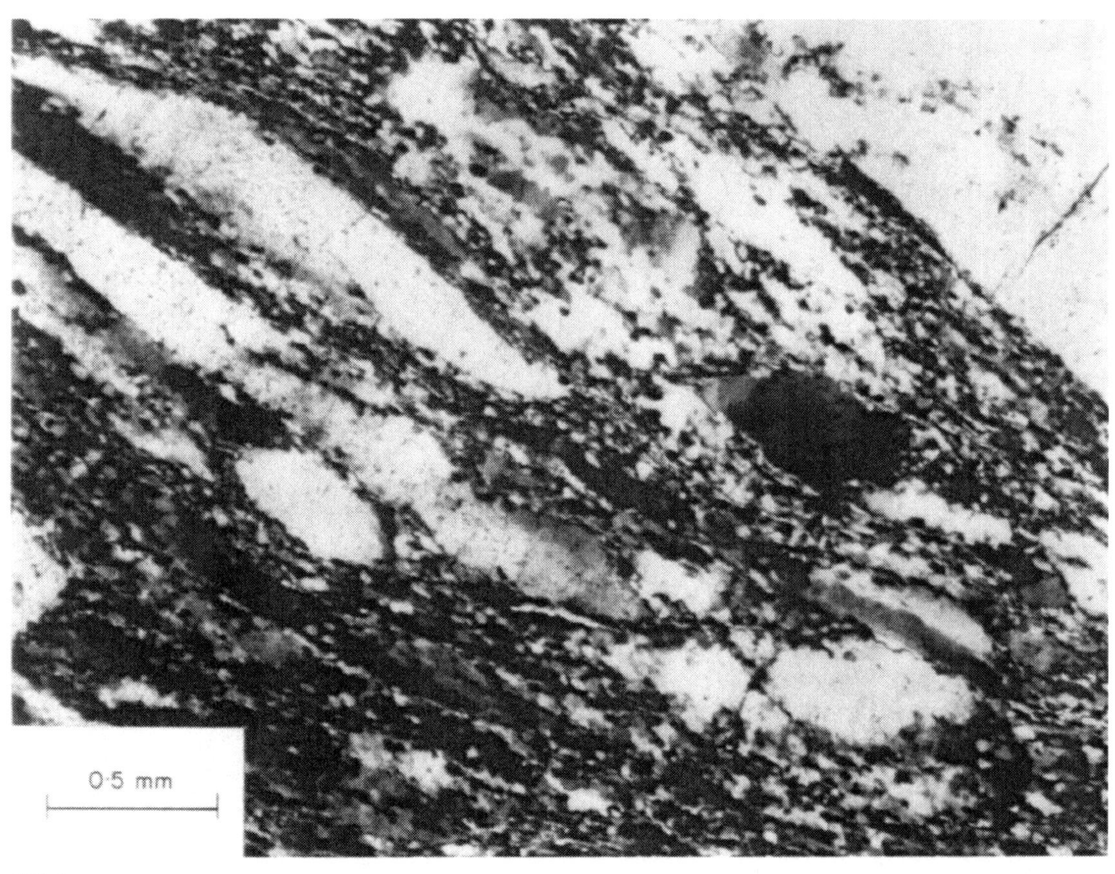

151 B

## 152. L-Tectonite Fabric in Deformed Sandstone

G. J. BORRADAILE

The specimen is from Proterozoic sandstones known as the Rhinns Group, in western Islay, Scotland (BORRADAILE 1979a). The sandstones have been regionally deformed, partly cataclastically (note fractured grains) and partly by ductile deformation (producing undulose extinction and subgrains). The deformation involved considerable grain boundary sliding (more properly termed particulate flow as this involves rolling etc. of grains) because the bulk strain of the rocks exceeds that of the grains (BORRADAILE 1979a, 1981). The fabric produced by the deformation is sometimes of the S-tectonite type (giving schistosity) but more usually here it gives L-tectonites (FLINN 1965) characterized by grain-shape lineation.

---

**A. Section perpendicular to lineation.** Note cracked grains (dilation) and subgrains around periphery of clastic quartz grains (core-and-mantle structure). NX

**B. Section parallel to lineation.** Note subgrains and undulose extinction of larger quartz grains. Several large quartz grains are ribbon-like (see also Plates 145, 146, 151). NX

152 A

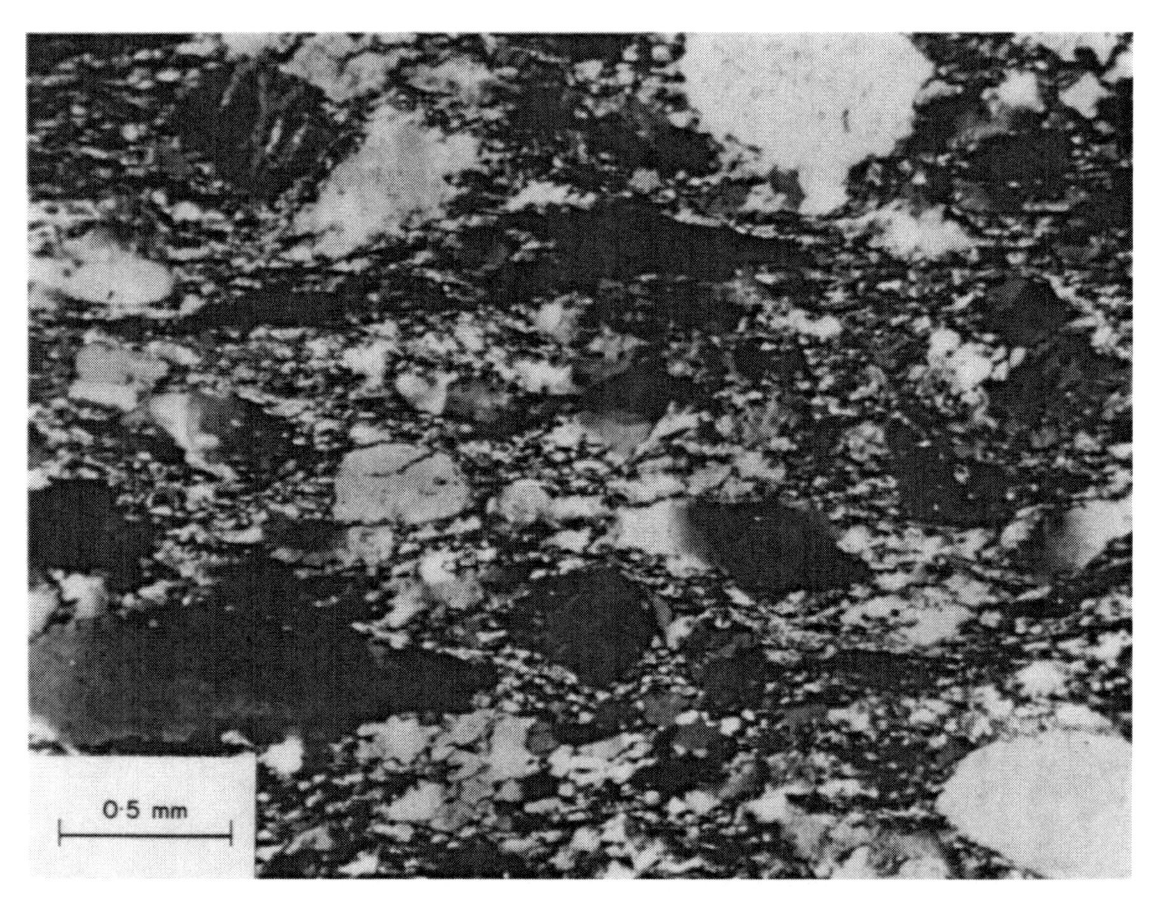


152 B

## 153. Pseudotachylite

A. by F. W. BREAKS and W. D. BOND

**A. Overview.** Sydney Lake Fault System, English River Subprovince, Sydney Lake, N.W. Ontario, Canada.
This bifurcating mass of pseudotachylite occurs in Archean migmatized metasediments which have been subjected to a Late Archean episode of faulting. The pseudotachylite is located at contact between medium to coarse-grained, protomylonitic, tourmaline-muscovite quartz monzonite leucosome and fine-grained, phyllonitic metapelitic paleosome (paleosome occupies upper one-third of photo). In terms of the morphological classification of FRANCIS and SIBSON (1973), this example would comprise a combination of subconcordant pinch-swell lensing (in lower and upper branches) emanating from a central discordant bleb. Note fragmentary material in the form of metapelitic paleosome and individual crystals, and wispy asymmetrical internal "onion-skin" structure which may have been induced by viscous flow or convective cooling. The pseudotachylite here post-dates the mylonitic schistosity which is observed in the granitoid leucosome. This indicates a change from SIBSON's (1977) Quasi-Plastic (QP) regime of aseismic flow to an Elastico-Frictional (EF) regime at higher crustal levels distinguished by brittle seismic deformation with the generation of pseudotachylite. Scale: coin diameter = 1.9 cm

B. by G. J. BORRADAILE and M. C. KENNEDY

**B. Detail.** Quetico Fault zone, Crowrock Inlet, Rainy Lake, N.W. Ontario, Canada.
This pseudotachylite occurs in rocks of the Rocky Islet Bay Intrusive Complex (HARRIS 1974). Here the pseudotachylite is developed in porphyritic quartz monzonite on the north side of the Quetico Fault zone. This thin section illustrates the structure in a veinlet branching off a structureless mass of pseudotachylite. The banding in the pseudotachylite is produced by porphyroclastic crystalline material of varying grain sizes. The fold pattern suggests flowage of the pseudotachylite. PPL

153 A

153 B

# 154. Veinlets in Mylonite

G. J. BORRADAILE

The sample illustrated is from a major transcurrent displacement zone known as the Quetico Fault at a location in Dance Township, N.W. Ontario, Canada. This fault is marked by a belt of mylonitic rocks cutting Archean granitic rocks.

---

**A. Overview of veinlets.** The mylonitic schistosity (*S*) in this hand specimen is cut by dark, gently curving veinlets (*V*) that cause sinistral displacements in the plane of the photograph. In some places the veinlets terminate in microfolds (*F*); elsewhere they curve until subparallel with the schistosity where they are dissipated, presumably by particulate flow along grain boundaries (BORRADAILE 1981).

**B. Detail of veinlet.** The veinlet curves from *X* to *W* in this thin section, cutting the horizontal schistosity and truncating the quartz ribbon (*Q*). Near (*W*) the veinlet terminates diffusely in micaceous laminae which may represent zones of grain boundary sliding. The veinlets are now composed of chlorite, epidote, white mica and very fine-grained quartz. They may be recrystallized pseudotachylite veins. NX

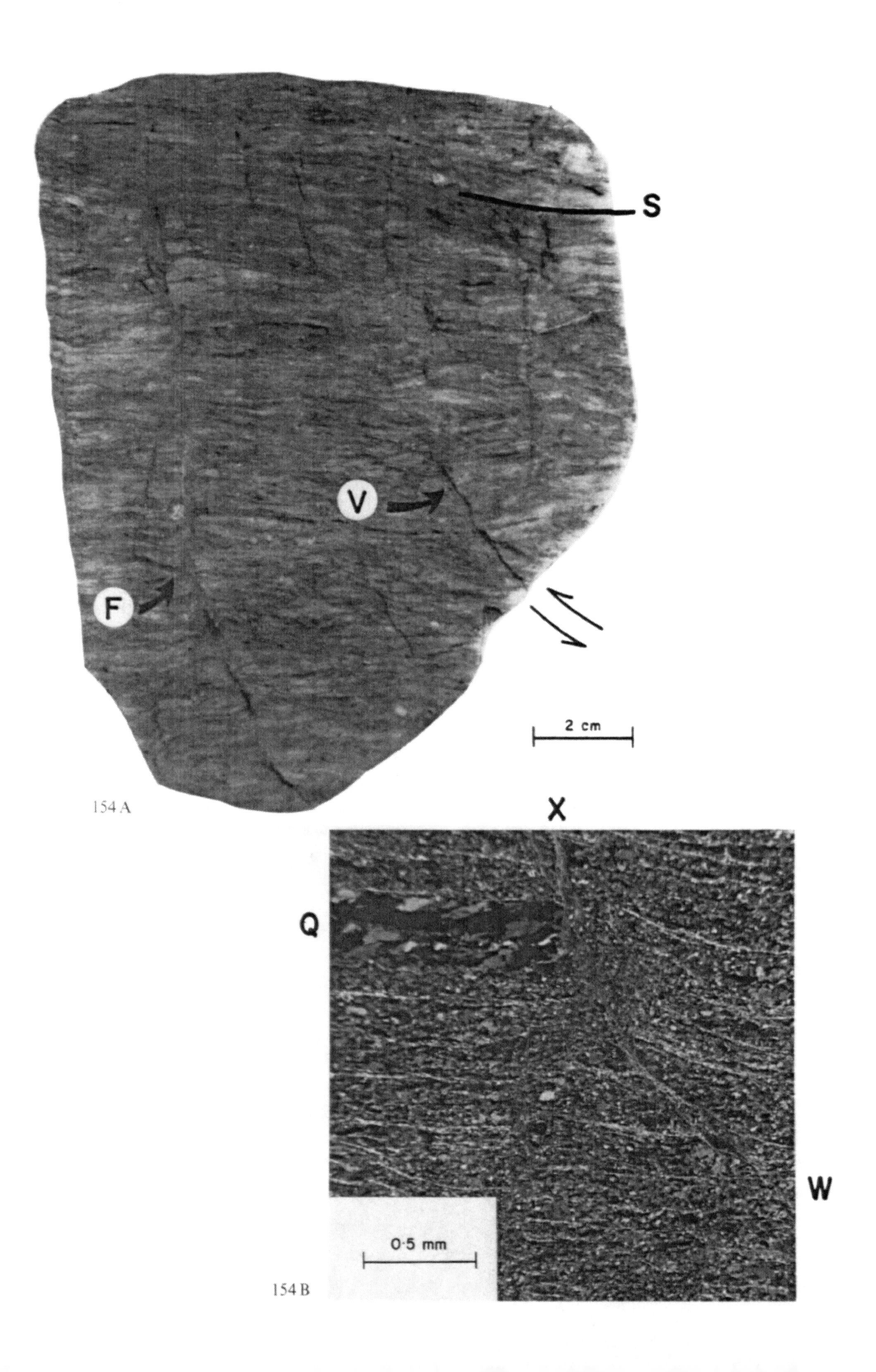


154 A

154 B

# 155. Pseudotachylites in Quartzo-feldspathic Rocks: Evidence for Intense Deformation and Recrystallization

H. R. WENK

Pseudotachylites are discordant, generally dark veins which have some aspects of an igneous rock that has intruded fractures (hence the name) and others of a sedimentary breccia cemented by fine-grained material. Most researchers agree that their formation is associated with deformation (see, e.g., the review by PHILPOTTS 1964).

Pseudotachylites are extremely fine-grained and commonly optically isotropic. Structures within them generally cannot be resolved with a petrographic microscope. Transmission electron microscopy (T.E.M.) is an obvious technique to use for the study of such material. This instrument permits acquisition of quantitative information about the internal structure and composition of these materials by combining brightfield and darkfield imaging with selected area diffraction, and energy dispersion elemental X-ray analysis. T.E.M. foils were prepared from standard petrographic thin sections by ion-beam thinning.

Pseudotachylites illustrated are in Precambrian granulite facies metamorphic rocks from Central Australia. Others, e.g., from the famous Hebrides locality (MACCULLOCH 1819), show similar features. All figures are T.E.M. brightfield micrographs.

---

**A.** In the host rock a quartz crystal displays evidence of high recovery which is typical of high grade metamorphic rocks. Dislocation densities are low ($\sim 4 \times 10\ cm^{-2}$) and generally arranged in low energy configuration such as subgrain boundaries where they form networks (*n*).

**B.** The vein consists of highly deformed material. The microstructure is complex with considerable distortion, fragmentation and tangled dislocations. Moiré fringes (*m*) spaced 5 to 22 nm are indicative of high dislocation densities or misorientation of superimposed fragments.

**C.** The strain energy is not reduced by recovery – i.e., dislocation movement to low energy arrangements – but by recrystallization through nucleation in the most heavily strained areas and subsequent growth, which is more effective if the strain is very large.

**D.** When the new grains coincide they have completely replaced the deformed material. Grain boundaries form a hexagonal network with abundant triple junctions. The grain size, however, remains small (less than 1 $\mu$m) and there is no evidence for additional surface energy reduction by coarsening

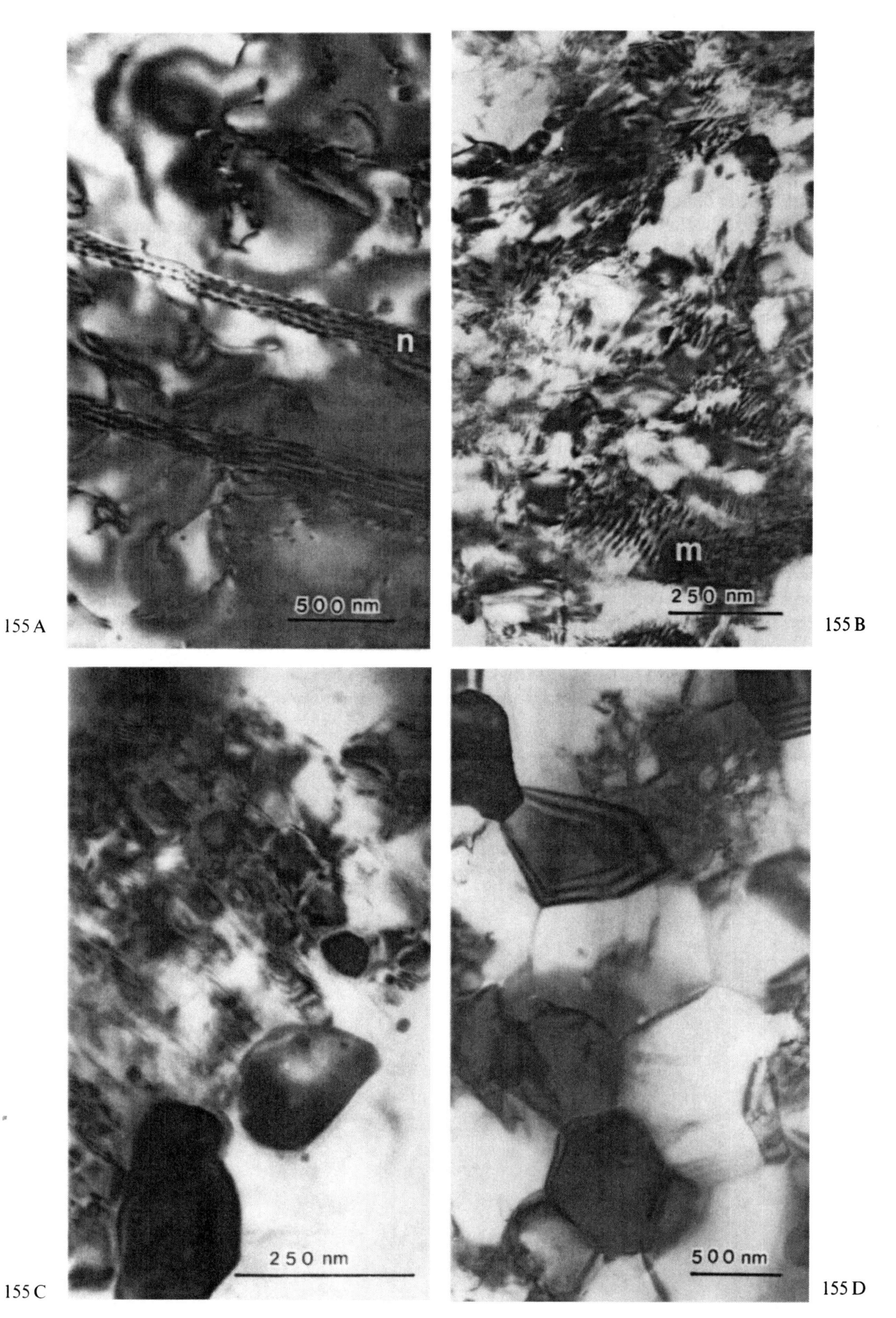
n
500 nm
155 A
m
250 nm
155 B
250 nm
155 C
500 nm
155 D

# 156. Glass in Pseudotachylites

H. R. WENK and J. AINSWORTH

On the previous page we describe the relationships between pseudotachylites and brittle deformation. Brittle deformation is locally heterogeneous and can result in very high stress concentrations over small volumes. Such concentrations may occur at the tip of a propagating microcrack during a seismic event. While average stresses, as determined from primarily geophysical evidence, are, at most, of the order of a few kilobars (KIRBY 1977), local stresses during extreme cataclastic deformation may reach magnitudes comparable to those produced under conditions of shock.

Crystals deformed by shock show a variety of responses which are dependent on the type of crystal, the environment, and the intensity of the shock wave. Crystals may behave in a brittle manner and be extensively fractured, or they may undergo penetrative plastic deformation during which dislocations are produced and move within the crystal. Microtwinning and phase transformations are also known to occur. Under extreme conditions the crystal structure collapses and crystalline material is transformed to an amorphous glass, known as diaplectic or thetomorphic glass (STÖFFLER 1974).

Some glass is found within pseudotachylite veins, but it comprises only a small volume fraction. Textural evidence precludes a mechanism of extensive melting on shear surfaces during deformation and subsequent intrusion of the melt into cracks. Glass is frequently isolated within a single crystal and sometimes occurs in subparallel bands similar to those seen in rocks shocked during meteorite impacts (KIEFFER et al. 1976). Glass has the same chemical composition as the parent crystal. It is sometimes found in small pockets between grains, but this may reflect the effects of fragmentation and mixing after the formation of the glass.

The two darkfield transmission electron micrographs illustrate the occurrence of glass in pseudotachylites.

---

**A.** Lamella of glass (*g*) in a highly deformed quartz grain from a pseudotachylite vein in a quartzo-feldspathic granulite from Ernabella, Central Australia. The *dark lines* are individual dislocations. Note the absence of Bragg contours in the glass.

**B.** Glass in a deformed pyroxene from a gabbro at Mt. Davies, Australia. The patchy appearance of the glass is the result of damage caused by the electron beam in the electron microscope. The glass (*g*) in this crystal occurs in roughly parallel bands

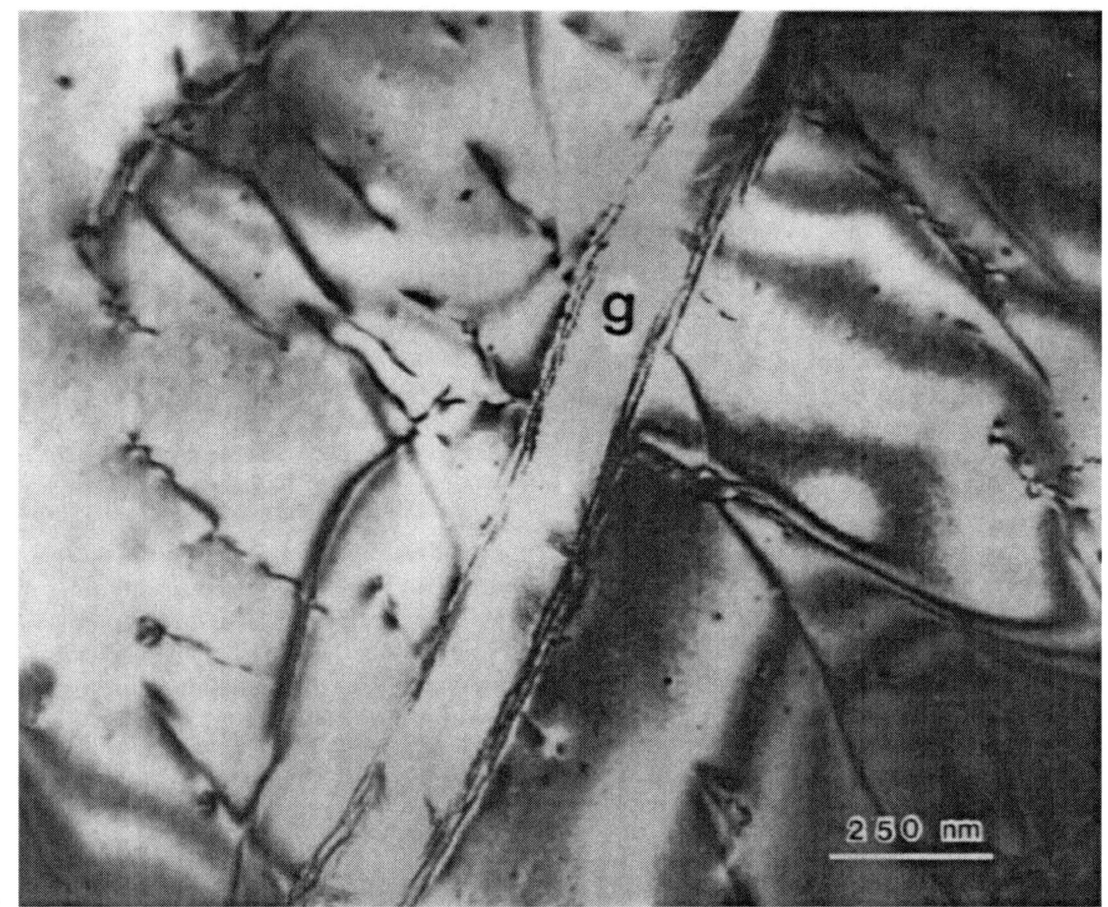

156 A

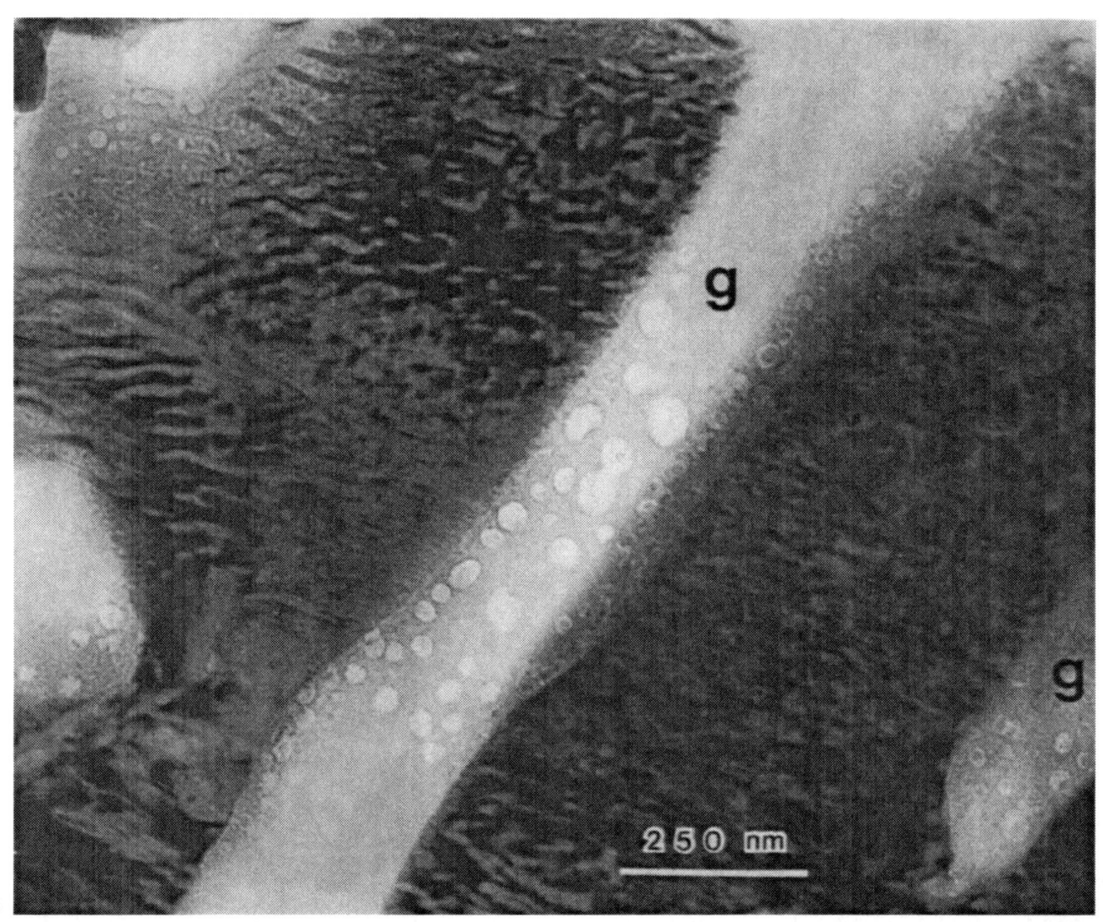


156 B

# 157. Schistosity in Galena: Recrystallization Mechanisms

K. R. McClay

Also see Plate 173

**A. Polygonization.** Sheared galena, Yerranderie, N.S.W., Australia. In this example the sheared galena has a strong foliation (*horizontal*) which is outlined by bands of small elongate grains. A large relict old grain in the center of the photograph shows the development of polygonal subgrains of approximately the same size as the recrystallized grains in the matrix. Progressive misorientation of these subgrains leads to dynamically recrystallized new grains.
Photograph of polished section (PPL) etched for 60 s at 65 °C in Brebrick and Scanlon (1957) etchant.

**B. Core-mantle microstructure.** Sheared galena, Halkyn, North Wales, U.K. This photograph shows the development of small new grains from progressively misoriented subgrains at the boundary of a large old grain. As the misorientation increases the subgrain – new grain boundaries become sharper and the different etching tones indicate orientations away from the parent grain orientation. Subgrain – new grain development can also be seen along vertical kink bands.
Photograph of polished section (PPL) etched 60 s at 65 °C with Brebrick and Scanlon (1957) etchant.

**C. Dynamic recrystallization along kink bands.** Sheared galena, Yerranderie, N.S.W., Australia. This photograph shows a segment of a large old grain traversed by two kink bands (*top* and *bottom*) which are slightly oblique to the main foliation (*horizontal*). A few subgrain walls are seen in the old grain but the kink band shows dynamic recrystallization into elongate new grains. In many examples of deformed coarse-grained galena, dynamic recrystallization along kink bands is the dominant mechanism which produces the foliation (McClay 1980).
Photograph of polished section (PPL) etched for 60 s at 65 °C in Brebrick and Scanlon (1957) etchant

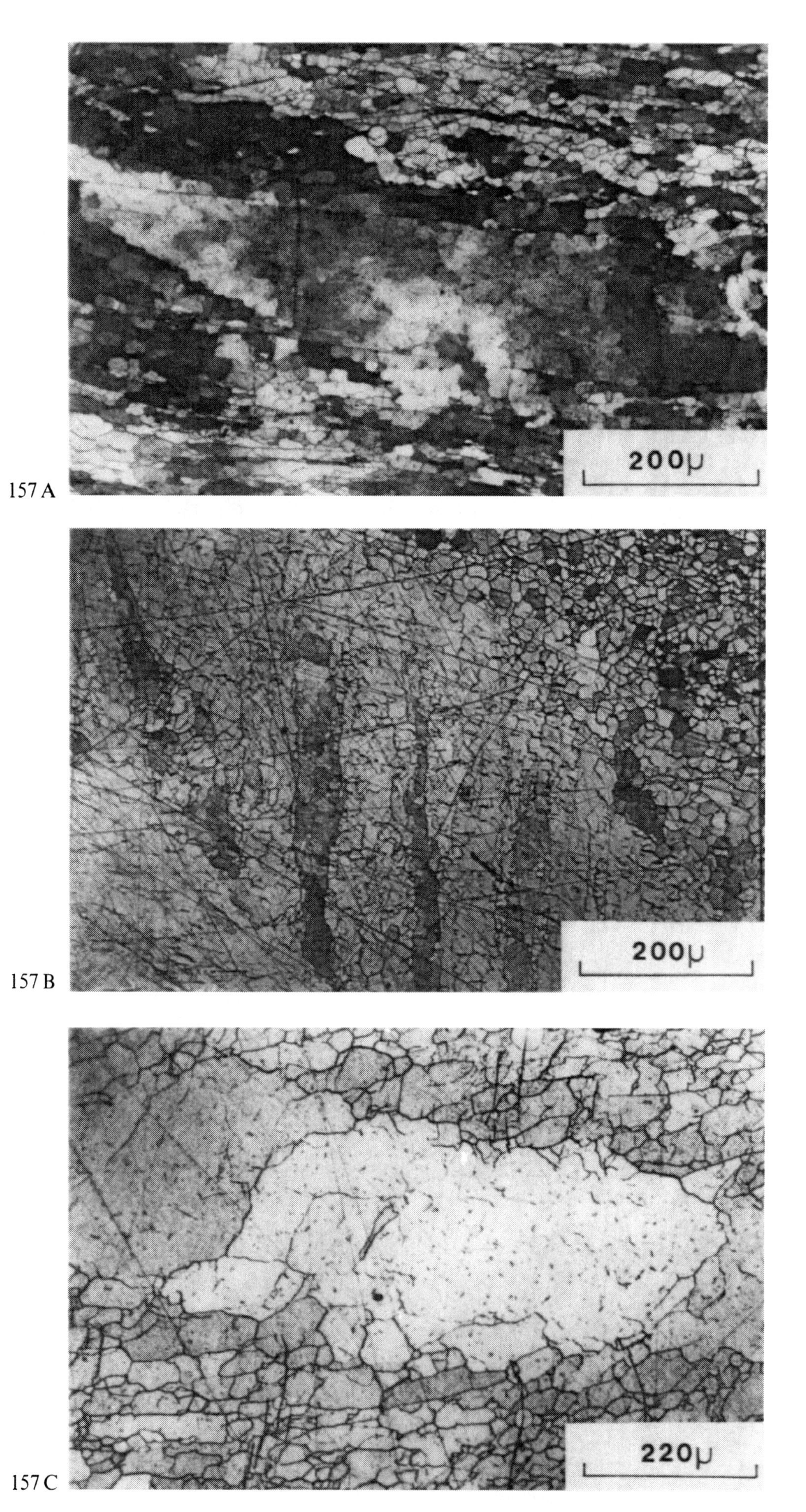


157 A

157 B

157 C

## 158. Schistosity in Galena: Recrystallization Fabrics

K. R. McClay

Deformed galena recrystallizes with ease as documented in experiments (Atkinson 1974, McClay and Atkinson 1977) and in studies of naturally deformed galena. Recrystallization may either be syntectonic dynamic recrystallization or static annealing. It is possible to distinguish between the microstructures resulting from these two processes (Clark et al. 1977).

---

**A. Dynamic recrystallization of sheared galena.** Ruth Hope mine, Slocan, British Columbia, Canada. Massive foliated galena occurs in a deformed vein deposit. In the photograph the foliation is horizontal but dynamic recrystallization along kink bands has produced two foliations slightly oblique to the primary foliation. The recrystallized grains exhibit a strong shape fabric. Photograph of a polished section (PPL) etched for 80 s at 65 °C in Brebrick and Scanlon (1957) etchant.

**B. Recrystallization and grain growth.** Galena from the silver-lead-zinc ore bodies, Mount Isa, Queensland, Australia. In this example the massive galena aggregate exhibits a foliation (*horizontal*), axial planar to minor folds. The grains have a slight shape fabric but the grain boundaries are straight, triple point junctions are developing and the boundaries are pinned by black pyrrhotite grains. Slight grain growth indicated by grain boundary movement has modified an originally dynamic recrystallization microstructure.
Photograph of polished section (PPL) etched for 60 s at 65 °C in Brebrick and Scanlon (1957) etchant.

**C. Annealing microstructure.** In this example sheared galena from the Yerranderie mine, N.S.W., Australia, has undergone a local post-tectonic annealing event which has produced equant grains with triple junctions. Fragmented calcite (*black*) marks the foliation and has inhibited grain growth by pinning grain boundaries during annealing. A strong preferred orientation is still preserved in this ore.
Photograph of polished section (PPL) etched for 60 s at 70 °C in Brebrick and Scanlon (1957) etchant

158 A

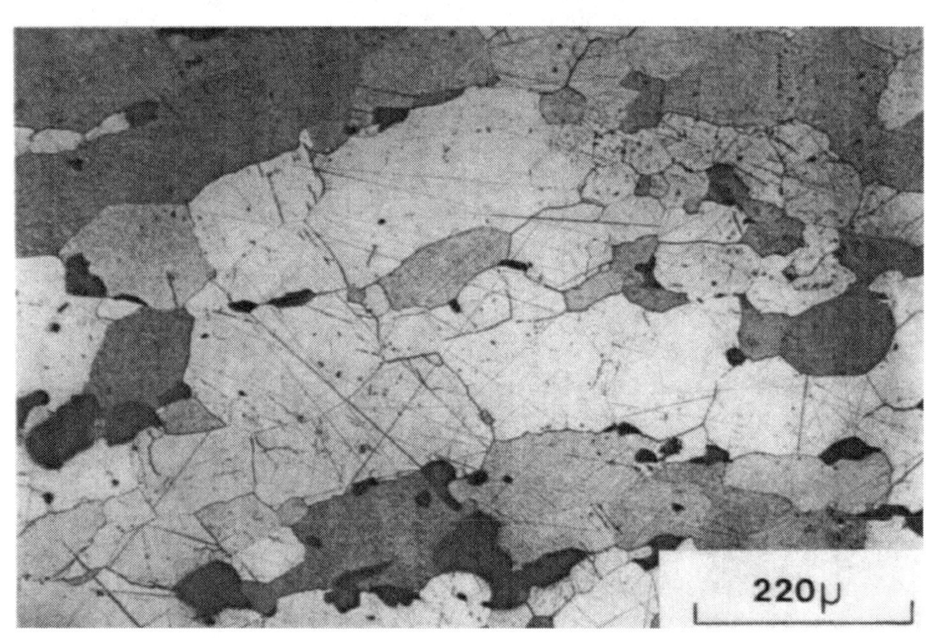


158 B

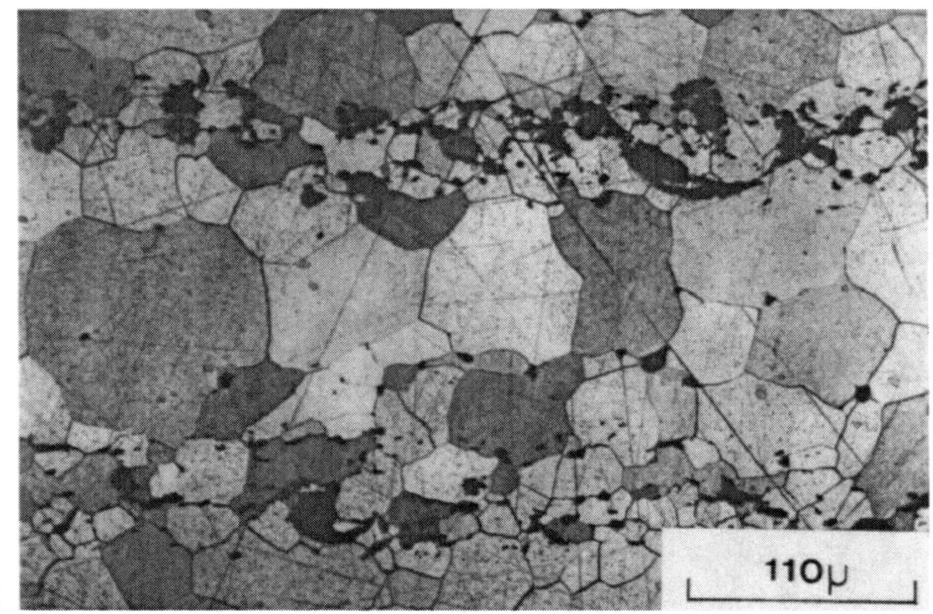


158 C

# 159. Schistosity in Sheared Galena: "Steel" Galena

K. R. McClay

See also Plate 160 B

Many lead sulphide ores have been tectonically deformed at low temperatures (< 350 °C) and they now exhibit a cleavage or schistosity ("Bleischweif"). This type of galena is characterized by a steely appearance and is usually fine-grained. A well-developed foliation and/or a stretching lineation parallel to the direction of shear is commonly present. "Steel" galena is often found in deformed vein deposits but is also found in deformed ore bodies such as Mount Isa, Australia, and Sullivan, British Columbia where it is often axial planar to folds.

---

**Sheared galena.** Yerranderie, N.S.W., Australia. This specimen of sheared galena comes from vein deposits of galena, Devonian in age, located at Yerranderie, N.S.W. Large, slightly deformed galena grains can be seen at the top of the photograph. As the amount of deformation increases toward the base of the specimen, abundant kink bands are first developed. Recrystallization along these kink bands produces a small (50–120 μm) grain-size matrix with occasional large augen galena grains (*bottom half of photograph*). As indicated by the curved kink bands at the top of the specimen, the sense of shear is dextral. The *white grains* in the photograph are fractured and brecciated pyrite grains. At the base of the specimen a strong dimensional preferred orientation is developed together with a strong crystallographic preferred orientation as indicated by the 220 pole figure. The features illustrated in this example of sheared galena are similar to those found in other sheared galena deposits.
The photograph is of an acetate peel of a polished slab of ore which was etched for 1 min with 48% HBr.

Parital 220 pole figure of Yerranderie galena. Contour levels are in steps of 20% of mean intensity. Projection plane is the foliation plane. Note the very strong, almost single crystal preferred orientation; maxima are indicated by stars

159

# 160. Schistosity in Galena: Microstructures

K. R. McClay

"Steel" galena or "Bleischweif" exhibits a strong dimensional orientation and commonly a strong crystallographic preferred orientation. Microstructural investigations (McClay 1980) have shown that dynamic recrystallization along kink and deformation bands and in core-mantle structures (White 1976) is responsible for the shape and crystallographic orientations. Examples of these microstructures are presented below.

**A. Steel galena.** Deformed vein deposit, Halkyn, North Wales, U.K.
Steel galena occurs in deformed vein deposits in Carboniferous limestones of North Wales. The galena is strongly foliated with relict augen of old grains traversed by kinks and surrounded by a fine grained matrix. The old grains exhibit a strong dimensional orientation.
Photograph of polished section (PPL) etched 65 s at 65 °C in HCl/Thiourea etchant (Brebrick and Scanlon 1957).

**B. Sheared galena.** Deformed vein deposit, Yerranderie, N.S.W., Australia.
This is a photograph of the sheared galena vein illustrated by the photograph of the hand specimen in the previous Plate. Elongate grains (100–200 μm) occur in well-defined bands. These are interpreted to indicate relict kink bands. A strong crystallographic preferred orientation is also indicated by the banding. The foliation is horizontal and the sense of shear is dextral.
Photograph of polished section (PPL) etched for 60 s at 65 °C in Brebrick and Scanlon etchant.

**C. Sheared galena.** Deformed vein deposit, Braubach, West Germany.
This sample of sheared galena from Braubach (Siemens 1977) also shows a well-developed banding of fairly equant galena grains along relict kink bands as in **B**. The main foliation is, however, horizontal and the sense of shear is dextral.
Photograph of polished section (PPL) etched 80 s at 65 °C in Brebrick and Scanlon (1957) etchant

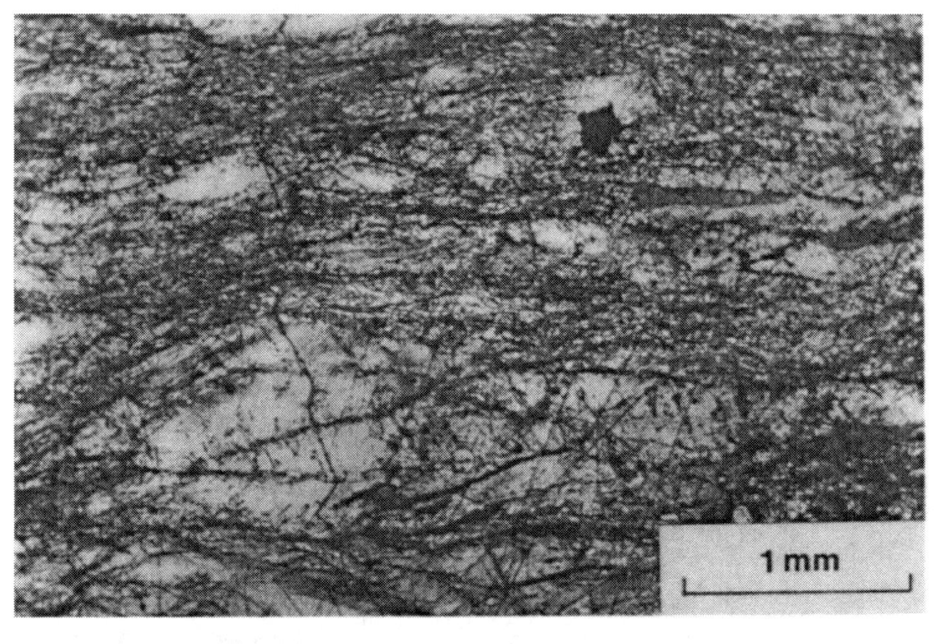


160 A

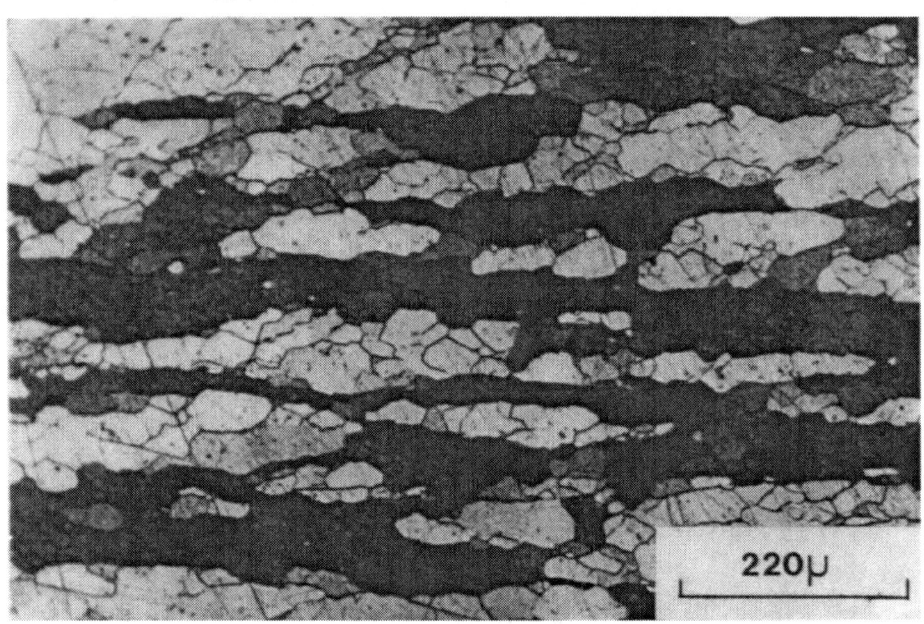


160 B

160 C

## 161. Fabrics in Ores: Sphalerite, Chalcopyrite, Pyrite

K. R. McClay

**A. Sphalerite.** Sullivan mine, Kimberley, British Columbia, Canada. Massive sphalerite occurs with galena in the Lower Proterozoic Sullivan lead-zinc orebody in British Columbia. Deformed under middle-upper greenschist facies metamorphic conditions, the galena commonly exhibits a well-developed foliation whereas the sphalerite does not. In the photograph, equant sphalerite grains exhibit broad annealing or growth twins and only a few thin lancelote deformation twins. Pyrrhotite (*white*) shows a preferred shape orientation.
Polished section (PPL) etched 40 s with sodium hypochlorite solution.

**B. Chalcopyrite.** Copper orebodies, Mount Isa, Queensland, Australia. The copper orebodies of Middle Proterozoic age at Mount Isa have undergone low grade metamorphism and deformation (Mathias and Clark 1975). Where the deformation is particularly intense a slaty cleavage is developed and chalcopyrite may take on a preferred dimensional orientation. Slightly elongate grains parallel to the foliation (*horizontal*) are found together with abundant thin polysynthetic deformation twins. The orientations of the twins also indicate a crystallographic preferred orientation.
Polished section (PPL), etched 2 min electrolytically in chromic acid.

**C. Pyrite: pressure solution.** Copper orebodies, Mount Isa, Queensland, Australia. This sample is from the same locality as photograph **B**. Pyrite deforms principally by cataclasis (Atkinson 1974) but in this case pressure solution of pyrite framboids can be observed. Small semicircular pyrite framboids show indentation and dissolution features and as a result a crude cleavage (slightly oblique) is developed. This emphasises the importance of pore fluids during the deformation of pyrite.
Polished section (PPL, oil immersion). Sample etched 3 min 30% nitric acid, 70% alcohol

161 A

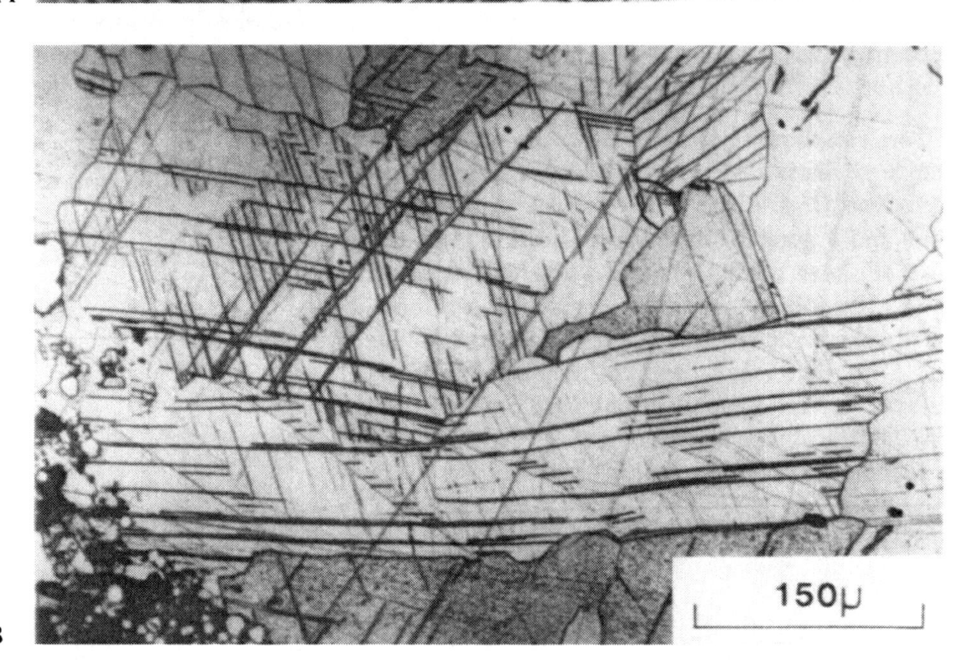


161 B

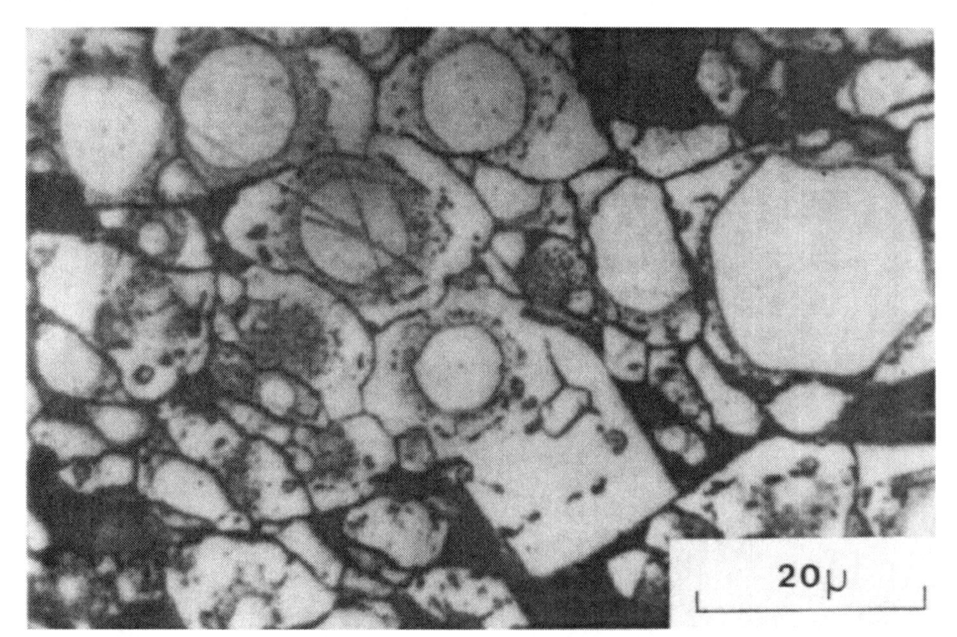


161 C

# 162. Peridotite Microtextures (I): Oceanic Environments

F. BOUDIER and A. NICOLAS

Textural types observed in cross-section of ophiolitic complexes have been recognized in specimens provided by oceanic dredging or drilling.

---

**A. Coarse granular type.** This has been observed in a few specimens drilled on the Mid-Atlantic Ridge. This texture shows little deformation and the crystallographic fabric is weak. As a result, recrystallization is not developed. Large olivine crystals (5 to 10 mm) present widely spaced subboundaries. Orthopyroxene is present as porphyroclasts in close association with spinel.
(Specimen from drilling DSDP 395, provided by H. Bougault) NX

**B. Porphyroclastic type.** This has been observed in dredgings from Pacific trenches. Olivine porphyroclasts (less than 5 mm) exhibit a strong shape fabric and a good lattice fabric indicating a dextral sense of shear on the photograph (NICOLAS et al. 1973). The olivine exhibits closely spaced subboundaries, and recrystallization of neoblasts is limited. Orthopyroxene porphyroclasts show a good preferred crystallographic orientation. Spinel is dispersed in the olivine matrix.
(Specimen from Tonga trench provided by D. E. Fisher) NX

**C. Mosaic type.** This is represented by mylonites from fracture zones. Sampling provided dunites, amphibole dunites and harzburgites. A tectonic banding is composed of bands of olivine with a grain size of the order of 0.05 to 0.10 mm and bands of amphibole or olivine plus orthopyroxene with grain sizes of the order of 0.01 mm. The olivine lattice fabric is composite, probably resulting from the activation of both the high temperature and low temperature slip systems. Typically the orthopyroxene porphyroclasts present a strong elongation and a marked preferred orientation. These characteristics suggest high deviatoric stress and relatively low temperature deformation conditions.
(Specimen from St. Paul islets, provided by W. G. Melson) NX

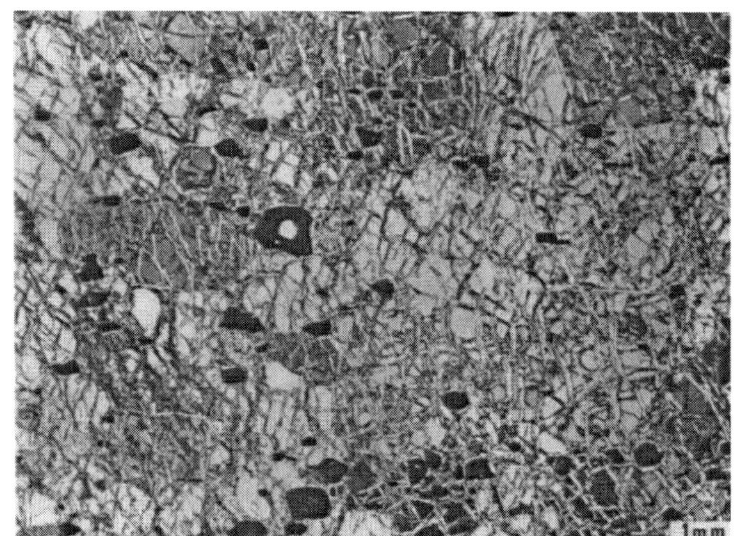

162 A

162 B

162 C

# 163. Peridotite Microtextures (II): In Tectonites from Ultramafic Sequences in Ophiolites

F. Boudier and A. Nicolas

From comparison with microtextures in oceanic peridotites (see Plate 162), and from the distribution of textures through the harzburgitic sequence in ophiolites, textural types have been correlated with specific geodynamic environments (Nicolas et al. 1973).

---

**A. Coarse granular type.** This characterizes peridotites from the upper part of the ultramafic sequence. Large polygonal olivine crystals (5 mm on the average) exhibit a generally weak tabular shape fabric and a moderate to very strong lattice fabric. Sub-boundaries are sharp and widely spaced. This indicates, along with polygonal grain boundaries, high recovery during deformation which operated at hypersolidus conditions. Orthopyroxene is recrystallized into aggregates of polygonal neoblasts replacing ancient porphyroclasts. Spinel is generally associated with orthopyroxene as observed in the protogranular texture of Mercier and Nicolas (1975). When present, clinopyroxene crystals have a concave shape. This textural type is interpreted as resulting from flow in a ridge environment (see figure below).
(Specimen from New Caledonia, provided by D. Cassard) NX

**B. Porphyroclastic type.** This is representative of the lower part of the harzburgitic sequence. For description, see previous plate. This textural type is considered to be representative of deformation due to flow in the upper mantle, at the base of the lithosphere (see figure).
(Specimen from Red Hill, New Zealand, provided by T. P. Thayer) NX

**C. The mosaic type** characterizes mylonites from the sole of the harzburgitic sequence, in contact with amphibolites. For description, refer to previous plate. This textural type is attributed to deformation accompanying suboceanic thrusting, predating emplacement onto the continent (see figure).
(Specimen from Samail nappes, Oman, provided by R. G. Coleman) NX

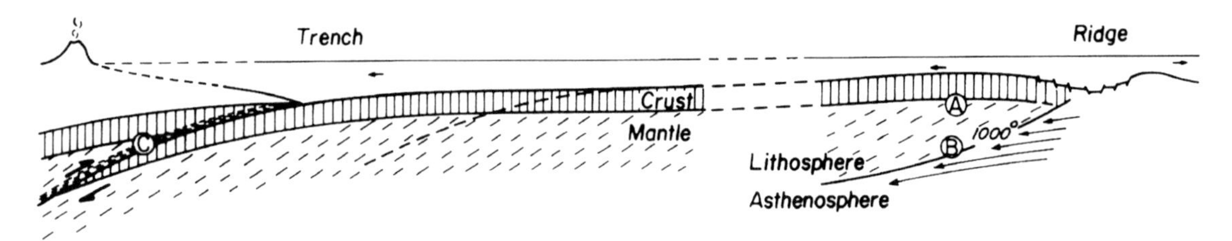

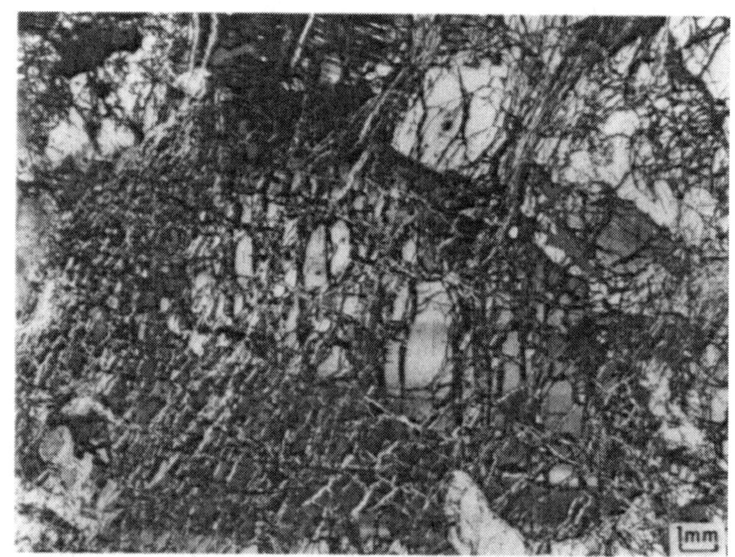

163 A

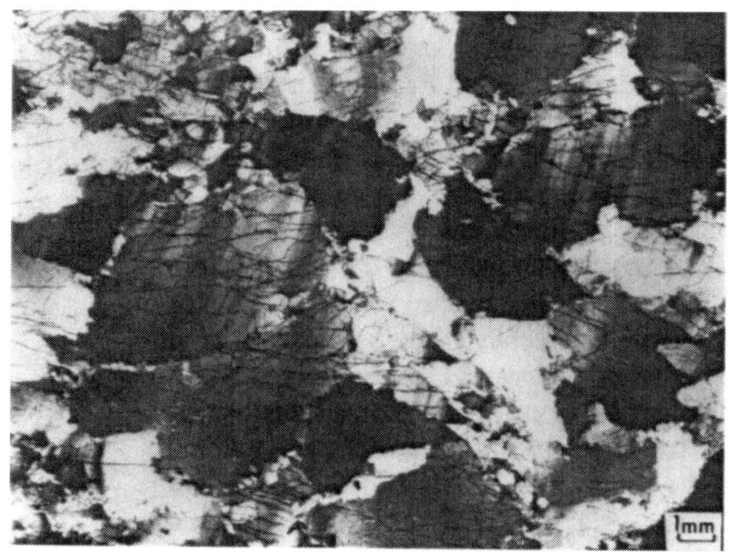

163 B

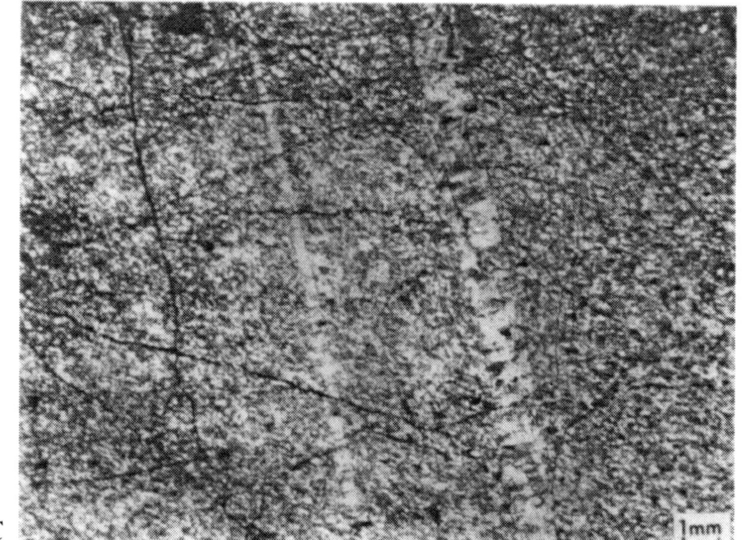

163 C

# 164. Color Banding and a Flow Fabric in a Salt Glacier

C. J. TALBOT

Infracambrian halite rises approximately 10 km up a salt diapir to form Kuh-e-Namak (=mountain of salt) at 28°17′N, 51°43′E in Dashti Province of S. Iran. The halite rises to 1500 m above sea level as an elliptical topographic dome (3×6 km) which stands well above the surrounding Mesozoic to Recent sediments. Two salt glaciers extend up to 3 km from the NE and SW slopes of the salt dome and the field photographs opposite illustrate the changes in the halite fabric from the dome down the lengths of the glaciers.

Wherever it is exposed the halite rock displays a color banding which tends to parallel the rigid boundaries imposed by the country rocks. Thus in the intrusive diapir the color bands have steep dips and in the extrusive salt glaciers they generally parallel the bottom and sides of the bedrock channels down which the glaciers flow. **A** shows halite typical of the dome. White efflorescence on the surface of a vertical face has picked out cubic grain boundaries in a steep band of transparent coarse grained halite (which appears *dark*). The coarse interlocking clean halite grains characteristic of the dome undergo progressive cataclasis down the length of each salt glacier. Transparent porphyroclasts of halite suffer continuous dynamic recrystallisation around their margins and their peripheral rotated subgrains contribute to a steadily increasing groundmass of fine-grained halite. In most areas of the glaciers the porphyroclasts are subcircular in outcrop and **B** illustrates a vertical face two-thirds of the way down the SW salt glacier where the subhorizontal color bands are alternating white and pink.

A flow fabric is general throughout the salt glaciers and is defined by the shapes and orientations of spicules of gypsum with or without specularite in the halite rich groundmass even where the halite porphyroclasts are subcircular in outcrop. The flow foliation and the color banding are parallel and almost planar where the bedrock channel is regular. However, near major obstructions in the channel, the flow planes diverge and the foliation becomes axial planar to trains of folds which develop in the color bands (TALBOT 1979). Even the halite porphyroclasts become ellipsoidal as they enter such zones of slowing and divergent flow and **C** shows clear orange porphyroclasts of halite (*dark*) three-quarters of the way down the length of the NE glacier. Soon after this stage differential strains (occasionally pure shear but usually a triaxial flattening) lead to rotation of the porphyroclasts which abruptly break up to several smaller subcircular daughter grains when they reach axial ratios of about $X/Z=4$ or 5. As the daughter porphyroclasts become ellipsoidal in their turn they define the axial plane of the folds in the color bands. Gentle asymmetric folds at the upstream end of each zone of divergent flow mature downstream to become isoclinal where the flow planes converge and the color bands and flow fabric again become parallel and planar. The color bands are drastically repeated and thinned by the folds developed in association with every one of the major obstructions down the length of both glaciers. **D** illustrates the results in the snout of the NE glacier of these repeated reworkings of the original broad bands of halite with subtle color differences exposed in the salt dome. In **D** most of the lumps in the thin bands of white, pale green and magenta fine-grained, halite-rich rock are aggregates of insoluble minerals. However, the medium-toned circular object just below the center of the picture is a porphyroclast of transparent halite which represents the largest visible remnant to survive this far, of the clean coarse-grained halite of the dome. The subvertical linear features in this photograph (and parts of the others) are sharp ridges between water run-off channels which are semicircular in section.

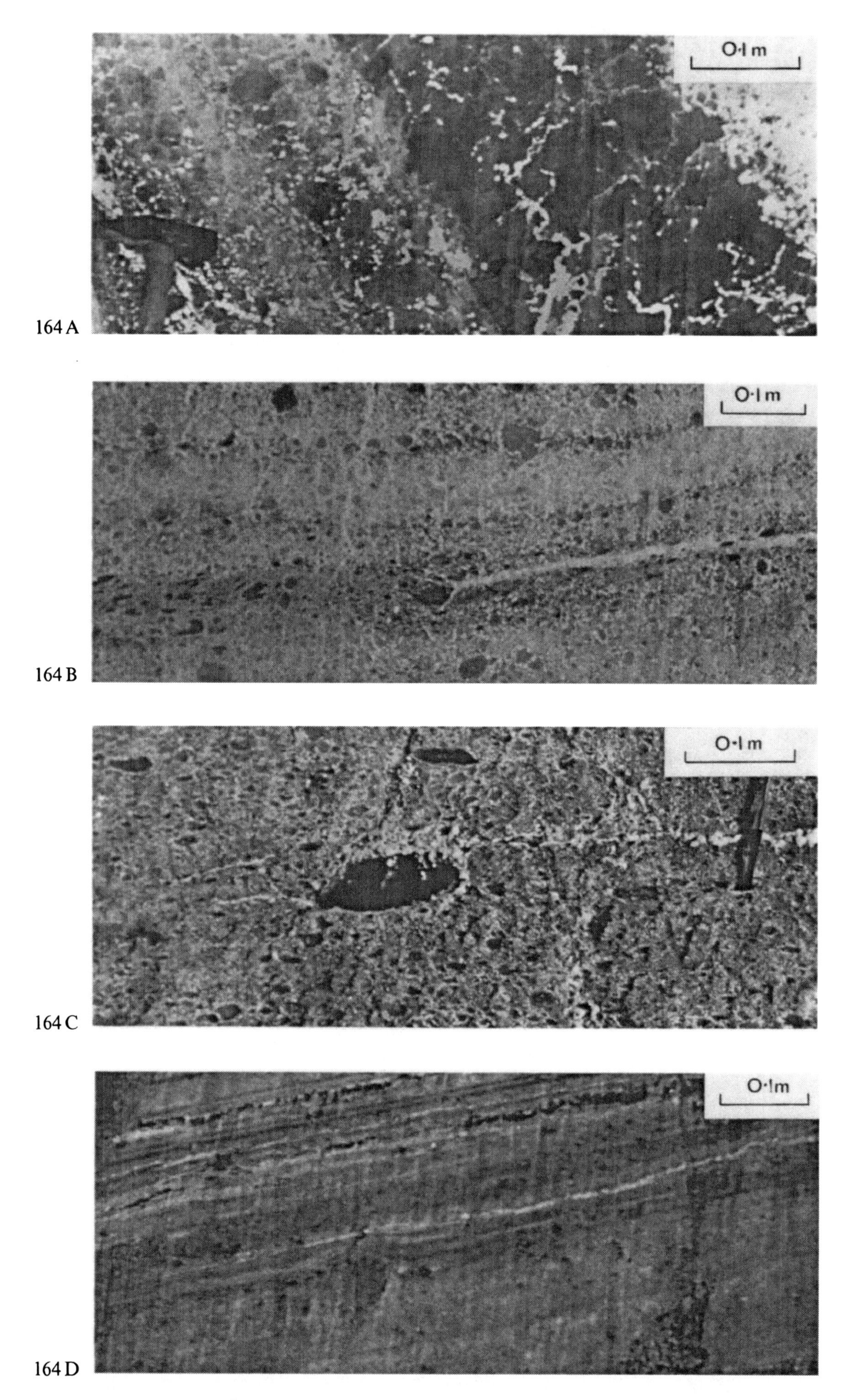

164 A

164 B

164 C

164 D

# 165. Load-perpendicular Cataclastic Fabric in Experimentally Deformed Limestone (I)

G. TWOMBLY and J. H. SPANG

This picture illustrates a load-perpendicular cataclastic fabric developed in experimentally deformed Indiana Limestone. Indiana Limestone is an undeformed, untwinned, indistinctly bedded micrite-cemented foraminiferal packstone of 1 mm mean grain size with 7% to 10% micrite and an initial porosity of 12.8%. The core was deformed dry, (approx. 0.04% volatiles by wt.) at room temperature, under 2000 lb $in^{-2}$ (13.8 MPa) confining pressure in a load-controlled experiment with a maximum strain rate of $3 \times 10^{-2}$ $s^{-1}$. The total bulk axial strain is 14.3% of which 1% was recoverable. Load axes are illustrated in the sketch. Porosity of the deformed core was 17.9% with a porosity increase during deformation of 5.1%. These same textures and features were observed in deformed cores at strain rates of $3 \times 10^{-5}$ $s^{-1}$ in which the displacement rate was controlled.

Note the load-perpendicular fabric, consisting of planar zones of grain-size and pore-size reduction. The dominant mechanisms of deformation are cataclasis, openings along mineral-cleavage planes, and twinning. In the deformed zone, grain long-axes are rotated to be normal to the load by cleavage or fracture (*arrow*). Because of grain-size reduction and porosity increase in these zones, solution or recrystallization would be favored.

The photo is a negative image of a thin section.

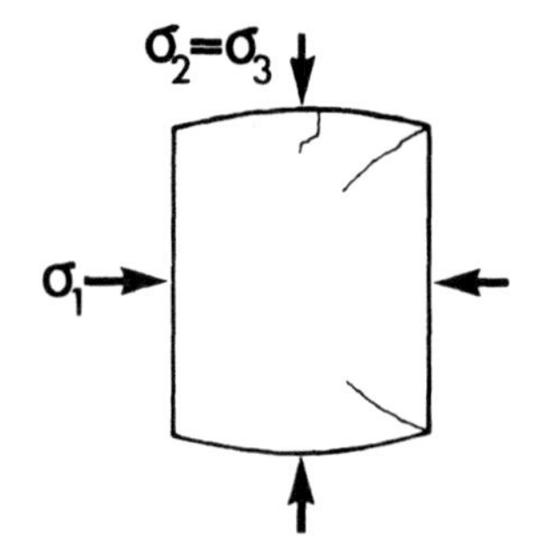

165

# 166. Load-perpendicular Cataclastic Fabric in Experimentally Deformed Limestone (II)

G. TWOMBLY, J. H. SPANG and N. C. WARDLAW

These pictures illustrate a load-perpendicular cataclastic fabric developed in experimentally deformed Indiana Limestone as in the preceding plate, for which the same experimental conditions apply.

These S.E.M. photographs are of etched pore casts of epoxy-impregnated samples. Pore space stands in relief. Photograph locations of the deformed specimens **B** and **C** are illustrated in the figure.

---

**A. Undeformed sample.** *C* Crinoid; *F* Foraminifera; *E* Echinoid spine; *M* Micrite cement.

**B. Deformed sample.** Note pervasive grain-size reduction; cleavage in grains (*G*); load-perpendicular shear zones wrapping around mechanically more competent micrite (*M*) and grains (*G*). Location (*2*) in the figure.

**C. Deformed sample.** Note shear zones wrapping around mechanically more competent grains (*G*). Note opening of (*G*) along cleavage at a high angle to load. Location (*3*) in the figure.

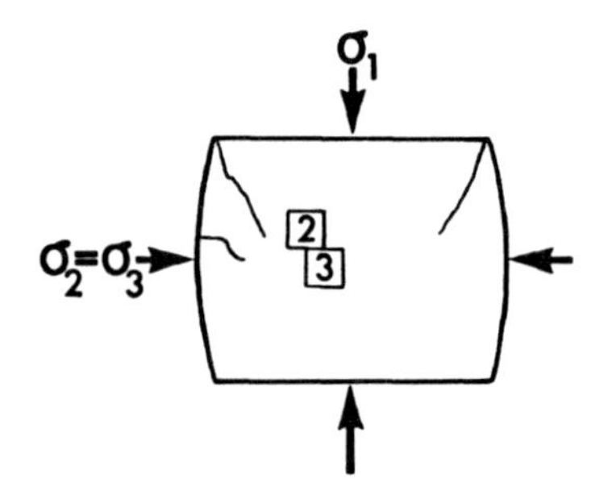

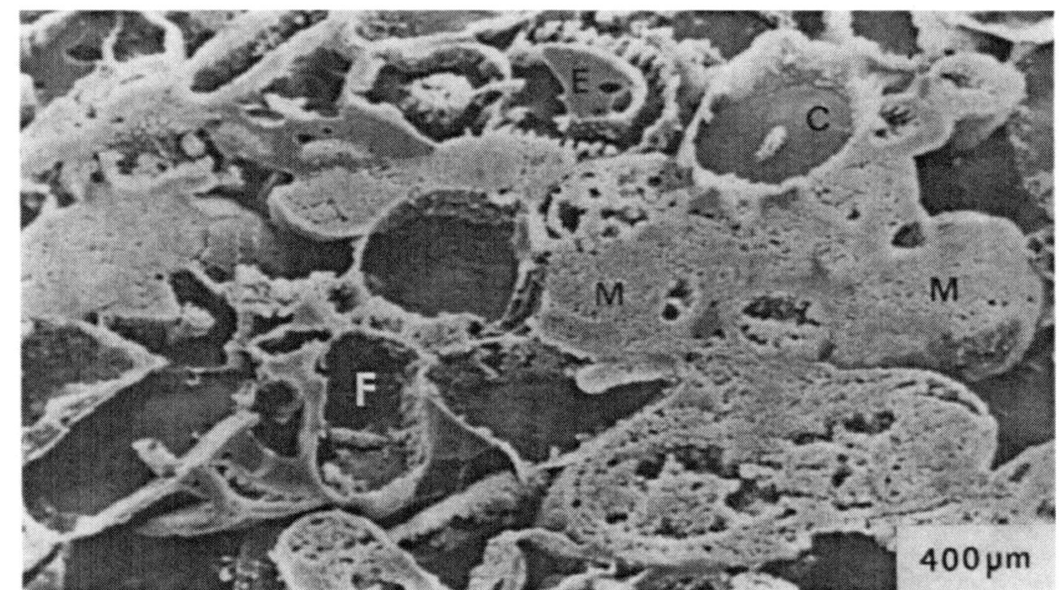

166 A

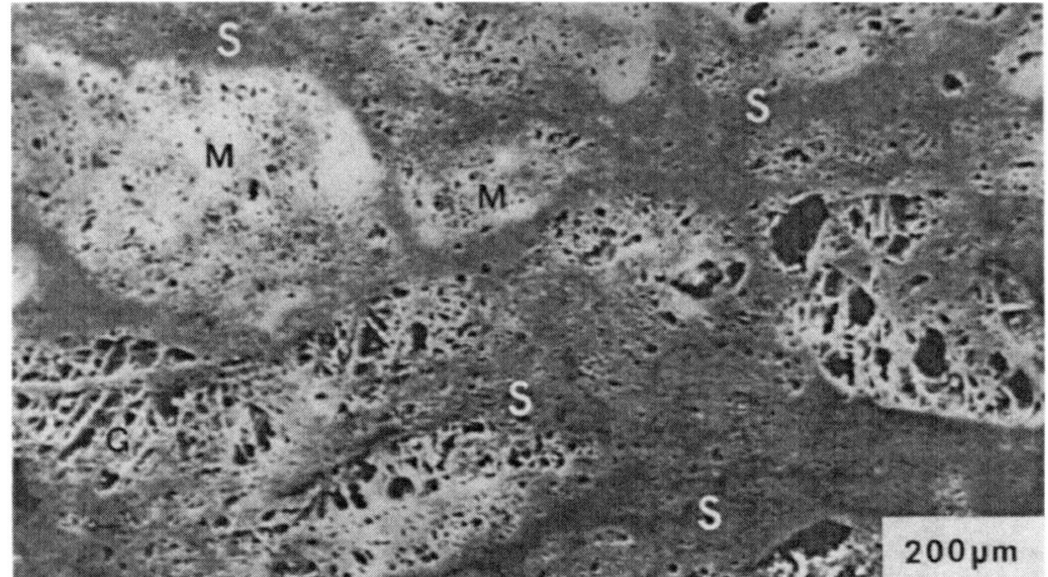

166 B

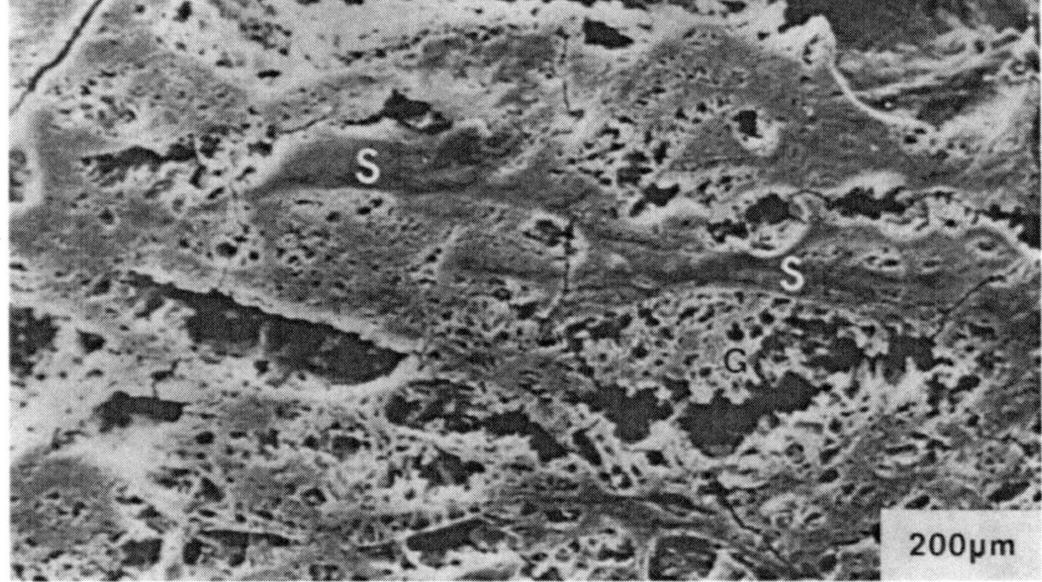

166 C

# 167. Foliation, Textural and Crystallographic Fabric Development in Glacial Ice

P. J. HUDLESTON

These are photographs of ice that has undergone variable amounts of (largely) simple shear. The samples are all vertical slabs (with flow towards the right) of superimposed ice from the Barnes Ice Cap, Baffin Island, N.W.T., Canada. This ice is formed at the margin of the glacier, which is frozen to its bed, and becomes overridden during an ice advance (HOOKE 1973, HUDLESTON 1977). Stages of deformation can be traced from completely undeformed ice, through a stage characterized by the development of shear zones (HUDLESTON 1977), to typically foliated ice that has undergone very high strains.

---

**A. Inhomogeneous strain and foliation marked by bubbles.** Air bubbles in ice are initially subspherical, becoming drawn out into ellipsoidal or cylindrical shapes during deformation, to form a component of foliation. The smallest remain spherical due to surface energy effects. The white band of very bubbly ice across the center marks primary stratification (Fig. **1**). A large strain gradient exists within this band, with ice above highly and uniformly sheared and that below weakly deformed. The bubbles can be used to estimate the magnitude of shear strain, $\gamma$ (HUDLESTON 1977), which varies from $\approx 0.2$ below to $\approx 3$ above the white band. A small component of horizontal shortening is required in this case (COWARD 1976), to account for the large angle between bubble trace and shear plane, which is parallel to the white band. Ice slab 6 mm thick.

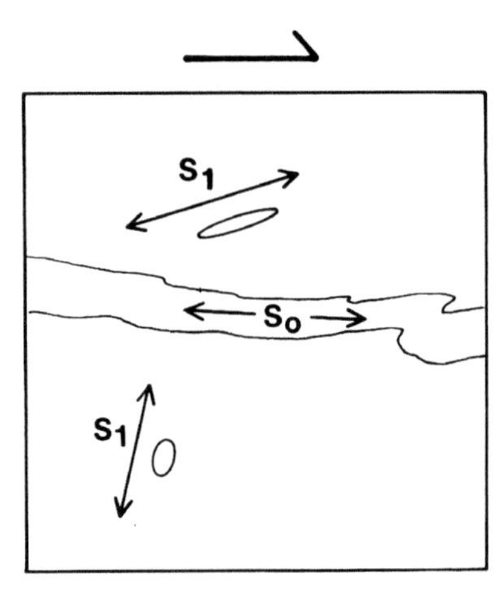

**Fig. 1**

**B. Textural changes associated with a strain gradient.** This is an enlargement of a thin section of the area outlined in photograph **A**. The small, equant, polygonal grains in the lower part of the photograph make up the ice below the white band in **A**: the large, irregular grains lie above it, and have grown due to recrystallization. Bubbles do not appear to have had a strong effect of pinning grain boundaries. The uniformity of elongation direction of the bubbles away from the white band indicates homogeneity of strain on the scale of several grains with unlike lattice orientations. In the coarse-grained ice the

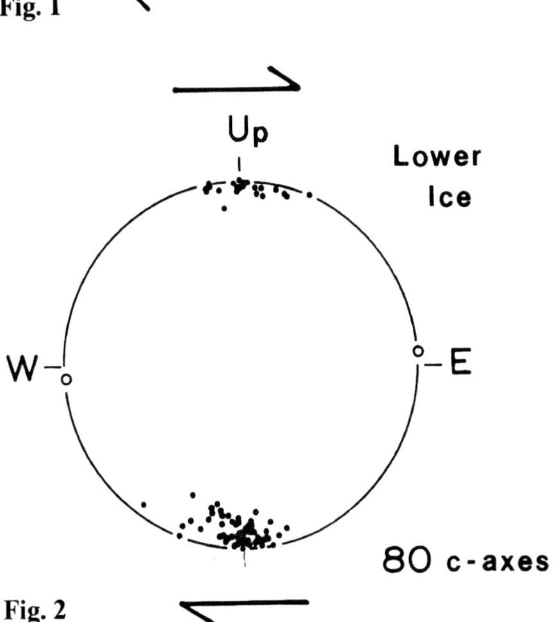


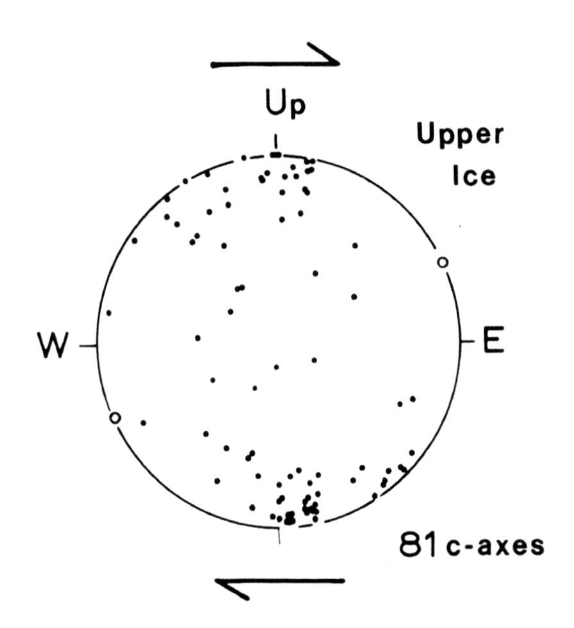

**Fig. 2**

$c$-axes are strongly clustered about the pole to the shear plane: in the fine-grained ice the $c$-axes show a diffuse pattern (fabric data not shown, but see caption for **C** below, and HUDLESTON 1977, fig. 5).

Section 0.5 mm thick

**C. Crystallographic fabric change across a shear zone boundary.** The plunge of stretched bubbles varies downwards from moderate to subhorizontal. Estimates of $\gamma$ are $\approx 1.5$ in the upper ice and $\gamma \approx 10$ in the lower ice. The latter displays a typical foliation, with bubble elongation direction parallel to bands of variable bubble concentration. Contrast this with **A**, in which banding and bubble traces are oblique. The $c$-axis fabric pattern here (Fig. **2**) varies from a rather weak double maximum to a strong single maximum as shear strain increases and bubble plunge decreases (HUDLESTON 1980).

Ice slab 3 mm thick

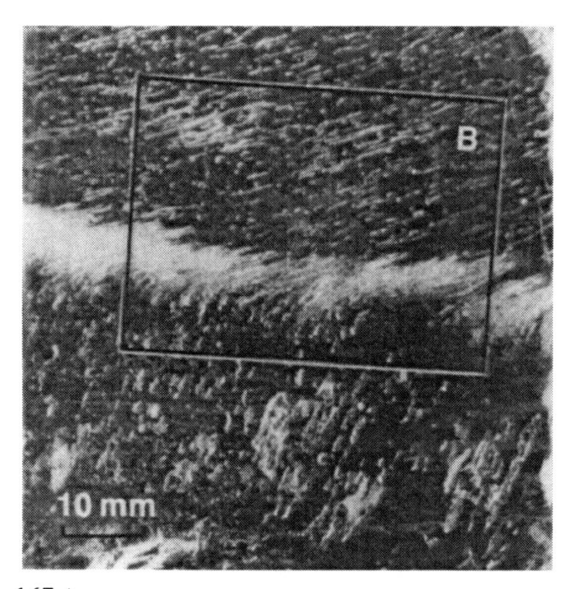


167 A

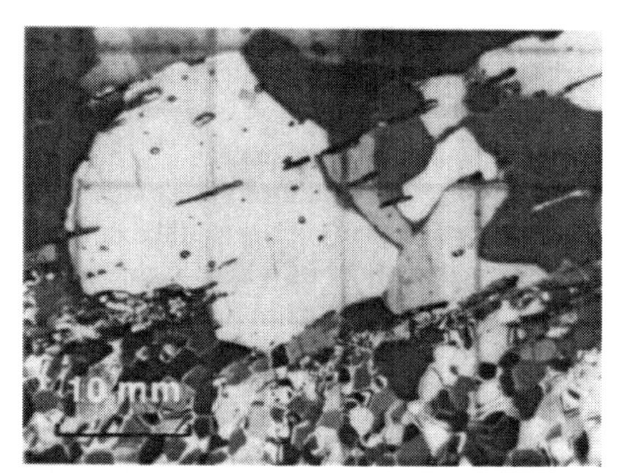


167 B

167 C

# 168. "Foliation" in Flowing Water

C. J. Talbot

The photographs on this and the next plate are of neither rock foliation nor rock schistosity but are included in this book as intriguing analogs of visual aspect if not of mechanisms.

Many others have remarked on the similarity between the flow phenomena indicated by the scum on river surfaces and deformation fabrics in rocks. The photographs are of white scum or foam floating on top of horizontal water surfaces which appear black.

The fabrics in the surface accumulations of foam are the result of deformations which involve translations which are immense compared to any lateral shortening. The strain markers have negligible thickness and the visible lines or bands of high shear strain do not represent outcropping planes.

**A** and **B** were both taken looking down (about 20° from the vertical) at the surface of the Claggain River just short of where it enters the sea on the east coast of Islay in Scotland.

Small irregular flecks of white foam or scum on the surface were being transported from upstream (i.e., from the left) for large distances (> 100 m) without any strain of their pattern. **A** shows about 4 m² of the edge of the zone where the foam was accumulating near a porous gravel bank on the outside of a bend in the river. The shear was induced in the patterns of foam as it accreted onto the almost stationary area of "foliated" foam by water flowing beneath at a slight angle to the shear zones which generated approximately parallel to the bank. **B** shows an area (approx. 4 m²) of the main area of "foliated" foam at the same location.

The river flow appeared to be steady and there was no obvious explanation for the spacing of the lines of greatest shear; this might relate in some way to the strength of the accreting foam. The outer zones of intense shear could be seen to involve differential movements of tens of meters at rates of about 1 cm/s

168 A

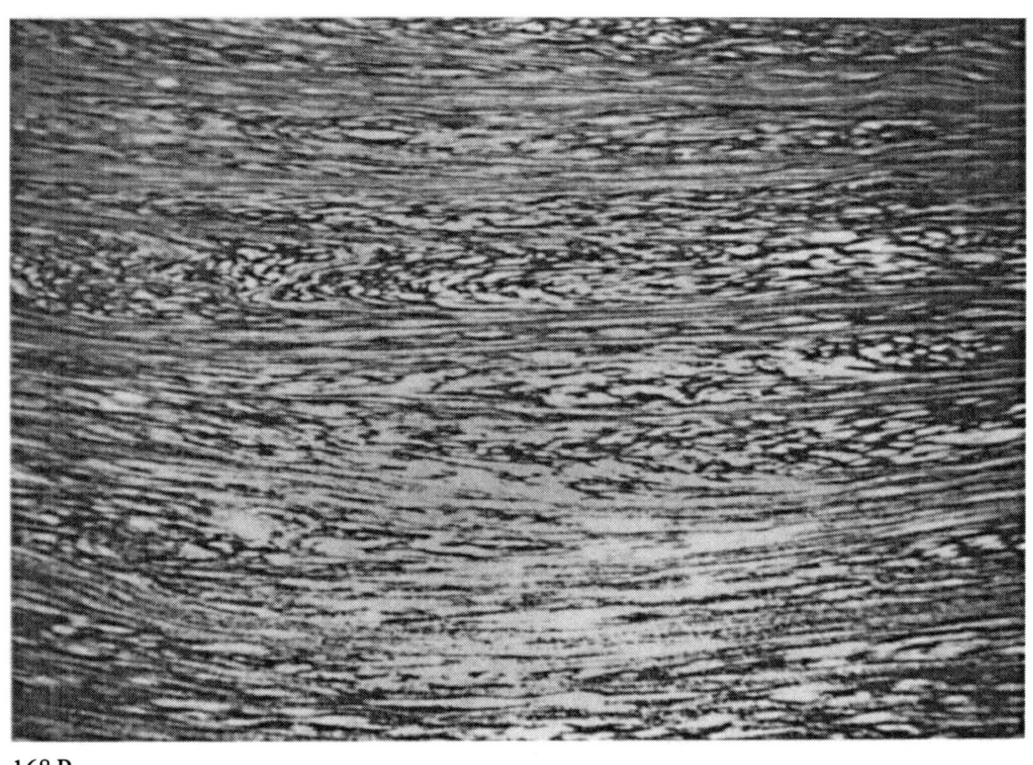

168 B

## 169. "Cleavage" and "Foliation" in Flowing Water

C. J. TALBOT

These two photographs were taken looking about 30° downward from the horizontal towards the surface of a pool in the River South Esk in Scotland. The main stream was flowing right to left beyond the top edge of **A** which shows a large patch of white foam circulating in an anticlockwise eddy (see sketch). Lines of foam were entering the picture top left and leaving the right hand edge from left to right in the foreground.

**B** is an enlargement of the area (~ 5 m²) outlined on **A**. The "crenulation cleavage" in **B** can be seen in **A** to represent healed tears in the layer of foam as it was being carried past small islands and branches breaking the water surface. The folded bands of foam were not very obvious where the "late crenulation cleavage" was imposed but became increasingly obvious where the "microlithons" were narrowed by small components of convergent flow as they were translated large distances.

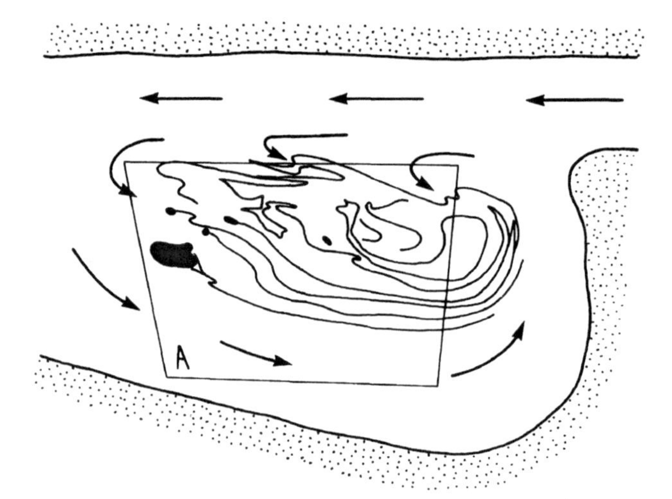


Sketch showing water currents (*arrows*) circulating in a pool between the banks of the River South Esk. Islands and branches breaking the surface are shown in *black* and the *thin lines* outline the patch of scum and indicate the main trails of foam. The outline of **A** is also indicated

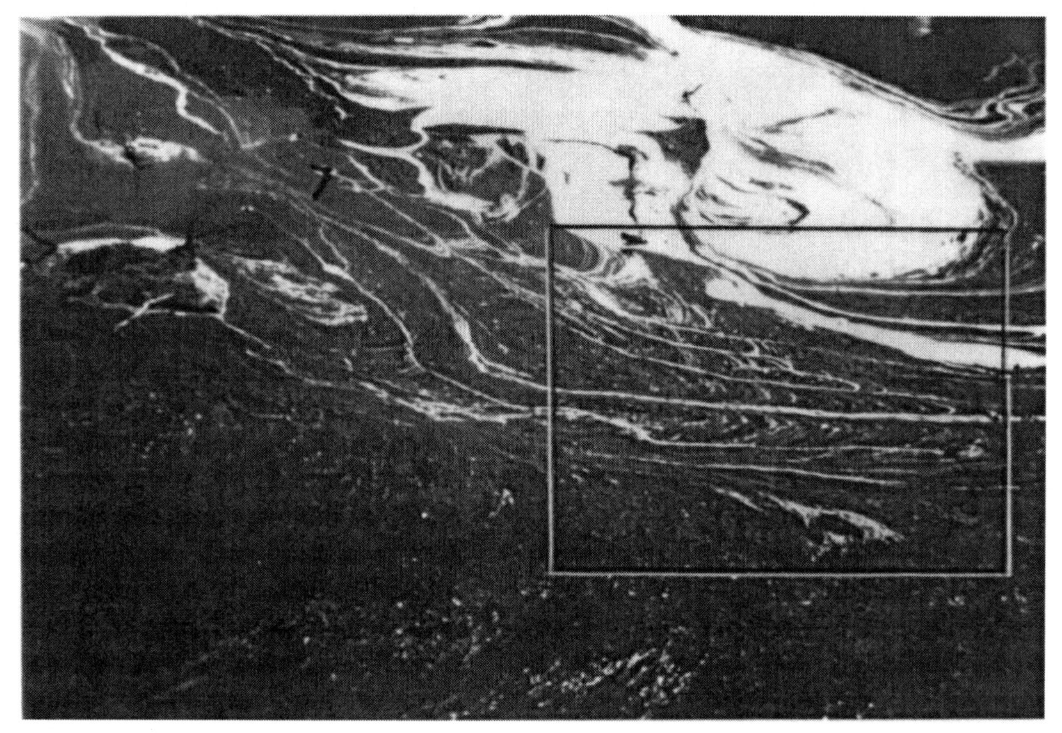

169 A

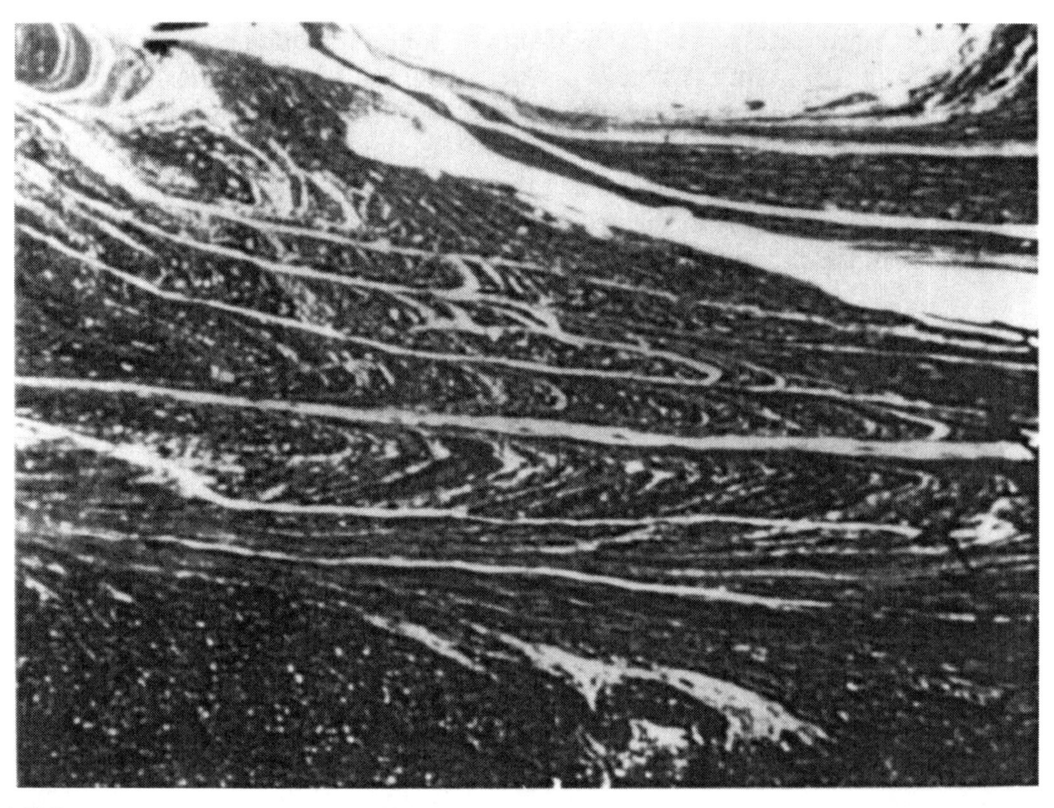

169 B

## 170. Deformation-induced Phase Transformation in Enstatite

H. R. WENK and J. AINSWORTH

The effects of deformation can be observed on small scales. This Atlas illustrates primarily macroscopic features of deformation. On this page we illustrate the other extreme; a change in the arrangement of atoms under conditions of stress. Metallurgists refer to such transformations as martensitic phase transitions, but for geologists, stress-induced twinning in calcite is a more familiar example.

TURNER et al. (1960) were the first to observe lamellae of different birefringence in experimentally deformed orthopyroxene. Optical properties indicated that orthoenstatite had transformed to clinoenstatite. Clinoenstatite was later observed in naturally deformed pyroxenites (TROMMSDORFF and WENK 1968) and is a good indicator for conditions of intense deformation such as that which occurs in mylonite zones.

The transformation from orthoenstatite to clinoenstatite is accomplished by shear on the (100) plane through an angle of approximately 13° in the [001] direction. The polymorphs retain the *b* and *c* crystallographic axes in common and are in contact across (100) planes (COE and KIRBY 1975, MÜLLER 1974). One-dimensional lattice images taken on the transmission electron microscope permit the differentiation of the polymorphs due to the fact that (100) lattice spacing is 18.2 Å in the orthopyroxene and 9.1 Å in the clinopyroxene.

The transformation is often incomplete and intergrowths of monoclinic and orthorhombic pyroxene appear. Intergrowths resulting from deformation can be distinguished from lamellar exsolution structures by counting the number of monoclinic (100) planes which appear between the orthopyroxene lamellae (CHAMPNESS and LORIMER 1974). If the clinopyroxene is the result of deformation, two clinopyroxene unit cells form from each orthopyroxene unit transformed so that only even numbers of clinopyroxene fringes should appear. Since there are no such constraints on intergrowths which form by chemical phase separation during cooling, the two modes of formation should be easily distinguished. The presence of numerous stacking faults or one-unit-cell-thick lamellae of orthoenstatite complicates the counting procedures, but if fringes are counted on both sides of a stacking fault and are then considered as one lamella, the technique may be successfully applied, particularly in the thinner area of the foil (*left side* of plate).

The crystal illustrated is from a sheared gabbro from a mylonite zone in the Precambrian basement at Mount Davies, Central Australia. It is a brightfield electron micrograph using (h00) reflections to image the fringe structure.

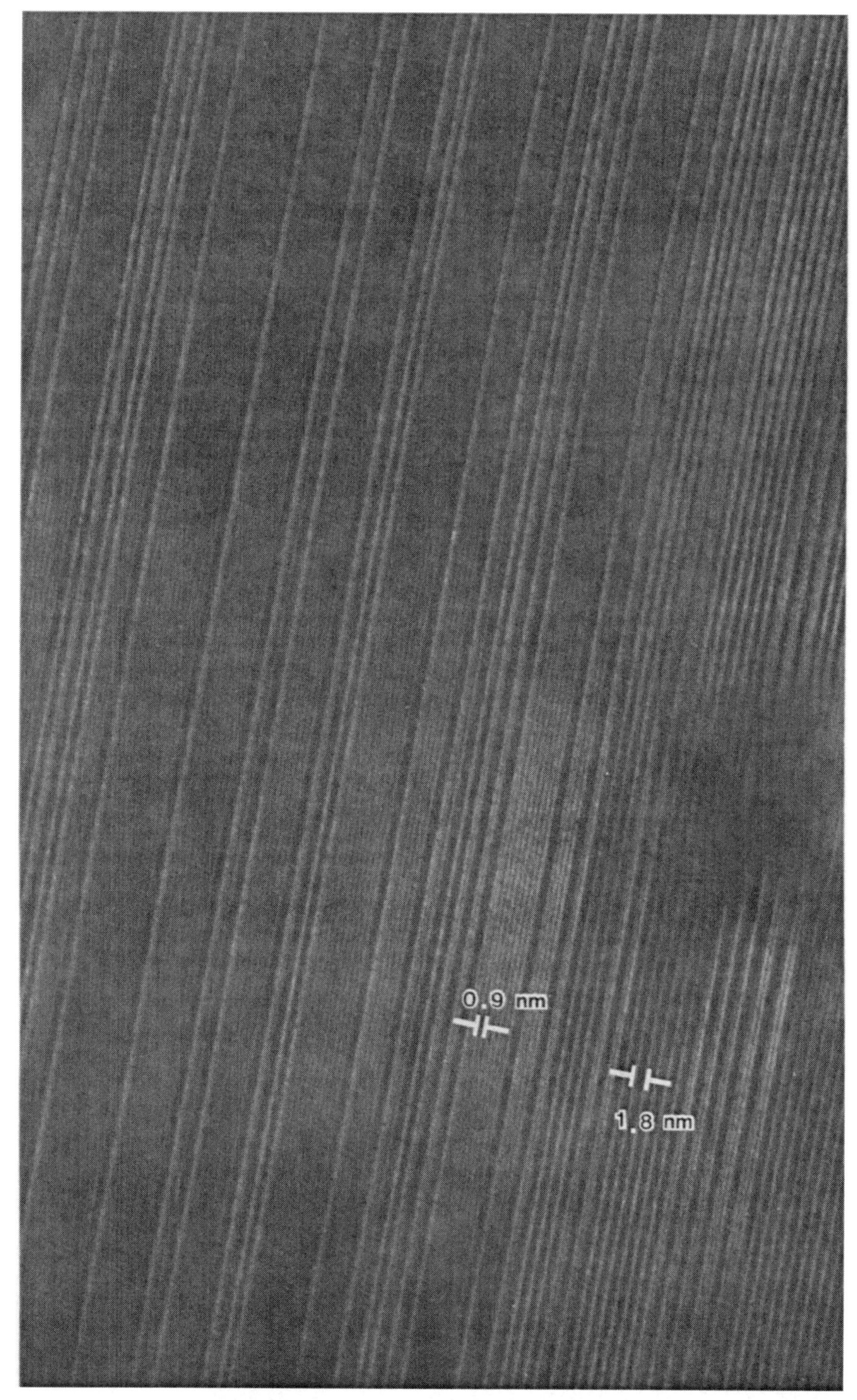


170

## 171. Shock Structures in Iron Meteorites (I)

S. A. KISSIN

Structures indicative of high shock pressures of various intensities occur in many iron meteorites. This topic has been reviewed by LIPSCHUTZ (1968), AXON (1969), and BUCHWALD (1975). These structures develop owing to the response of two minerals kamacite ($\alpha$-Fe, Ni) and troilite, the former comprising the bulk of iron meteorite and the latter occurring principally as nodules within a complex kamacite ($\alpha$-Fe, Ni) – taenite ($\gamma$-Fe, Ni) intergrowth. The shock pressures were originally believed to have been generated by the impact of the meteorite upon the earth; however, terrestrial impact is now believed to cause considerable shock only in instances of large crater-forming, explosive impacts. Various evidence indicates that shock was of cosmic origin created by collision-generated break up of meteoritic parent bodies and/or subsequent cosmic collisions. In meteorites the effects of shock are not uniform, even over small distances, so that the intensity of shock varies not only from meteorite to meteorite but within a single specimen. The following photographs illustrate structures commonly found in order of increasing shock. Abbreviations for source collections are as follows: NMNH = National Museum of Natural History, Smithsonian Institution; ROM = Royal Ontario Museum; WNSE = Wards Natural Science Establishment.

---

**A. Neumann lines.** Toluca meteorite (NMNH 3378). Neumann lines are shock-generated twins parallel to (121) in kamacite. They generally range from 1 to 10 $\mu$ in width, although they are considerably wider in some classes of iron meteorites. The twin law indicates that 12 orientations of Neumann lines are possible in one single crystal of kamacite. Neumann lines are generated by low shock pressures, generally < 130 kbar, and are ubiquitous in iron meteorites, except where obliterated by structures due to more intense shock. In this figure, Neumann lines occur in massive kamacite on the right side, but are interrupted on the left by an intergrowth of kamacite and taenite called plessite. Note that the Neumann lines do not occur in the darker taenite, but appear to be developed in the kamacite lamellae within the taenite. Reflected light PPL

**B. Neumann lines.** Goose Lake meteorite (NMNH 1332). Neumann lines above and below its boundaries can be seen to be interrupted by the taenite (*dark*) of a plessitic intergrowth. Note that the Neumann lines also occur in the kamacite of the plessitic intergrowth. Reflected light PPL

**C. $\varepsilon$ Structure.** Trenton meteorite (WNSE 30.103). The phase diagram for iron indicates that $\alpha$-Fe (kamacite) undergoes a transition from the body-centered cubic structure to the hexagonal close-packed $\varepsilon$-Fe structure at a pressure of 130 kbar. The transition is reversed after release of pressure resulting in a hatched microstructure designated $\varepsilon$ structure. The structure is actually a distorted $\alpha$-Fe structure and is not continuous through the taenite of a plessitic intergrowth on the *left*.
Reflected light PPL; Nital etched

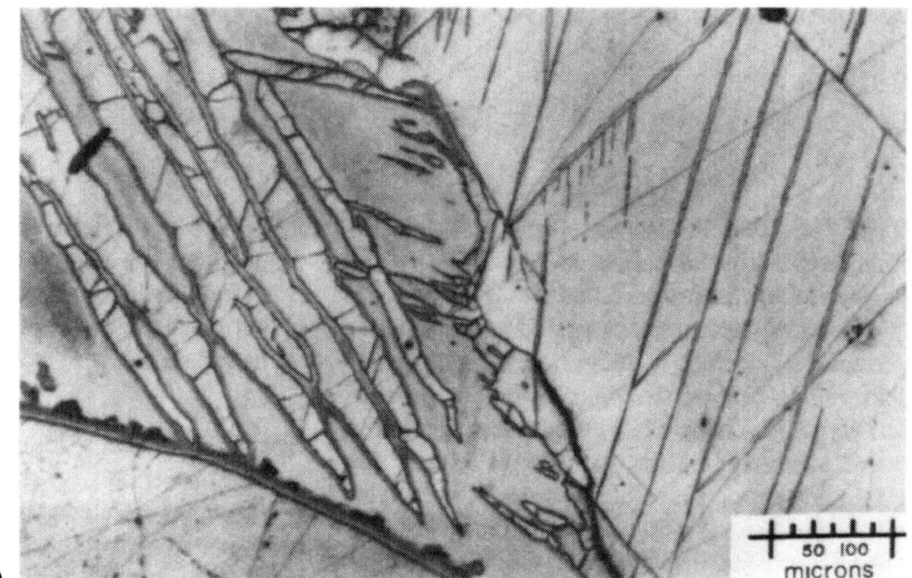


171 A

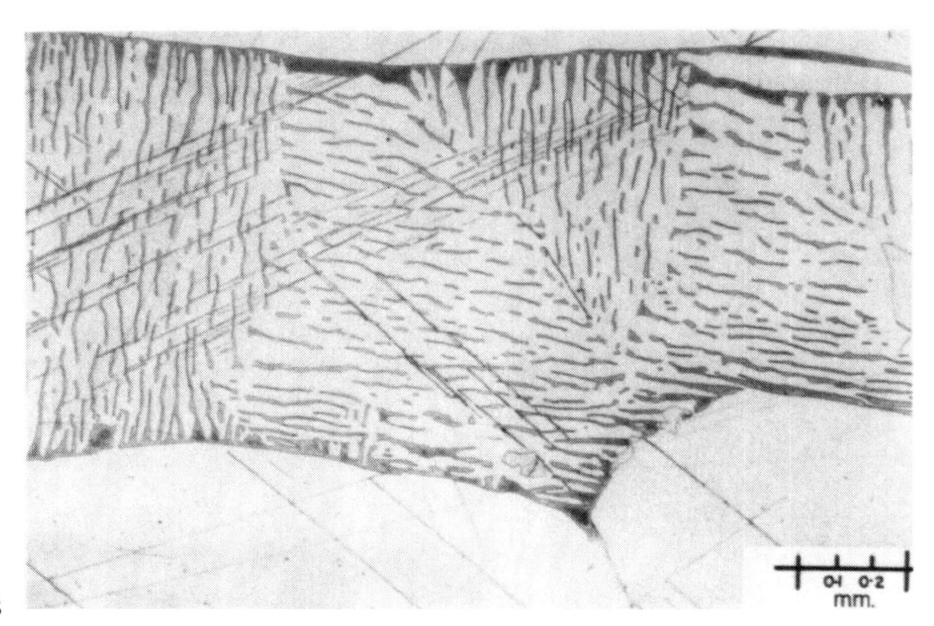


171 B

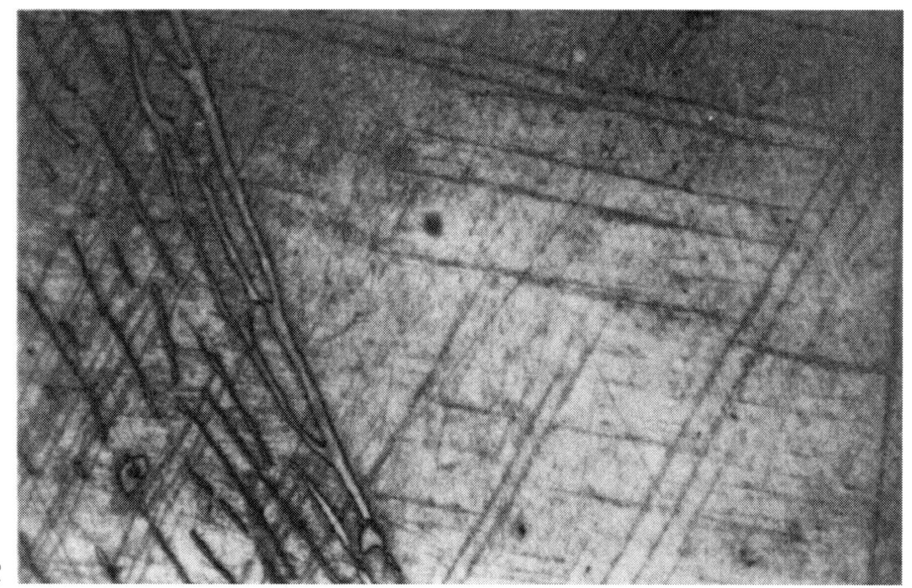

171 C

## 172. Shock Structures in Iron Meteorites (II)

S. A. KISSIN

**A. Type I troilite.** Toluca meteorite (ROM 3923). Troilite is sensitive to shock with a series of deformational features proportional to the intensity of the shock pressure. BUCHWALD (1975) differentiated five stages of troilite deformation ranging from undeformed monocrystals (type 1) to shock-melted troilite (type 5). LIPSCHUTZ's (1968) classification is, however, based upon more quantitative criteria. His type I (lightly shocked) troilite includes BUCHWALD's type 1, 2, and 3 categories. The figure illustrates troilite with shock-induced multiple twinning (BUCHWALD's type 2). This type of deformation is generated by shock pressures of less than 400 kbar.
Reflected light NX

**B. Type I troilite.** Sardis meteorite (NMNH 1381). Shown here is a large sphalerite grain (*dark*) bounded sharply by lightly shocked troilite (*light*). Note the highly oriented trains of minute troilite exsolutions in the sphalerite. Reflected light PPL

**C. Type III troilite.** Sardis meteorite (NMNH 1381). Elsewhere in the specimen illustrated in **B**, shock-melting generated by shock pressures greater than 800 kbar has affected sphalerite with exsolved troilite such that the sphalerite (*dark*) is mobilized into ribbons and swirls in troilite. The adjacent areas of troilite show areas of undulatory extinction (*upper right*) characteristic of BUCHWALD's type 3 with some areas of recrystallized troilite (Type II of LIPSCHUTZ and type 4 of BUCHWALD). Reflected light NX

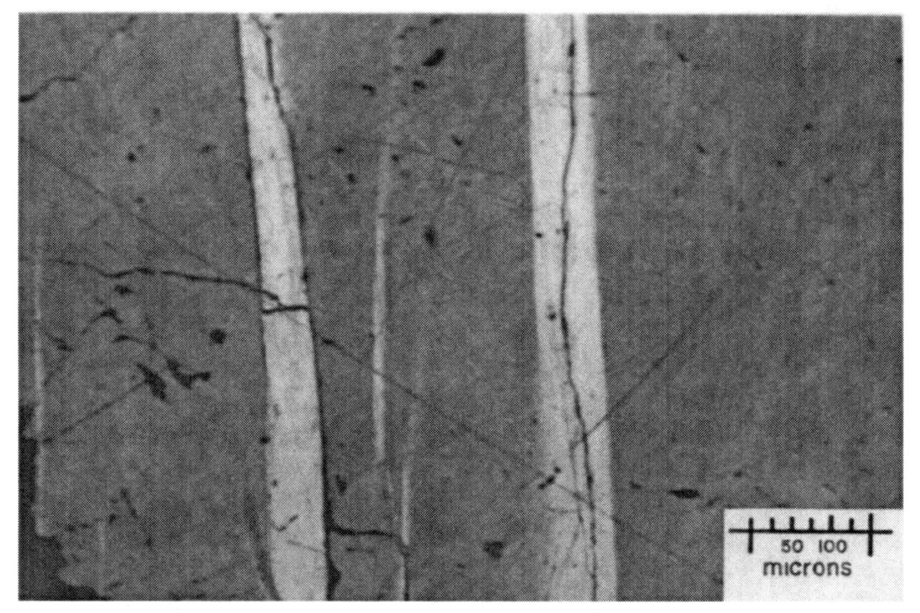


172 A

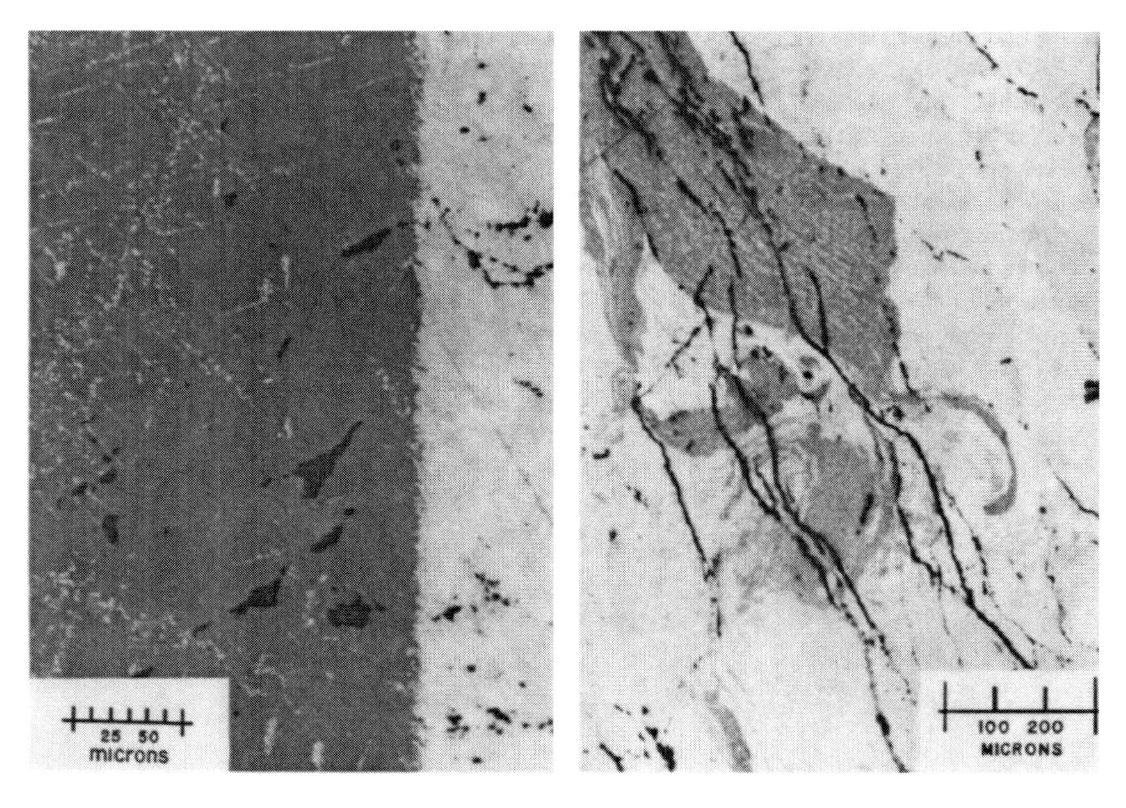


172 B

172 C

# 173. Schistosity in Galena: S.E.M. Studies

K. R. McClay

S.E.M. studies of etched polished sections and etched cleavage fragments (McClay 1977) have thrown light on the recrystallization mechanisms of galena (see also Plate 157).

**A. Kinks in galena.** Yerranderie N.S.W., Australia. A Devonian vein deposit of coarse-grained galena exhibits a strong dimensional fabric and strong crystallographic preferred orientation. Recrystallization from coarse (cm-size) to fine (50–100 $\mu$m) grains is achieved principally along kink and deformation bands. S.E.M. studies show kink bands (*kb*) as outlined by bent cleavage steps. The *large arrows* indicate the amount of rotation in the kink band. Elongate subgrains (*s*) form in the kink band with walls of dislocations (*arrowed*) making the subgrain boundary.
S.E.M. photograph of cleavage fragment etched 50 s at 65 °C in HCl-Thiouria etchant (Brebrick and Scanlon 1957).

**B, C. Core and mantle microstructure.** Halkyn, North Wales, U.K. Fine-grained "steel" galena occurs at Halkyn, North Wales, in deformed post-Carboniferous veins. Large old grains recrystallize into a fine-grained equi-dimensional matrix (1–5 $\mu$m grains in **C**) with no shape or crystallographic preferred orientation (McClay 1981). The recrystallization mechanism is illustrated in **B**. On the *right hand side* can be seen the core structure of equant subgrains (*s*) of similar orientation (as indicated by the flat cleavage faces).
Progressive misorientation of these subgrains, probably by rotation, produces a mantle of rotated subgrains (*rs*) in the center of the photograph. Further misorientation of these rotated subgrains leads to new, recrystallized grains (*ng*). The final product is seen in **C**. This is a mosaic of small equant grains with no crystallographic nor shape orientation. This core and mantle microstructure is indicative of dynamic recrystallization and is similar to that found in quartz (White 1976) as indicated in the sketch.
The sketch (after White) illustrates dynamic recrystallization and the core-mantle microstructure. Subgrains form in deformation bands, and undergo progressive rotation until new grains are formed

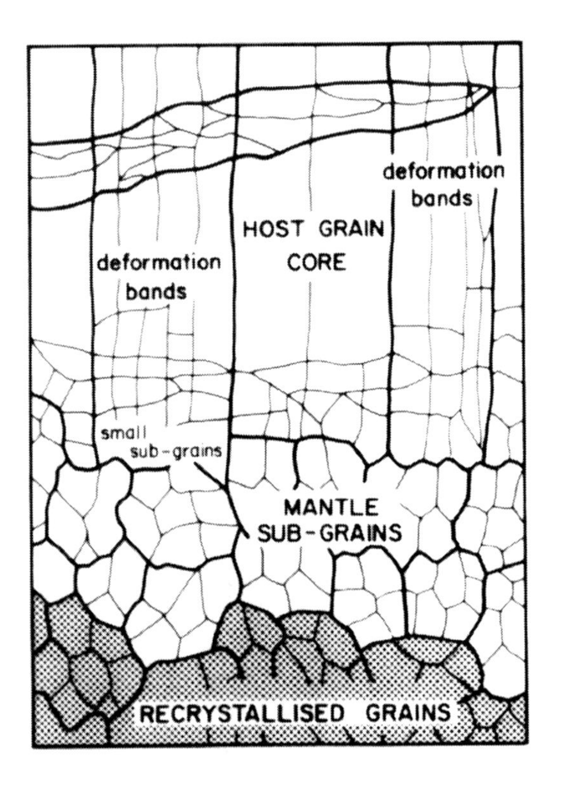

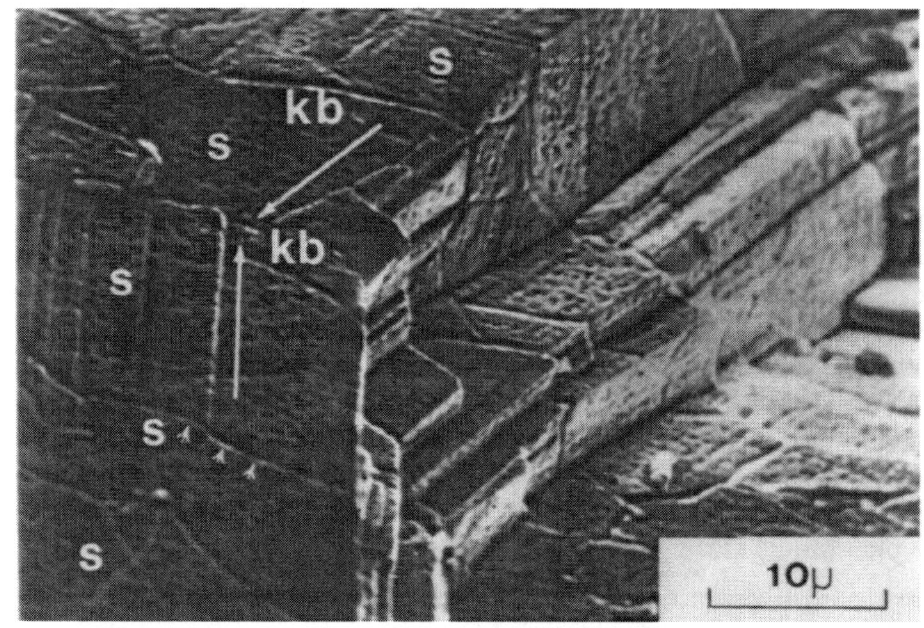


173 A

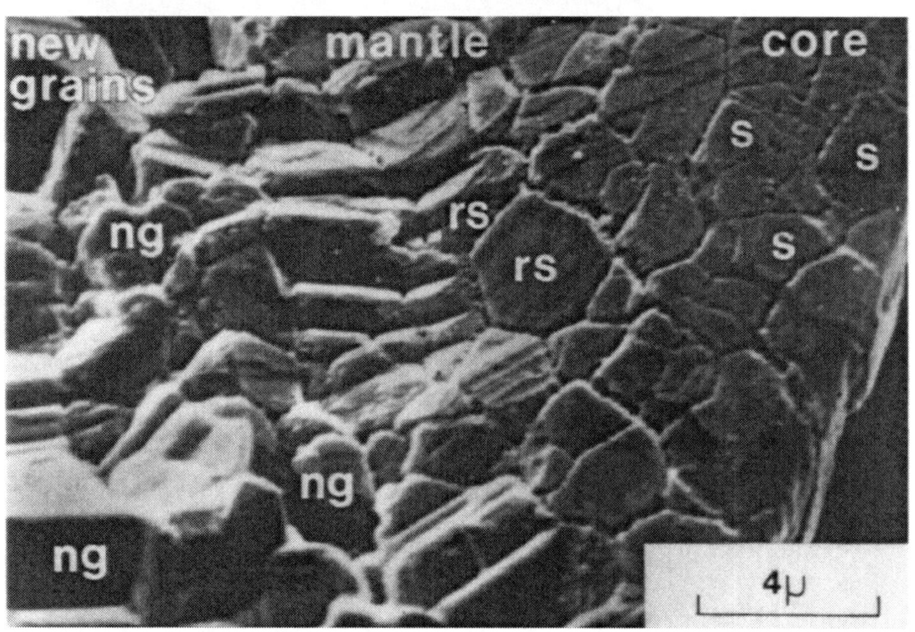


173 B

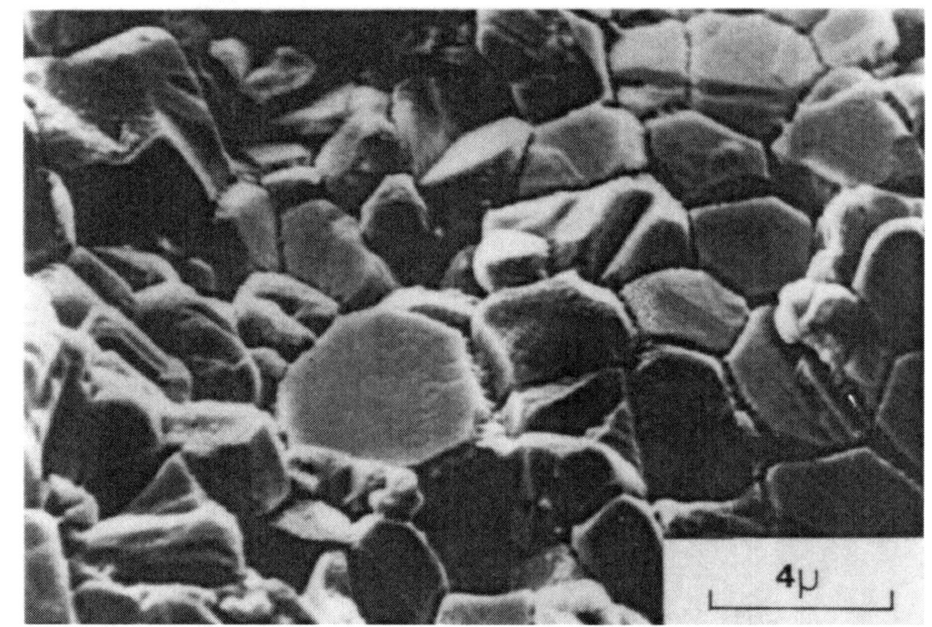


173 C

## 174. Foliation Formed by Alignment of Kink-band Boundaries in Biotite

R. H. VERNON

Foliation is produced by the subparallel alignment of kink-band boundaries in deformed biotite occurring in a mica-staurolite schist near the Nairne Pyrite Mine, Mount Lofty Ranges, South Australia. The deformation is restricted to narrow, bifurcating zones up to about 4.0 mm wide (averaging about 1.5 mm), and is very heterogeneous, especially in more weakly deformed areas. The metamorphic conditions appear to have been those of the lower part of the hornblende hornfels (low-pressure amphibolite) facies, although partial replacement of garnet and staurolite by chlorite near these deformation zones suggests lower-grade conditions after the deformation or during its later stages. The foliation is similar to that formed experimentally in strongly kinked mica by MEANS (this volume Plate 175; also Plate 179). The *dark circular spots* in the photographs are pleochroic haloes.

---

**A.** Early stage of the deformation process, in which a biotite grain in suitable orientation has been kinked much more strongly than its neighbors, involving a shortening of about 30%. This has resulted in a local foliation produced by the subparallel alignment of kink-band boundaries in the deformed grain. PPL

**B.** Rough alignment of kink-band boundaries in a deformed grain of biotite, producing a foliation approximately parallel to the long dimension of the photograph. The grain also shows open folding at one end and tight folding with some kinking at the other. PPL

174 A

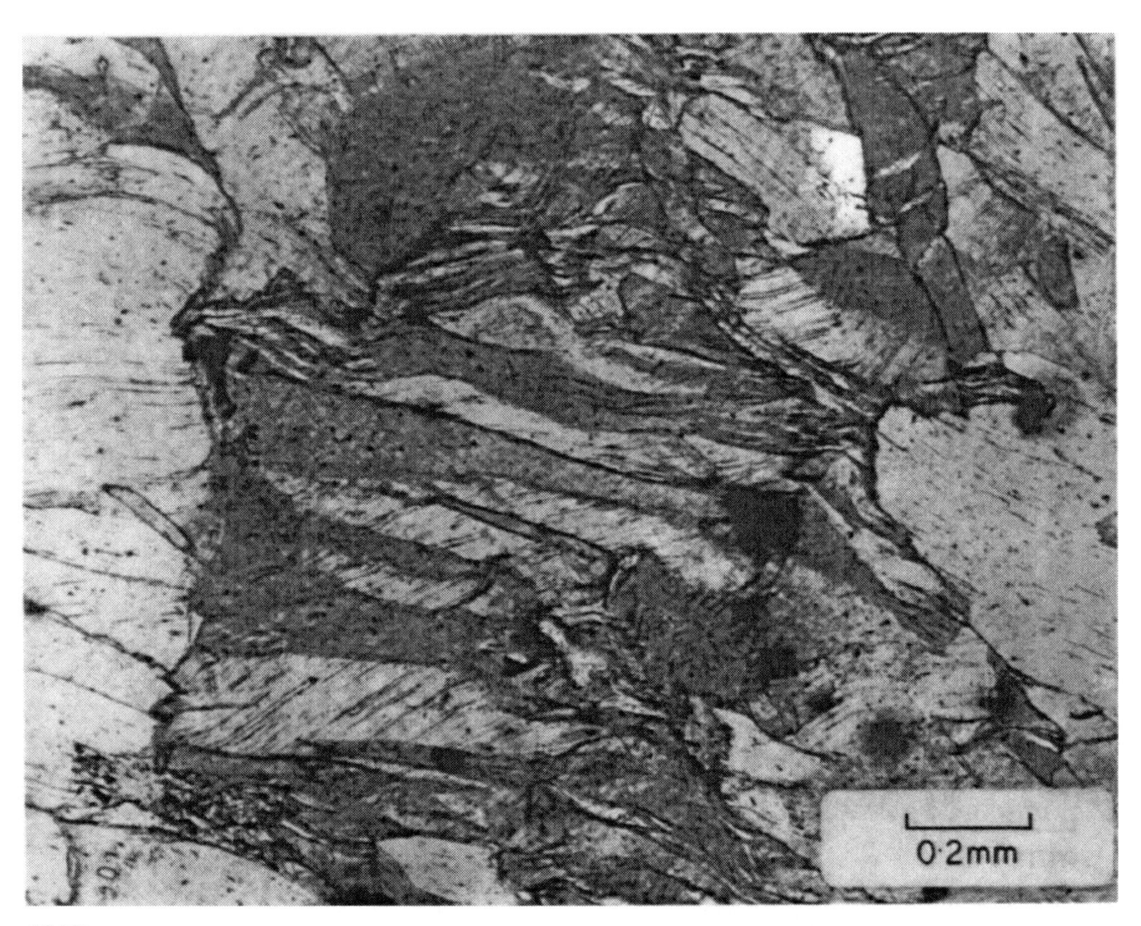


174 B

## 175. Foliation Defined by Kink-band Boundaries in Experimentally Deformed Phlogopite Crystals

W. D. MEANS

Part of a specimen of an initially schistose, glass-bonded mica ceramic shortened experimentally at about one kbar confining pressure, room temperature. The specimen was shortened parallel to the long dimension of the photograph, and parallel to the initial schistosity.
The mica crystals (synthetic fluorine phlogopite) were shortened about 15%, with prolific development of intracrystalline kink bands and chevron folds. The boundaries of these regions (*fine, dark, horizontal lines* in the photograph) define a kind of secondary foliation that is also occasionally seen in rocks, and that may more often be a stage in the development of certain slaty cleavages and schistosities (WILLIAMS et al. 1977, WHITE and KNIPE 1978). Specimen TP-46 from experiments described briefly by MEANS (1968).

175

# 176. Incipient Mechanical Cleavage in an Experimentally Deformed Biotite Single Crystal

M. FRIEDMAN and H. C. HEARD

The photograph shows incipient cleavage formed by extension fractures (*curved arrow*) in feldspar lath and kink-bands (*straight arrow*) in the enclosing biotite crystal. Specimen shortened parallel to (001) in biotite at 3.0 kbar confining pressure, 300 °C, and at a shortening rate of $10^{-4}$ $s^{-1}$. An overview of the total specimen is shown in CARTER and RALEIGH (1969, Plate 4a). The specimen illustrates two principles:
(a) that the mechanical response of a body is an interaction between the intrinsic properties of the material and the physicochemical environment at the time of deformation – here ductile kink-banding of the biotite and brittle fracture of the feldspar; and (b) that apparent polyphase deformation may in fact result from the single deformation of a heterogeneous body. Note here that (1) extension fractures in the feldspar accurately reflect the bending stresses within the lath (FRIEDMAN and SOWERS 1970, Fig. 14), and depending upon location, these are perpendicular, parallel, and inclined to the long axis of the lath; (2) the kink-band boundaries are inclined at about 90° to the end loads (parallel to the long axis of the photograph) so that the local maximum stress $\sigma_1$, if taken to be normal to the kink-band boundaries, parallels the externally imposed $\sigma_1$ which also produced the bending; and (3) dynamic petrofabric analyses of kink-bands and fractures might be interpreted to indicate polyphase deformation unless one recognized the operation here of principles (a) and (b) above.

NX

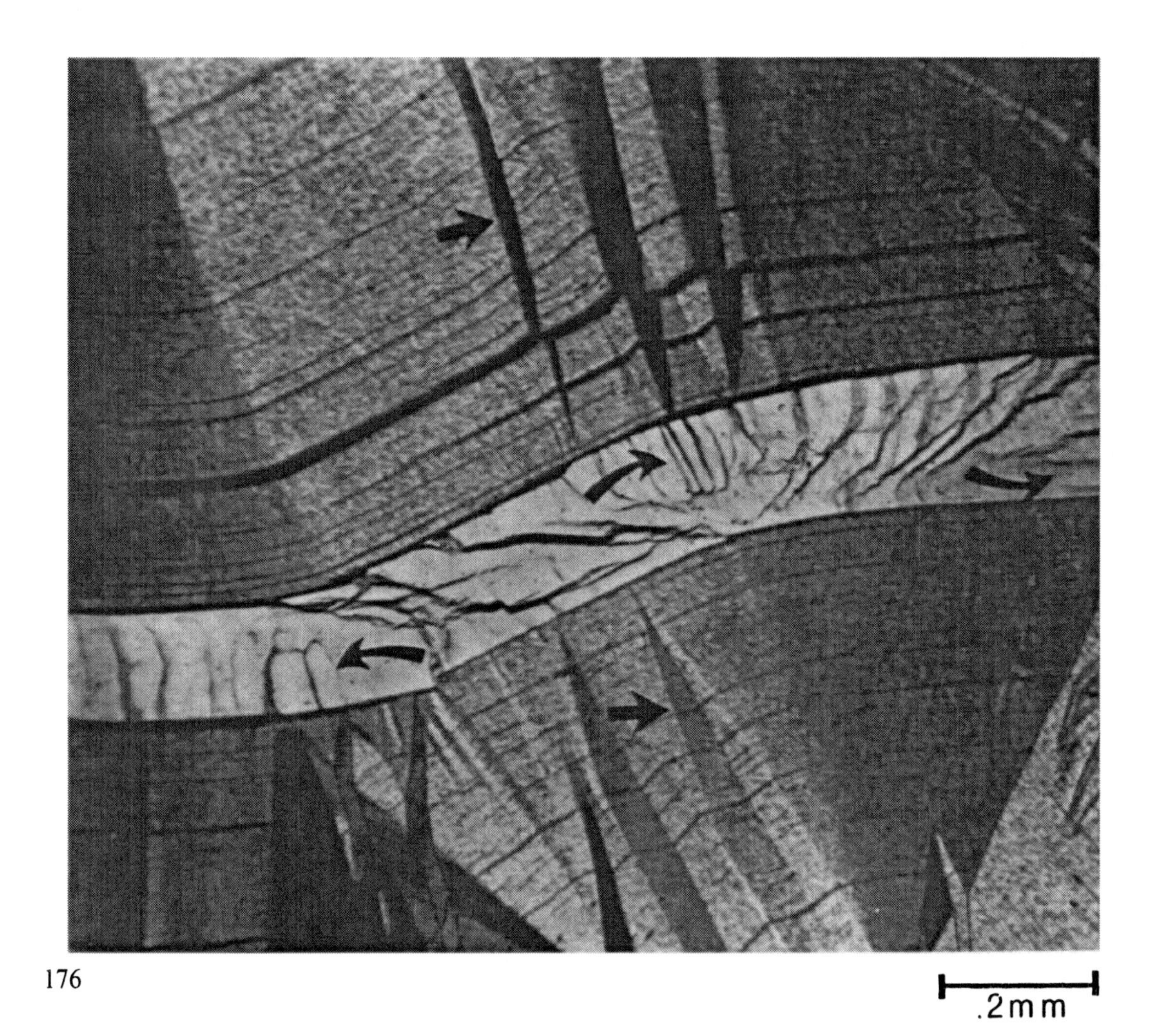

176

# 177. Deformed Detrital Mica (I)

P. J. Conaghan and C. McA. Powell

The specimen is from a road outcrop in the lower to middle part of the Late Miocene to Early Pleistocene Siwalik Group, near Udhampur, N.W. India, and is located south of, and adjacent to, the Himalayan Main Boundary Fault (Gansser 1964) (see map, next plate). Macquarie University specimen MU5562A.

Clastic sediment in the molasse apron on the SW side of the Himalayas reflects the progressive rise and unroofing of the metamorphic core of the High Himalaya during the late Cenozoic. Lithologically, the rocks are feldspathic litharenites with rock fragments of metamorphic provenance, detrital micas and heavy minerals, as well as quartz and feldspar, derived from the rising Himalayan chain and deposited in the wedge-shaped Himalayan foredeep which locally contains in excess of 5000 m of sediment. Johnson and Vonda (1972) provide some petrographic analyses of these sandstones from a nearby area to the east.

In this specimen detrital muscovite and altered biotite lying both parallel and oblique to bedding have been deformed into angular kink folds. The section is cut perpendicular to bedding and to the trend of regional folds and thrusts. Strain indicated by the geometry of the kink folds is a bedding-parallel shortening around 15%. This is accomplished in the body of the rock with no visible signs of intragranular deformation of the quartz and feldspar grains, and only moderate deformation of the less rigid rock fragments. There is no secondary cleavage in outcrop. Presumably, much of the bedding-parallel shortening that kinked the detrital mica grains has been absorbed by intergranular movements between the smaller equidimensional quartz and feldspar grains. The rock is free of clay and is cemented by ferroan calcite.

---

**A. Conjugate kinks in deformed detrital muscovite grain.** The sense of rotation on the kink bands corresponds to a bedding-parallel shortening with the obtuse angle between the kink axial surfaces (105°) bisected by the bedding. PPL

**B. Overview.** Broad-scale view of kinked mica grains within a host of more rigid clastic (dominantly quartz) grains. Some grain impingement has occurred, as well as slight to moderate bulk deformation of rock fragments. The detrital mica grains have ragged ends indicative of their detrital origin. PPL

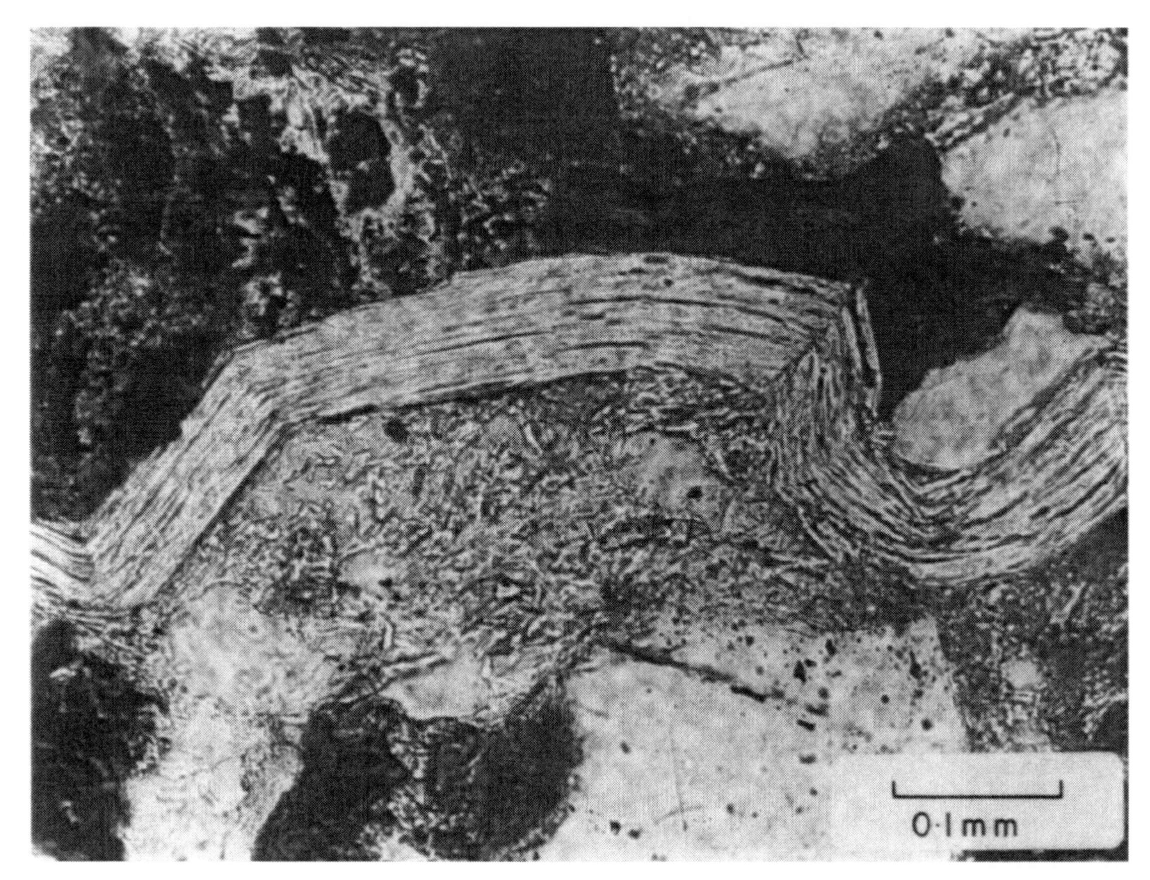


177 A

177 B

# 178. Deformed Detrital Mica (II)

P. J. CONAGHAN and C. MCA. POWELL

The mica grain is from the same thin section as Plate 177. Although a detailed cross-section of this part of the Himalayas is not available, the locality probably lies within GILL's (1951) Structural Zone 3 of the Sub-Himalayas – a zone characterized by closely spaced strike faults and severely compressed folds. There is no cleavage in outcrop.

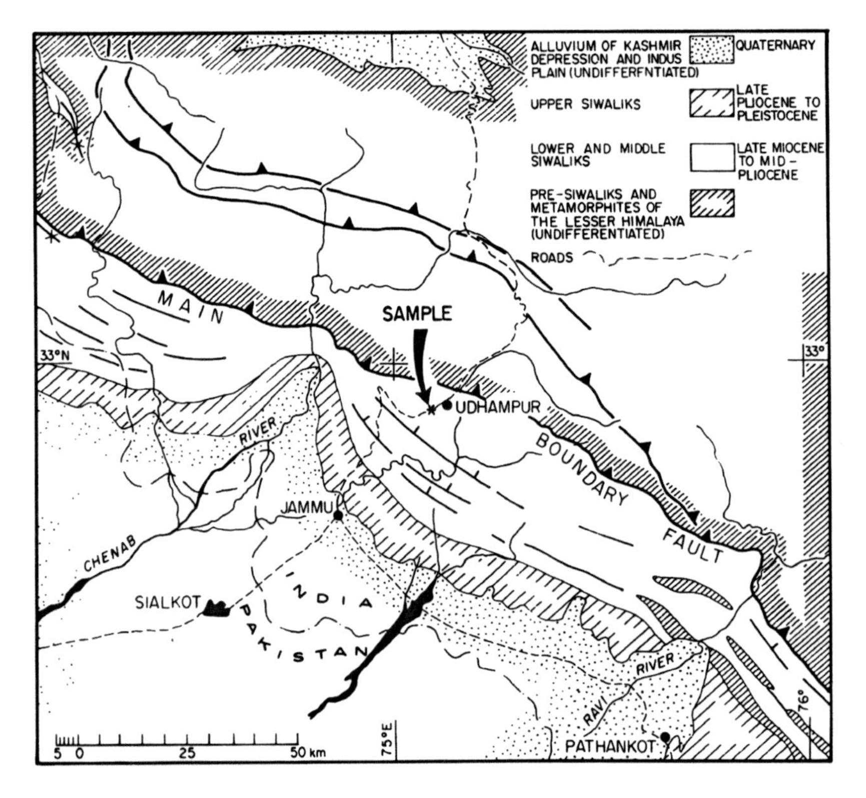


The figured mica grain shows two conjugate kink bands converging to form a box fold. The geometry is the same as mesoscopic folds commonly seen in phyllites and schists, and the mechanism of deformation is presumably by (001)-parallel slip in the mica. Individual (001) laminae may bend gradually, as on the right-hand side of the specimen, or change orientation abruptly as at the main kink boundaries. A poorly defined deformation zone parallel to the dextral kink boundary shows up in crossed nicols (**B**). The ends of the muscovite grain are ragged. The box fold has developed between detrital quartz grains that have moved relative to one another to accommodate the strain. The sense of shortening in this mica grain, which lay oblique to the bedding, is consistent with strain during compaction.

---

**A.** Box fold formed by conjugate kinks in a deformed detrital muscovite grain. PPL

**B.** Same grain NX

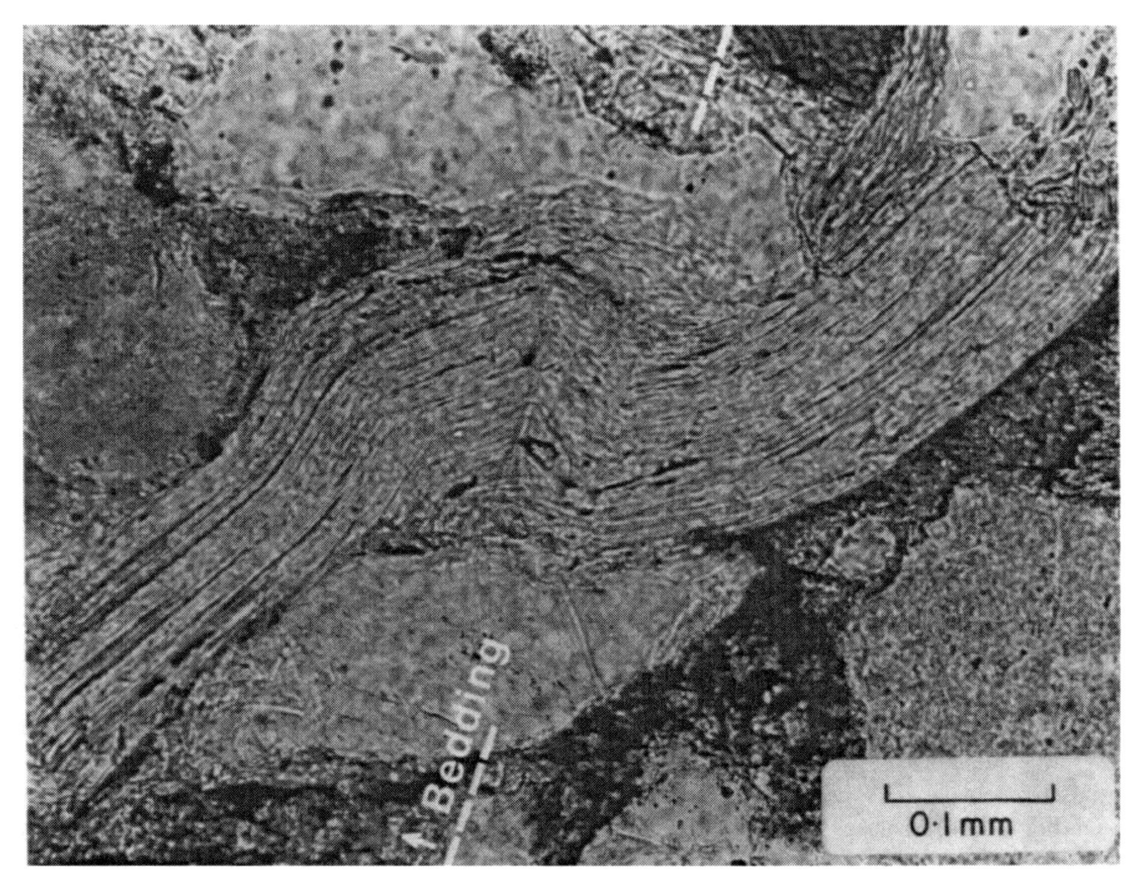


178 A

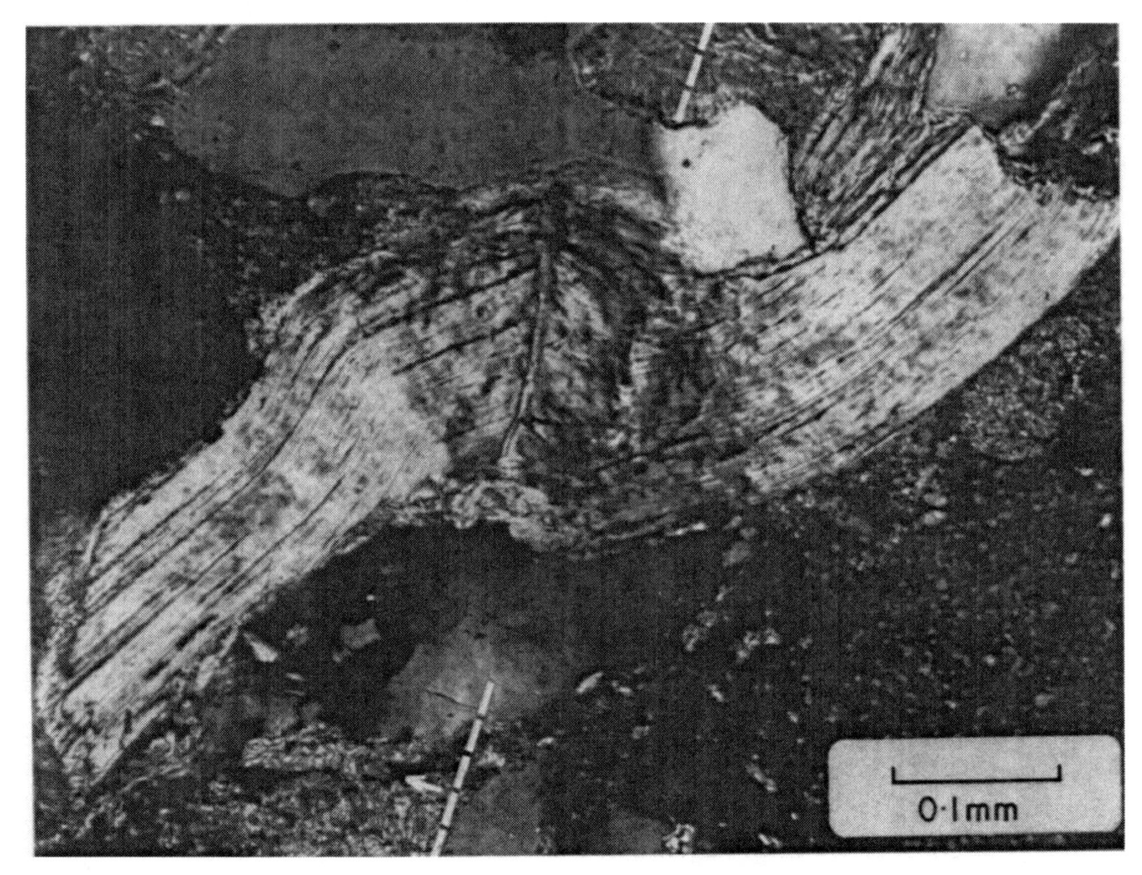


178 B

## 179. Foliation Development in Biotite Schist

C. J. L. WILSON

A tectonic foliation of variable intensity is associated with closely spaced shear zones in a pegmatite deformed under amphibolite facies conditions (at Cava di Mica near "I Mondei", Valle d'Antrona, Italian Alps). A heterogeneous foliation is characterized by alternating layers of quartz, albite, biotite, and/or muscovite together with albite-mica mixtures. The foliation is defined by the alignment of mica and the elongation of quartz and albite grain aggregates. An examination of the biotite-rich portions suggests that the foliation is controlled by four parameters: the amount of bulk strain, influence of the surrounding material, the extent of grain growth and grain nucleation, and the operation of diffusive processes (WILSON and BELL 1979).

---

**A. Biotite schist.** This is characterized by a heterogeneous foliation development outlined by two biotite types:

1. Domains composed of large deformed 1M and $2M_1$ biotite grains. These are relic pegmatitic biotite, now substantially reduced in grain size. They exhibit extensive kinking (*A*), gentle bending (*B*) and dissolution features (WILSON and BELL 1979). The parallelism and degree of preferred orientation of the deformed biotite is achieved predominantly by mechanical rotation, and is therefore strain-dependent.
2. Small elongate undeformed biotite grains concentrated in discontinuous lenticular aggregates (*C* and *D*). The biotite grains have a marked preferred orientation (east–west) and are characterized by 1M and $1M_r$ (120°) polytype structures.

NX

**B. Individual biotite grains in schist.** This photograph shows in greater detail the relationship between the deformed and undeformed biotite grains. The undeformed biotite is characterized by numerous planar defect structures (BELL and WILSON 1977) and is nucleated by either a syntectonic crystallization process (VERNON 1977) or the formation of strain-free grains by a segmentation process, or by both of these (WILSON and BELL 1979). The segmentation process breaks down large biotite grains into smaller grains by a series of misorientations across boundaries parallel and subnormal to (001) and achieves microstructures similar to that produced by a recovery-recrystallization process. NX

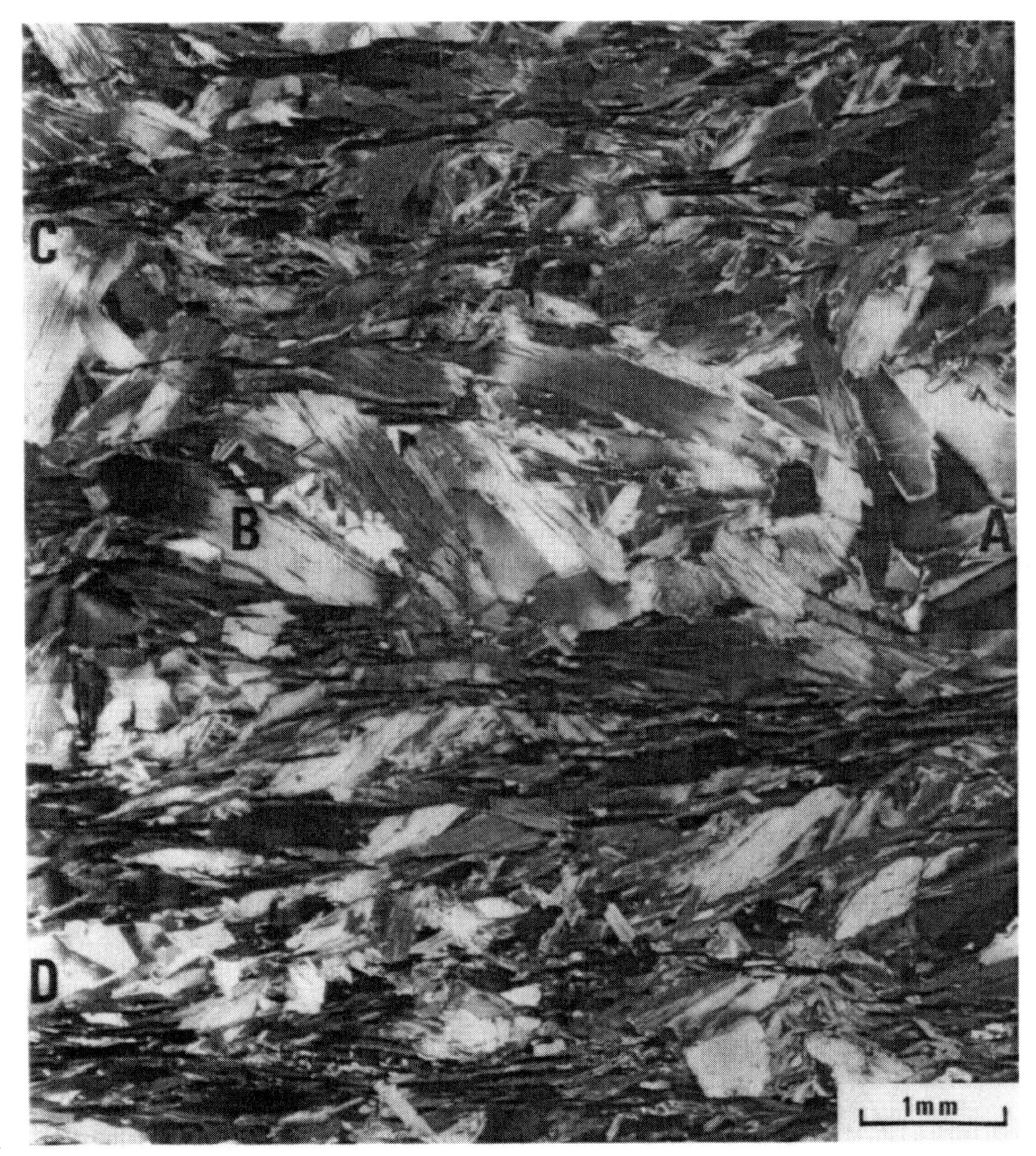

179 A

179 B

# 180. Foliation Development in Deformed, Hydrothermally Altered Acid Volcanic Rocks (I)

S. F. Cox and M. A. Etheridge

The following series of illustrations are from the Cambrian Mt. Read Volcanics near Queenstown, Tasmania, Australia. The sequence of hydrothermally altered acid volcanics in this area has been folded and metamorphosed at low greenschist-facies conditions during the Middle Devonian. Foliation is defined largely by layer silicate "films" and fibrous microstructures in a range of extension sites (also see Plate 126).

---

**A. Spaced layer-silicate film development.** Chlorite which pseudomorphs a biotite phenocryst has been gently crenulated. Chlorite films up to 20 $\mu$ wide with (001) subparallel to the film length have developed approximately parallel to crenulation axial surfaces. The films extend out of the host chlorite grain into the surrounding quartz + chlorite + phengite matrix. NX

**B. Detail of chlorite films.** (001) of host chlorite is abruptly truncated against the films. The difference in composition of film and host chlorite favors the interpretation that films have grown by a solution-transfer mechanism involving ion-exchange reactions. NX

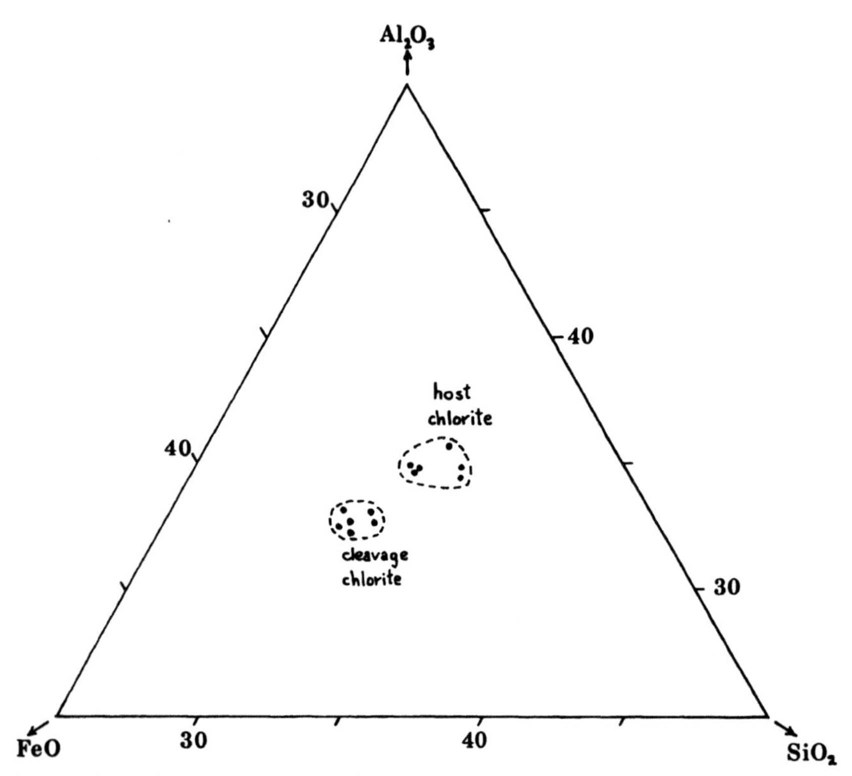


Compositional variation between host and film chlorite

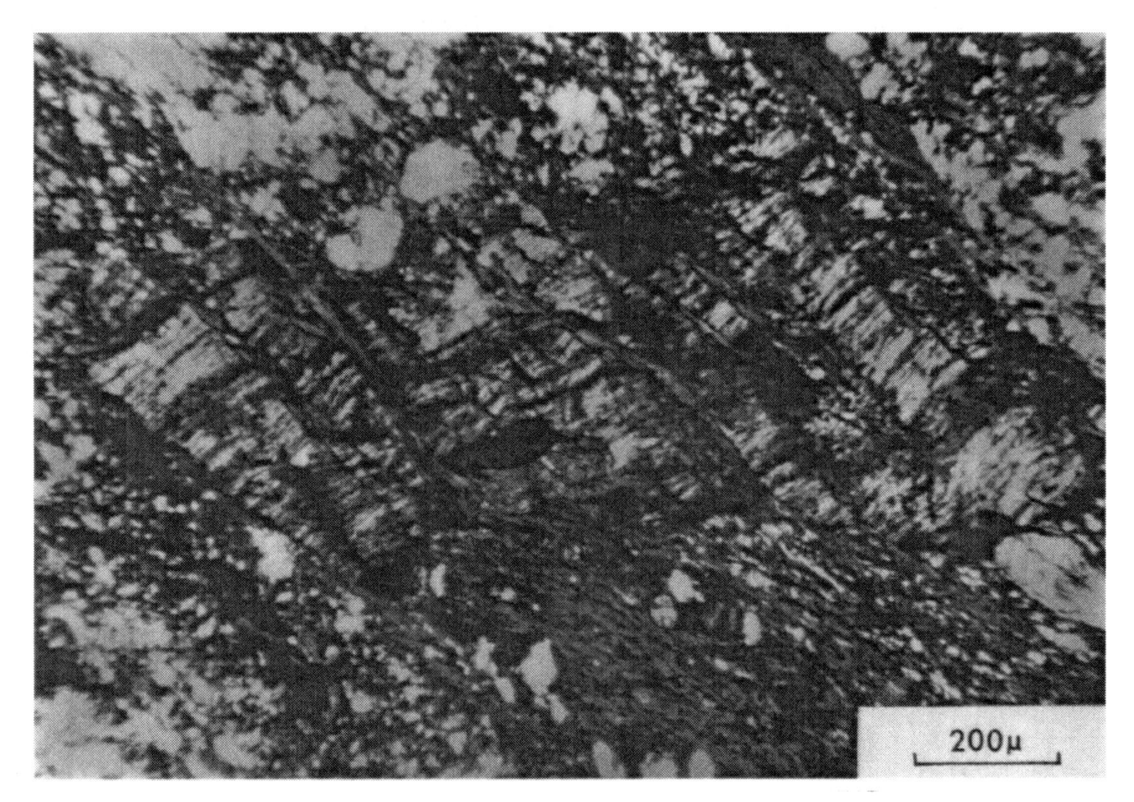


180 A

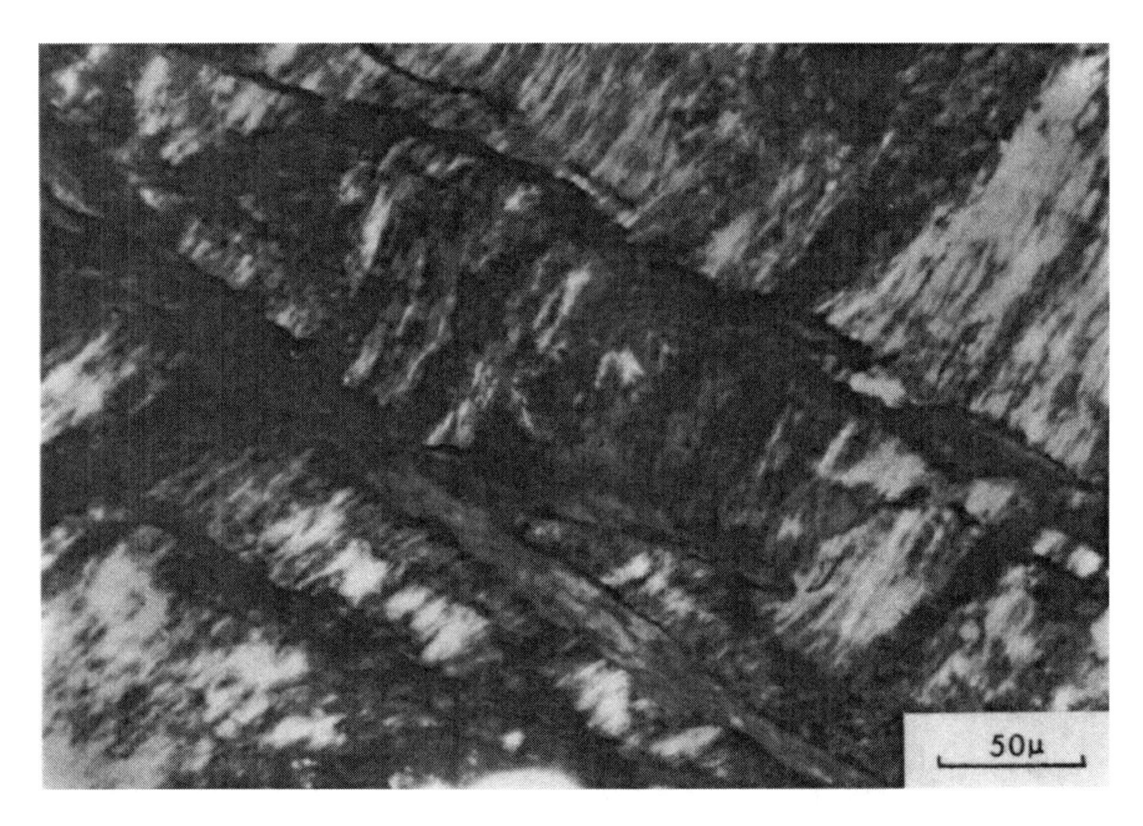


180 B

# 181. Foliation Development in Deformed, Hydrothermally Altered Acid Volcanic Rocks (II)

S. F. Cox and M. A. Etheridge

**A. Domainal layer-silicate film development in a pure layer-silicate aggregate.** A fine-grained muscovite aggregate has anastomosing mica films up to about 20 $\mu$ wide in which (001) is subparallel to film length. Films are spaced up to 70 $\mu$ apart, with the interfilm domains being only weakly deformed sheaf-like aggregates of muscovite. NX

**B. Detail of films and interfilm domains.** NX

**C. Transmission electron micrograph.** From the same rock showing small domains of undeformed micas inclined to the foliation direction between zones of elongate and undeformed foliation-parallel micas

181 A

181 B

181 C

# 182. Kinking in a Rotating Stress Field

B. Bayly

A common problem in studies of cleavage is as follows: suppose a stress field operates for a sufficient time to create an appropriate texture and then the stress field rotates with respect to the rock. Will the texture rotate, or will a new texture develop and destroy the old, or will a new texture develop superimposed on the old, both remaining visible? Of course the answer depends on the rock, but there is no rock for which the answer is direct and obvious. Often when we see a texture, we suspect that some peculiarities may be due to stress rotation, but it is usually difficult to be confident. This plate illustrates a rock where the geometry seems definitely due to stress rotation.

Cambrian Hoosac formation from north shoulder of Worth Mountain, by Middlebury Gap, 4 miles SE of Ripton, Vermont, U.S.A. Chlorite schist. Specimen by courtesy of P. Smith

---

**A.** and **B.** Crossed nicols

**C.** and **D.** Same fields of view in ordinary light

In the sketch *abab* shows an earlier set of kinks and *1–2–1–2* shows a later cross-cutting set. Although they cannot be seen in the plate, the rock in fact seems to contain a third set, earlier than the set marked *abab*, with orientation farther round toward east–west

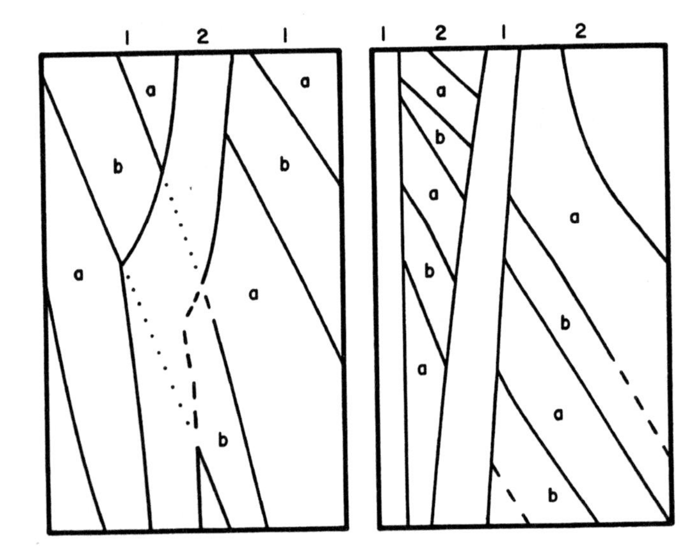

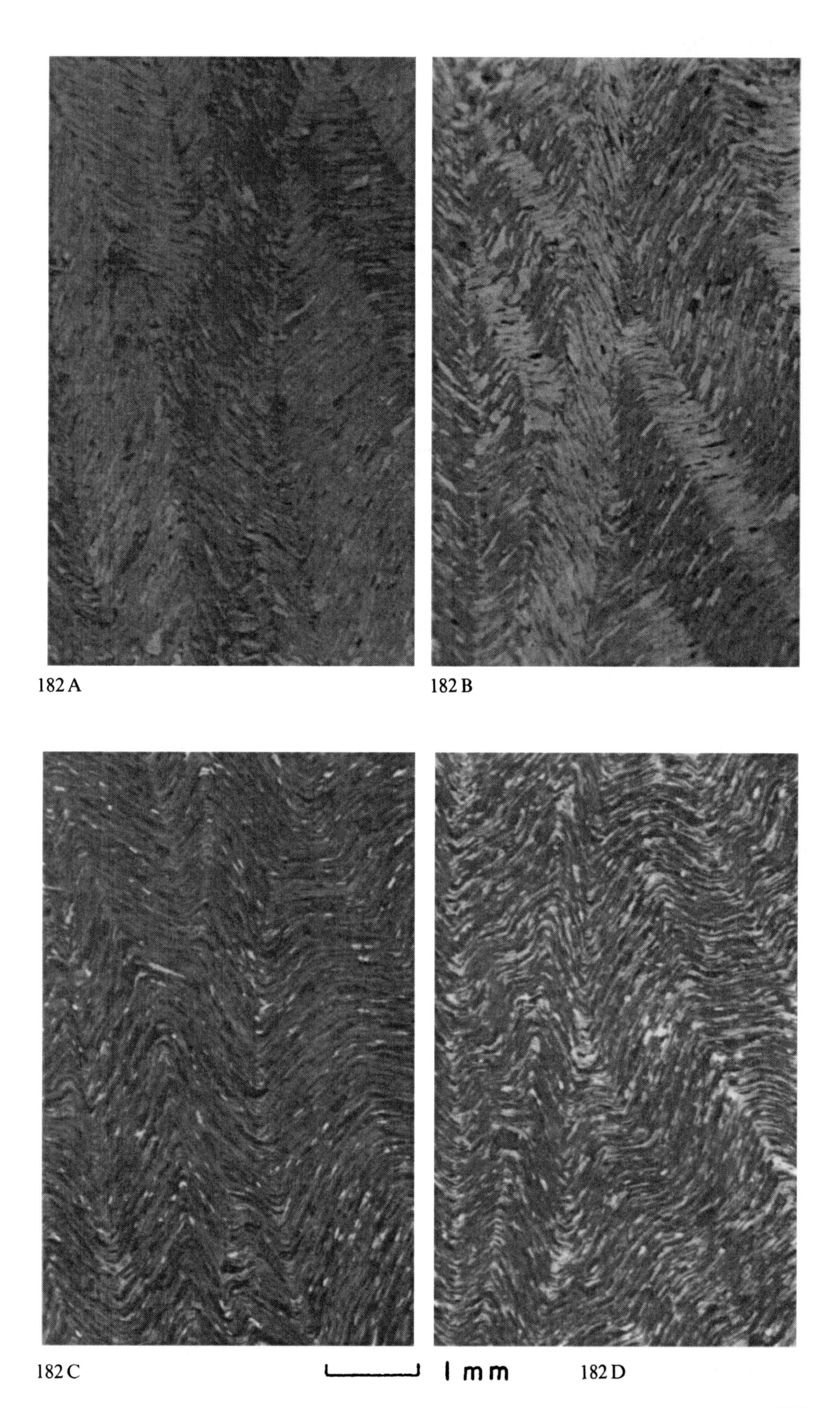
182 A
182 B
182 C
1 mm
182 D

## 183. Experimentally Deformed Clay (I): Cleavage-like Microstructures

A. J. MALTMAN

Photographs at different scales (**A** through **C**) of ball clay deformed in simple shear. The "primary settling fabric" (*p.s.f.*, oriented as shown) has been sharply crenulated in a periodic but highly asymmetric style (the "creases" of MALTMAN 1977). PPL

*Bulk conditions of deformation.* Sinistral distributed simple shear parallel to *p.s.f.*; shear strain (tan $\psi$) = 0.83; strain rate = $3.38 \times 10^{-5}$ $s^{-1}$; stress normal to shear direction = 3 bar; moisture content of clay = 30%; room temp.

*Notes on techniques* (further information in MALTMAN 1977):

*Specimen preparation.* Clay/sea-water slurry was allowed to settle under gravity; superjacent water was syphoned off, and material 'air-dried' to the required moisture content. Material was then cut to dimensions required for test specimen.

*Thin section preparation.* Slice (~ 1 cm thick) was cut from the central portion of specimen and impregnated with melted Carbowax P.E.G. 6000. After immersion for about one week at 70 °C, slice was cooled to room temperature and ground (paraffin coolant) to interior portion to avoid artefacts. Thin section was ground in the conventional way.

*Optical examination.* Individual clay crystals were not discernible with an optical microscope, but viewing under crossed nicols revealed preferred orientations of clay aggregates. Without experimental deformation and away from the disturbed edges of sections, an even "primary settling fabric" is always apparent.

*Shear tests.* "Direct shear" deformation is the conventional shear box test of Soil Mechanics. The arrangement is as below except that the strain is concentrated along a narrow, predetermined, central zone. In "distributed shear", a block 60×60×20 mm is held in a thin rubber sleeve surrounded by thin metal plates, the whole of which is subjected to simple shear. Loads may be added perpendicular to the shear direction. Dimensional change is negligible, except under normal stresses greater than about 5 bar, which may cause major expulsion of pore water and volume loss.
Please note that the *experimental conditions are not necessarily homogeneous* throughout the specimen. The bulk conditions quoted are a comparative guide, but may not be accurate quantities for the portions illustrated.

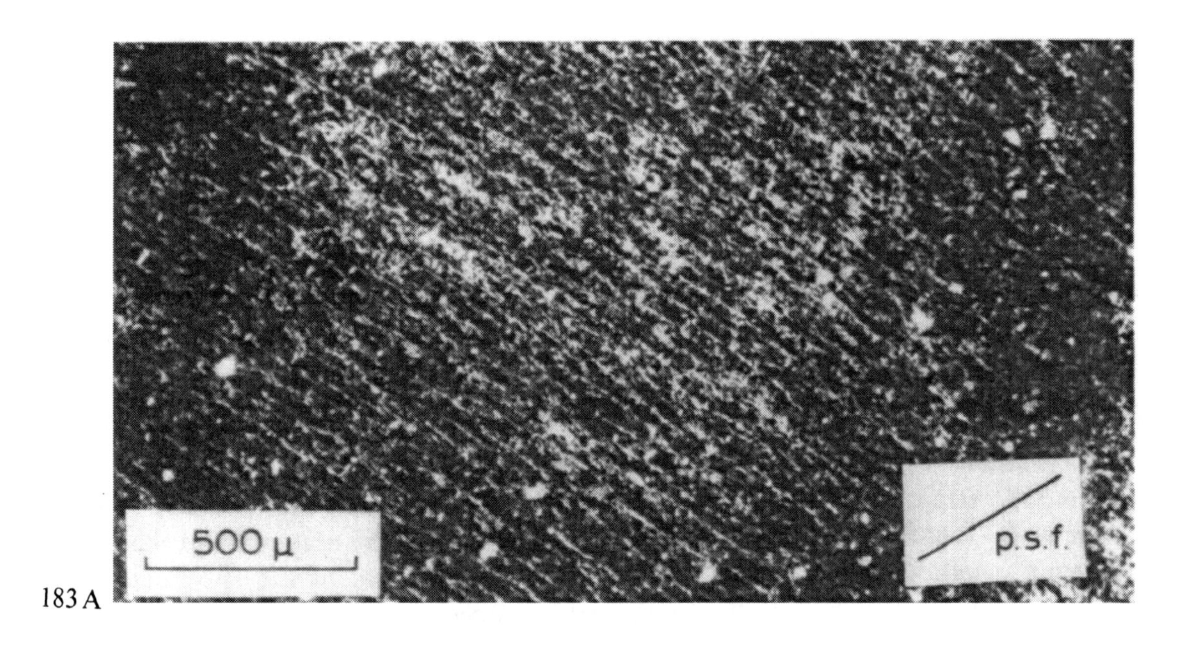


183 A

183 B

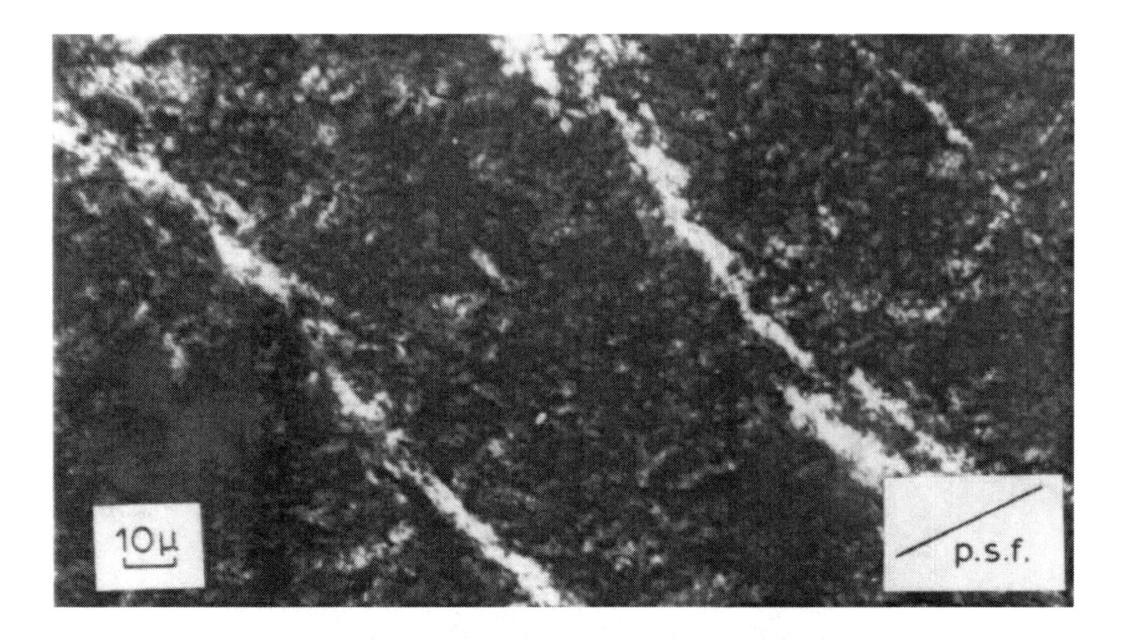


183 C

# 184. Experimentally Deformed Clay (II): Cleavage-like Microstructures

A. J. Maltman

Photographs at different scales (**A** through **C**) of ball clay deformed in triaxial compression. "Creases" (sharp, periodic, highly asymmetric crenulations) of the primary settling fabric. PPL

*Bulk conditions of deformation.* Compressive strain = 30%; strain rate = $5 \times 10^{-6}$ $s^{-1}$; effective pressure = 0 (confining pressure = 7 bar, no drainage); moisture content of clay = 30%, p.s.f. parallel to length of test cylinder (parallel to bulk principal compression).

*Notes on techniques.* Specimen and thin section preparation as outlined.

*Triaxial compression tests.* Cylinder 107 × 27 mm of settled clay, typically with p.s.f. either parallel or perpendicular to length of cylinder. Specimen held in thick rubber sleeve with oil confining fluid. Tested in the way conventional to Soil Mechanics.

Please note that the bulk conditions given are not necessarily homogeneous throughout the specimen, and that because of the mechanical nature of the material and the ubiquity of small irregularities in the specimen, the orthogonal geometry of the test is not assured.

*Ball clay.* Test material is a proprietory brand of clay marketed for the ceramics industry. It is a grey, high alumina, Devon ball clay, consisting typically of about 45% kaolinite, 30% quartz, 20% mica, 5% organic matter, with the clay particles mainly being less than 1 $\mu$ e.s.d.

184 A

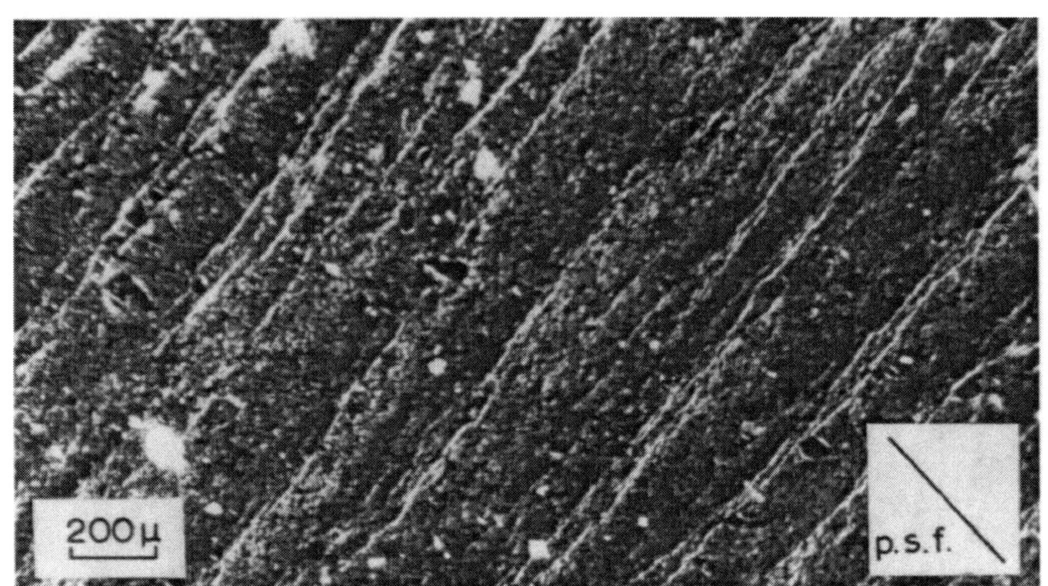

184 B

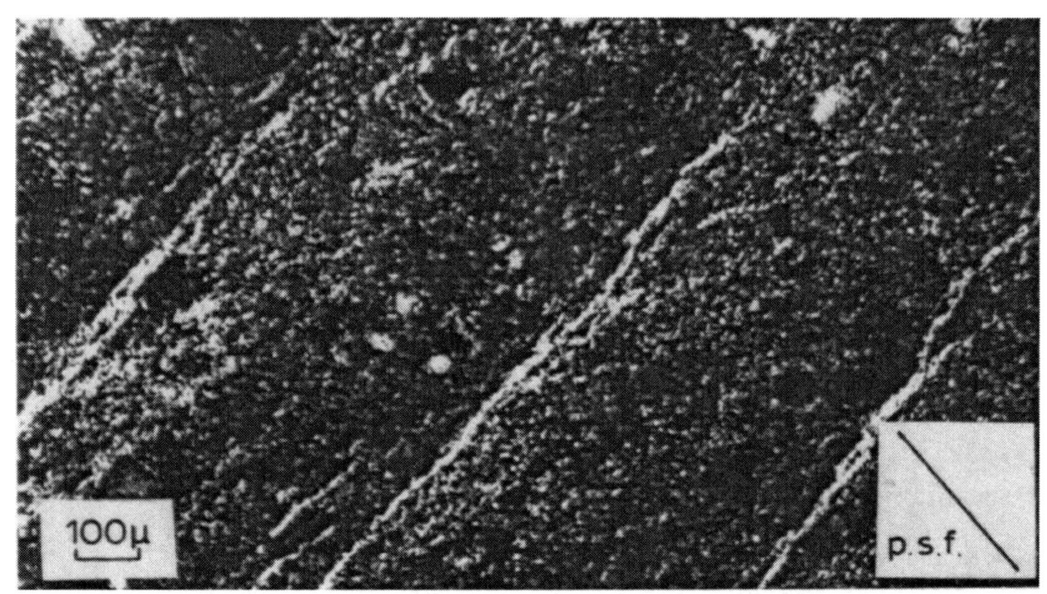

184 C

## 185. Experimentally Deformed Clay (III): Crenulations

A. J. Maltman

Photographs at different scales (**A** through **C**) of ball clay after "hydrostatic" dewatering. PPL

*Bulk conditions.* Arrangement as for triaxial compression test; confining pressure 300 bar, fluctuating pore pressure as pore water drained from base of specimen; *no deviatoric force* applied.

*Notes on techniques.* Specimen and thin section preparation as outlined.

*"Dewatering" tests.* Arrangement as for triaxial tests, with specimens allowed to drain, constantly or intermittently, from top and/or bottom of cylinder, while under hydrostatic confining pressure only, or during triaxial compression.

[The results illustrated here were from a test unsuccessful in that failure of the rubber sleeve allowed the confining fluid to touch the outside of the clay cylinder. However, the interior of the specimen revealed crenulations (atypically photogenic) of the primary settling fabric, the latter being parallel to the length of the cylinder].

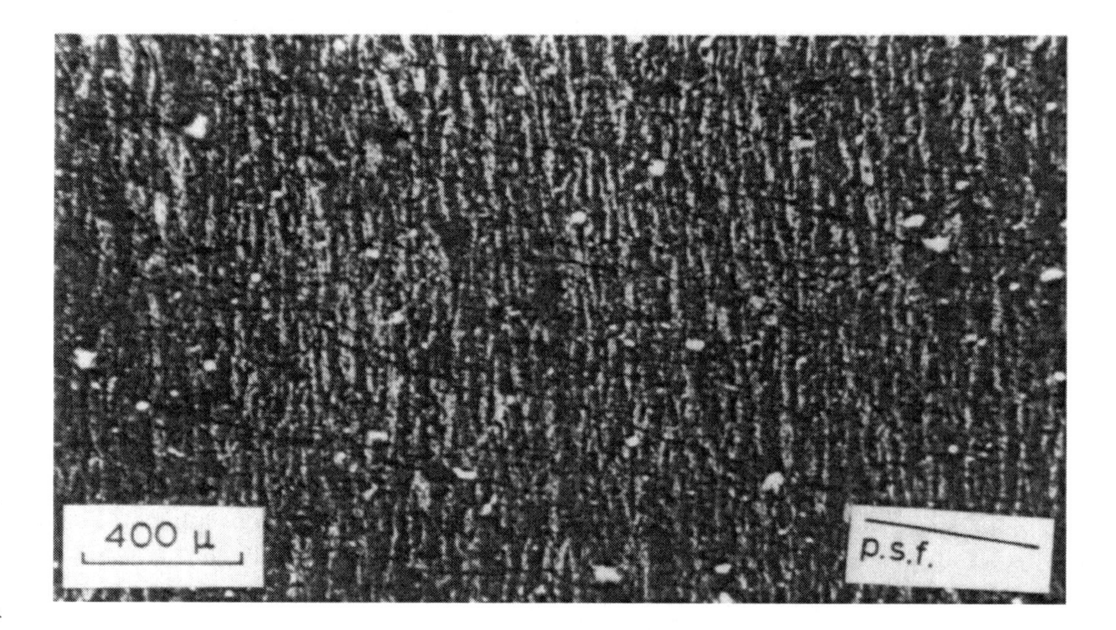


185 A

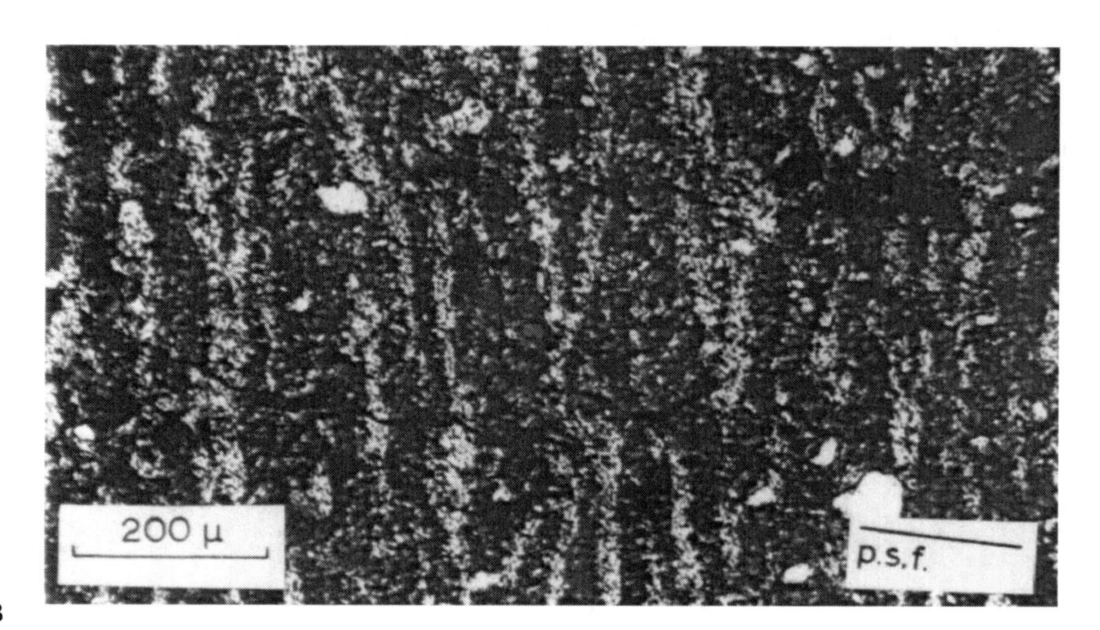


185 B

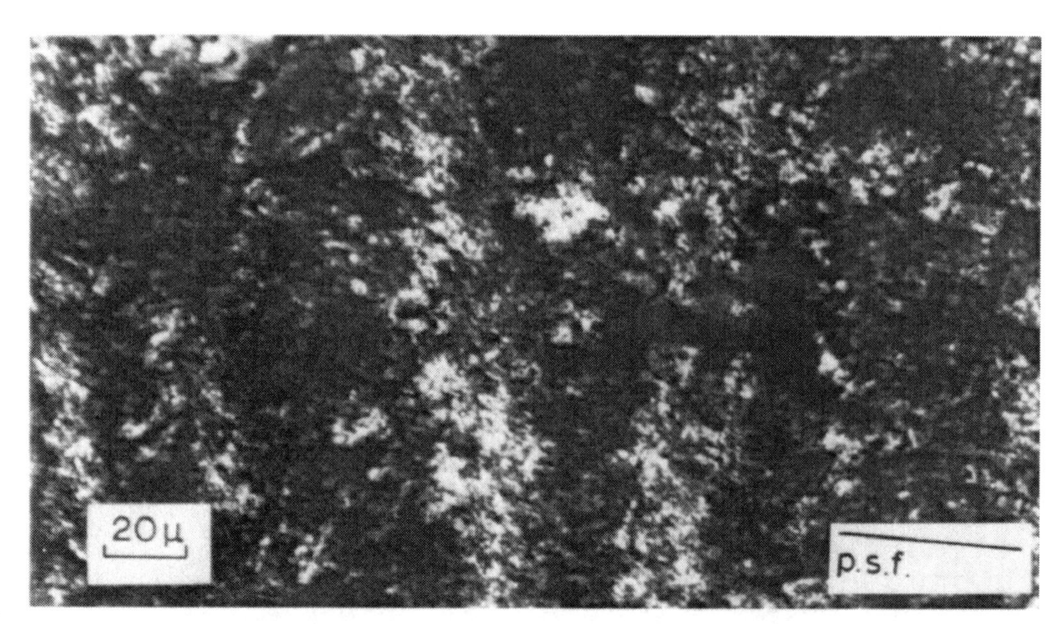


185 C

# 186. Experimentally Deformed Clay (IV): Other Microstructures

A. J. MALTMAN

**A. Layer Boundary Crenulations.** This type of crenulation varies in appearance with the deformation conditions, but is restricted to the zone adjacent to a plane of high mechanical anisotropy. In this example, broad, tapering crenulations in kaolinite are adjacent to a band of a different clay. Preparation and procedures as for ball clay experiments, except several slurries involved. PPL

*Bulk conditions.* Layered kaolinite block in sinistral distributed simple shear, parallel to p.s.f.; shear strain = 1.082; strain rate = $3.38 \times 10^{-3}$ $s^{-1}$; normal stress = 0.03 bar; moisture content of clay = 35%.

**B. "Shear zones".** Narrow, spaced, commonly anastomosing, zones of re-orientation of the p.s.f., which arise at failure of the clay. Band of dark clay and p.s.f. approximately vertical in photograph, displaced by shear zones. PPL

*Bulk conditions.* Layered kaolinite block in sinistral direct shear, perpendicular to p.s.f.; displacement = 8.5%; strain rate = $2 \times 10^{-5}$ $s^{-1}$; normal stress = 0.3 bar; moisture of clay = 35%.

**C. "Kink bands".** Angular, spaced crenulations, comparable to "conventional" kink bands, which are usually assumed to be restricted to lithified material. PPL

*Bulk conditions.* Kaolinite block in sinistral distributed simple shear, parallel to p.s.f., shear strain = 0.72; strain rate = $1 \times 10^{-5}$ $s^{-1}$; normal stress = 0.03 bar; moisture content of clay = 34%.

186 A

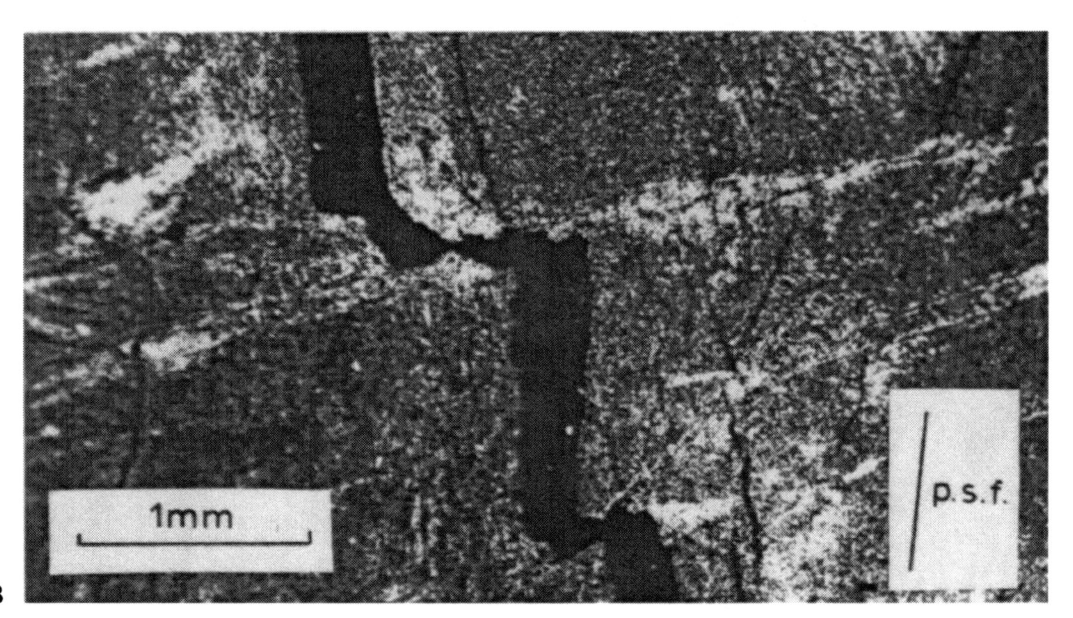


186 B

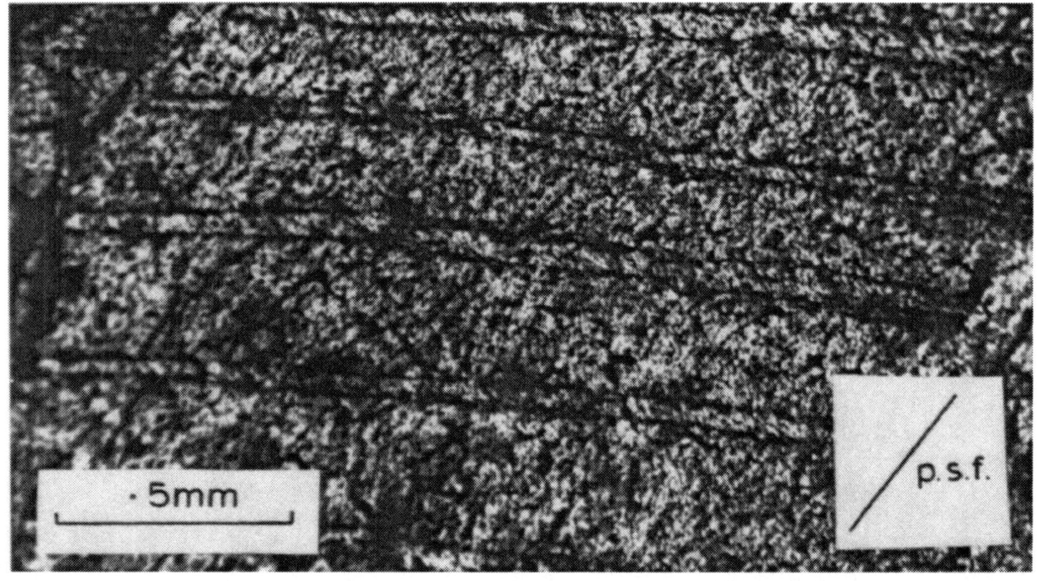


186 C

# 187. Primary Crenulation Cleavage in Clay Shale

R. P. NICKELSEN

**A. Class 3 fold** (RAMSAY 1967, p. 365). Llewellyn Formation, Pennsylvanian age, Anthracite Coal Region, Bear Valley Strip Mine, near Shamokin, Pennsylvania, U.S.A. (NICKELSEN 1979). This fold contains the cleavages illustrated in **B** (*left limb*), **C** (*crest*), and **D** (*right limb*). Upward converging fractures approximately parallel the crenulation planes. All specimens on this plate were collected east of the 2.5 CAI (Conodont Alteration Index), indicating a temperature range 110–140+ °C (EPSTEIN et al. 1976, fig. 17). Percentage of fixed carbon in associated anthracite coal is 92% (DAMBERGER 1974) suggesting a temperature of approximately 160 °C. Negative print of thin section.

**B. Primary crenulation cleavage, left limb of fold in (A).** Bedding ($S_0$) dips left. Crenulation cleavage is shown by Z-shaped crenulations of originally bed-parallel phyllosilicates. No known recrystallization of detrital phyllosilicates has occurred, and the limbs of crenulations are not broken by shear or pressure solution. As strain incremented on the fold limb, cleavage was internally rotated in a clockwise direction, opposite to the counterclockwise external rotation of the fold limb. PPL

**C. Primary crenulation cleavage, crest of fold in (A).** Symmetrical crenulations which manifest 15–25% shortening perpendicular to the axial plane of the crenulations. NX

**D. Two intersecting primary crenulation cleavages from right limb of fold in (A).** Bedding ($S_0$) dips right. Crenulation cleavage $S_{1a}$ formed first and has been externally rotated in a clockwise sense with bedding as the fold formed. Crenulation cleavage $S_{1b}$ overprints $S_{1a}$ and bedding and was formed after rotation of $S_{1a}$ to its present attitude. Gypsum plate NX

187 B

187 A

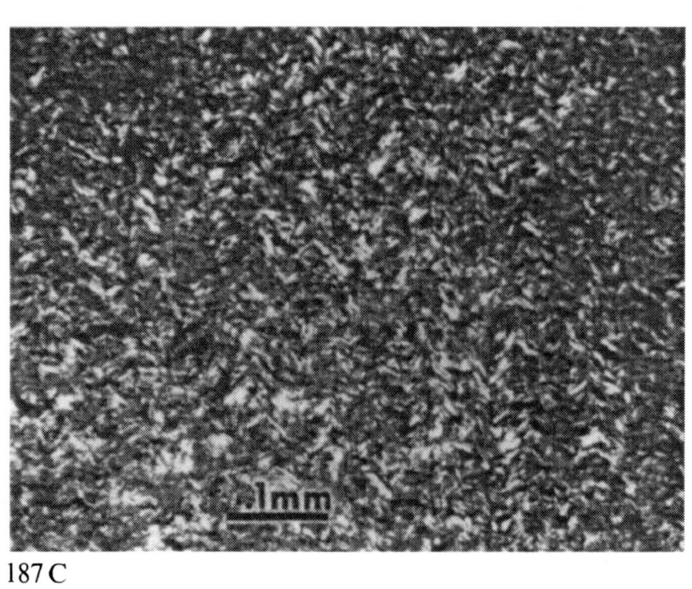

187 C

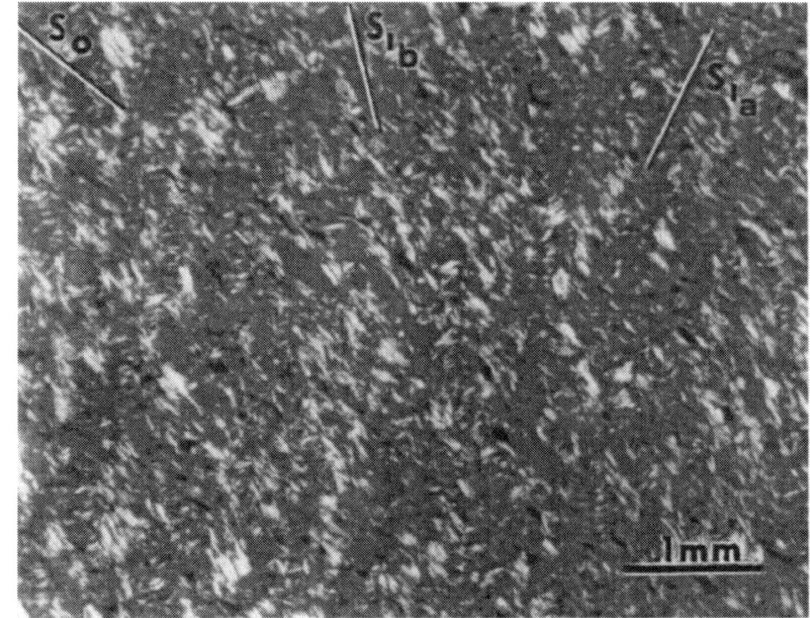

187 D

# Other Topics

## Section 15: Nonplanar Differentiation and Blastesis

(Plates 188 to 200)

Thirteen contributions are here grouped to illustrate porphyroblast development and strain shadows and overgrowth structures associated with nonplanar fabric components. The lithologies represented are schists, phyllites, or slates in all cases.

---

**Plates**

# 188. Prekinematic Porphyroblasts (I)

R. H. VERNON

**A.** A strongly kinked porphyroblast of biotite is illustrated. It came from a mica schist from the Robertson River metamorphic complex, north Queensland, Australia (BLACK et al. 1979). Some of the kinks in the porphyroblast are continuous with kinks of similar size in micaceous aggregates in the adjacent matrix. (See also Plates 174, 175 and 179.) The porphyroblast is prekinematic with respect to the kinking of the matrix foliation, but earlier relationships cannot be inferred from the available evidence, except for a suggestion that the matrix foliation has been deflected around the porphyroblast at some stage. This suggests that the porphyroblast is also prekinematic with respect to shortening across matrix. An alternative interpretation is that growth of the porphyroblast pushed the schistosity of the matrix aside, but this idea is not generally favored (VERNON 1978, p. 297). PPL

**B.** A bent porphyroblast of chloritoid is shown in a mica schist from the Robertson River metamorphic complex, north Queensland. The porphyroblast is highly oblique to the schistosity in the matrix which tends to be deflected around the porhyroblast, forming a relatively quartz-rich "pressure-shadow". The matrix to the right of the porphyroblast is crenulated, the axial surfaces of the crenulations being approximately parallel to the main schistosity in the matrix. These relationships suggest that the porphyroblast is prekinematic with respect to the shortening across the schistosity and also with respect to the crenulation of that schistosity in the "pressure-shadow" zone. The bending of the porphyroblast probably occurred during the matrix flattening, although this is uncertain. The strongly oblique orientation of the porphyroblast suggests, but does not prove, that it grew after the initiation of the schistosity, but not necessarily later than the minerals that now delineate the schistosity (VERNON 1978). PPL

188 A

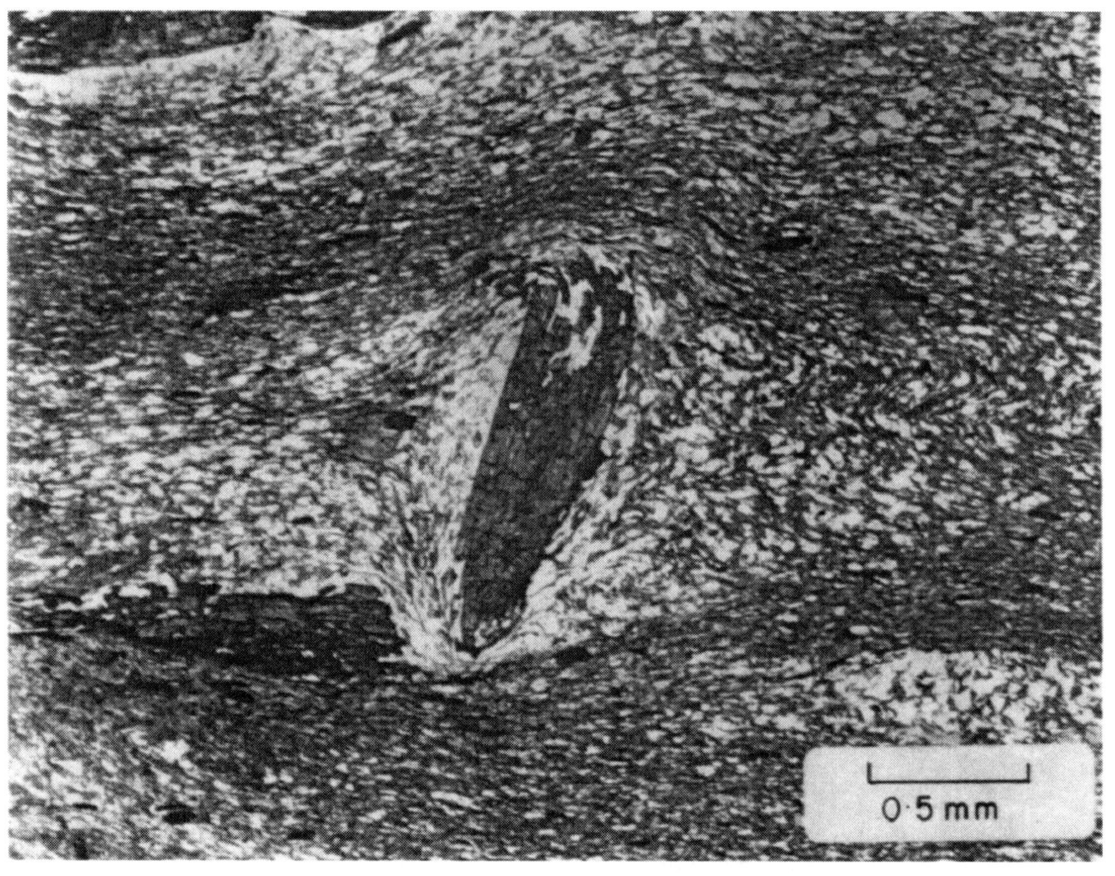


188 B

## 189. Prekinematic Porphyroblasts (II)

R. H. Vernon and C. McA. Powell

**A.** Illustrated are concentrations of white mica and graphite, relative to quartz, adjacent to a garnet porphyroblast in a mica schist (amphibolite facies) from the Karakorams, Pakistan. The concentrations occur in portions of matrix folia adjacent to garnet-matrix boundaries lying approximately parallel to the schistosity. Boundaries approximately normal to the schistosity do not show these concentrations, but are adjacent to poorly schistose "pressure-shadow" regions related to the deflection of the schistosity around the porphyroblast. The concentrations are inferred to have been formed by the preferential solution and removal of quartz and the resultant concentration of less mobile material, in zones of relatively high mean stress. The porphyroblast is inferred to be prekinematic with respect to a flattening deformation of the matrix. The sample was collected from near Hini in the Hunza valley (Powell and Vernon 1979, fig. 2). PPL

**B.** Staurolite porphyroblast, partly replaced marginally by white mica, in a schist (hornblende hornfels or low-pressure amphibolite facies) from the Bosost area, Pyrenees, showing straight $S_i$ formed by trails of quartz inclusions. The $S_i$ passes continuously into the matrix schistosity ($S_e$). In view of the folded $S_e$, compared with the straight $S_i$, and in view of the obliqueness of $S_i$ to $S_e$, the porphyroblast is inferred to be prekinematic with respect to the folding of the schistosity and postkinematic with respect to the initiation of the schistosity (although the growth of the porphyroblast and the minerals now occupying $S_i$ and $S_e$ could have been simultaneous). PPL

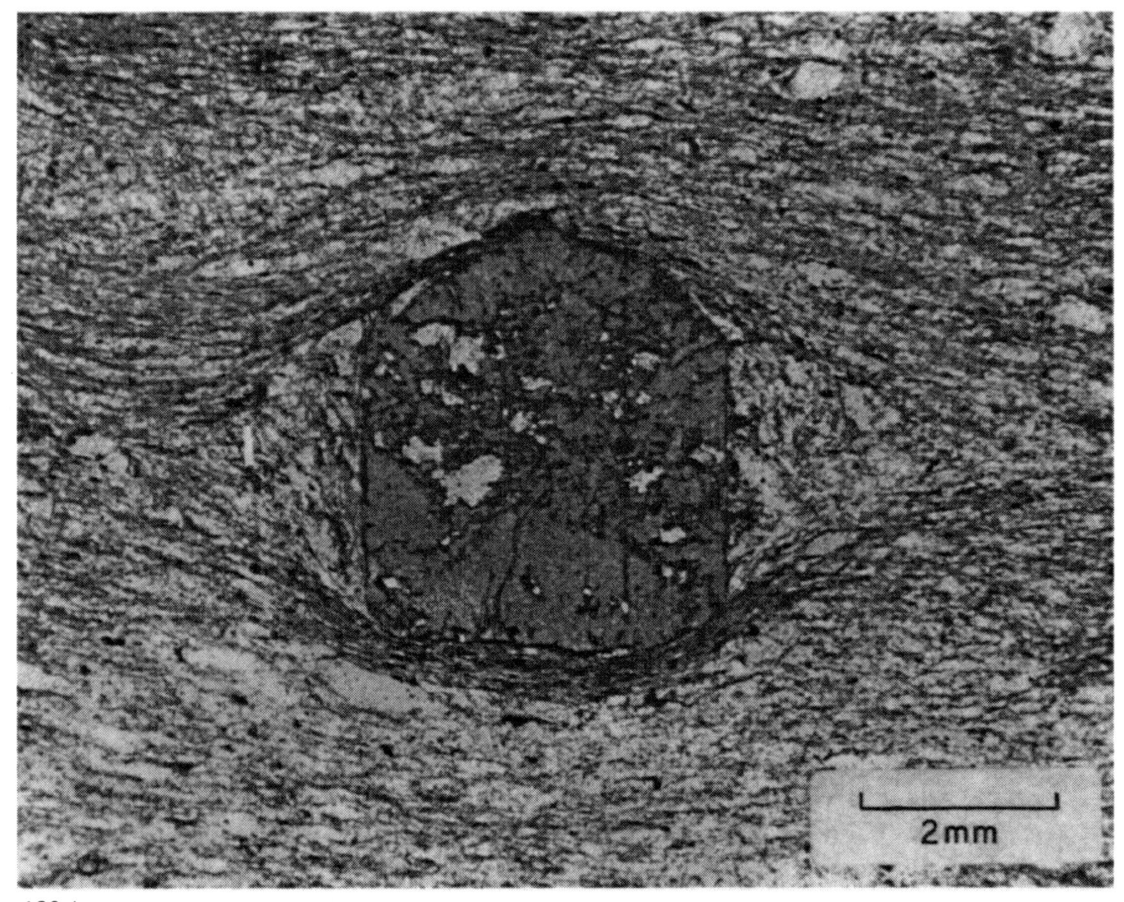

189 A

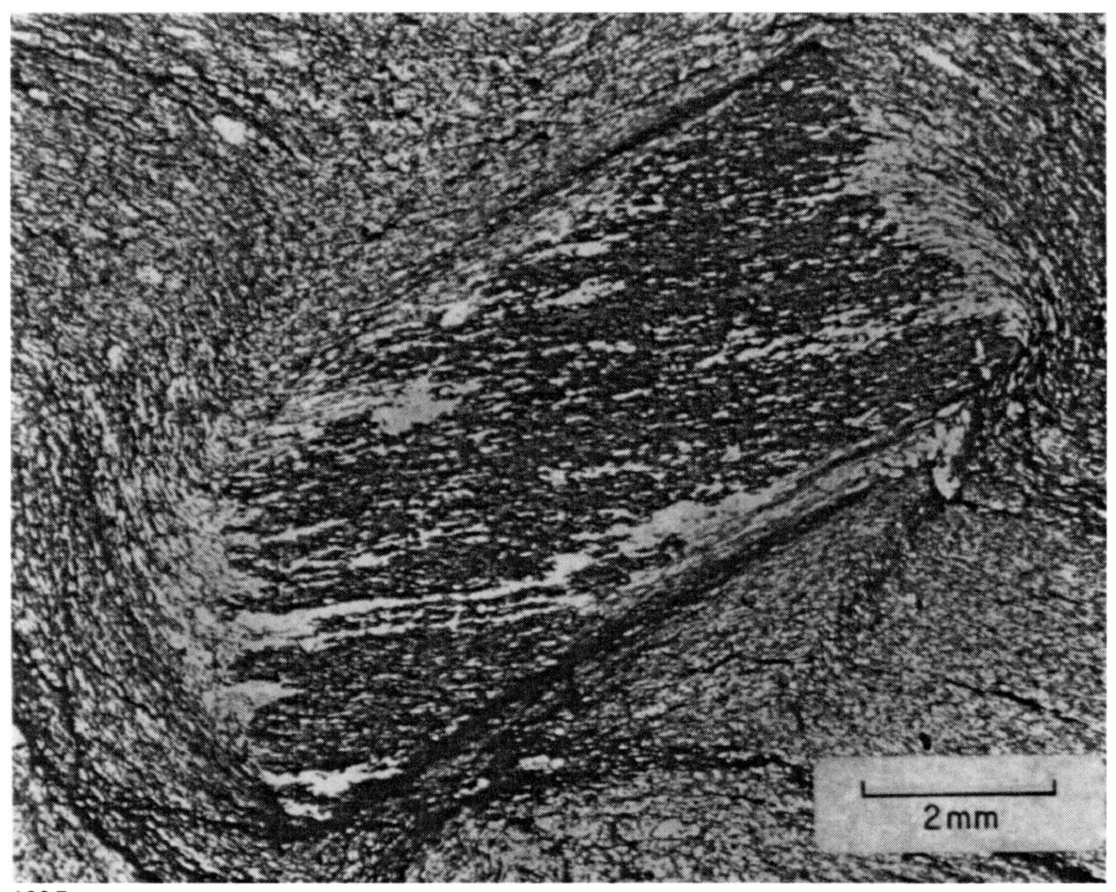

189 B

# 190. Synkinematic Porphyroblasts (I)

R. H. VERNON and C. MCA. POWELL

**A.** A porphyroblast of garnet in Karakoram schist shows concentric, idioblastic growth zoning (marked by varying concentrations of minute graphitic inclusions), an inclusion spiral ($S_i$, marked by trails of opaque inclusions, mainly graphitic), and a spiral of quartz inclusions inferred to have been incorporated from overgrown "pressure-shadows". The intersecting arrangement of $S_i$ and the quartz spiral is identical to the observations and experiments of SCHONEVELD (1977), from which we infer that the porphyroblast grew synkinematically with respect to rotational strain in the matrix; i.e., the garnet crystal grew while being rotated relative to the matrix. The growth zoning indicates that the crystal maintained an idioblastic shape throughout its growth history. This is Fig. 5A of POWELL and VERNON (1979, p. 31). Criteria for recognizing porphyroblasts that grew while rotating relative to the schistosity in the matrix include the following (VERNON 1978, p. 300):

1. Each porphyroblast, irrespective of size, contains only one $S_i$ sigmoid.
2. $S_i$ is symmetrical about the center of the porphyroblast.
3. The inferred axis of rotation is parallel to the axis of folds formed by the deformation episode concerned.
4. Spirals of quartz inclusions (incorporated from "pressure-shadows") and spirals of inclusions of other matrix minerals (true $S_i$) intersect and show geometrical relationships described by SCHONEVELD (1977).

We could add two other criteria (illustrated in Plate 191 B), namely:

5. In small rock volumes, the sigmoids in adjacent porphyroblasts indicate the same sense of rotation.
6. Microfolds of similar size to the sigmoids are absent from the matrix. PPL

**B.** Same field as **A**, but with crossed polars, showing the quartz spiral more clearly. NX

**C.** Enlargement of the central part of **A**, to show $S_i$ more clearly and the crenulation preserved where $S_i$ intersects the quartz spiral (POWELL and VERNON 1979). This photograph is rotated counter-clockwise by about 45° with respect to **A** and **B**. PPL

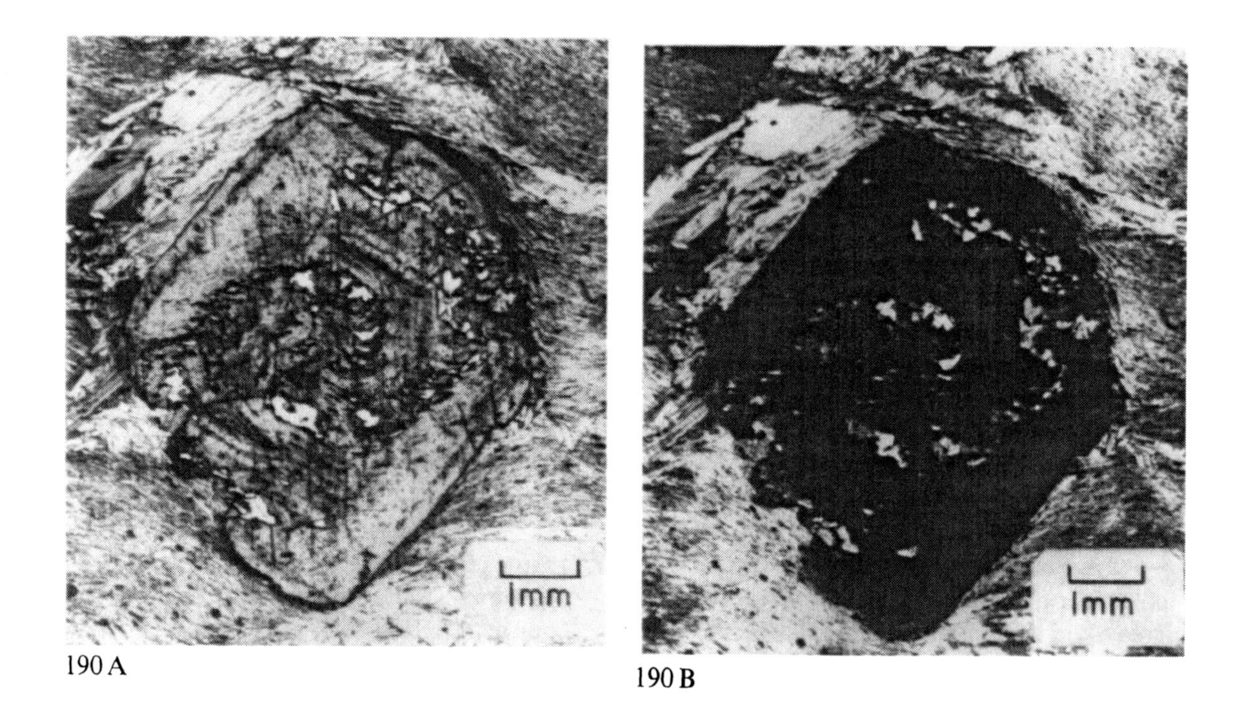

190 A
190 B

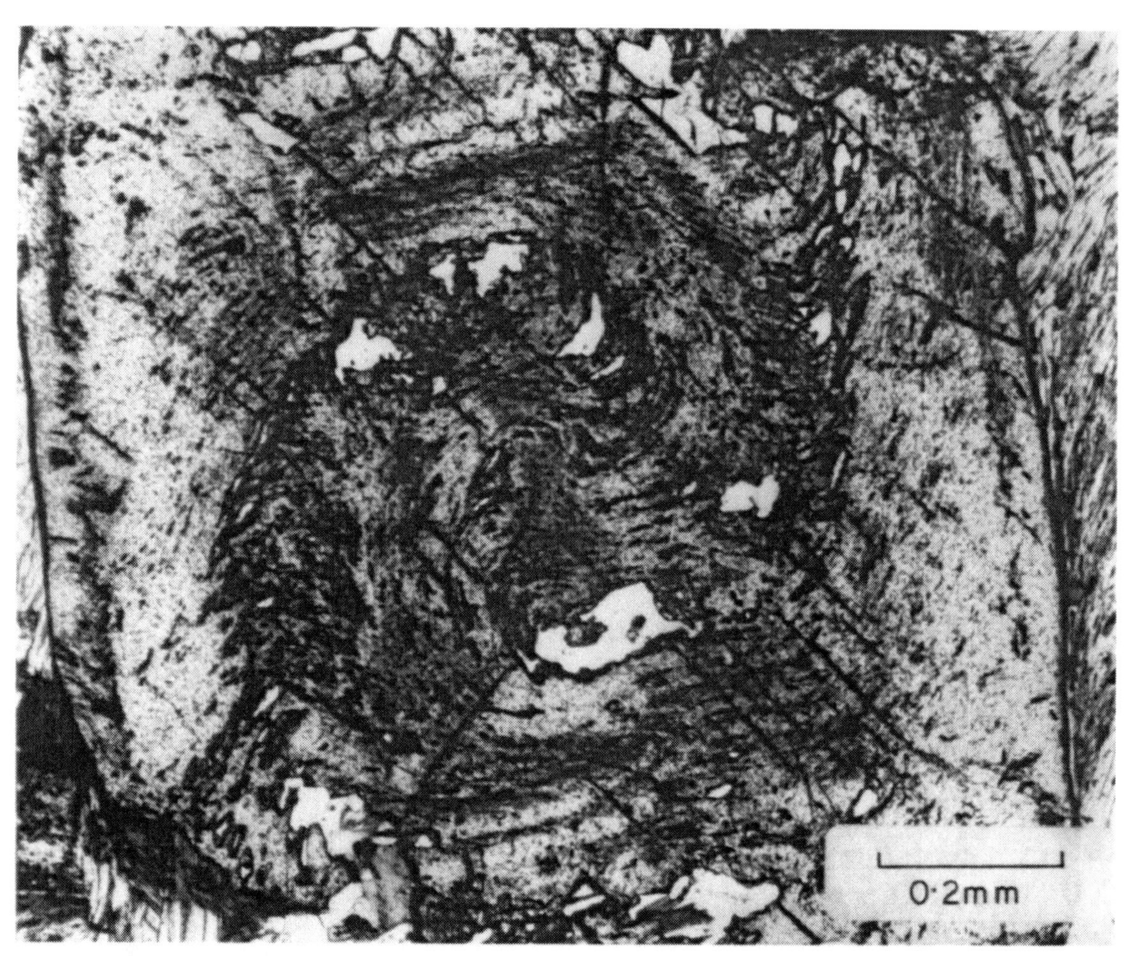

190 C

# 191. Synkinematic Porphyroblasts (II)

R. H. VERNON

**A.** An irregularly shaped porphyroblast of garnet in a schist from the Robertson River metamorphic complex, north Queensland, Australia (BLACK et al. 1979), shows slightly curved $S_i$ (marked by elongate inclusions of quartz). $S_i$ is continuous with, but oblique to the main trend of $S_e$, and a relatively quartz-rich "pressure shadow" occurs where $S_e$ is deflected around the porphyroblast. The absence of folds in $S_e$ suggests that the porphyroblast has been rotated relative to the matrix, rather than having grown over a pre-existing microfold. The fact that $S_i$ is smoothly curved, and continuous with $S_e$ suggests that the porphyroblast grew while the schistosity of the matrix was being progressively rotated. Thus the porphyroblast is inferred to be synkinematic with respect to the deformation of an existing schistosity that gave rise (with concomitant recrystallization or neocrystallization of framework and layer-silicate minerals) to a new orientation of $S_e$. The included quartz grains are much smaller than those in the matrix, which is generally taken to imply that they were included when the rock was much finer grained (presumably at a lower metamorphic grade). PPL

**B.** Several garnet porphyroblasts in the same rock show a single $S_i$ spiral and quartz-rich "pressure shadows". Some of the spirals are no longer continuous with $S_e$, but each spiral indicates the same sense of rotation relative to $S_e$, which is one of the criteria that can be used to indicate a porphyroblast that grew while undergoing rotation relative to the matrix, as discussed previously. PPL

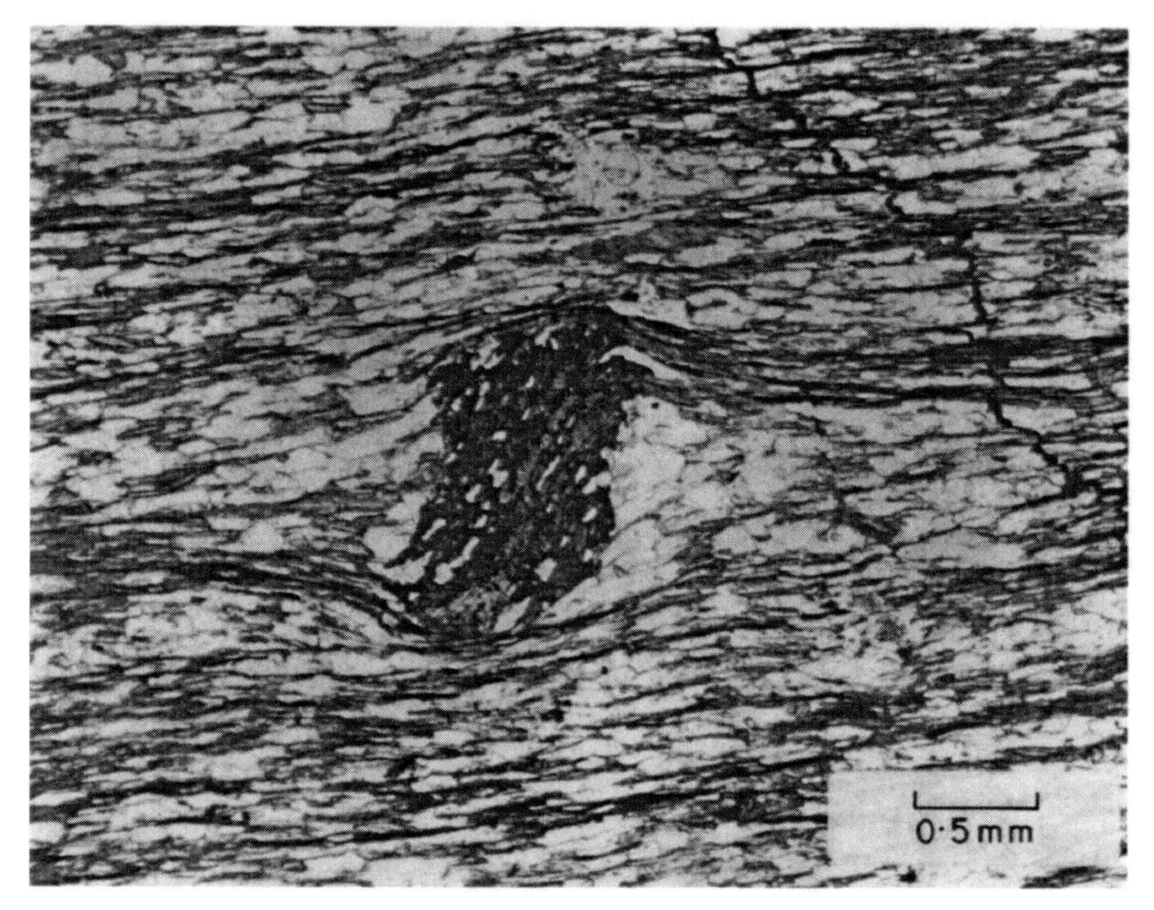


191 A

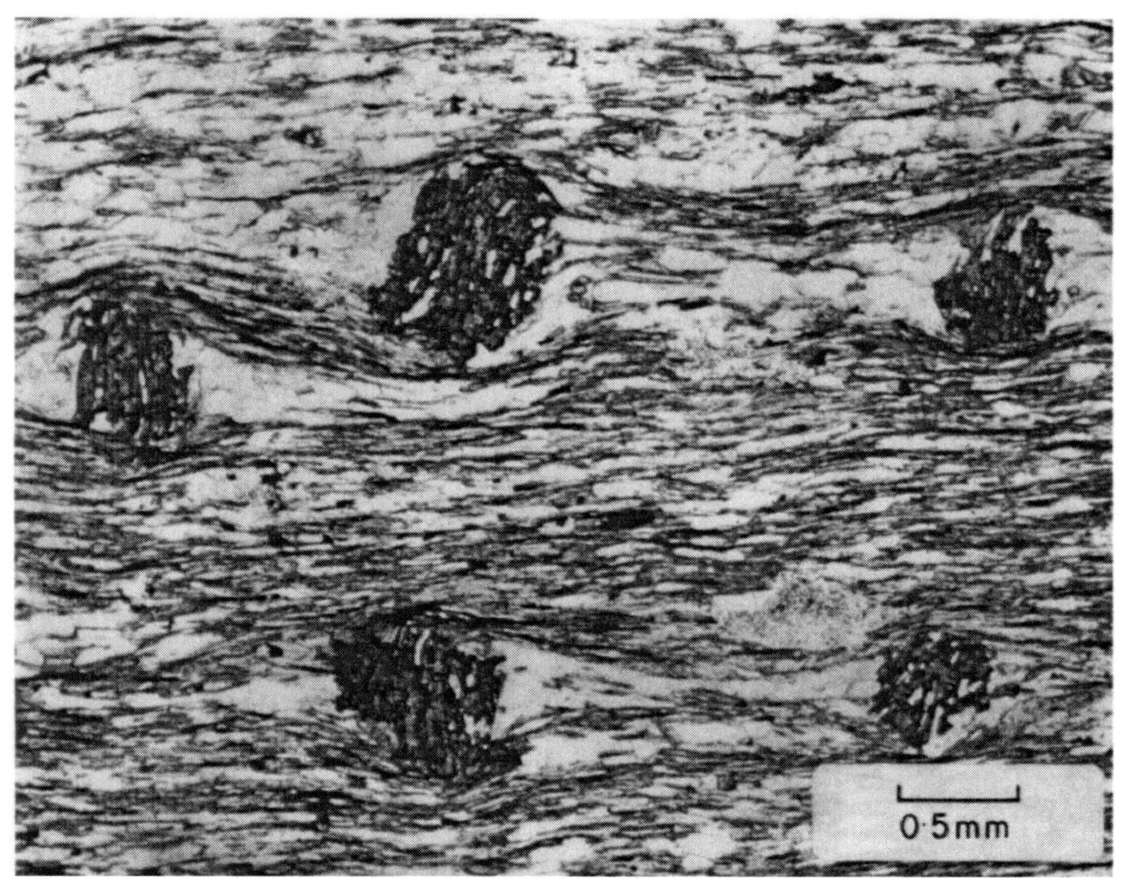


191 B

# 192. Postkinematic Porphyroblasts (I)

R. H. Vernon

**A.** A porphyroblast of biotite in a schist from the Pyrenees shows $S_i$ (marked by curved inclusion trails of quartz and opaque material) continuous with $S_e$, implying growth of biotite after the formation and weak folding of $S_e$. However, growth of the biotite and the minerals now defining $S_e$ could have been simultaneous (Vernon 1978). Folds of similar size and style occur in the matrix nearby. The porphyroblast also contains a "false foliation" formed by partial replacement by quartz along (001) crystallographic planes. PPL

**B.** Two porphyroblasts of hornblende are illustrated in a quartz-feldspar-amphibole schist (amphibolite facies) from an unknown locality. The $S_i$, marked by essentially straight trails of quartz inclusions, passes continuously into a largely straight $S_e$. This implies that the porphyroblasts grew after the initiation of $S_e$, though not necessarily after the mineral grains now outlining $S_e$. The slight outward curvature of $S_i$ toward the ends of the porphyroblast, passing continuously into a similarly bowed $S_e$, suggests that the porphyroblasts grew during incipient flattening of $S_e$ about them. That is, the porphyroblasts may be postkinematic with respect to the initiation of $S_e$, but synkinematic with respect to this particular slight deformation of $S_e$. PPL

192 A

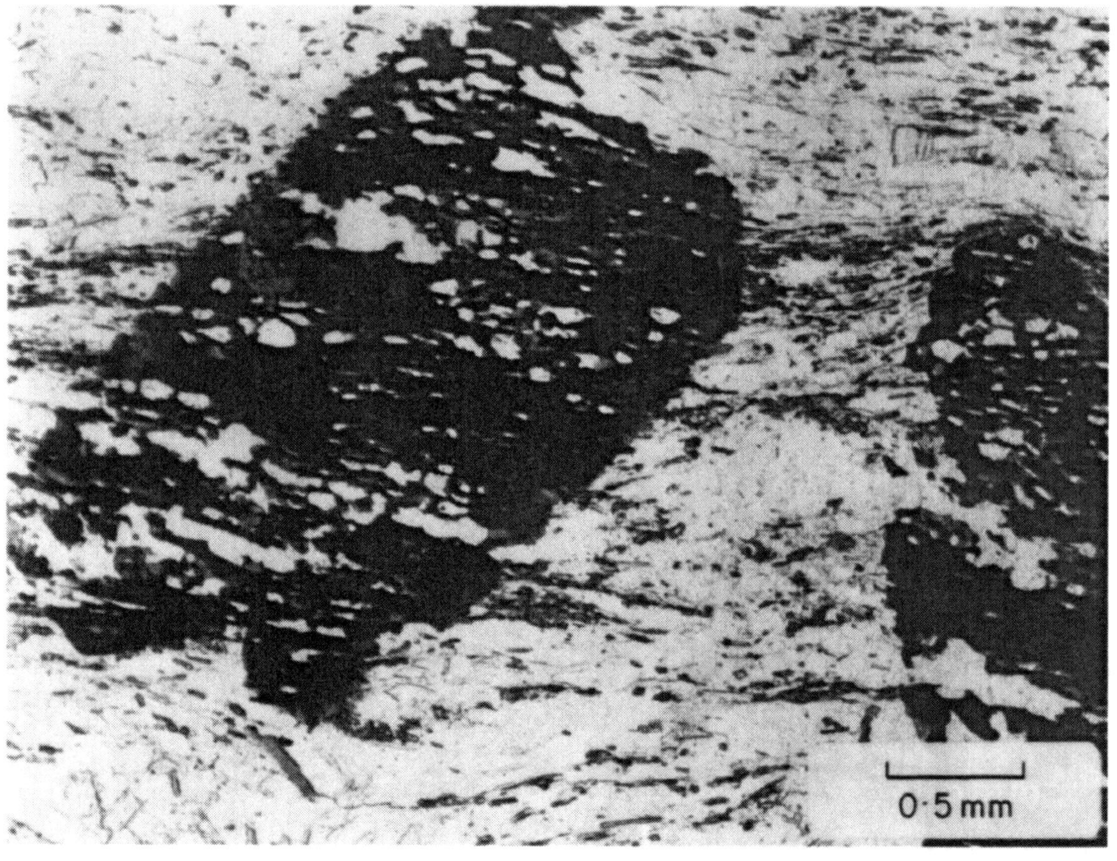


192 B

## 193. Postkinematic Porphyroblasts (II)

R. H. VERNON

**A.** Layers rich in fine grained graphitic material are interfolded with light-coloured layers containing relatively coarse-grained biotite, in a kyanite-plagioclase-mica-quartz schist (lower amphibolite facies) of the Dalradian, Glen Clunie, 1 mile SSW of Auchallater, Scotland. In the right-hand side of the photograph an isoclinal fold in graphitic material has been refolded into an open upright fold. PPL

**B.** Same field as **A**, showing that the folded layering is preserved as an incorporated $S_i$ ("helicitic" structure) in a large plagioclase porphyroblast with multiple twinning. The plagioclase grew postkinematically with respect to the folding of the layering, although not necessarily with respect to the nucleation of the minerals now outlining the folds. NX

193 A

193 B

# 194. Porphyroblasts with Complex Histories (I)

C. McA. Powell and R. H. Vernon

**A.** A porphyroblast of garnet is illustrated in a Dalradian schist from an outcrop on the A93 road approximately 0.8 km SSW of Devils Elbow, Perthshire, Scotland. It shows sigmoidal inclusion trails that may have been due to synkinematic growth of the garnet during rotation relative to the matrix, or may be due to postkinematic overprinting of an earlier microfold. Criteria for distinguishing the two interpretations have been given in the caption to Plate 190, but, although the porphyroblast contains only one spiral, it shows insufficient criteria to make a confident inference of growth during rotation. In particular, it does not show the intersecting double spirals of Schoneveld (1977). The porphyroblast is prekinematic with respect to the deformation that has produced a segregated crenulation cleavage ($S_e$) in the matrix, which truncates $S_i$. PPL

**B.** Large-scale view of the above, showing the segregated foliation ($S_e$) and later microfolds of this foliation. PPL

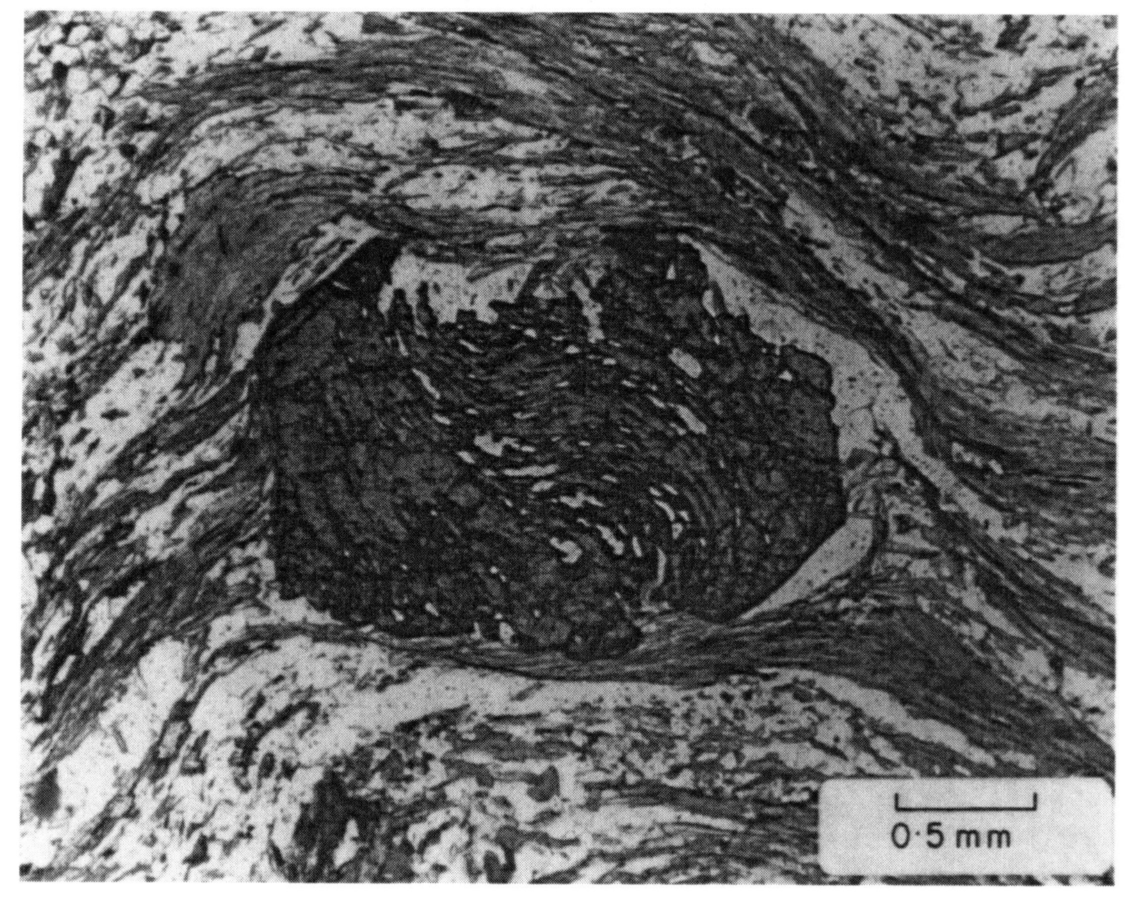


194 A

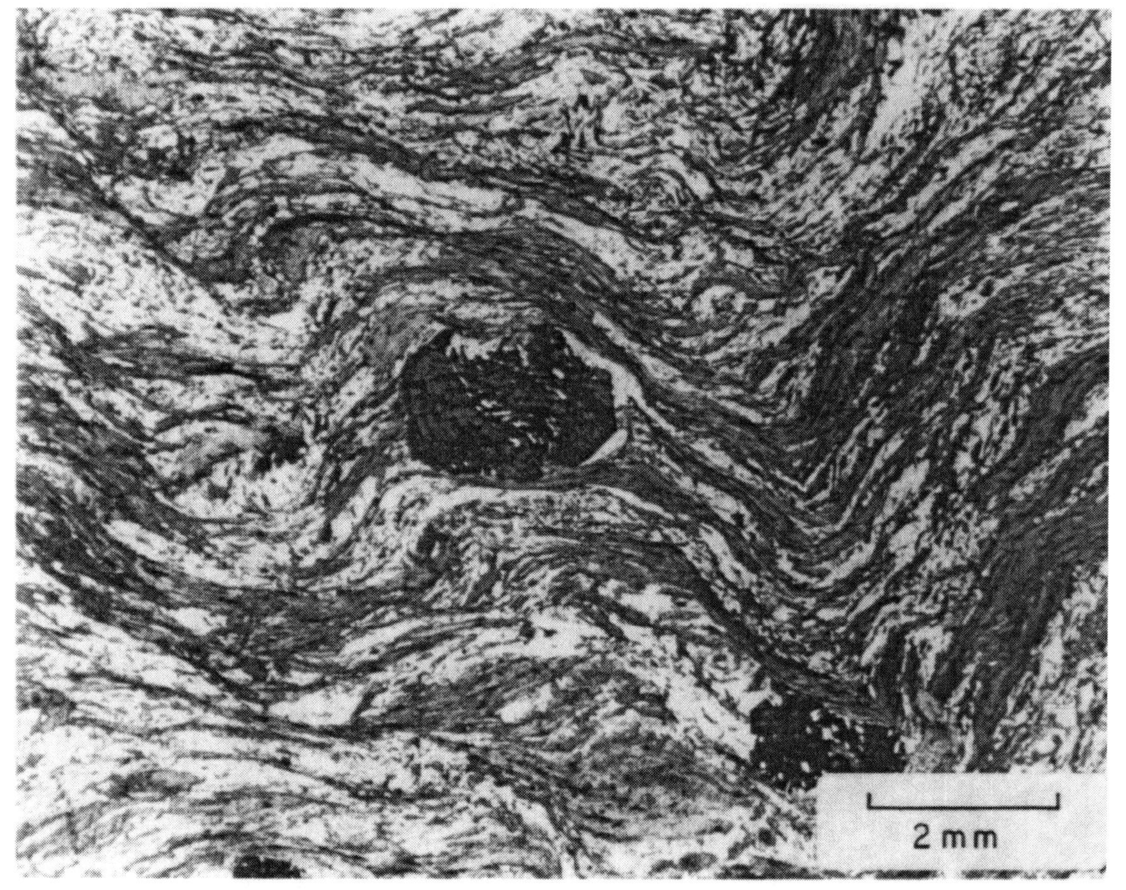


194 B

## 195. Porphyroblasts with Complex Histories (II)

R. H. VERNON

**A.** An idioblastic porphyroblast of staurolite in a Pyrenean schist (hornblende hornfels or low-pressure amphibolite facies) shows continuity between $S_i$ and $S_e$. This implies that the staurolite grew postkinematically with respect to the initiation of $S_e$ (but not necessarily after the growth of the minerals now delineating $S_e$). $S_i$ shows a slight tendency to bow outward toward the top end of the porphyroblast, in conformity with $S_e$, in which case the porphyroblast could be interpreted as having grown synkinematically with respect to the flattening of $S_e$ about it. However, in view of the folding in $S_e$ and its absence from $S_i$, the porphyroblast is prekinematic with respect to most of the matrix folding and draping of $S_e$ about the porphyroblast. PPL

**B.** Apparently conjugate kink folds in $S_e$ appear to have nucleated on the end of a pre-existing idioblastic porphyroblast of staurolite in the same rock as in **A**. $S_i$ is again continuous with $S_e$, having been inherited when $S_e$ was straight. That is, the porphyroblast is postkinematic to the initiation of $S_e$ and prekinematic with respect to the deformation of $S_e$ (flattening about the porphyroblast and presumably later kinking). PPL

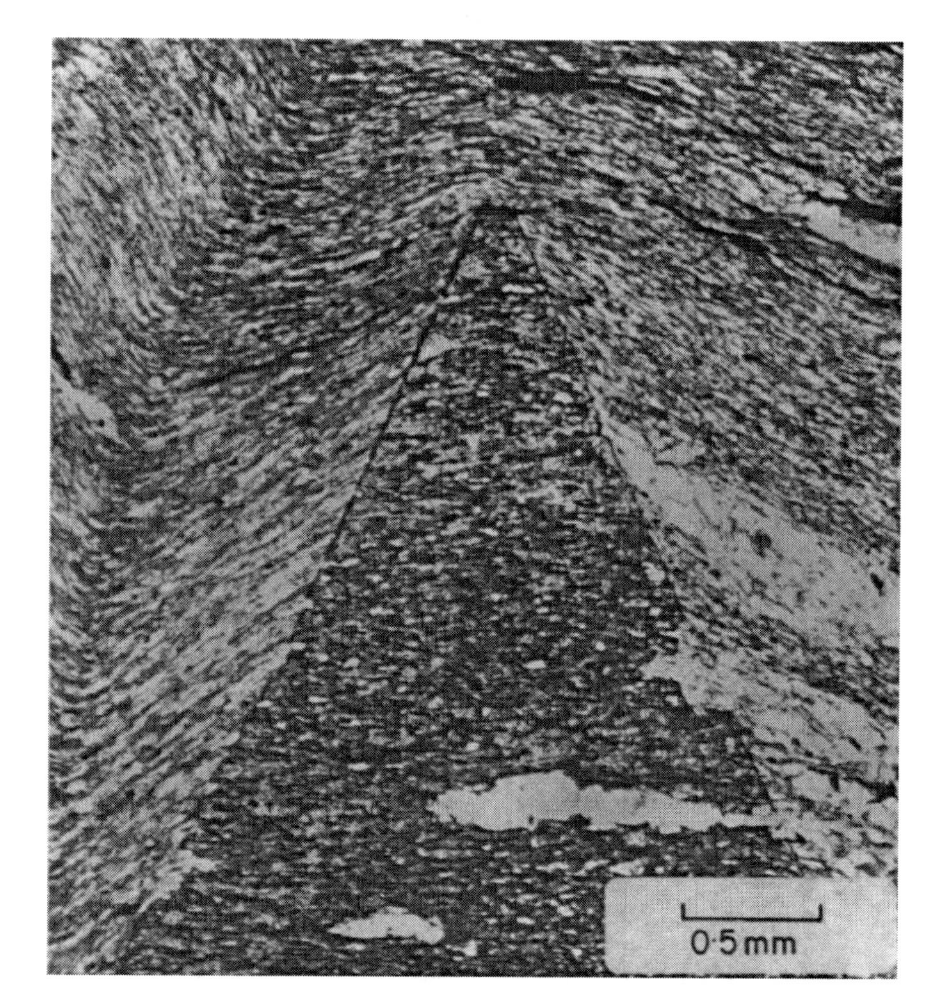

195 A

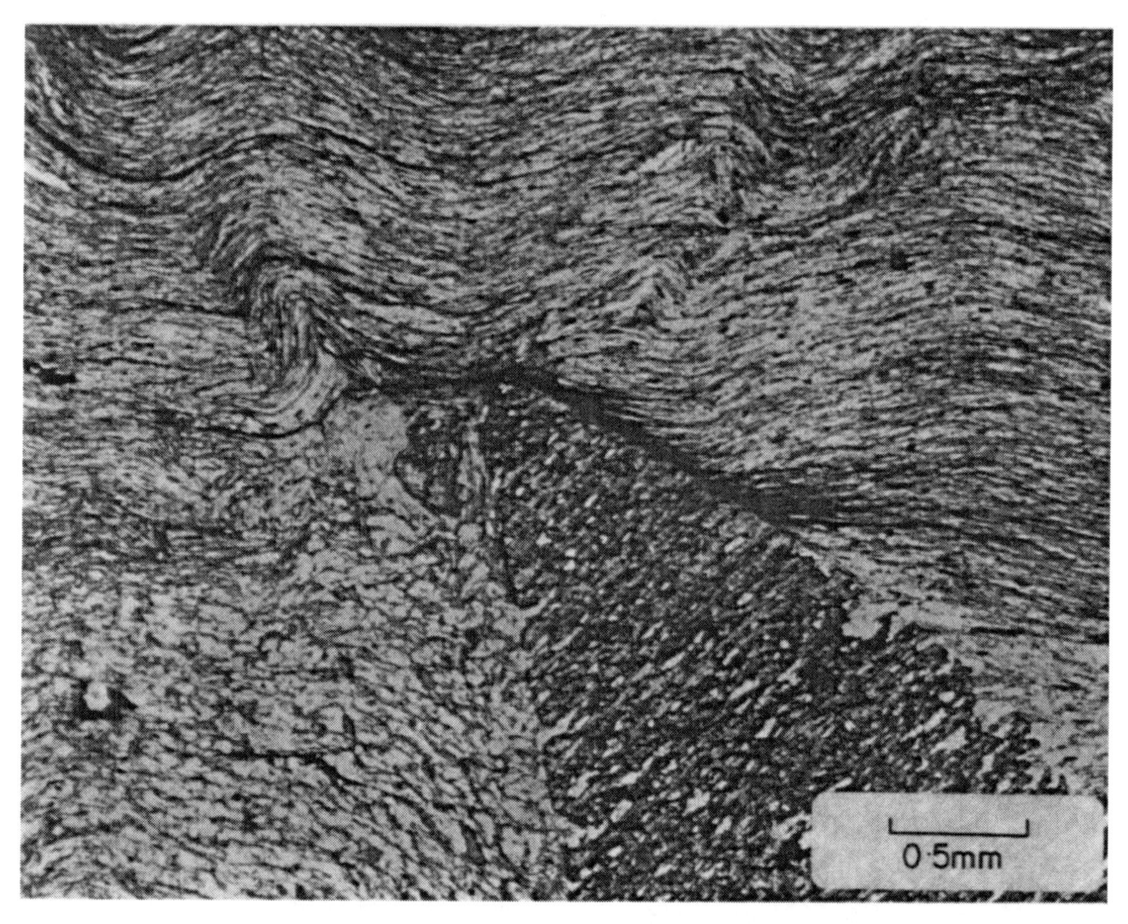

195 B

# 196. "Millipede" Structure in Porphyroblasts

T. H. BELL and M. J. RUBENACH

**A. Millipede microstructure in a plagioclase porphyroblast.** The plagioclase porphyroblast consists of an An22–30 core and an An11–18 rim. The curved disposition of $S_1$ inclusion trails in the rim indicates that it grew synchronously with $D_2$. However, the core, where inclusion trails are planar, could have grown pre- or syn-$D_2$. The unique geometry of this microstructure may indicate that it formed by bulk inhomogenous shortening (BELL and RUBENACH 1980). NX

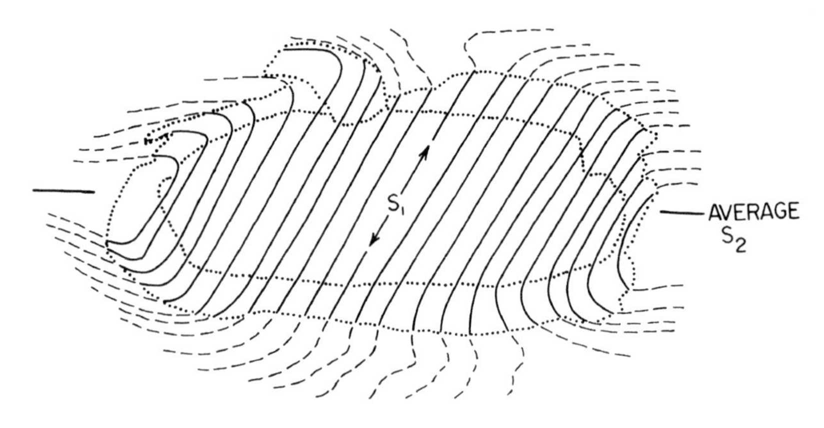


**B. Millipede microstructure in crenulated schist.** The photograph shows a crenulated cleavage ($S_1$) bounded by crenulation cleavage ($S_2$) produced during the second deformation in these rocks. The $S_1$ geometry is similar to that of the inclusion trails in **A** and indicates deformation involving progressive bulk inhomogeneous shortening. The porphyroblasts are plagioclase. NX

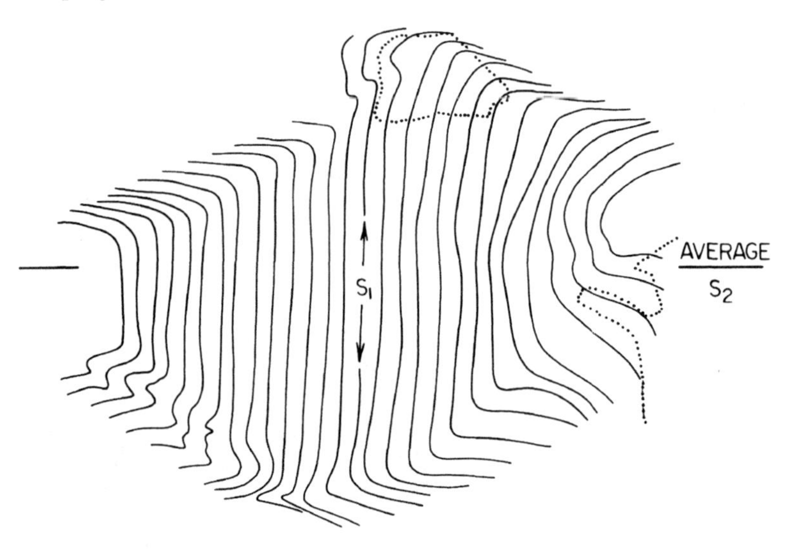

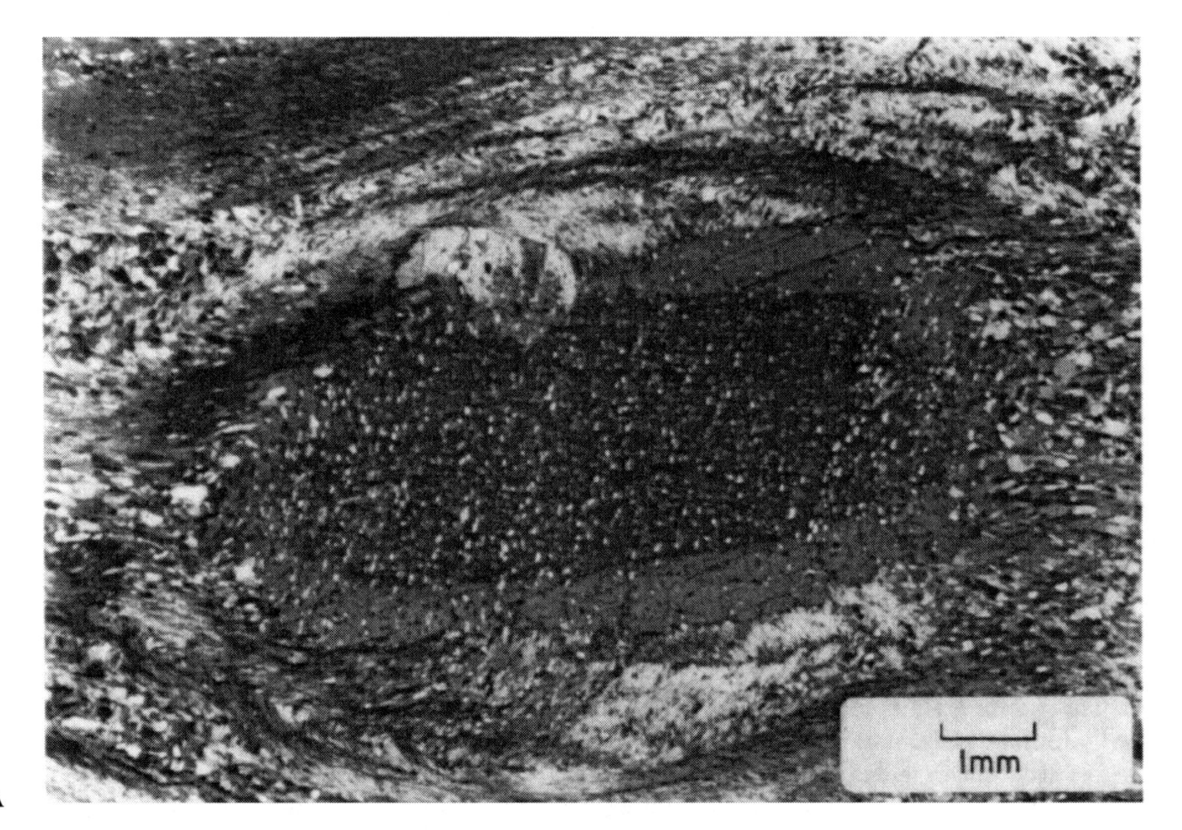


196 A

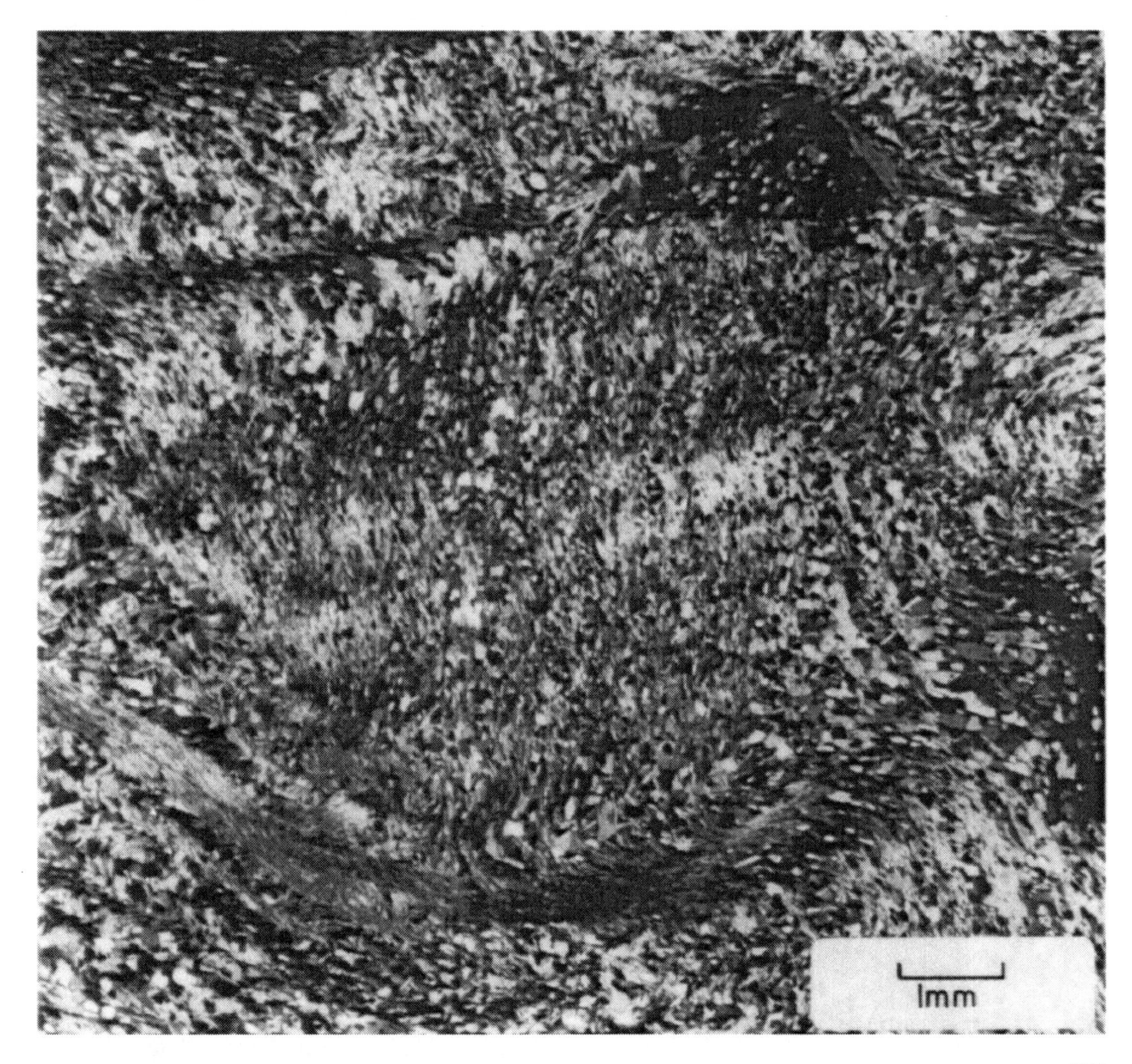


196 B

# 197. Overgrowth Microstructures Associated with Slaty Cleavage Development in the Adelaide Geosyncline

T. H. BELL

**A. Inclusions parallel to $S_0$ and $S_1$ preserved in overgrowth on quartz.** Tapleys Hill Formation, Orroroo, South Australia (BELL 1978). Photograph of a siltstone with slaty cleavage trending NE–SW and bedding oriented NW–SE. The slaty cleavage is defined by aligned (001) of muscovite and chlorite, ellipsoidal quartz and feldspar grains and rutile needles. The large quartz grain in the center of the photograph contains mica, rutile and smaller unidentified inclusions. These inclusions are aligned to give an $S_i$ fabric in the overgrowth. They are parallel to $S_0$ closer to the center of the grain but parallel to $S_1$ in the long axis extremities (*A*). The center of the grain and the edges to its right and left are relatively inclusion-free. The quartz grain is interpreted to have grown initially during diagenesis and/or the very early stages of slaty cleavage deformation ($D_1$). This preserved the inclusions parallel to $S_0$. It continued to grow during $D_1$ and thus preserved the inclusions parallel to $S_1$. The lack of inclusions on the short-axis margins of the elliptical quartz grain is attributed to pressure solution of these edges by the micas aligned parallel to them. Note that the regions of the quartz grain marked *A* contain no inclusions oriented at intermediate attitudes between $S_0$–$S_1$. This tends to indicate that those aligned parallel to $S_1$ crystallized and grew in that orientation rather than that they were rotated toward it. NX

**B. Inclusions parallel to $S_0$ and $S_1$, preserved in an albite "porphyroblast".** Tapleys Hill Formation, Orroroo, South Australia (BELL 1978). Photograph of a twinned albite grain which has grown in metasiltstone. The regions marked *A* are twinned portions of the albite. Note that the $S_i$ inclusion trails and orientation are parallel to $S_0$ in the albite center and that they change abruptly to $S_1$ orientations near the regions marked *A*, with no intermediate attitudes. The albite is interpreted as having grown initially during diagenesis and/or early slaty cleavage deformation and has thus overgrown inclusions parallel to $S_0$. It continued to grow during $D_1$ and consequently grew over inclusions oriented mainly parallel to $S_1$ with some still parallel to $S_0$. This indicates that rotation was not as important as crystallization in development of the $S_1$ preferred orientation. NX

197 A

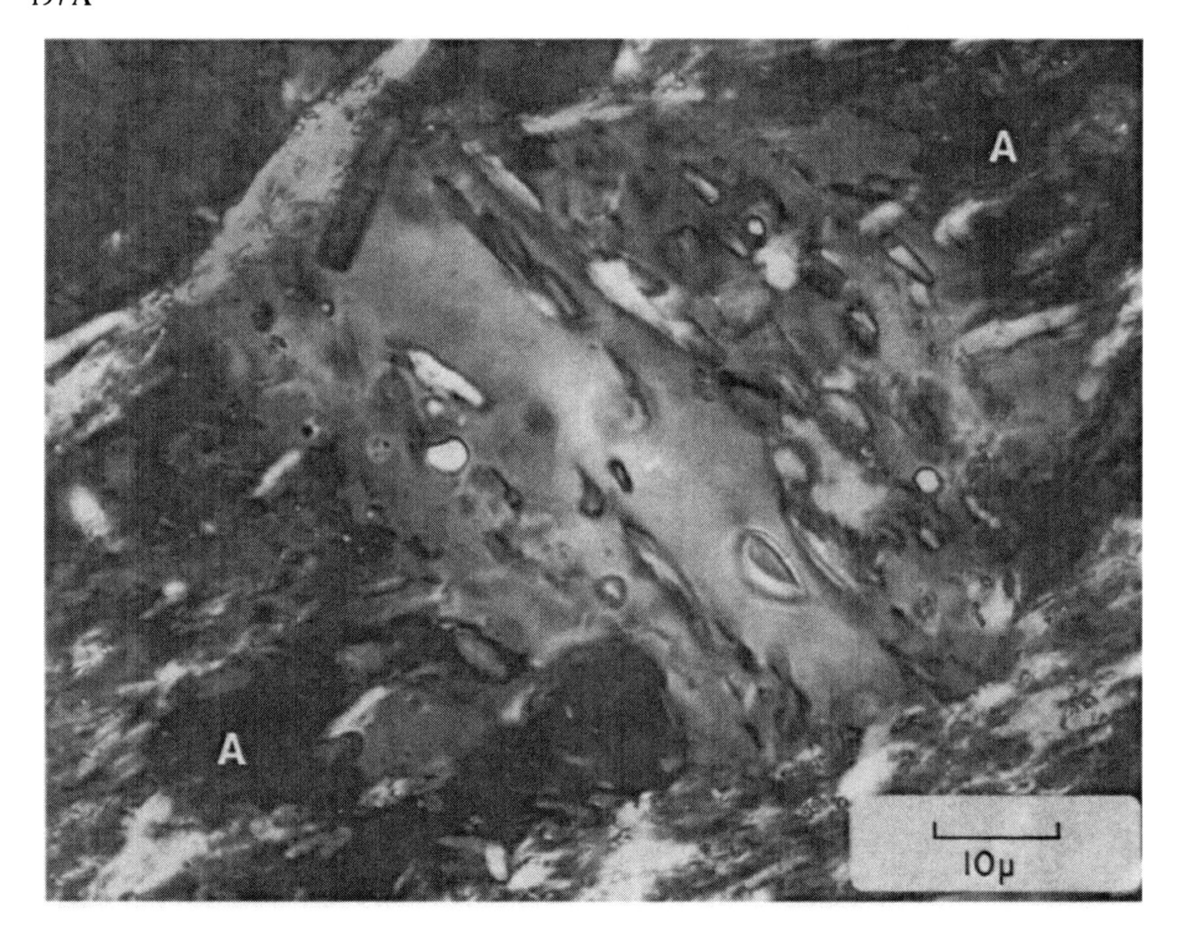


197 B

## 198. Chlorite-Mica Blasts in Slate with Continuous Cleavage

G. J. Borradaile

Jura Slate from the Late Precambrian Dalradian Supergroup at Tarbert House, Jura, Scotland (Borradaile 1979a, b).

This lower greenschist facies slate is composed of quartz, chlorite, white micas, graphite and opaque minerals. The excellent preferred orientation of the phyllosilicates and quartz blobs in the matrix provides an example of continuous cleavage. The rock was quarried as a roofing slate. The cleavage shown is the first tectonic fabric in the rock and its formation has accompanied or partly post-dated the growth of the blasts of intergrown chlorite and white mica. These have mineral cleavage oriented at a high angle to rock cleavage (so-called "cross micas"). The blasts have strain shadows or "beards" of quartz and chlorite.

Somewhat similar chlorite-white mica blasts are illustrated in Plates 64 and 105. (See also Beutner 1978). PPL

198

# 199. Experimental Production of Strain Shadows

W. J. GREGG

Experimental study of the development of strain features in rock materials and rock analogues has been greatly facilitated by the development of the Transmitted Light Apparatus (MEANS 1977). This apparatus consists of a steel frame enclosing a thin (1–2 mm) specimen between thick glass blocks. These blocks allow light to pass through the thin specimen and thus continuous photographic recording of microstructural detail is possible. A thin piston deforms the specimen at strain rates variable from $10^{-3}$ to $10^{-7}$ $s^{-1}$ and confining pressure to a few hundred bars (for illustration and description, see MEANS 1977).

The sequence of photographs shows the production of strain shadows (similar to the pressure-shadows of SPRY 1969, p. 246) around a galena "porphyroblast" in a rock analog composed of halite grains (*white*) and fine mica (*m*) in saturated NaCl brine. The specimen has been flattened from the top at a strain rate of $10^{-6}$ $s^{-1}$. In addition to strain shadows, the process of "grain neighbor switching" has been observed. One of the examples from area (*n*) is sketched below in sequence. Initially neighboring grains *b* and *c* have been displaced and close contact has been "switched" to grains *a* and *d*. ASHBY and VERRALL (1973) have shown that large overall strains can be achieved by this process without equivalent intragranular deformation, and WILLIAMS (1976, p. 190) and BORRADAILE (1981) have commented on its possible significance in geological materials.

---

**A.** After 30% overall shortening only a few grains have been flattened. Most of the strain has been accommodated by pore-space reduction. *Dark area* is galena crystal.

**B.** At 63% shortening mica films (*f*) have developed in regions of high strain and halite grains have suffered considerable flattening due to crystal plasticity. The galena crystal has been highly strained on the left side but remains nearly intact on the right. Halite grains on the right hand side (*s*) show less strain due to the shadowing effect of the galena and mica film development is poor.

**C.** At 70% shortening the strain shadow at (*s*) is well developed, but where the galena has been deformed on the left side the grains are highly strained and mica films are well developed. Note also the curvature of the fabric around the galena "porphyroblast".

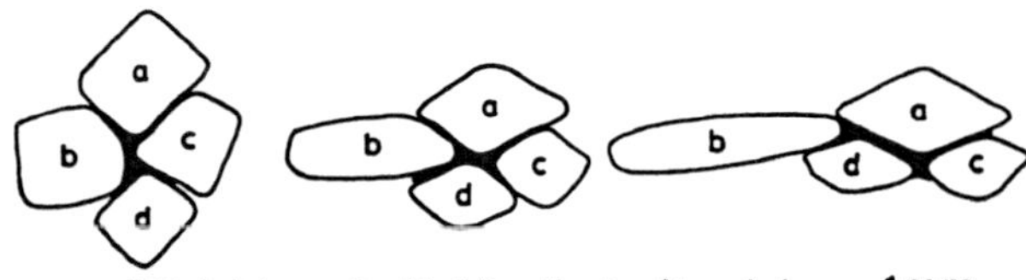

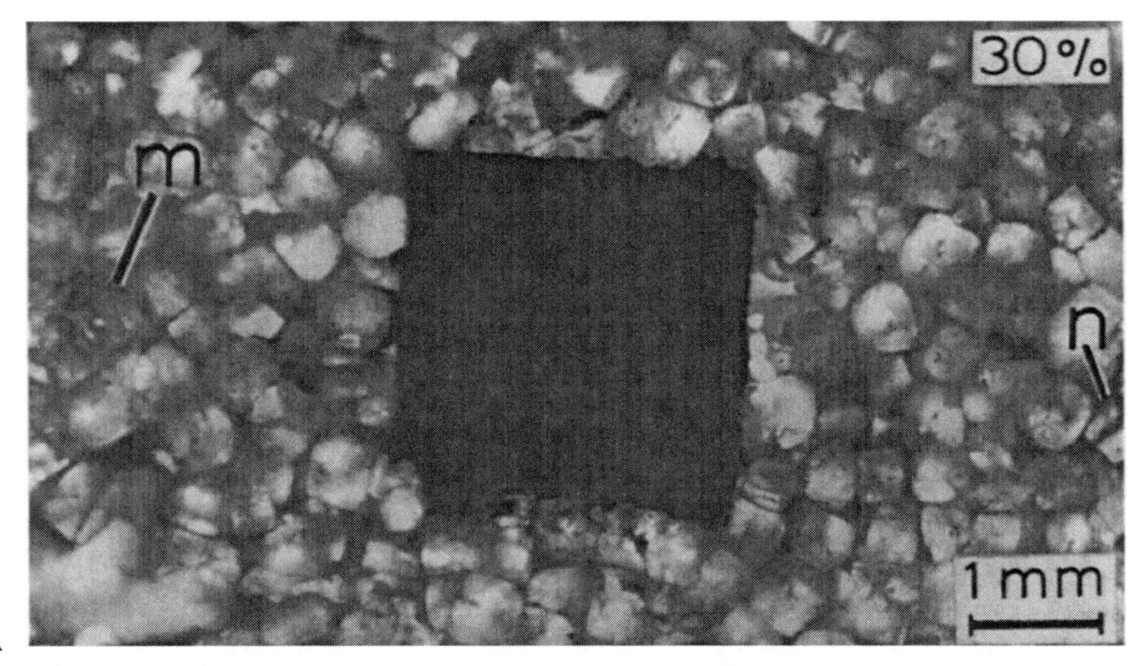


199 A

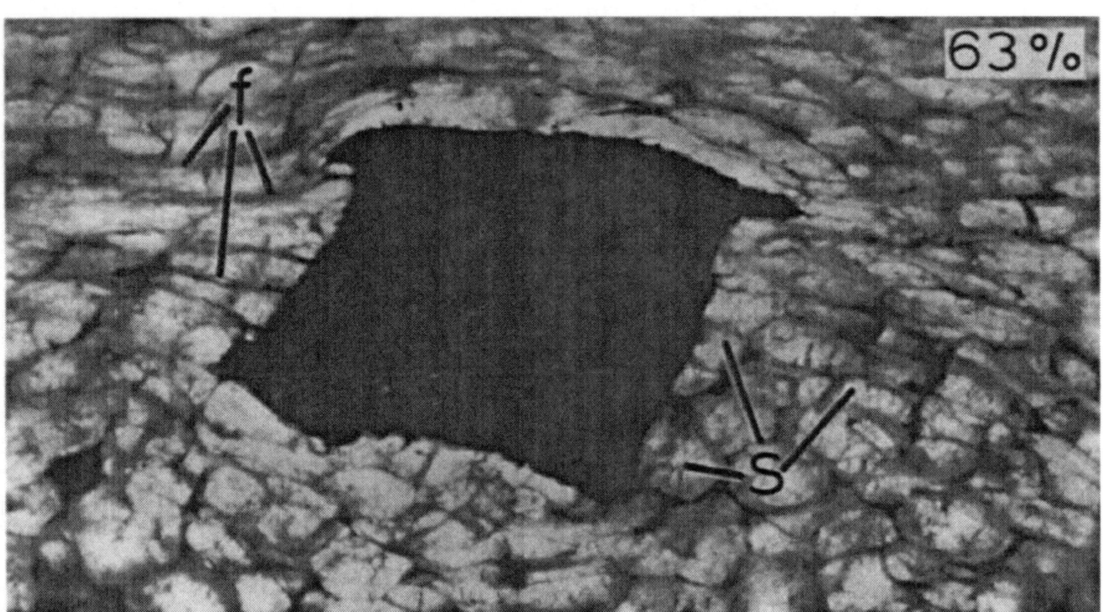


199 B

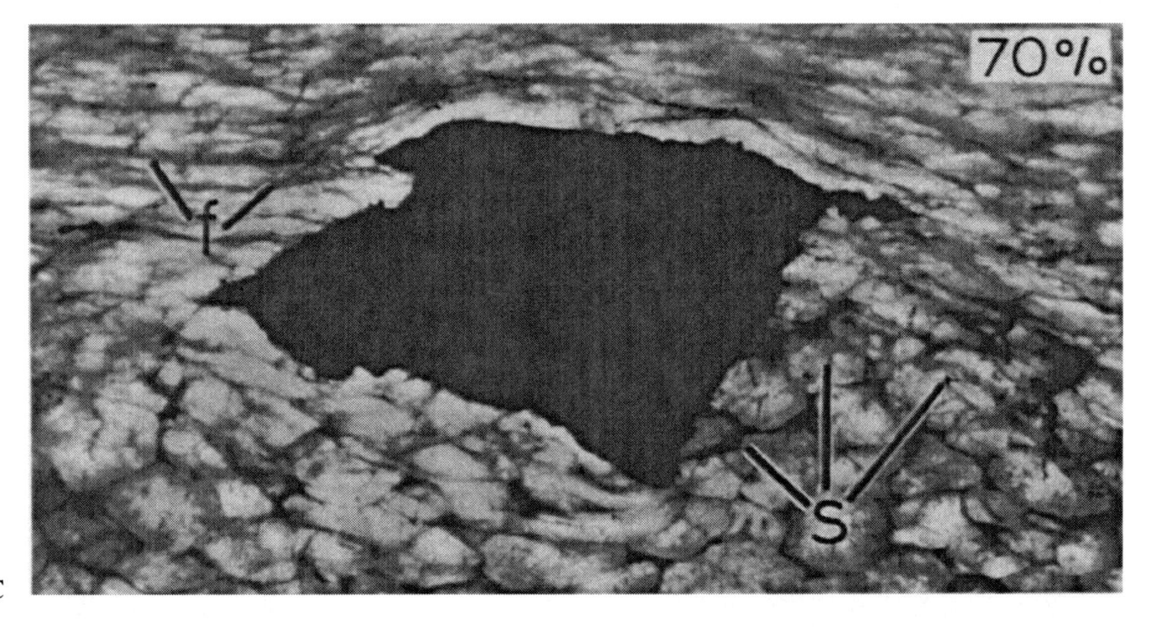


199 C

# 200. Strain Shadows in Phyllite

B. Z. LINCOLN

Phyllite from the Middle Ordovician Partridge Formation, collected five miles north of Fairlee, Vermont, U.S.A.

---

**A. Strain shadows.** In this section, which is normal to the two schistosities and the lineation which results from their intersection, the older schistosity is horizontal. The large opaque grain is rutile, with many inclusions. The mineralogical composition of the rock in the strain shadows around the rutile is the same as that in the matrix; in both, the preferred orientation is defined by fine-grained chlorite, muscovite, opaque material, and elongate quartz and feldspar grains. The long axes of the coarse rutile grains are commonly at an angle to the penetrative early schistosity, although the long axes of their strain shadows lie subparallel to it. Within these strain shadows, the orientation of the grains fans from parallel to the rutile, near the rutile, to parallel to the schistosity at the ends of the shadows. PPL

**B. Strain shadows.** The mineralogical composition is the same as in **A**, although there is less opaque material in the strain shadows. This section is from the same hand sample as **A**, but is cut parallel to the lineation and perpendicular to the older fabric. The coarse rutile grains appear subparallel to the penetrative schistosity in sections of this orientation. PPL

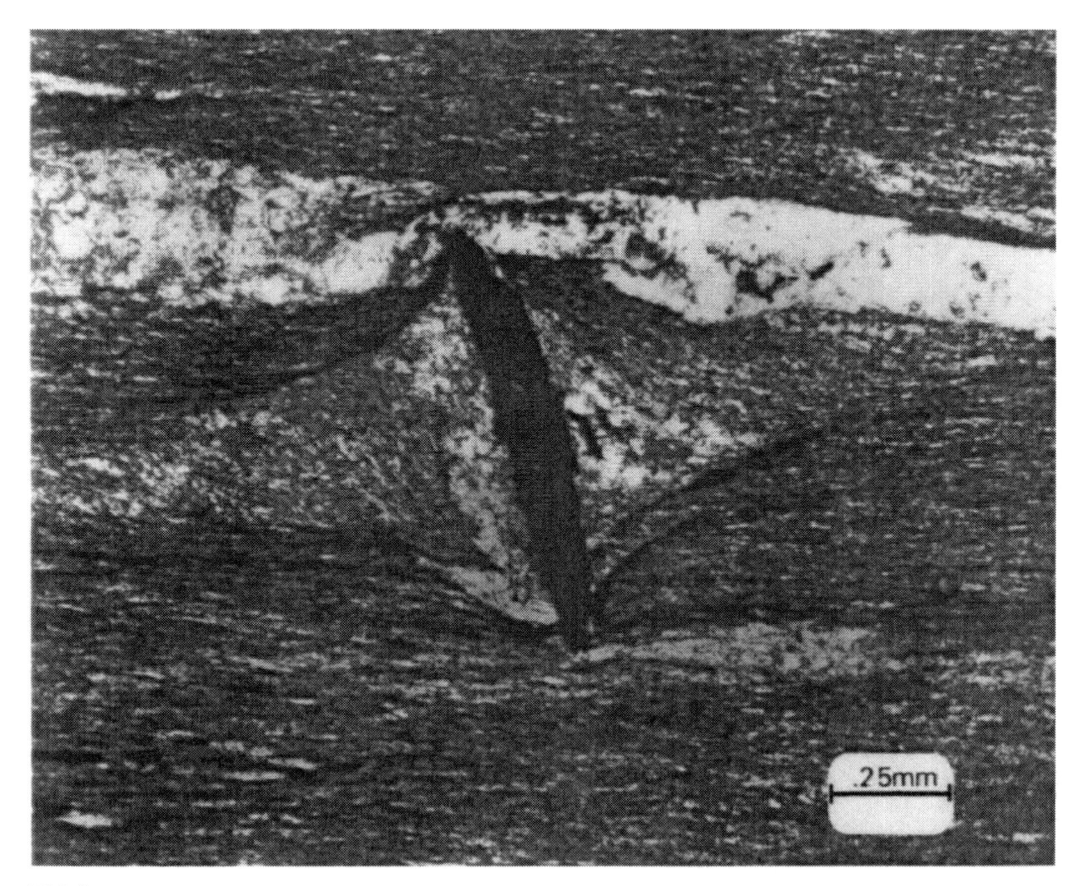


200 A

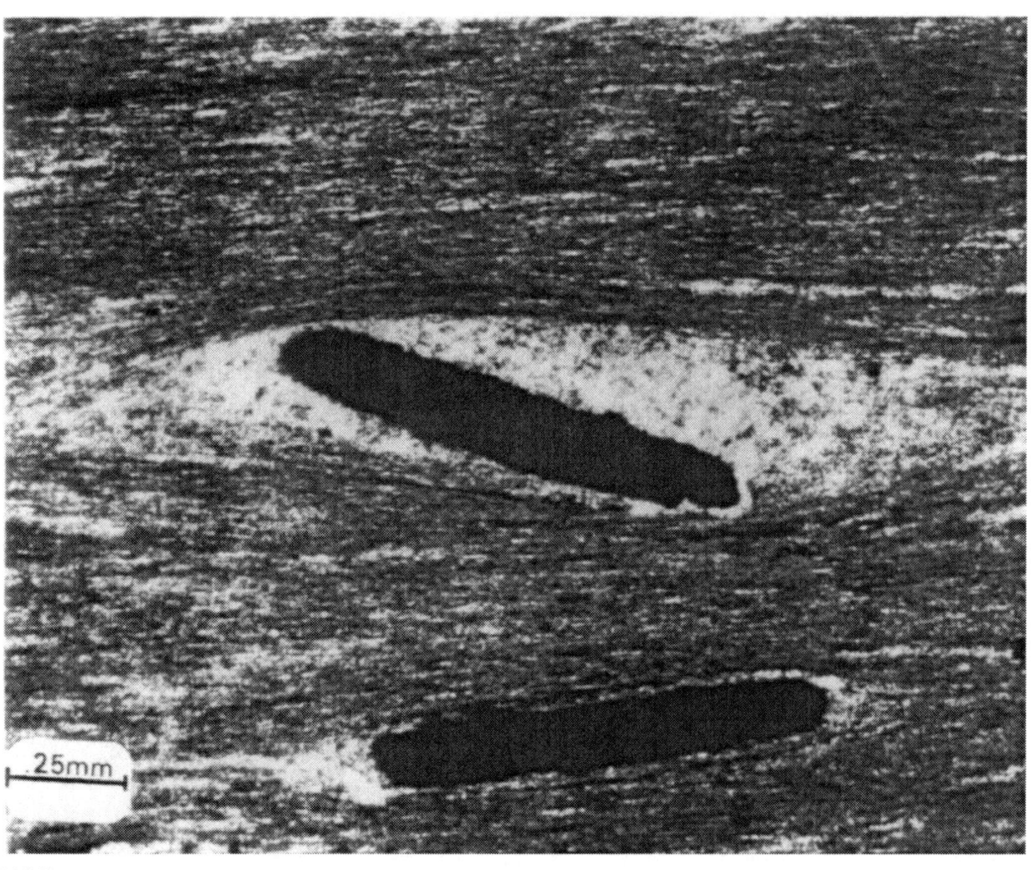


200 B

# Other Topics

## Section 16: Cleavage: Indications of Genesis and Strain

(Plates 201 to 212)

The first five contributions in this section are concerned with clastic dikes and cleavage. Interpretations of such dikes fall into two categories. Some dikes appear to be of tectonic origin or are perhaps related to a dewatering event associated with (early) cleavage formation. In other cases clastic dikes clearly predate deformation and are sedimentary structures deformed during the cleavage-forming event. MAXWELL and GREGG even discuss the opposing interpretations for the same outcrop (Plates 201, 202). Other clastic intrusions, generally smaller in scale and different in character, may be more readily associated with physical disaggregation and are discussed in Section 12 (Plates 132 to 135).

The fifth and subsequent contributions in this section (Plates 205 to 212) are concerned with strain in cleaved rocks. The initial two contributions (Plates 205 and 206) are concerned with the angular relationships between first cleavage and the principal plane of the strain ellipsoid. In particular, it is possible to show that cleavage may or may not be parallel to the principal plane within the limits of the test described. The last eight contributions taken together illustrate some useful strain markers which allow strain ratios to be determined, though Plate 210 provides a rare example in which the absolute strains can be determined, and therefore allows an estimate of volume-loss accompanying cleavage formation. Strain estimates for cleaved rocks are also reported elsewhere in this book, e.g., Plates 3–5, 60, 61, 66, 82–84, 89, 93, 94, 96, 113, 131, 165–167, 175, 183–186, 199 and 228.

---

### Plates

## 201. Martinsburg and Hudson River Slates, U.S.A.

J. C. Maxwell

Features suggesting a relationship between the origin of slaty cleavage and the deformation of water-saturated, unconsolidated silty mudstone and sandstone.

---

**A.** Passive folding and associated slaty cleavage. Martinsburg slate, road cut 2 miles south of Columbia, N.J. along U.S. 46. Silty mudstone with thin layers of siltstone (*white*). Bedding trends approximately orthogonally to axial surfaces of microfolds.
Scale: 10 mm PPL

**B.** Sandstone dikes in slate. Hudson River slate, low road cut 1.8 miles east of New Paltz exit, New York Thruway. Mudstone and interlayered siltstone cut by dikes of fine sandstone. Fine parallel cracks mark trace of slaty cleavage. Note faint orientation of clay flakes approximately perpendicular to walls of larger dike and small apophysis parallel to cleavage on right side of dike. Small dike visible in *upper left corner.*
Scale: 10 mm PPL

**C., D.** Sandstone dikes in slate. Field photographs of Martinsburg slates in road cut on U.S. 46, about 2 miles south of Columbia, N.J.

**C.** Sandstone dike intruded downward from 7 cm thick bed of fine, graded sandstone into underlying silty mudstone, approximately parallel to trace of slaty cleavage. Note second, thinner sandstone dike near *middle of finger.*

**D.** Detail of area at source of dike. Note that sand was eroded preferentially from left (west) side of fissure and injected into mudstone along similar trajectory, close to present trace of slaty cleavage. Sandstone dike apparently was not rotated appreciably after emplacement, as indicated by absence of bending of dike at base of sandstone. Slaty cleavage is continuous from overlying mudstone into fissure through sandstone bed.
Scale: Sandstone bed is 7 cm thick

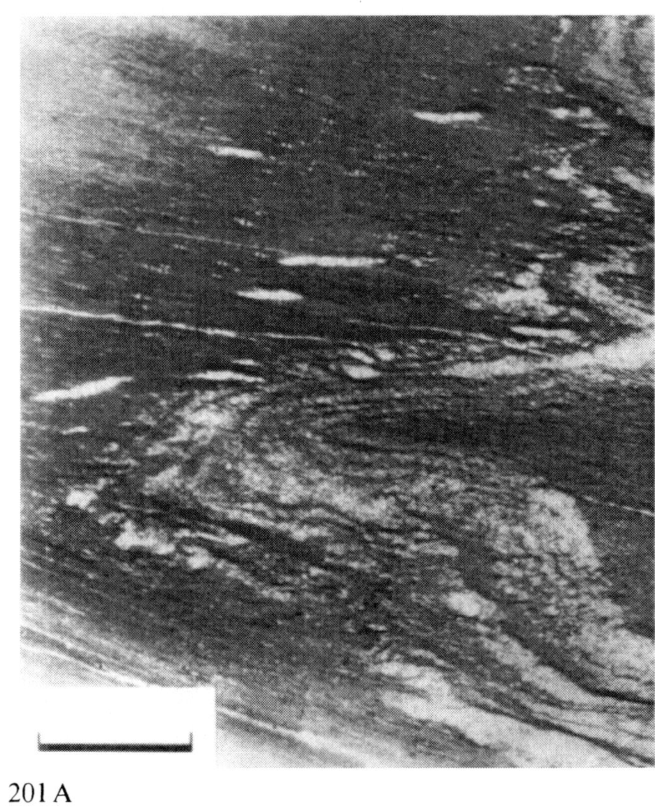

201 A

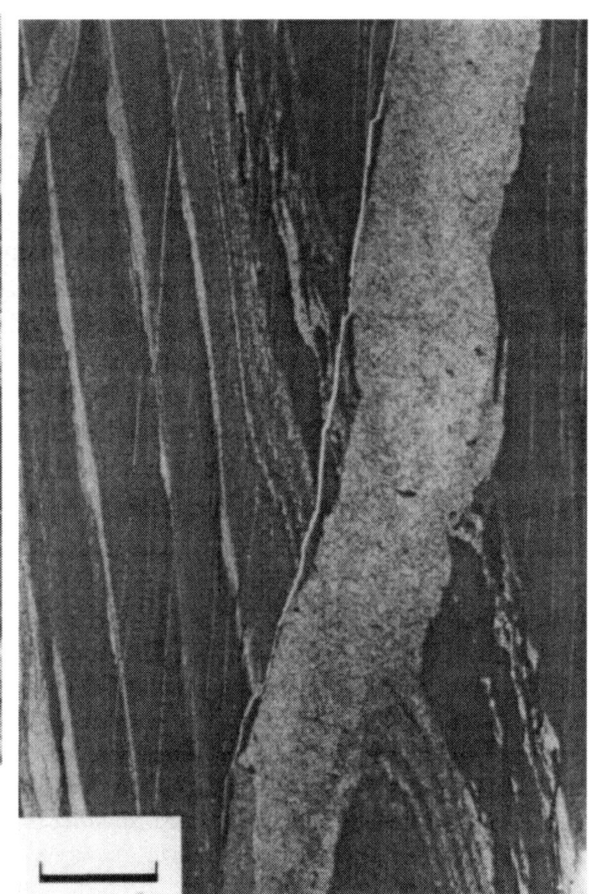

201 B

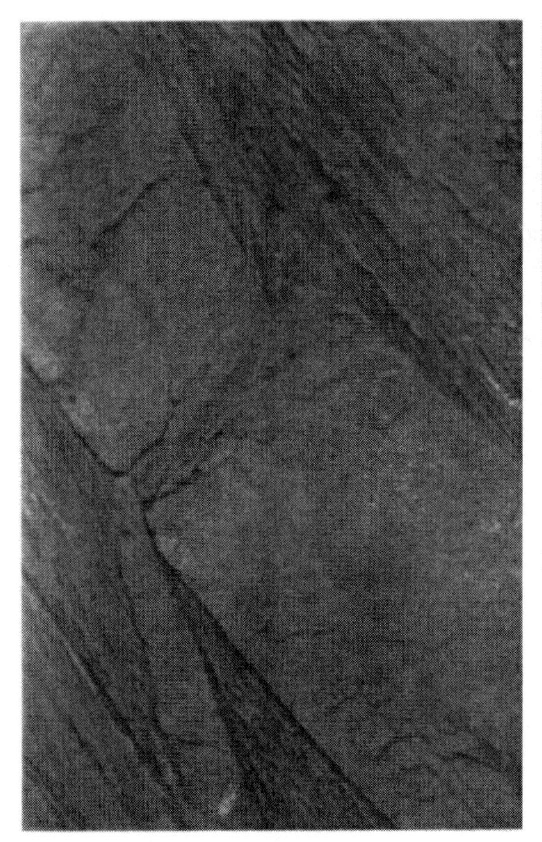

201 C

201 D

# 202. Pre-cleavage Clastic Dikes at New Paltz, New York, U.S.A.

W. J. GREGG

An important line of evidence often used by supporters of MAXWELL's (1962) hypothesis for cleavage formation by tectonic dewatering has been the parallelism of soft sediment clastic dikes and slaty cleavage. The parallelism of these structures indicates, to many supporters of this hypothesis, that the dikes were injected along cleavage planes, and thus indicate the "soft" condition of the sediment during cleavage formation. A number of investigators have objected to the claims of clastic dike-cleavage parallelism cited by MAXWELL and his supporters (for a review, see GREGG 1979). Reinvestigation of key outcrops has shown that dikes and cleavage are not parallel in most cases and the discovery of deformed fossils associated with slaty cleavage at these outcrops (GEISER 1975) has essentially invalidated MAXWELL's hypothesis.

---

**A.** Deformed precleavage clastic dike. Martinsburg Formation at New Paltz, New York. This specimen is taken from precisely the same outcrop discussed by MAXWELL (1962). The surface is normal to the clastic dike-cleavage intersection lineation and is shown in the accompanying sketch as the "cut surface." The relationships shown here are considerably different from the apparent parallelism displayed if one observes only the typical outcrop surfaces. Slaty cleavage is vertical in the photograph and passes through the dike. The dike is disposed in folds which contain the cleavage as an axial surface cleavage.

**B.** Uncleaved segment of the dike illustrated in **A** from area *b* shows typical clastic grains and lack of cleavage films.

**C.** Well-cleaved segment *c* in **A** shows development of mica films continuous with the pelitic slaty cleavage and also illustrates shape changes of clastic grains due to corrosion along cleavage planes

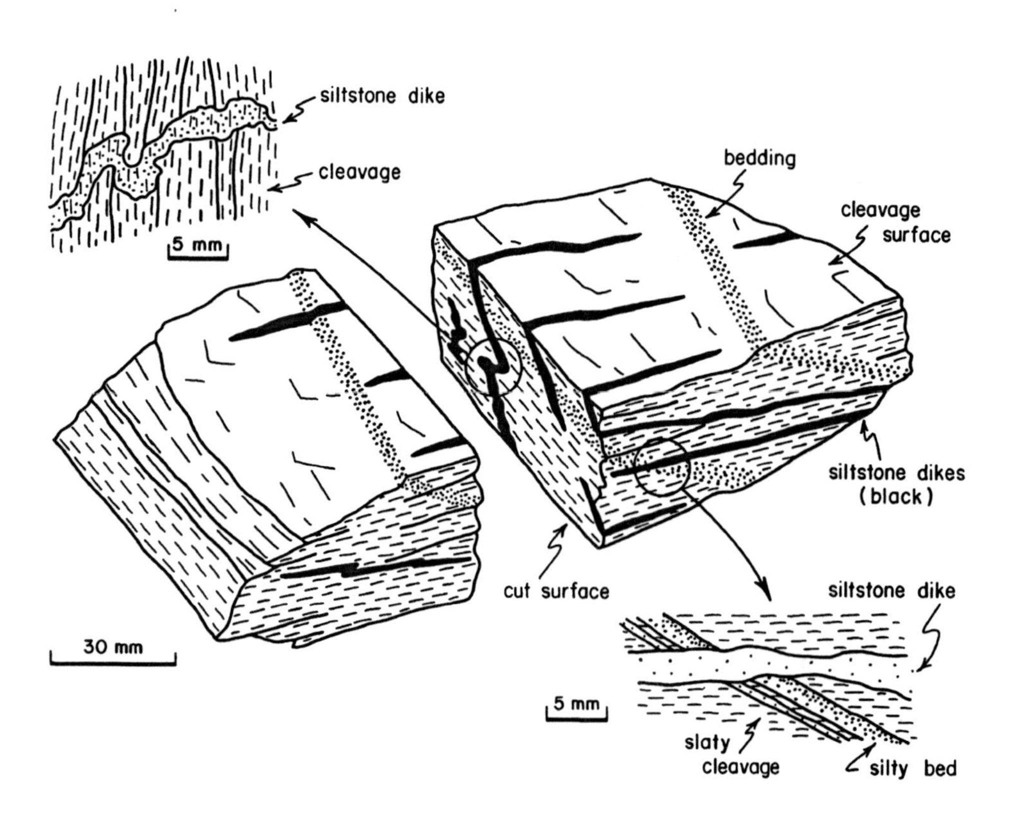

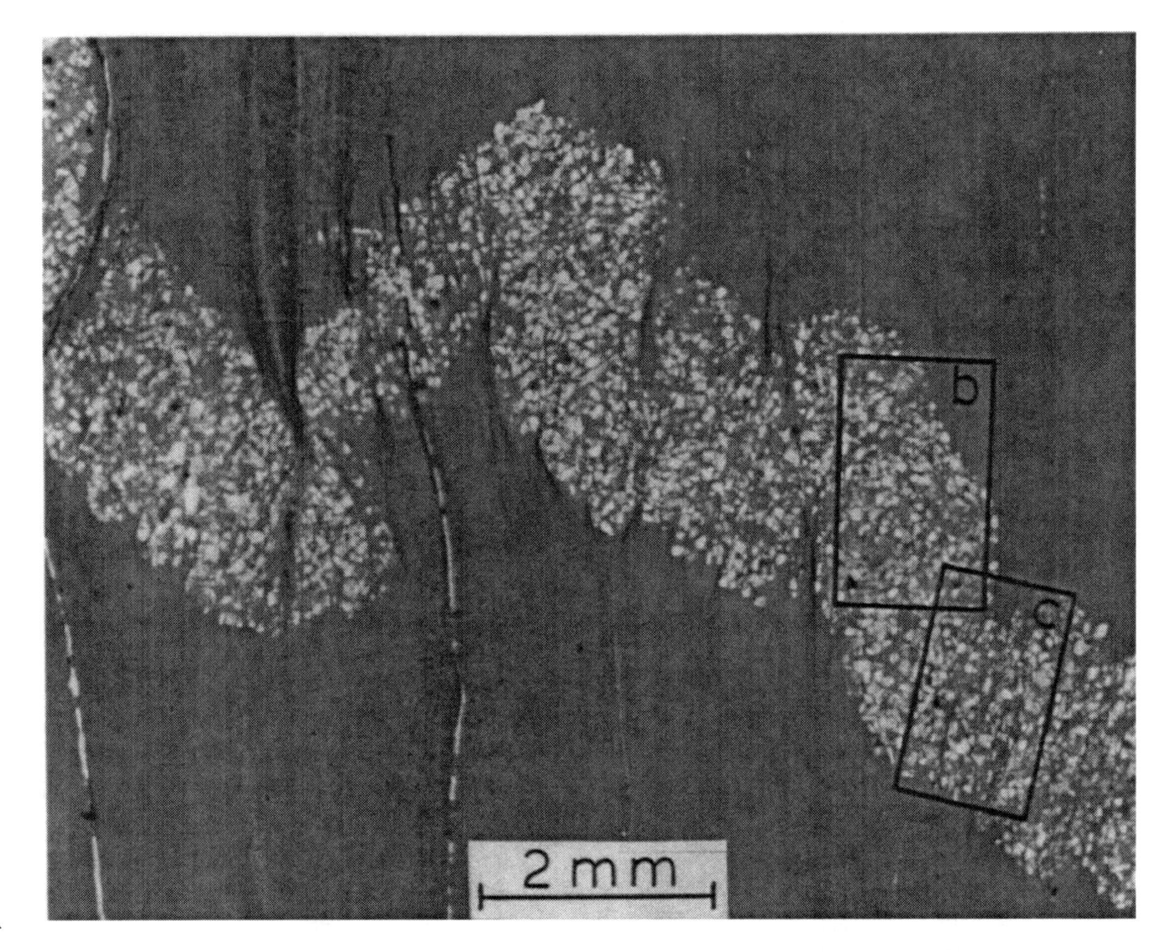


202 A

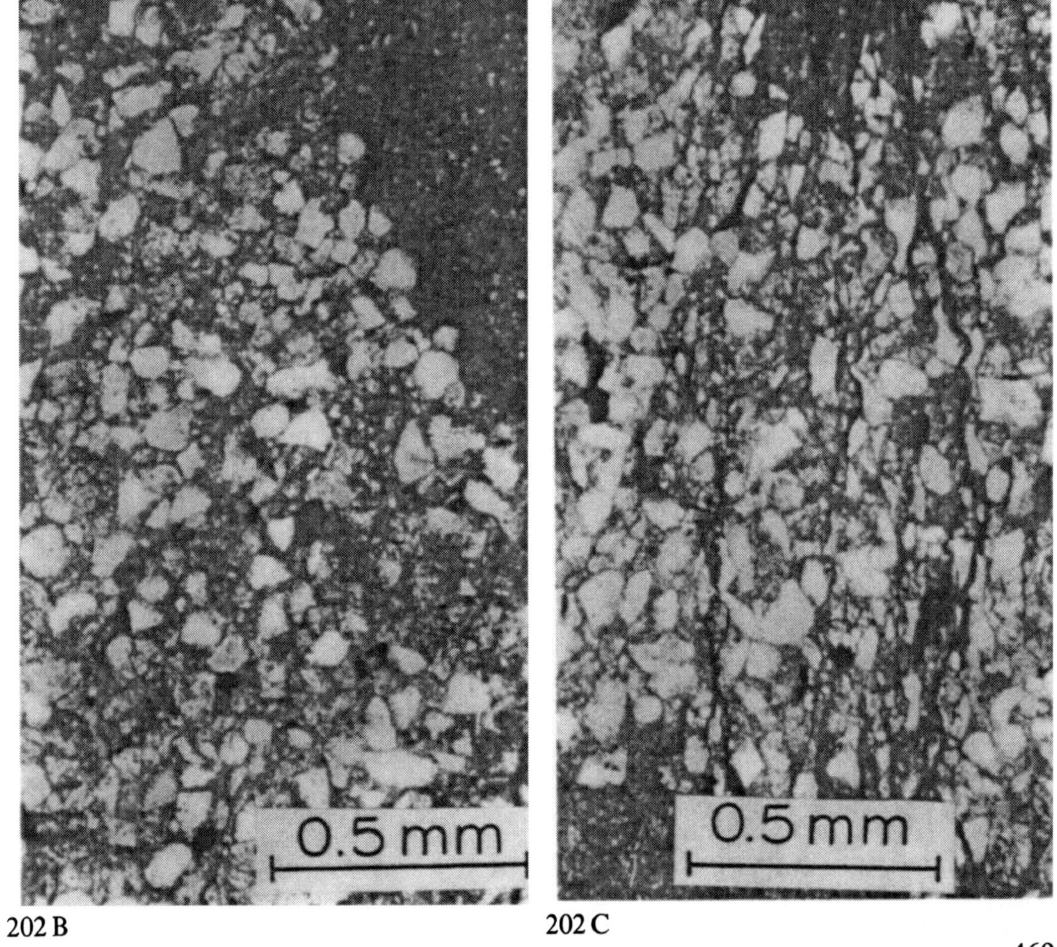


202 B

202 C

# 203. Slaty Cleavage and Clastic Intrusions

R. H. MOENCH

Upper Ordovician (?) Quimby Formation near Rangeley, Maine, U.S.A. Specimen from roadcut on route 4 & 16, 2.8 km W of Rangeley (MOENCH 1966, locality 1). Upper limb of northerly overturned syncline, but right way up in upper photographs; near chlorite-biotite isograd. Mineral assemblage is quartz, albite, chlorite, sericite, and sparse biotite, calcite, and opaques; higher proportions of albite and quartz in metasandstone (*light, graded*). $S_0$ bedding; $S_1$ cleavage; $D$ and $L$ dike and lobe intrusions.

---

**A.** Hand specimen. Etched with HF.

**B.** Close-up showing thinnest dikes.

**C.** Thin section normal to common intersection of $S_1$ and $D$ with $S_0$. PPL

**D.** Thin section normal to $D$ and parallel to common intersection of $S_1$ and $D$ with $S_0$. PPL

*Observations:* Intersections of bedding with cleavage and the dikes and lobes are parallel and form a single lineation. Normal to lineation the dikes and lobes are subparallel to $S_1$, but on the hidden rear face, $S_1$ crosses one tabular dike at 22°. The lobes extend for at least the length of the specimen (15 cm), but their shapes in cross-section change greatly, from features that mimic sedimentary casts to irregular features, clearly of intrusive origin. The tabular dikes extend downward from well-graded source beds or from lobes; thinnest dikes are the thickness of single clastic grains (0.1 mm).

*Interpretation:* The dikes and lobes support MAXWELL's (1962) dewatering mechanism for at least the earliest stages of slaty cleavage formation. The lobes are formed by downward extrusion of liquefied sand from source beds into mud. The dikes formed by continued downward extrusion of mobile sand when the underlying shale was semi-compacted, having a poor $S_1$ that guided but did not strictly control dike orientations. As $S_1$ improved metamorphically, it was impressed on the intrusions and locally crossed the dike contacts

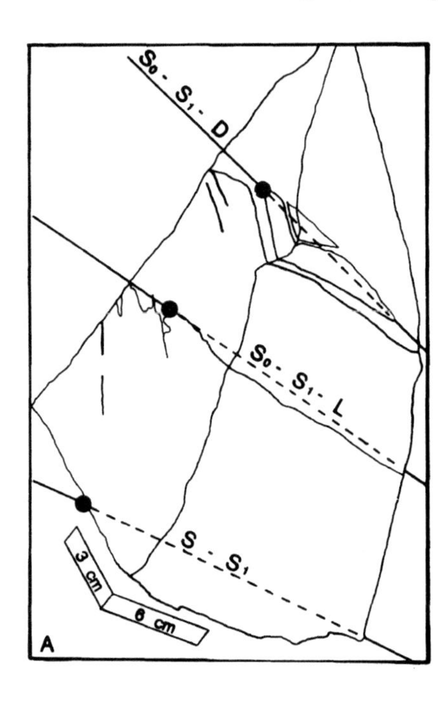


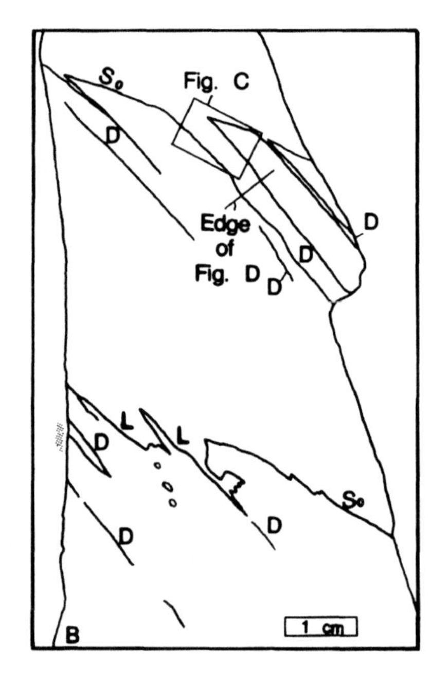

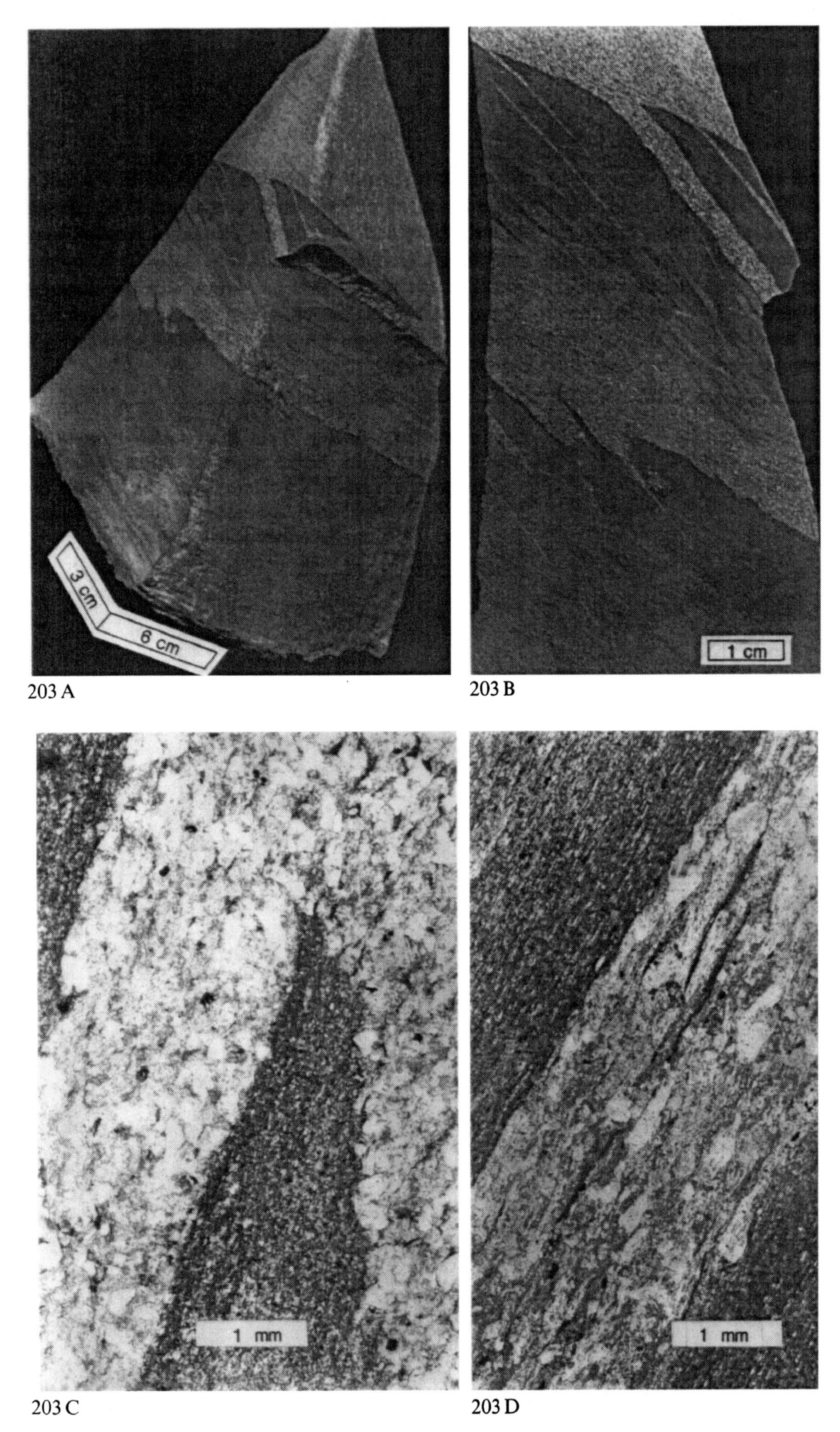

203 A

203 B

203 C

203 D

# 204. Sandstone Dikes and Cleavage in Slates

D. E. B. Bates

Field photographs from the Llanbedr Slates (Lower Cambrian) exposed in a small slate quarry (British National Grid SH/60281908) 4 km north of Barmouth, Gwynned, U.K. (Bates 1975).
In this outcrop there are developed both slaty and crenulation cleavages parallel to one another, and the pelites are divisible into mudstones, crenulated shales, and slates. It is suggested that the slaty cleavage was initiated in unlithified clays, and that its subsequent development has been confined to these, leaving the mudstone horizon uncleaved, except where the crenulation cleavage was developed in shales. The sandstone dikes were probably intruded after initiation of these cleavages, but before the lithification of the rocks.

---

**A. Sandstone dike in slate.** The dike is injected downward, with a sharp termination, from sandstone bed some 30 mm thick. See sketch for cleavage-dike relations.

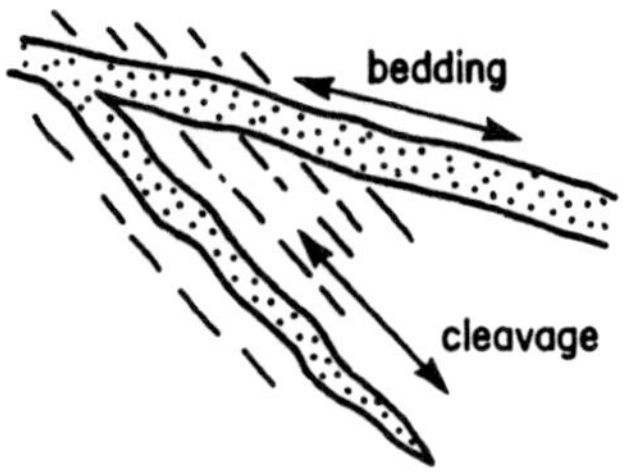

**B. Sandstone dike parallel to slaty cleavage.** Note the fretted edges of the sand grains in the dike, where the chlorite grains have grown in the pressure-shadows, contrasting with the grains in the sandstone at the *bottom* of the picture. PPL

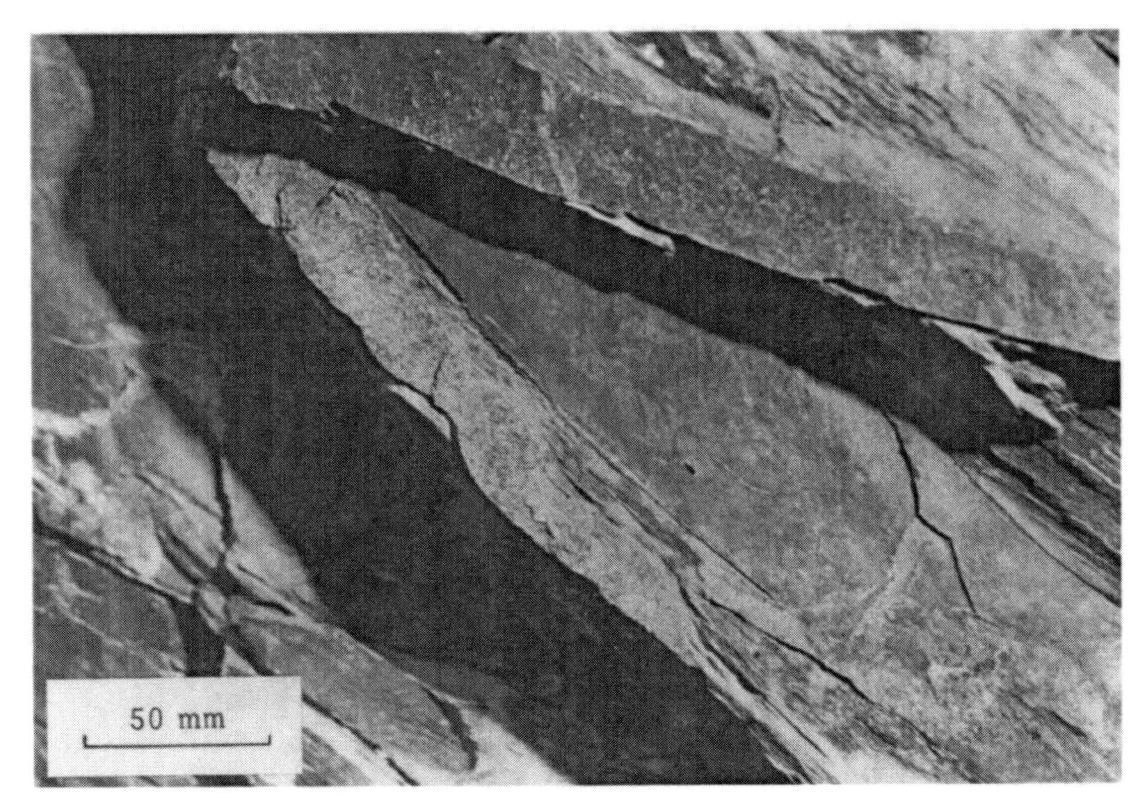


204 A

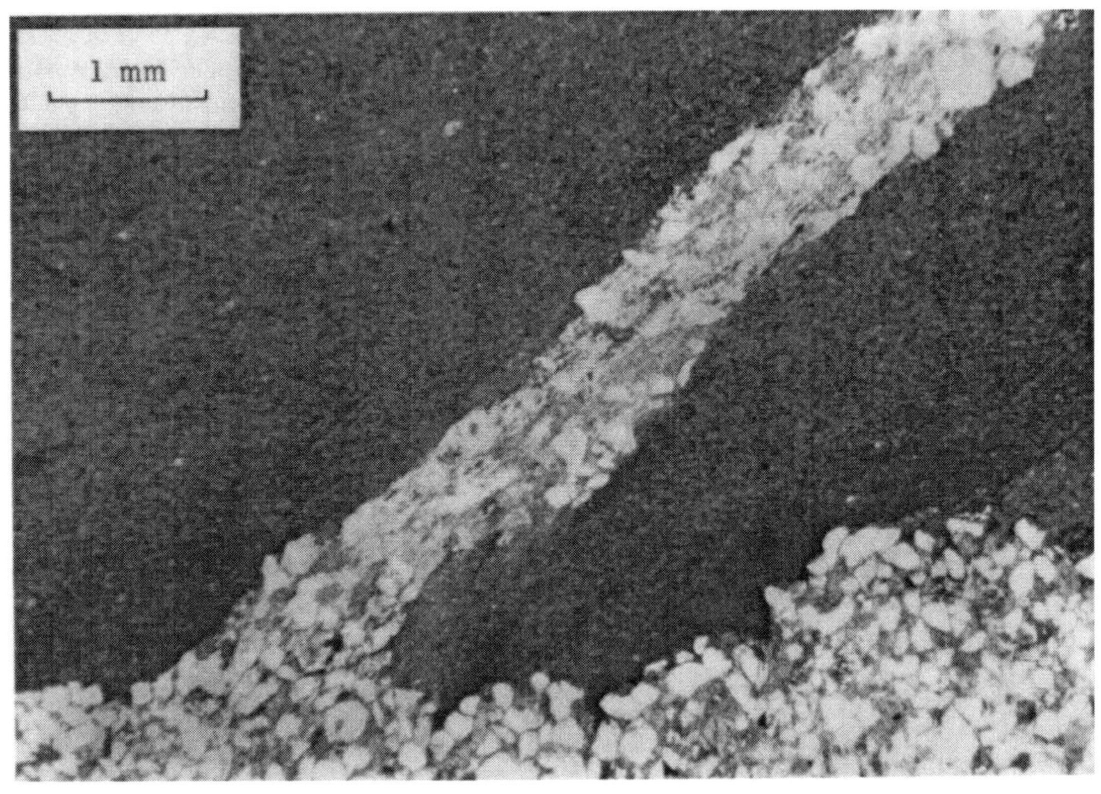


204 B

# 205. Cleavage and Right-angle Strain Markers

G. J. BORRADAILE

Strain markers which involve an initial right-angle are useful for determining the magnitudes of the principal strains. They may also be used to fix limits on the approximate orientation of the principal strain directions. The figure indicates, in a two-dimensional analogy, that a deformed right-angle must always contain the principal stretch ($x$) in the acute angle after deformation:

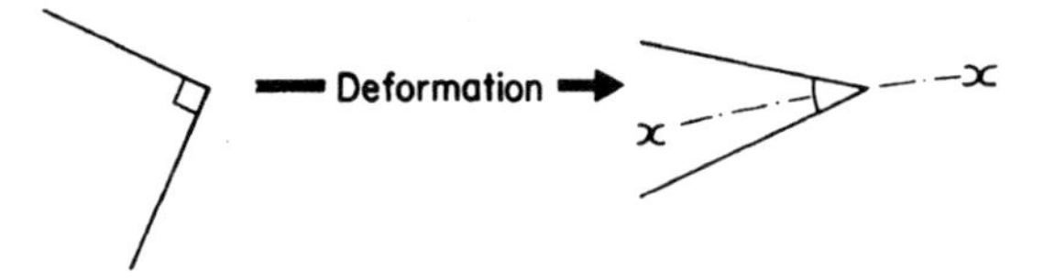


In three dimensions the acute dihedral angle brackets the trace of the $XY$ plane of the total strain ellipsoid: see BORRADAILE (1977). The method has now been applied in several regions and has been developed to include testing for the validity of assuming an original right-angled relationship (BORRADAILE 1977 and 1979). Two examples are illustrated (see also Plate 206).

---

**A. Deformed sedimentary sand dikes and continuous cleavage.** Late Precambrian Dalradian Supergroup, Port Ellen Phyllite, Islay, Scotland (BORRADAILE 1974, outcrop 3).

Bedding is subhorizontal and right way up; dike dips to left down from 5 cm thick sand bed which is interbedded with pelites. Cropping out best in the *top left* and *left* is the trace of the continuous $S_1$ cleavage. This is a preferred orientation of fine-grained muscovite and chlorite and preferred dimensional orientation of quartz. Sand dikes have been rotated from their original, near orthogonal, relationship with bedding. A full three-dimensional analysis indicates the following principal strains if $XY$ is taken to be parallel to $S_1$:

| $X$ | $Y$ | $Z$ |
|---|---|---|
| $2.09 \pm 0.25$ | $1.63 \pm 0.10$ | $0.29 \pm 0.02$ |

A stereoplot of sand dikes, bedding, and $S_1$ shows that $S_1$ lies in the "acute angle" field formed by sand dikes of varying azimuth and bedding. $S_1$ may therefore approximately indicate the orientation of the $XY$ plane of the total strain ellipsoid (see BORRADAILE 1979 a, b for further details). Scale: 25 cm

**B. Deformed worm tubes and disjunctive cleavage.** Lower Carboniferous shales, Edersee near Nieder-Werbe, Rheinisches Schiefergebirge, West Germany (PLESSMANN 1965).

Bedding is horizontal and right way up; pair of worm tubes dip to right on *right-hand side of coin.* Rough, widely spaced, disjunctive cleavage planes dip to right ($S_1-S_1$ on outcrop). The $S_1$ trace is contained in the acute angle ($\sim 70°$) between bedding and worm tubes. The $XY$ plane of the strain ellipsoid passes through this angle, but because it is large and defined by a line-element and a plane, rather than by two planar elements as in **A**, it is not possible to make a precise statement about the relationship of cleavage to the strain ellipsoid.

Strain estimates using the cross-sectional shapes of the worm tubes on the bedding surfaces indicate a strain ellipse with axes 1 : 0.65 (PLESSMANN 1965) and the shearing of worm tubes from an orthogonal position indicates a shear strain, $\gamma = 0.40$. Because the principal strain directions cannot be determined for this outcrop it is not possible to specify the shape of the strain ellipsoid. Scale: 2 DM coin

205 A

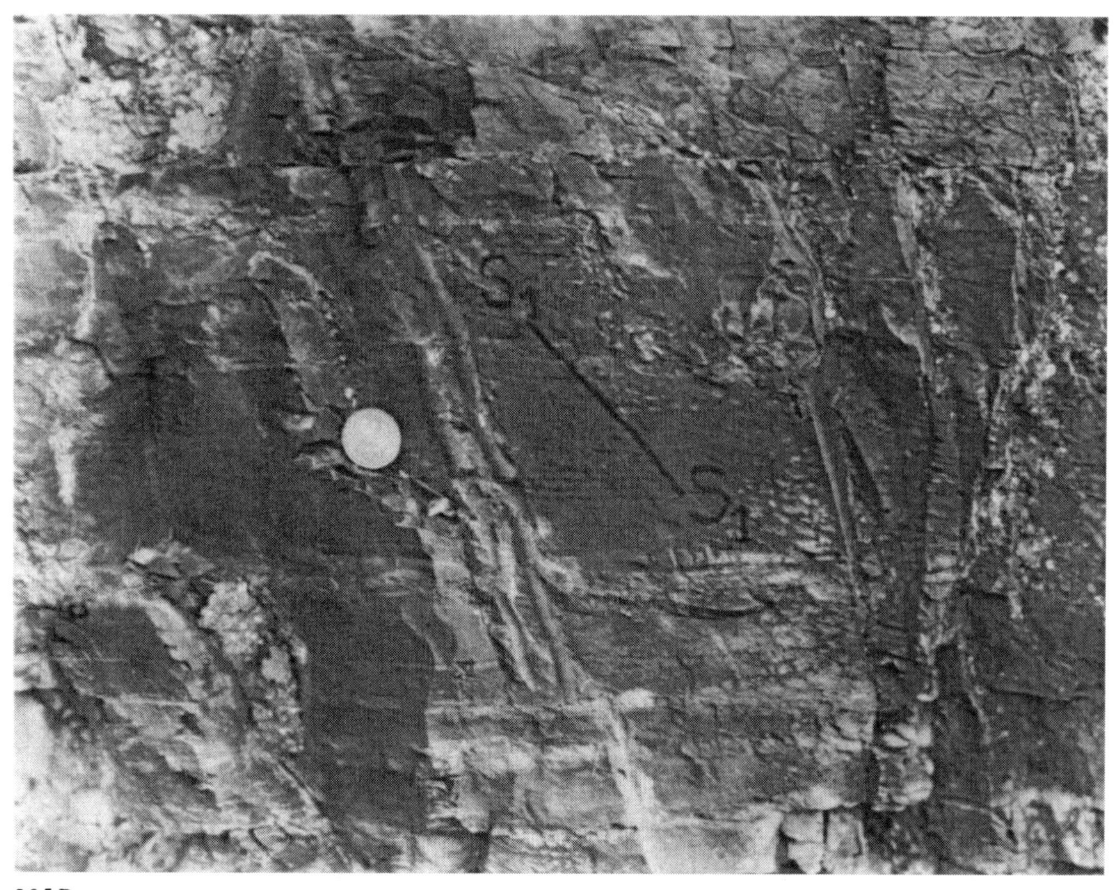


205 B

# 206. Tectonic Deformation of Pillow Lava

G. J. BORRADAILE and K. H. POULSEN

From a small reef in Lake Superior near Jackfish in the Archean Marathon Greenstone belt in Northwest Ontario, Canada.

The deformed basaltic pillow lavas are exposed on a horizontal glaciated surface. The "patchy" appearance of the pillows is partly superficial and partly due to the uneven distribution of varioles. The rocks have been metamorphosed in the lower greenschist facies and bear a single tectonic schistosity (*S*) of the continuous type.

Deformed pillow lavas are of importance to cleavage studies for two reasons. Firstly, it has been shown that the ratio of the principal strains is given by the ratio of the thickness of pillow selvages at points of maximum thickness (*A*) and minimum thickness (*B*) (BORRADAILE and POULSEN, 1981). The orientation of the strain ellipse is also given by the direction in which the selvages are most thickened. (This method might also be employed to estimate strain from envelopes or mantles on ooids, e.g. Plate 94, or lapilli e.g. Plate 209). In this instance the strain ellipse has an axial ratio of 2.15± (s.e.) 0.12 (n = 42) and is oriented parallel to the schistosity trace. Secondly, it is possible to show that if the schistosity is formed in an approximately coaxial strain history, its trace should lie in the acute angle formed by the long axis of the deformed pillow and the cusp-midpoint line of the pillow (see *sketch*). The argument is essentially the same as that involving sand dikes and bedding (see Plate 205).

Scale: Card in photograph is 5 cm long

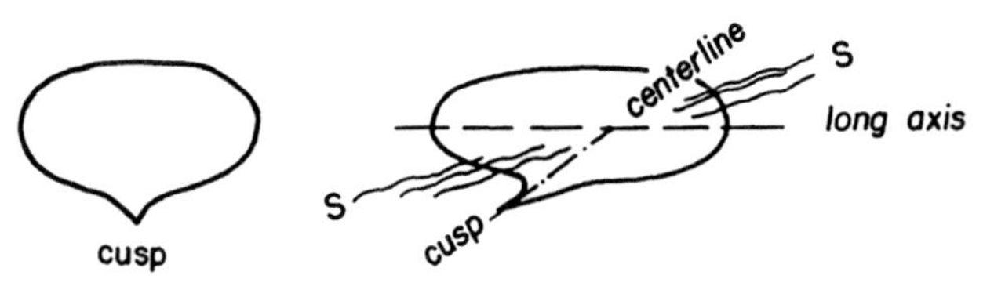

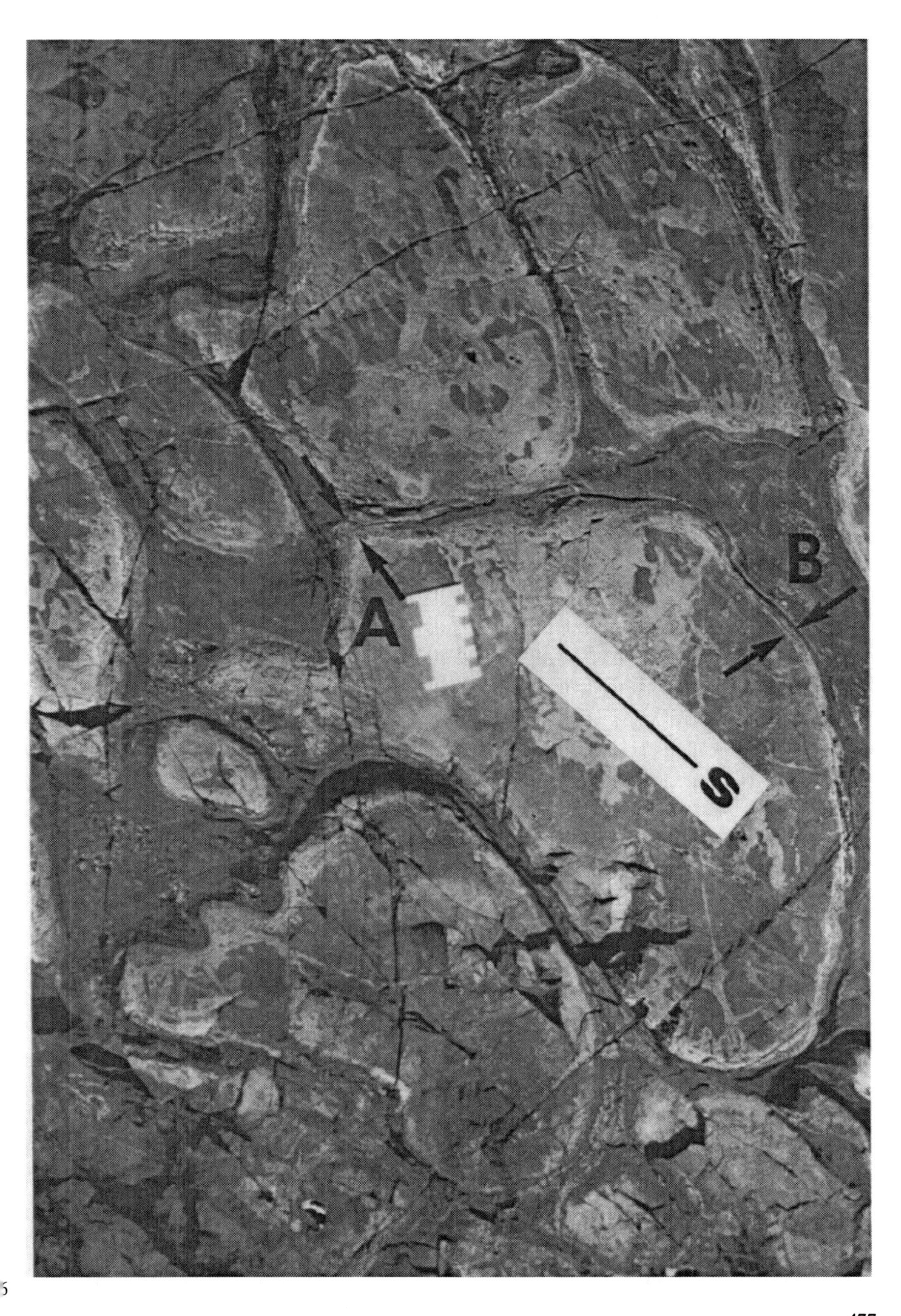


6

## 207. Deformation of Conglomerate (I)

G. J. Borradaile and P. A. Jackson

This plate and the succeeding one are of an Archean conglomerate cropping out in the Seine River Area west of Atikokan, NW Ontario in the Canadian Shield.

This photograph is normal to cleavage and is of a joint surface approximately parallel to the *YZ* plane of the finite strain ellipsoid. A simple method of strain analysis suggested by Lisle (1977, 1979) indicates that the approximate minimum strain experienced by the number of pebbles is

| *X* | *Y* | *Z* |
|---|---|---|
| 2.24 | 1.64 | 0.27 |

assuming constant volume strain. (This result was derived from 24 pebble shapes on an *XZ* surface and 31 pebble shapes on a *YZ* surface).

The clasts are of three main types: acid volcanic fragments (*light-colored*), tonalite-diorite (e.g., pebble *A*) and basic volcanic fragments (extremely elongated so that they are difficult to distinguish from the chloritic/pelite matrix). The acid volcanic fragments were initially angular and developed "fish-mouth ends" (*F*) as do boudins occasionally (Ramsay 1967, pp 105–106).

Some pebbles indent others (*arrows*). The indentation is achieved by plastic deformation rather than pressure solution. Usually a thin film of matrix persists between the indenting pebble and the indented pebble.

The schistosity shows marked strain shadows about rigid clasts such as *A*. By analogy with the results of Shimamoto (1975) it may be possible to determine the strain of the matrix from the geometry of such pressure shadows.

207

## 208. Deformation of Conglomerate (II)

G. J. Borradaile and P. A. Jackson

This is an outcrop surface parallel to the $XZ$ plane of the strain ellipsoid, from an outcrop adjacent to that shown in the previous plate. In this plane the pebbles are much more elongate, they show much less fluctuation and the strain shadows are more slender. The tonalite/diorite pebbles ($P$) are fractured and filled with fibrous vein quartz.

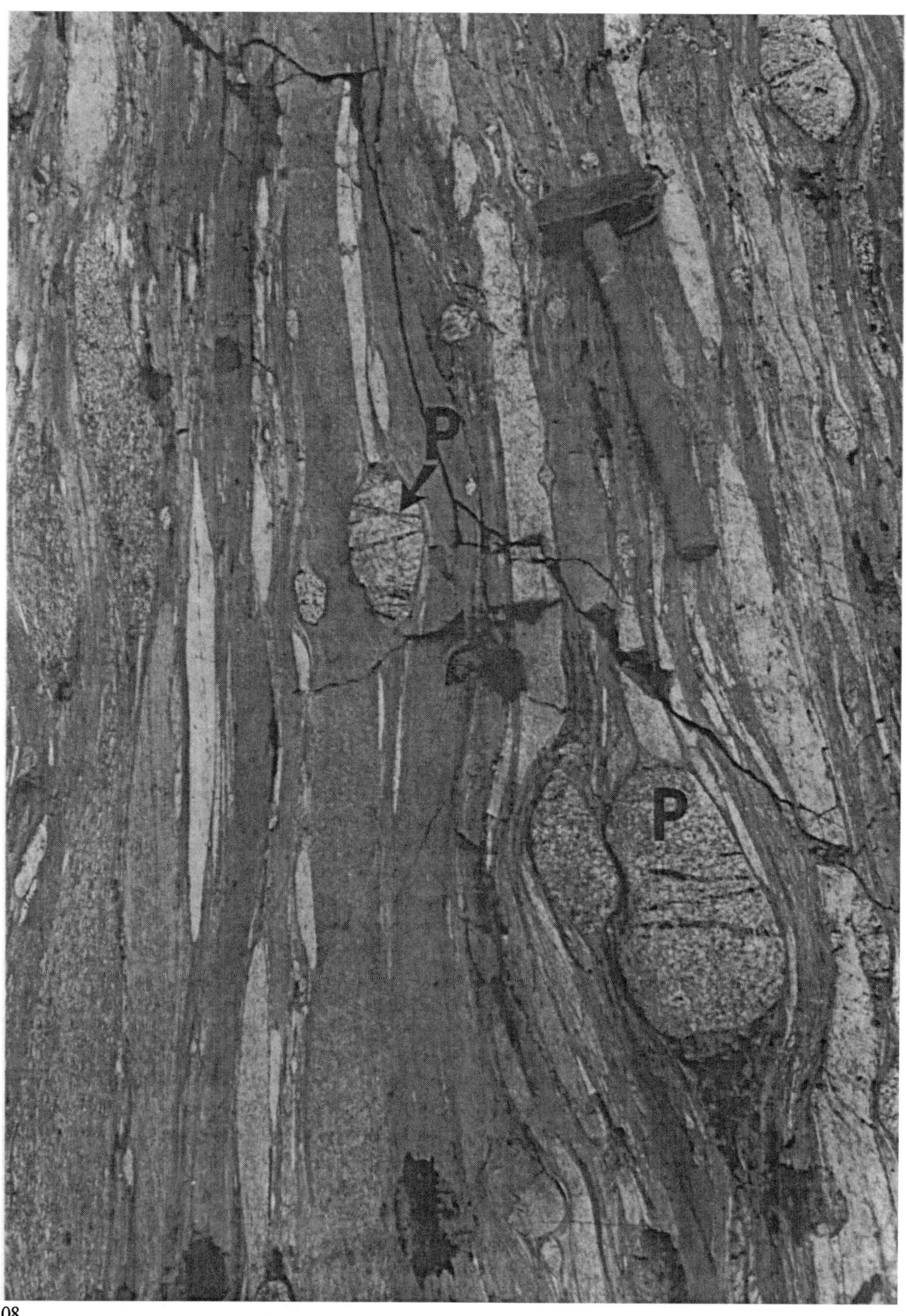


208

## 209. Fabrics Produced by Compaction Strain of Accretionary Lapilli

C. A. BOULTER

The accretionary lapilli are from the Triassic Brisbane Tuff and were collected on the bank of the Tingalpa Creek at Castra, 19 km ESE of Brisbane, Australia. This sedimentary succession is flat-lying and shows no sign of tectonic modification.

---

**A. Section normal to bedding.** The lapilli are not perfectly elliptical in this photograph. There is evidence for moulding of lapilli around their neighbours and the serrated contact in the *top left-hand corner* may be a stylolite. The departures from ellipticity are considered to be small and do not hinder strain analysis using axial ratios and long-axis orientation. Measurements of 76 lapilli from this face give a mean axial ratio of 2.00 to 1.00. Assuming that the precompaction fabric of this specimen is random, the Rf/Ø curve technique of DUNNET (1969) shows the compaction strain in this section to be 1.95 to 1.00. For another specimen with a lower volume concentration of lapilli a compaction strain of 1.50 to 1.00 was obtained. Measurements of fluctuation (long-axis orientation) were referred to the bedding trace and a range of $+8°$ to $-16°$ included all but one of the lapilli, indicating a strong preferred orientation which would markedly influence any tectonic fabric that could subsequently develop in such material. The distribution of long axes was, however, found to be asymmetric about bedding with 20 to one side, 52 to the other with four being coincident with the reference. This could be the result of slightly oblique compaction. If a fabric of this type is subjected to pure shear with the $XY$ plane of the tectonic strain ellipsoid at a moderate angle to bedding, the long axes of the lapilli would converge on the tectonic $XY$ plane. Any principal section at a high angle to bedding would show a mean preferred orientation of lapilli long axes at an angle to the tectonic $XY$ plane.

**B. Section parallel to bedding.** The majority of axial ratios are less than 1.2 to 1 and long axes are random in orientation. This specimen has a lower volume concentration of lapilli than the first and the compaction strain is lower. However, the strain ellipsoid shape $X = Y > Z$, is consistent with that expected for compaction. Rule shown has centimeter divisions

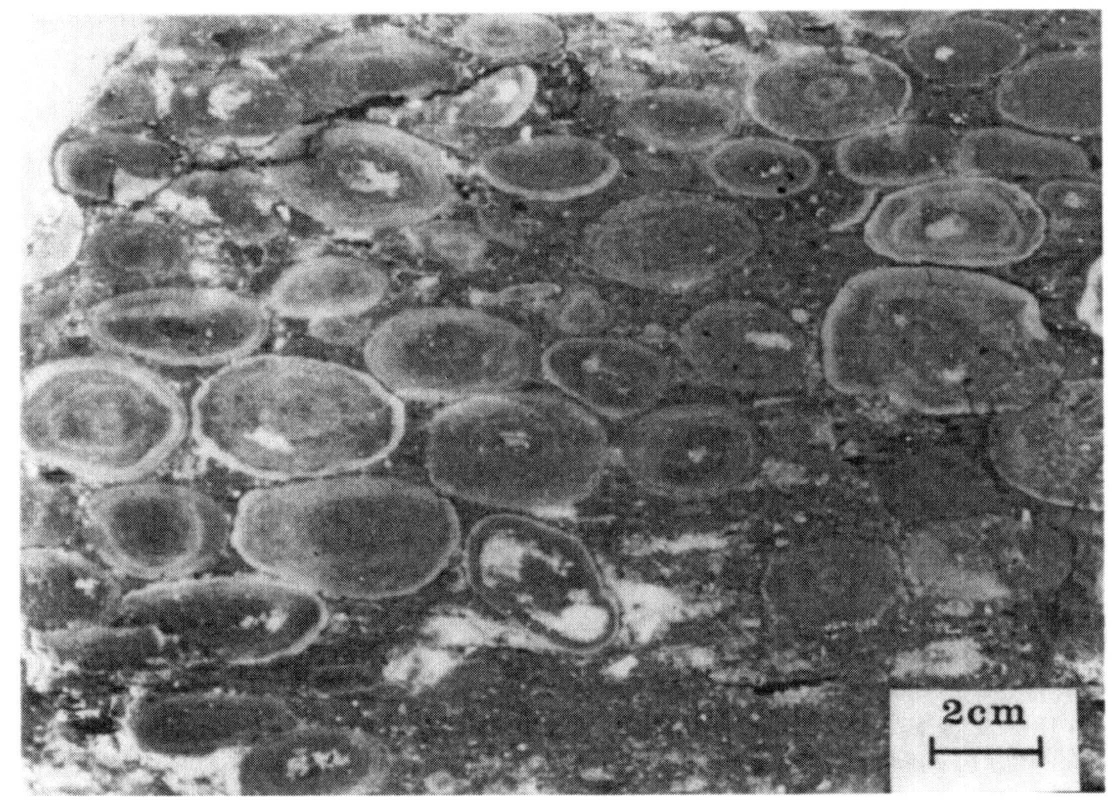


209 A

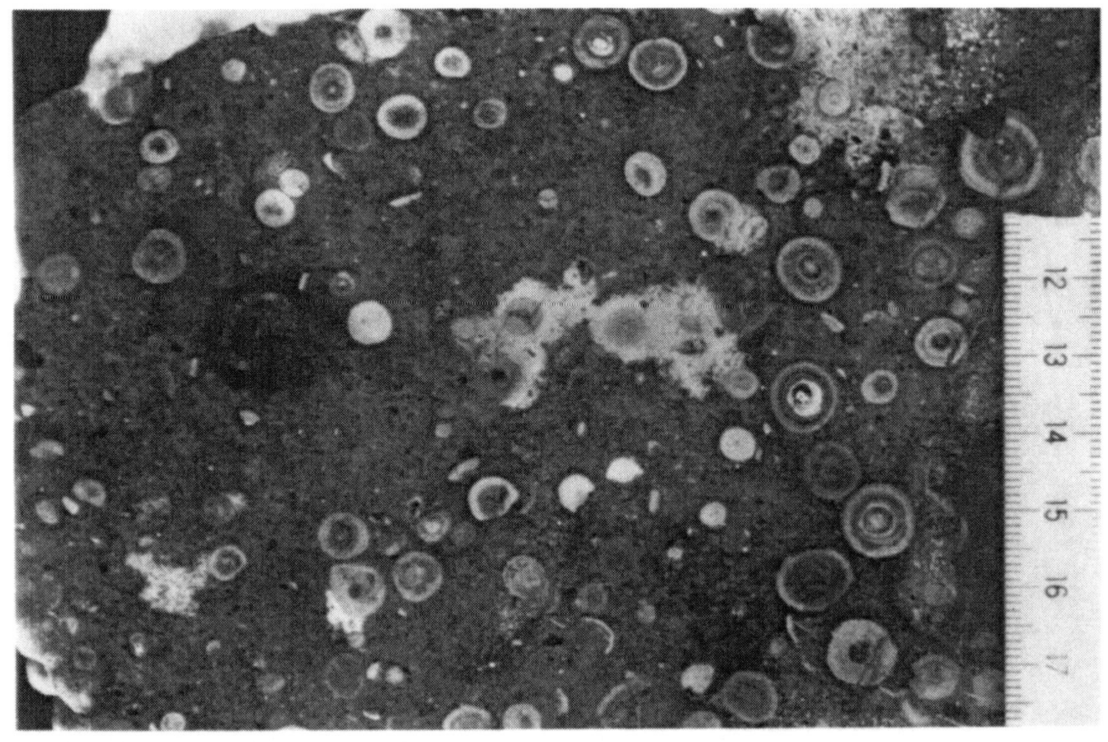


209 B

## 210. Cleavage and Strained Graptolites

T. O. Wright and L. B. Platt

Graptolites from the Martinsburg Formation in Pennsylvania, U.S.A., show distortion caused by pressure dissolution shortening during cleavage formation.

The trace of cleavage on the bedding plane is indicated by the horizontal lines on the photographs. Graptolites oriented parallel to the cleavage trace have undistorted thecal spacing, but are narrower than normal. The graptolites oriented perpendicular to the cleavage trace have normal widths, but the thecae are crowded together. The maximum shortening is always perpendicular to the trace of cleavage. This apparent shortening of the graptolites in the Martinsburg Formation is greatest where bedding and cleavage are at high angles. The apparent deformation in the bedding plane decreases to near zero when bedding and cleavage are parallel. These observations indicate a volume-loss mechanism for cleavage formation that removes material so that the direction perpendicular to cleavage is shortened. The apparent bedding plane deformation follows this relationship: $d = \text{sine}\ a \cdot \text{sine}\ b \cdot D$, where $d$ = the apparent bedding plane deformation measured along a specific direction relative to the trace of cleavage on the bedding; $a$ = the angle between the trace of cleavage on the bedding plane and the direction along which the deformation is measured; $b$ = the angle between bedding and cleavage in the specimen; $D$ = the actual deformation, perpendicular to cleavage planes.

---

**A. Apparent strain: large bedding to cleavage angle.** Several *Orthograptus amplexicaulis* specimens in silty shale from approximately 3 km southwest of Chambersburg, Pa., U.S.A. The angle between bedding and cleavage is 60°. The apparent strain (35%) represents a loss of 40% of the original rock volume by pressure solution.

**B. Apparent strain: small bedding to cleavage angle.** Several *Glyptograptus* specimens in limy shale. The bedding/cleavage angle is 40°. Apparent strain normal to the cleavage trace on the bedding plane (40%) represents a loss of 60% of the original rock

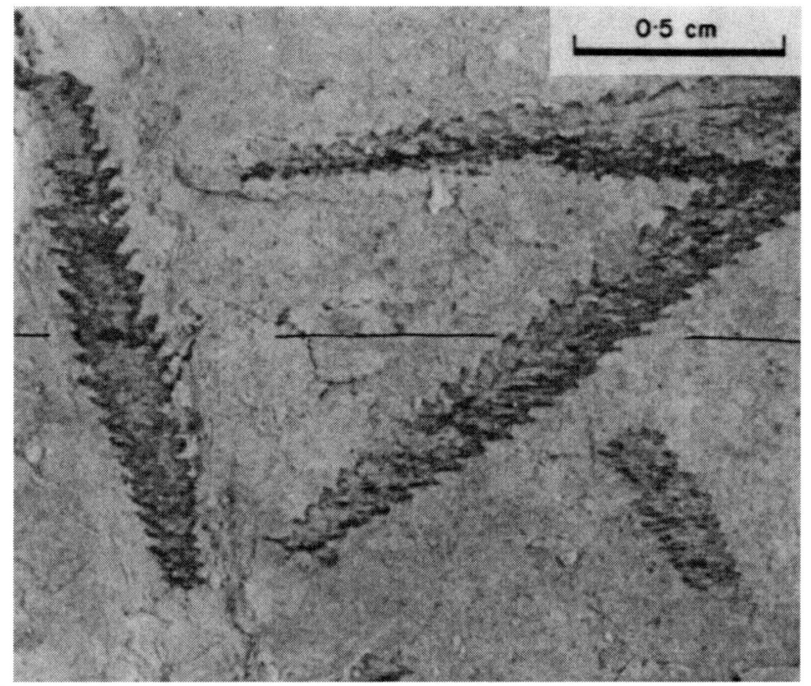

210 A

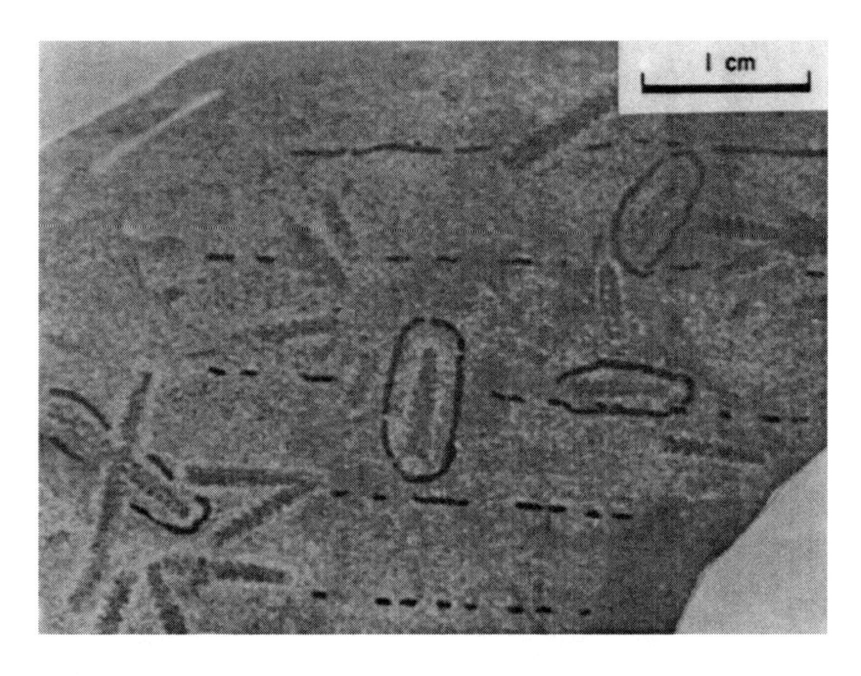

210 B

# 211. Cleavage and Strain in an Oolitic Limestone

O. A. PFIFFNER

Middle Jurassic (Upper Bajocian – Lower Callovian) Blegi oolite from Infrahelvetic complex (eastern Switzerland). The oolite consists of chamosite or hematite ooids, embedded in a fine grained calcitic matrix. Undeformed specimens show initially uniformly oriented elliptical ooids; deformation results in a planar fabric defined by the shape preferred orientation of flattened ooids (PFIFFNER 1977). Rock cleavage is controlled by preferred shape orientation of the calcite crystallites in the matrix. The fabric is of post-early Oligocene to pre-Miocene age (Cavistrau and Calanda phase; MILNES and PFIFFNER 1977, PFIFFNER 1977). Deformation occurred during increasing burial and progressively more ductile behavior of the rocks. Metamorphic mineral assemblages indicate P-T conditions of 3–4 kbar and 350° C at the end of the Calanda phase (FREY 1978; PFIFFNER 1977). Strain, as calculated from ooids, and whole rock strain are roughly the same, whereas strain indicated by the preferred shape orientation of the calcite crystallites of the matrix is much less; this means that either grain boundary sliding must have been active as the grains were deforming, or that some specific type of grain boundary migration occurred at some stage during the deformation (PFIFFNER 1977, 1981). Strains were determined by graphical methods ($R_f$-Ø diagrams and polar graphs) and by the reduced means method (PFIFFNER 1977, 1981). The long axes of strain ellipsoids are oriented at high angles to the fold axes, and the flattening plane coincides with the axial plane. Strains in these rocks are generally of the flattening type and include a component of plastic flow (movement of dislocations) and pressure solution/redeposition, both mechanisms operating simultaneously (PFIFFNER 1981).

---

**A. Flattened ooids.** Middle Jurassic Blegi oolite at Vättis (north of Calanda, eastern Switzerland). The sample was collected on the normal limb of a large scale Calanda phase fold with a basement core (external massif; Vättis inlier of Aar massif). *Reference line:* cleavage trace. Weathered surface of hand specimen.

**B. Calcite matrix of oolite.** Middle Jurassic Blegi oolite at Val Frisal (north of Cavistrau, eastern Switzerland). The sample was collected near a Cavistrau phase thrust fault, which follows the basement-cover contact, situated on the normal limb of a large scale Calanda phase synform into which this thrust is folded. Strains in this sample are extremely high (the axial ratio of the finite strain ellipse in this principal plane is larger than 12 : 1, possibly 100 : 1 or more). The calcite crystallites show a preferred shape orientation and relatively straight grain boundaries often meeting at 120° triple-points. *Reference line:* fabric trace. NX

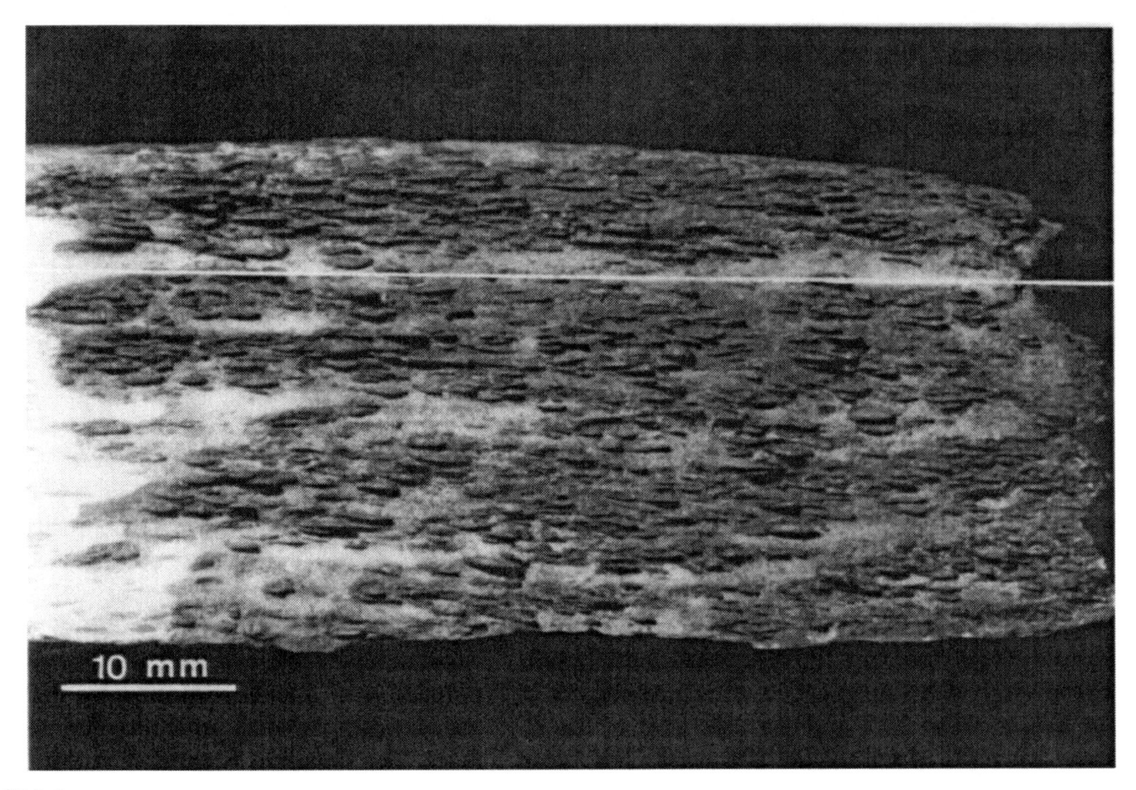


211 A

211 B

## 212. Relative Cleavage Development in Relation to Deformed *Beaconites Antarcticus* Burrows

P. L. Hancock

Reduction spots and other sedimentary structures provide well-known strain markers in deformed continental sequences but the trace fossil *Beaconites antarcticus* has not been used for this purpose. The illustrated examples occur in sandstones of the Milford Haven Group (Allen and Williams 1978) of the Old Red Sandstone (Siluro-Devonian) within the Variscan fold belt of southwest Wales, U.K. (Hancock 1973, Hancock et al. 1981).

---

**A. Weakly deformed *Beaconites* in an uncleaved sandstone.** Milford Haven Group, Freshwater West, Dyfed, Wales. The plate illustrates the underside of a steeply inclined bedding plane which exposes a transverse section through two *Beaconites antarcticus* burrows at the base of a fining-upward fluvial cycle. The average width/length ratio of burrow ellipses in the bed is about 0.75, and the long axes of the ellipses are oriented subparallel to a nearby fold hinge line. Scale: rule 6 inch (15.3 cm)

**B. Moderately deformed beaconites in a cleaved sandstone.** Milford Haven Group, St. Ann's Head, Dyfed, Wales. A transverse cross-section through the burrow is exposed on the upper surface of a bed of sandstone. The width/length ratio of several elliptical sections is 0.50, and long axes are parallel or subparallel to a lineation formed by the intersects of anastomosing spaced, pressure solution cleavage surfaces. The lineation is parallel to the hinge lines of nearby parasitic folds. Scale: 20 mm diameter coin

212 A

212 B

## Section 17: Cleavage and Polyphase Deformation

(Plates 213 to 220)

Eight contributions have been selected for this section on polyphase deformation. Many plates elsewhere in the book feature polyphase deformation. For example, most of the crenulation cleavages (Plates 24 to 64) are not primary tectonic fabrics and are thus polyphase deformation fabrics. Some examples of those sections show multiple crenulation cleavages which might be more appropriate here (e.g., Plates 34, 37). Nevertheless, it was considered necessary only to group a small number of plates here to illustrate the comparatively well known phenomenon of polyphase deformation. Three contributions in this section illustrate what could be described as conjugate cleavage (Plates 215 to 217). However, the components of the cleavage show sequential development and the plates are retained here for that purpose. Conjugate cleavage is also described in Section 10 (Plates 106, 107).

---

**Plates**

## 213. Folded Crenulation Cleavage

J. W. Cosgrove

The New Harbour Series, Treaddur Bay, Holy Isle, Anglesey, North Wales, U.K.
An unfolded example of this cleavage is shown on Plate 29. If this folding had continued another crenulation cleavage would have formed parallel to the axial planes of the folds.

3

## 214. Two Schistosities

B. Z. LINCOLN

Phyllite from the Middle Ordovician Partridge Formation, collected five miles north of Fairlee, Vermont, U.S.A., by D. RUMBLE and G. OERTEL. The following photographs are details of the textures seen in samples from the same outcrop.

---

**A.** The older schistosity, which is approximately horizontal in this photograph, is axial planar to a set of isoclinal folds. The younger, nonpenetrative cleavage, seen in the upper third of the photograph, is related to the refolding of these folds. This second fabric is observed most readily in the pelitic layers but can be found in the psammitic layers as well. This view is perpendicular to the cleavages and to the lineation which results from their intersection. PPL

**B.** Detail of **A**, taken from the area indicated by the *arrow* in the upper left of **A**. Note that the field of view is rotated approximately 90° counterclockwise relative to **A**, and that the older penetrative schistosity can be traced from the *bottom left* to the *upper right* of this photograph. The discrete traces of the younger cleavage cut across the penetrative older schistosity which is locally deflected. PPL

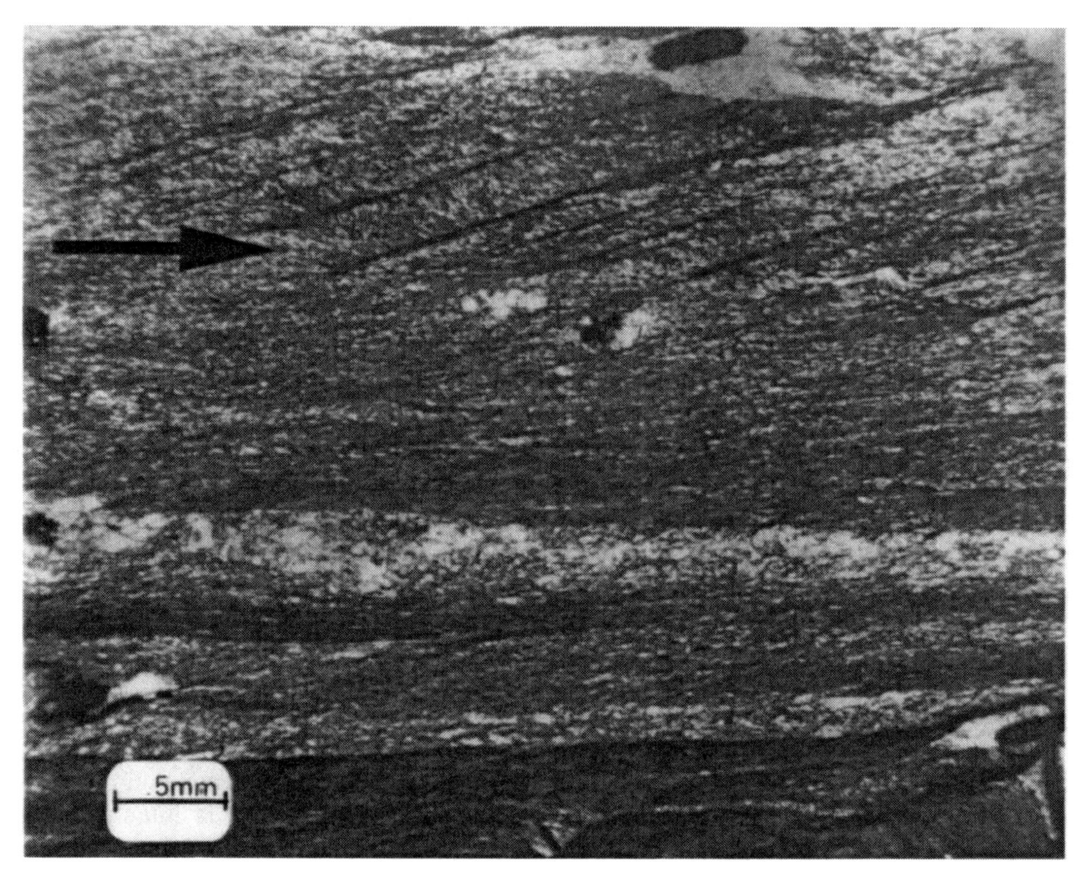


214 A

214 B

# 215. Two Inclined Cleavages Formed in One Folding Event (I)

C. A. BOULTER and K. C. HUGHES

This and the following two plates show cleavages from the low-grade, folded foreland of the Late Proterozoic Stirling Range Beds, SW Western Australia. The characteristic feature of the region is that many arenite or wacke layers have two inclined cleavages whose intersection lineations on bedding are fairly closely related to the local fold-hinge line. Cleavage-related folds are markedly asymmetric with long-limb to short-limb ratios usually in excess of 5 to 1. Common limbs up to 50 m in length have been recorded.

BOULTER (1979) proposed that well spaced micaceous bands were initiated during an early phase of the folding dominated by layer-parallel shortening. Until dihedral angles of about 140° were achieved fold amplification was weak, and the cleavage ($S_{IB}$) defined by the micaceous bands was able to maintain a nearly axial-planar geometry. Body rotation of the limbs then took over, dispersing the bedding and $S_{IB}$, but maintaining their approximate 70° angular relation. When the folds reached dihedral angles of about 100°, flattening became important producing a pervasive axial-planar mica-film cleavage ($S_{IG}$).

---

**A. In outcrop.** The mica-band cleavage in this arenite unit at Mondurup has a spacing between 2 and 5 cm and in weathered outcrop appears to be a joint set. Closer examination reveals up to 2 mm of mica coating each $S_{IB}$ plane. The mica forms a zone of very weak cohesion and the micas may be easily removed with a razor blade. $S_{IG}$ is a grain-scale fabric inclined at between 15° and 20° to $S_{IB}$. In the field it appears to be a dimensional preferred orientation of quartz grains. However, in thin section the fabric is seen to be of a mica-film type with little preferred orientation of quartz grains. This fabric varies considerably in character as arenite grades into wacke.

**B. Detail.** The sandstone unit is close to the arenite/wacke contact and is interlayered with units that were originally siltstone and mudstone. Bedding and the mica-band cleavage are the most obvious features in the thin section, with $S_{IB}$ spacing varying from 1 to 2 cm. $S_{IB}$ is a zone of nearly pure mica together with minor zircon, tourmaline, and opaques. Quartz grains are truncated against $S_{IB}$ and show markedly scalloped faces. In some cases $S_{IB}$ surfaces die out gradually as they pass from pure mica zones into coarsely anastomosing seams with quartz becoming more important and seams fading out. $S_{IG}$ in the wacke is subparallel to the cleavage seen in the finer-grained units. PPL

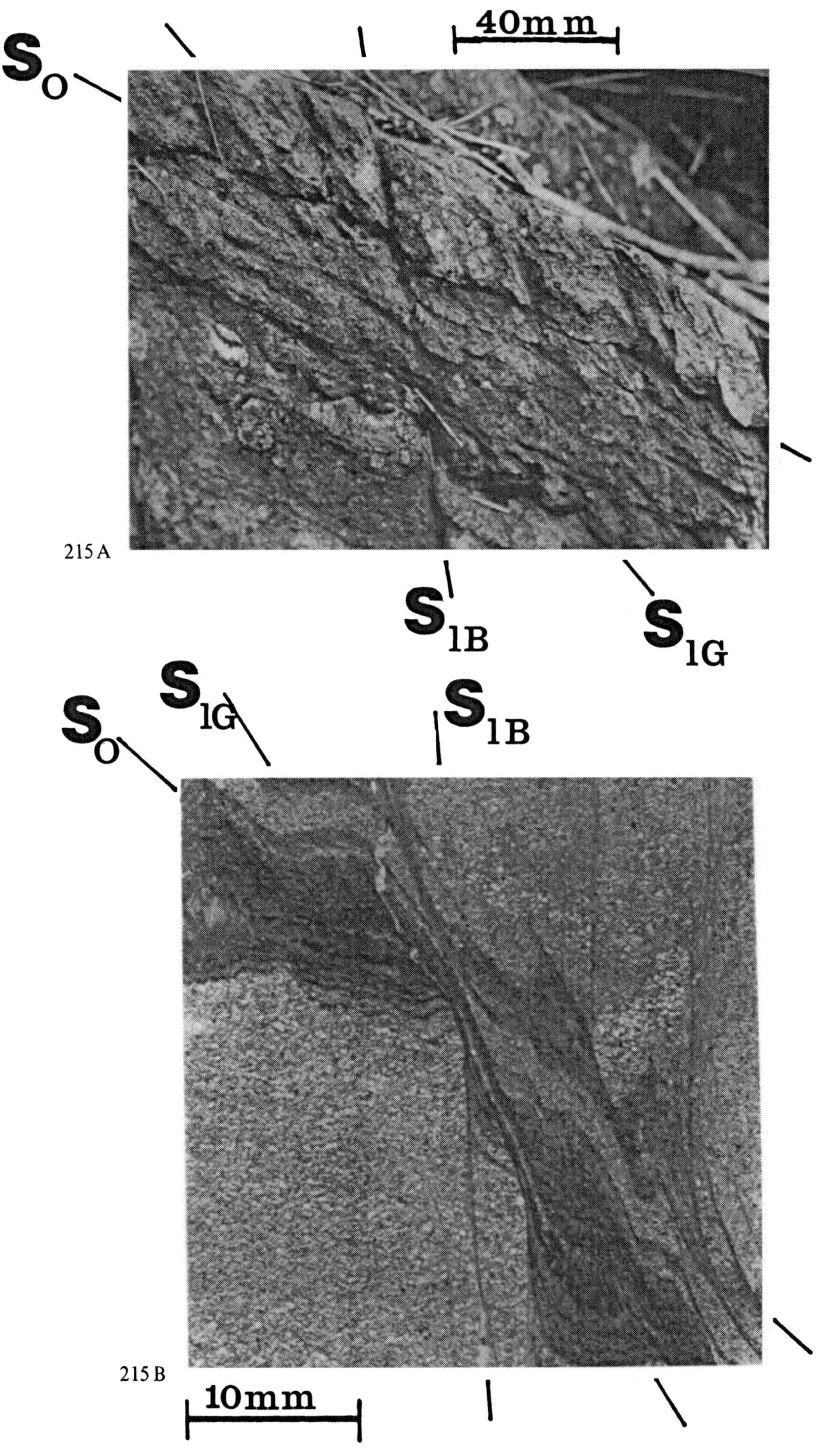

40mm
$S_O$
$S_{1B}$
$S_{1G}$
215 A
$S_{1G}$
$S_O$
$S_{1B}$
215 B
10mm

# 216. Two Inclined Cleavages Formed in One Folding Event (II)

C. A. BOULTER and K. C. HUGHES

**A. The $S_{1B}$ microlithon fabric.** In wacke layers with more than 20% mica content the microlithons between $S_{1B}$ micaceous bands are dominated by a continuous to nearly continuous film-plus-beard fabric ($S_{1G}$) that is inclined to $S_{1B}$. In mica-poor wacke and micaceous arenite two fabrics are found in microlithons; one is subparallel to $S_{1B}$ and the other is parallel to $S_{1G}$. Remarkably, there is little fluctuation about either of these two trends and very little evidence of overprinting of one by the other. Quartz grains within the microlithons approximate to parallelograms being bounded by faces roughly parallel to $S_{1B}$ and $S_{1G}$.

The angle between $S_{1B}$ and $S_{1G}$ in this example is about 60°, but this angle varies in a single thin section from 61° to 39°, with lower angles being the most common. The angular variation of the $S_{1B}$ surfaces is due to their curvature, their refraction into more silty layers and by some fluctuation of the individual micas in both the $S_{1B}$ and $S_{1G}$ trends. The specimen illustrated was taken from a tight fold (dihedral angle of 22°) where higher than average $S_{1G}/S_{1B}$ angles occur (BOULTER 1979). Two outcrops show a pattern similar to the conjugate cleavages with a bisecting mica-band cleavage as illustrated by POWELL (Plates 106 and 107) from the Hervey Group, N.S.W., Australia. NX

**B. Detail of mica-band cleavage.** This area is transitional into a mica-band ($S_{1B}$) and contains 20% to 30% quartz, some of which is labeled *Q*. Mica-bands in this specimen are between 0.5 mm and 1 mm thick and are spaced at 6 to 7 mm intervals. The large grains are intergrowths of white mica and chlorite which, away from $S_{1B}$ zones, form only 20% to 30% of the silty layers. The mica grains show much internal kinking on two scales; one affecting the chlorite/white mica multilayers and the other a finer kinking of single white mica layer. Mica rotated by kinking into the rock-cleavage direction is labeled *K*. Other mica grains in the same orientation may owe their position to kinking but their orientation mechanism is obscure. Grain-boundary readjustments have taken place and simple kink-band boundaries are not common. Many mica grains parallel to the rock cleavage are found within large mica intergrowths with apparently no relation to any kinks: their origin is enigmatic. NX

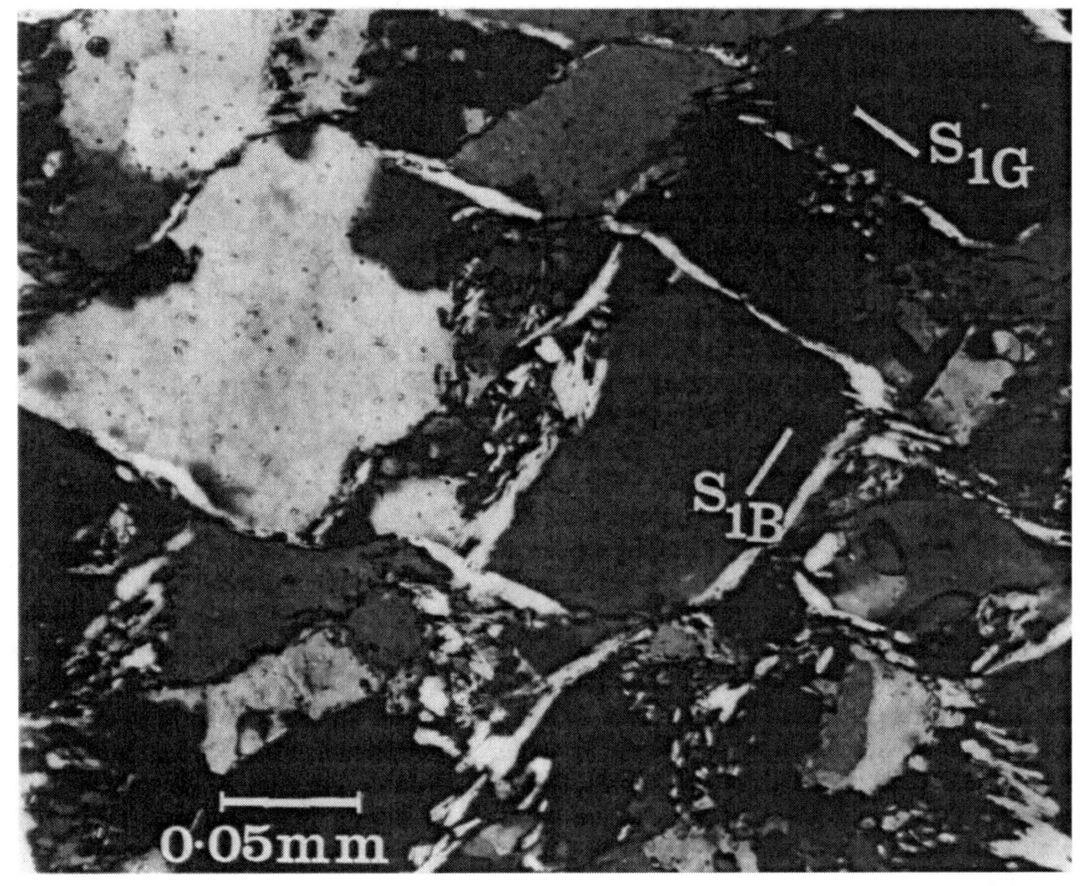


216 A

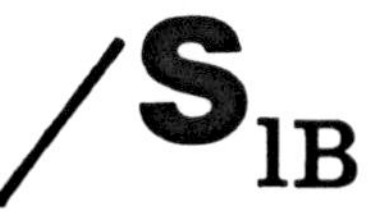


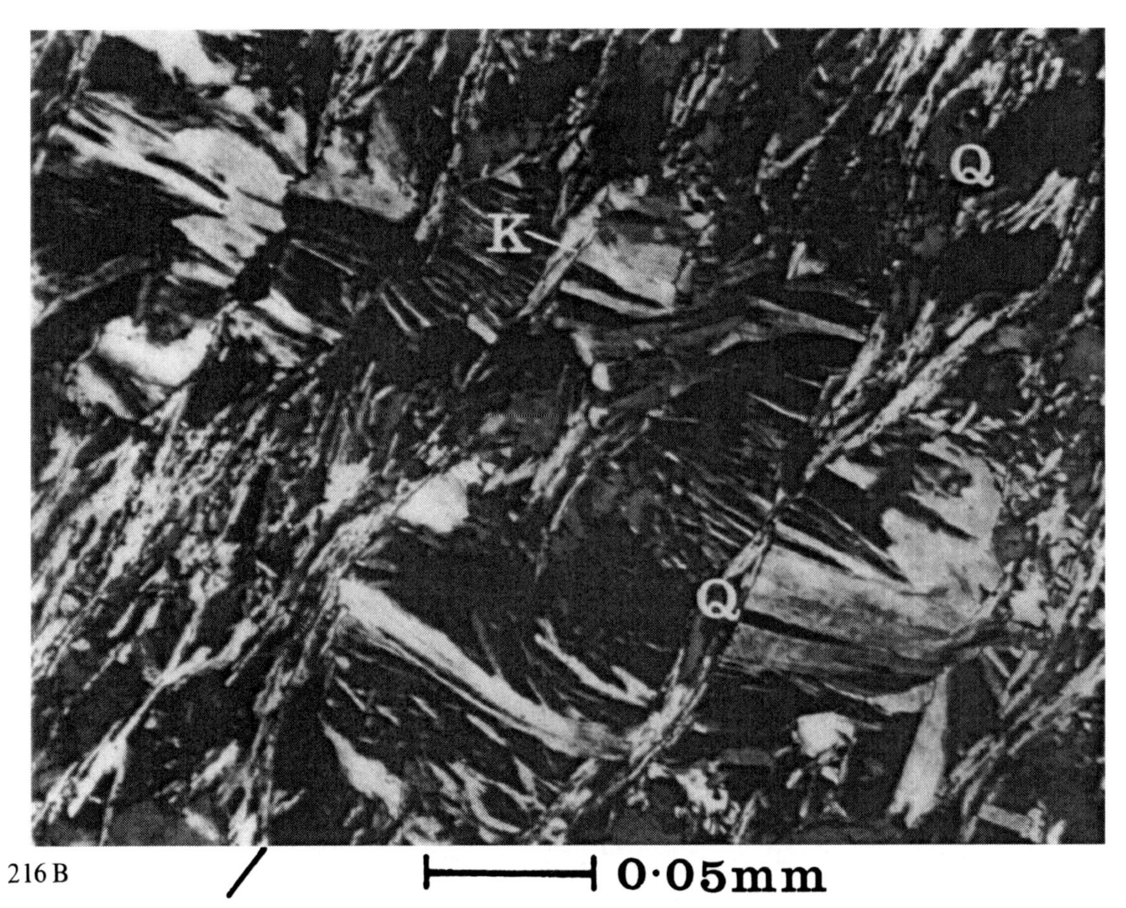


216 B

0·05mm

## 217. Two Inclined Cleavages Formed in One Folding Event (III)

C. A. BOULTER and K. C. HUGHES

**A. Detail of a mica-band** (Same specimen as photograph **B** on the previous plate). Within the mica-bands ($S_{IB}$) exposed at Bluff Knoll, the interlayered chlorite/white mica grains are markedly elongate and lozenge-shaped. Aspect ratios of these remnant detrital or diagenetic mica grains are usually greater than 4 to 1. White mica grains define the rock-cleavage direction and chlorite is virtually absent in this orientation. In the remnant grains the white mica is clear, whereas most of those in $S_{IB}$ are very pale green and pleochroic, probably indicating a phengitic composition. The increase in opaque grains in the $S_{IB}$ zones may reflect the release of iron in the transformation of *white mica + chlorite* to *phengite.*
Kinking of the pre-$S_{IB}$ mica grains is an important mica-orienting mechanism but the role of mass transfer as it affects the mica is difficult to assess. Metamorphic conditions are not well known, though the upper limit could not have been above lowest greenschist facies. Mass transfer of quartz plays a major role in the production of the $S_{IB}$ zones. The $S_{IG}$ direction in adjacent lithons is marked in **A**. Mica within $S_{IB}$ shows a secondary peak of dimensional orientation in this direction. It is not known whether this represents a fortuitous alignment of kinked mica grains or is the result of mica reorganisation during $S_{IG}$ formation. NX

**B. Overprinting of $S_{1B}$ by $S_{1G}$.** Mica grains in $S_{IB}$ show a persistent open crenulation with an axial surface subparallel to $S_{IG}$ mica grains in the lithons. In **B** a rare example of a lithon mica (*M*) is shown curving from the $S_{IG}$ trend to the $S_{IB}$ trend. Without the crenulation of the mica band it would be impossible to use the curved mica to determine the sequence of overprinting. Because of the crenulation, it is believed that $S_{IG}$ began to form when $S_{IB}$ was already well established. Because $S_{IB}$ intensifies towards fold closures, the mica bands must have been accentuated during fold flattening, though $S_{IG}$ appears to have outlasted $S_{IB}$ development.
Polars not quite crossed; $1/4\,\lambda$ plate used

217 A

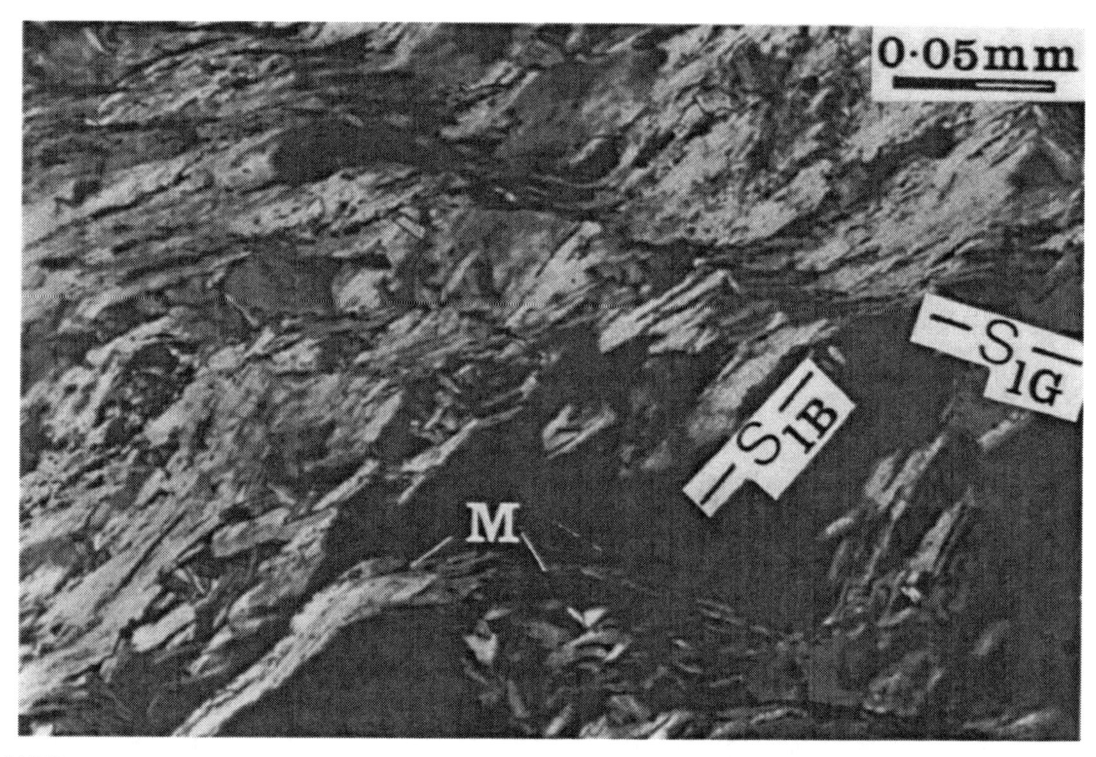


217 B

# 218. Cleavage Regeneration, Western Maine (I): Chlorite and Staurolite Zones

R. H. Moench

Three steps of structural and metamorphic superposition are shown in low- to middle-grade pelite exposed near Rangeley, Maine, U.S.A. Three additional steps in high-grade rocks are shown in Plate 219. Early slaty cleavage in **A**, the first recognized generation, is coarsened to schistosity and overprinted by staurolite in **B**, and is crossed by late crinkles in **C** in rocks that have been retrogressed to the chlorite zone. (In Plate 219 early schistosity is replaced by coarse-grained late schistosity and is preserved only as the poikiloblastic inclusions in staurolite.) The examples were selected from a thick succession of Lower and Middle Paleozoic rocks that were regionally deformed and weakly metamorphosed early in the Devonian, and deformed and metamorphosed again during Devonian plutonism. (See Moench 1970, Moench and Zartman 1976.) All photographs are at the same scale to illustrate changes in coarseness.

---

**A.** Typical regional slaty cleavage in chlorite zone, first metamorphism (M-1). Axial planar to major northeast-trending folds of early deformation. Photograph is of dark, somewhat graphitic pyrrhotite-bearing, chlorite-sericite slate of Upper Ordovician (?) Quimby Formation. PPL

**B.** Early schistosity overprinted by staurolite porphyroblast of second metamorphism (M-2), which was static at this locality. Isograds of M-2 are related to granodiorite in a nearby pluton. Photograph is of garnet-staurolite, two-mica schist of Silurian Rangeley Formation. PPL

**C.** Early schistosity and late crinkles in chlorite zone of third metamorphism (M-3), downgraded from M-2 staurolite zone. Interpreted to represent the most distant expression of metamorphism associated with the granite of the Mooselookmeguntic batholith. Photograph is of chlorite-muscovite schist of the Rangeley Formation. Coarse-grained M-3 chlorite and minor muscovite have replaced M-2 staurolite, and euhedral porphyroblasts of M-3 chlorite overprint early schistosity and have grown subparallel to the late crinkles. Nicols not completely crossed

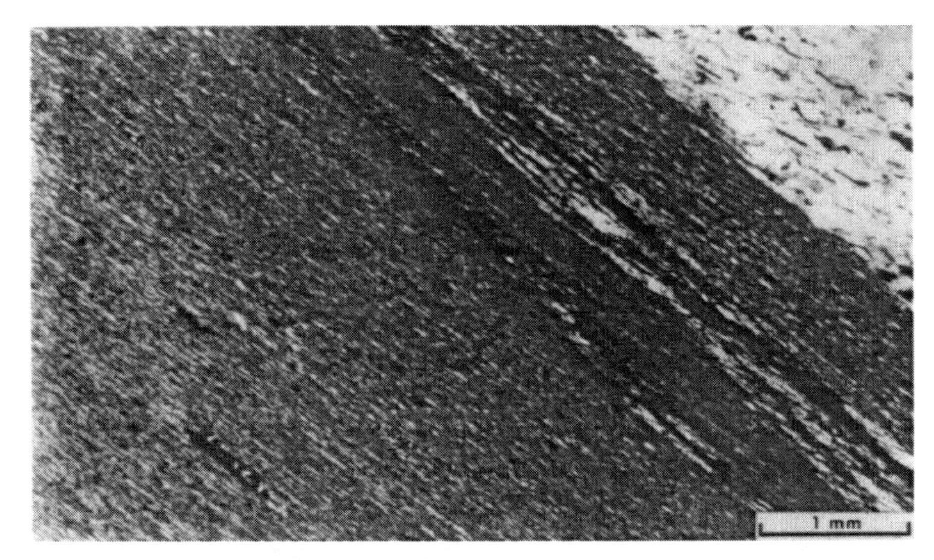


218 A

218 B

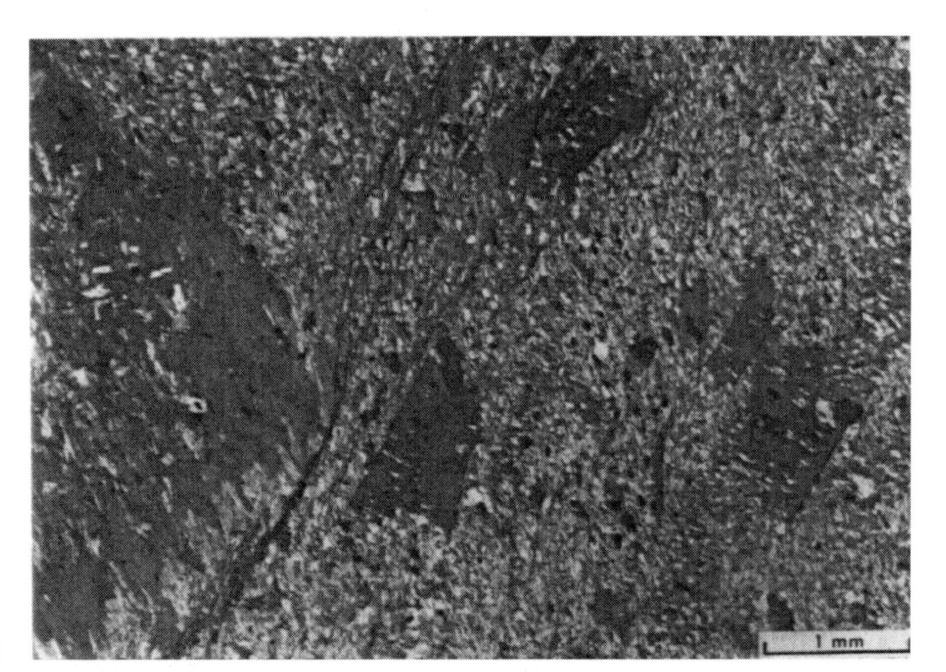


218 C

# 219. Cleavage Regeneration, Western Maine (II): Lower Sillimanite Zone

R. H. MOENCH

Three steps are shown of structural and metamorphic superposition in the lower sillimanite zone (GUIDOTTI 1974), which commonly contains staurolite, but not chlorite. This metamorphism was the third event (M-3) of MOENCH and ZARTMAN (1976). In the M-3 sillimanite zone, early cleavage is almost entirely replaced by coarse-grained late schistosity and is preserved only as the poikiloblastic texture in the staurolite porphyroblasts. The late schistosity is axial-planar to recumbent folds that are superposed on the regional early folds and cleavage. It is also subparallel to the gently dipping contact of a tonalite/granite batholith. The batholith is interpreted as a subhorizontal sheet emplaced at a depth of about 12 km (MOENCH and ZARTMAN 1976). The M-3 metamorphism is directly related to the granitic sheet with decreasing grade of M-3 eastward from the eastern contact of the sheet; late schistosity passes gradationally to crenulation cleavage (previous plate), much as described by WHITE (1949) in Vermont.

---

**A.** Typical late schistosity in sillimanitic, garnet-staurolite, two-mica schist of the Silurian Rangeley Formation; sillimanite is sparse and fibrolitic. Late schistosity is expressed by muscovite-rich laminations about 0.1 mm thick, alternating with thicker laminations of quartz, plagioclase, and randomly oriented biotite. This interdigitation is overprinted by staurolite porphyroblasts that grew, without evident rotation, over crenulated early cleavage. The early cleavage in the porphyroblasts is preserved as crenulated trains of quartz and scattered platelets of ilmenite. PPL

**B.** Similar to **A**, except that banding is much thicker, and that the poikiloblastic texture of the porphyroblasts is 40° or so out of alignment with the late schistosity of the matrix. The staurolite evidently grew over early cleavage that was deformed by late crenulation cleavage, and then was rotated as the crenulation cleavage of the matrix was transformed to schistosity. This rotation took place on the limbs of a small recumbent fold (MOENCH and ZARTMAN 1976, fig. 6). PPL

**C.** Coarsened late schistosity in sillimanite-garnet, two-mica schist of the Rangeley Formation; sillimanite is sparse and fibrolitic. Staurolite and andalusite of M-3 are present nearby, but they are reconstituted and do not preserve crenulated early cleavage. All evidence of the early cleavage is thus obliterated in these and higher grade rocks of the region. PPL

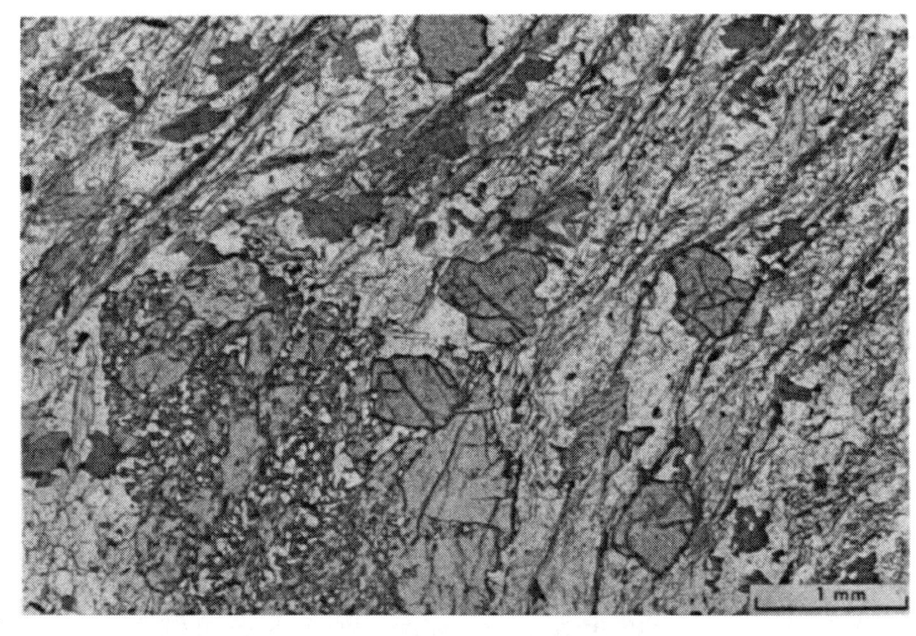

219 A

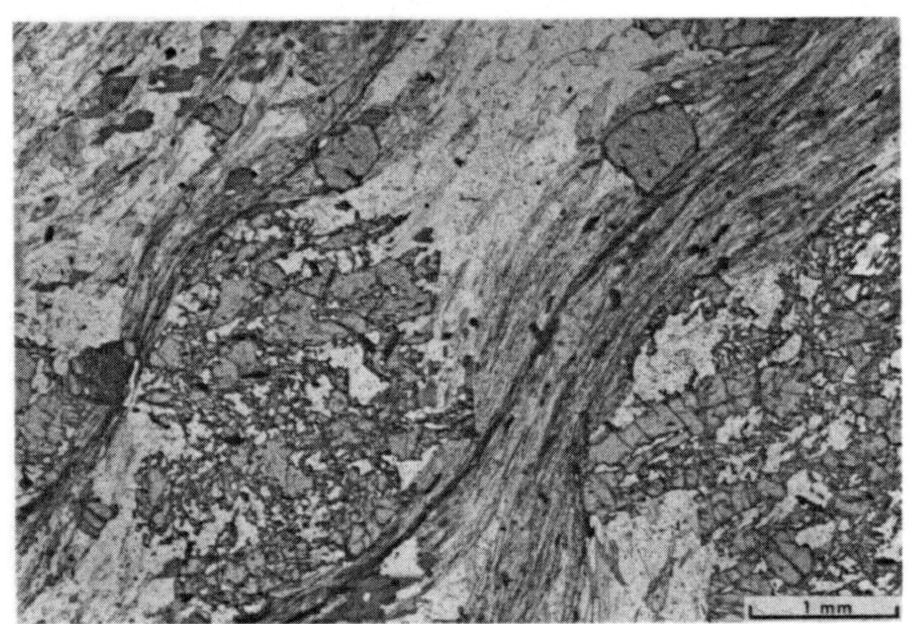

219 B

219 C

# 220. Development of Long-limb Schistosity

R. A. GLEN

$F_4$ Crenulations of $S_2$. Andalusite-grade mica schists, Precambrian Willyama Complex, Broken Hill, N.S.W., Australia. *Dashed line* $S_0$ indicates bedding and separates upper psammite from lower psammopelite. $F_4$ crenulations are markedly asymmetrical (areas *A, B, C, D*), with a well developed vertical long limb defined by rotated $S_2$. The alignment of $F_4$ long limbs forms a new penetrative surface ($S_4'$), best seen in the lower part of the photograph. $S_4'$ (= rotated $S_2$) lies oblique to axial surfaces of $F_4$ crenulations ($S_4$) which are inclined from *upper right* to *lower left* (areas *A, B, D*).

A thin-section sketch from the psammopelite shows that $S_2$ is defined by the alternation of *M* (mica) and *QM* (quartz & mica) domains. In the *sketch, A* indicates andalusite and *S* indicates a sericite aggregate. (The *scale bar* in the sketch is 1 mm.) Bent micas are restricted to hinge zones only. Note coalescence of several short limbs into a single long limb. This is accompanied by the formation of axial-surface discontinuities and "pressure" solution.

Scale (sketch): 1 mm
Scale (photograph): matchstick 4.2 cm

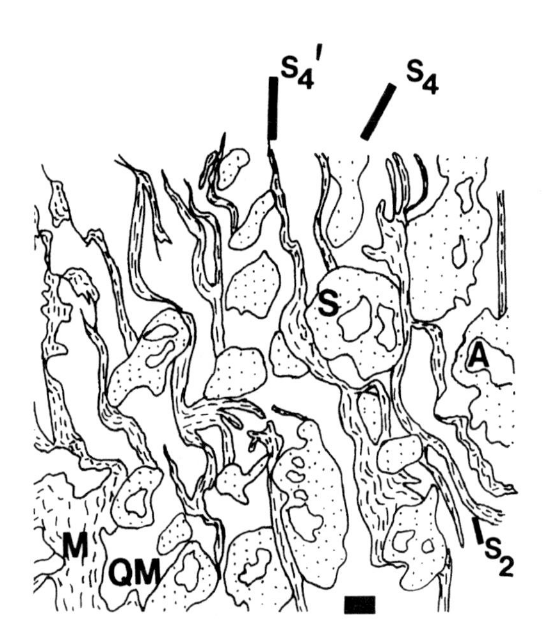

220

# Other Topics

## Section 18: Cleavage Refraction and Cleavage-fold Relationships

(Plates 221 to 234)

This final group of fourteen contributions indicates some of the special geometrical relationships which can exist between cleavage and rock layers. Two topics are treated: refraction of cleavage across layering (Plates 221 to 223) and the relationship of cleavage to folds. The latter covers fanning cleavage, axial-planar cleavage (Plates 224 to 229) and nonaxial-planar (transecting) cleavage (Plates 230 to 234).

---

**Plates**

## 221. Refraction of Cleavage

D. PUPPOLO and B. M. BAYLY

Ordovician Moretown formation at West Cummington, Massachusetts, U.S.A., roadcut on Route 9 beside Westfield River (PUPPOLO 1979, HATCH 1975).

The layer across the center of the photograph is about 1 cm thick and is rich in quartz; it is set between thick units of more micaceous rock whose character is better shown in Plates 42 and 43 (as seen here in the *upper left* and *lower right corners*). The micaceous rock has shortened by crenulation and dissolution; total shortening is at least 20% parallel to the layering. The plate allows us to see how a more quartz-rich lithology responded to the same shortening and rotation.

Some features are shown in the simplified diagram:

i. sharply defined upper boundary, crenulated

ii. less well defined lower boundary, with more open crenulation

iii. most quartz-rich central zone (mica content 10%)

iv, v. transition zones (mica content 19%–21%)

vi. angle of cleavage to bedding changes with quartz/mica ratio: more mica permits more shear parallel to the layering as these layers rotate in the limb of a larger fold.

vii. the cleavage surfaces are closely spaced and well defined in the mica-rich rock. As the mica percentage diminishes, the surfaces become diffuse zones. Average spacing in the quartz-rich material is greater – not all cleavage surfaces in the mica-rich material lead into dissolution zones in the quartz-rich material, some just die out. (The natural periodicity for quartz dissolution seems not to match the natural periodicity for mica crenulation – there is no reason to expect that it would.)

Comparison with the following plate reveals many similarities yet the conditions of hydration were rather different. The rock shown opposite had been folded into recumbent folds with limbs in tens of kilometers and metamorphosed to garnet grade long before the crenulation process began. Much of the chlorite is retrograde: that is, the rock was absorbing water by mineral reaction rather than releasing it. But the geometry of the dissolution surfaces, and the contrast between flake-rich and flake-poor material seem similar despite the rather different metamorphic states and histories. NX

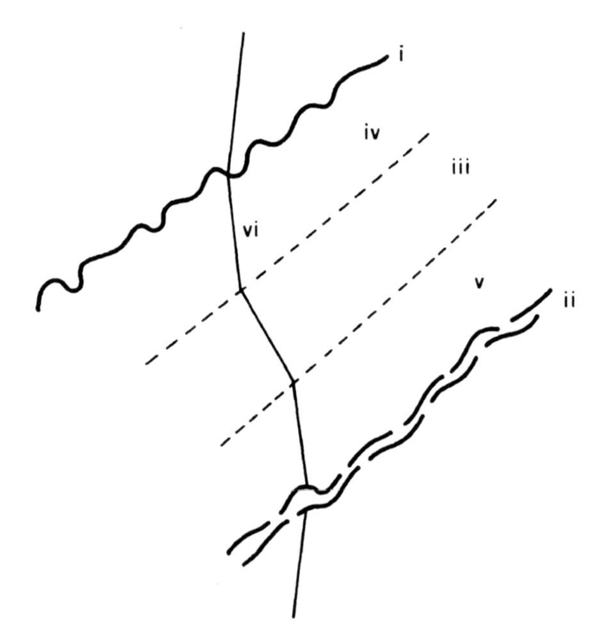

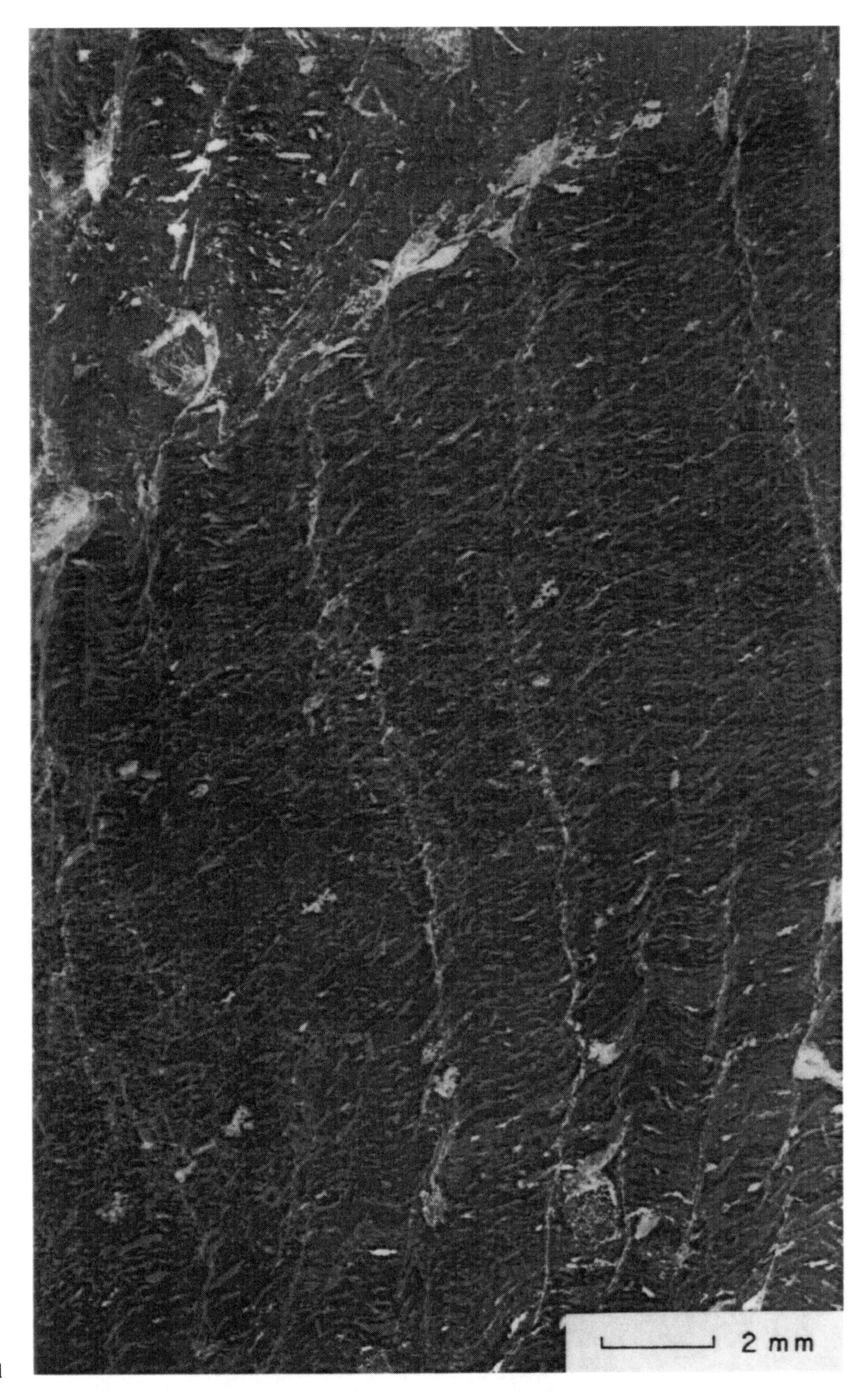


221

# 222. Solution Cleavage in Turbidites of Central Maine, U.S.A.

H. R. Burger

**A. Solution cleavage and suggested bedding transposition.** Silurian Kenduskeag Formation exposed on Route 169 near I-95 Etna Interchange (Griffin and Lindsley-Griffin 1974). Greenschist facies. This formation is characterized by turbidite features, chaotic zones in which bedding is totally disrupted, and extreme variation in bedding thickness and fold style. Rock types vary from quartzites to zones of thin, alternating metasiltstones and phyllite. Sedimentary and slump structures clearly visible at this locality include graded beds, cross-laminations, slump folds, sedimentary breccia, balls and pebbles, and flame structures. Weathering of the outcrop surface enhances contrast between lighter-colored silt layers and darker pelitic material. Bedding trends *from left to right* across photograph; cleavage trends *from top to bottom*. Dark pelitic "seams" disrupt the silt layers and define the dominant cleavage. At many places on the outcrop surface apparent offset of silt layers by cleavage is best explained by removal of material from zones now containing a strong cleavage. Disrupted layers cannot be matched, even roughly, by removing the apparent offset across the cleavage which should be possible if the offset is due to slip along cleavage planes. In addition, the cleavage zones contain pelitic material in much greater abundance than elsewhere in the rock, suggesting a "stylolite-type" accumulation of relative insolubles.

Scale: cm

**B. Nature of the cleavage.** Negative print of thin section taken from sample collected from a homogeneous siltstone layer a few centimeters thick at the outcrop pictured in **A**. Bedding trends *from left to right.* Note that the bedding details are still preserved. Cleavage is concentrated in narrow sinuous zones separated by domains marked by a virtual absence of cleavage although careful observation reveals that a few thin, discontinuous cleavage folia are present.

**C. Cleavage-bedding relations.** Negative print of thin section from same area showing much thinner siltstone layers interbedded with phyllite. Note the preservation of bedding laminations and structure. Bedding trends *from upper left to lower right*; cleavage trends *from top to bottom.* In light colored areas pelitic material is in abundance and cleavage is closely spaced and barely discernible in the photograph. In dark-colored, silty layers cleavage is more widely spaced and is marked by light colored, pelitic-rich folia disrupting the silt layers. This pattern of cleavage spacing and variations in relative disruption of closely spaced layers (as at *Q*) is more satisfactorily explained by solution than by slip on the cleavage.

**D. Cleavage-bedding relations.** Enlarged portion of area in **C**

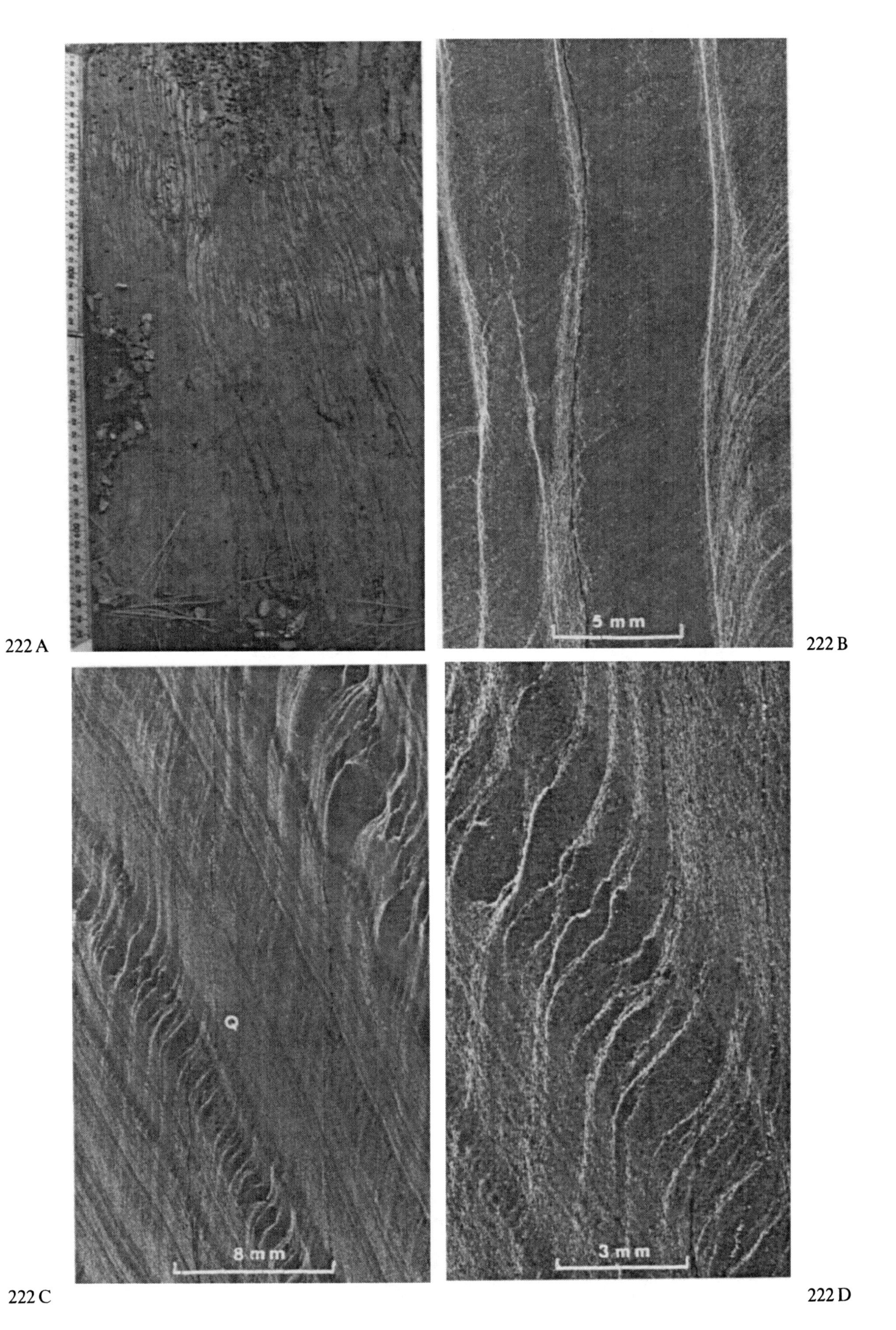

222 A

222 B

222 C

222 D

# 223. Refracted Continuous Cleavage

G. J. BORRADAILE

Devonian slate, from the Zeller Stadtwald, Zell, Mosel Valley, West Germany.

Alternating slate (*S*) and silty layers (*SP*) exhibit refraction of cleavage from a planar continuous one in slate to a sigmoidal, less continuous cleavage in the siltier layers. The sketch is of the lower part of the photograph.

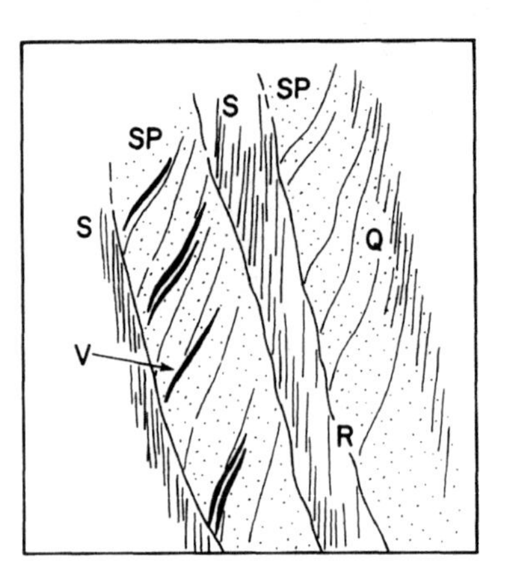


In many locations the beds appear so similar that they can only be identified in a cursory examination because of the cleavage refraction which they cause. The refraction is sharp where there is a strong lithological/grain-size contrast (*R*) and more gentle where the transition from silt to slate is more gradual (*Q*).

Quartz infilled veins (*V*), which have been recrystallized, show a sigmoidal form consistent with dilatation caused by dextral shearing of the beds. A heterogenous shear component of the same sense could have contributed to the sigmoidal form of the refracting cleavage in the silty beds. The quartz veins are slightly oblique to the cleavage in several instances.

Scale: ball-pen

223

## 224. Axial-Planar Cleavage: Soft-Sediment Deformation

B. C. Burchfiel

Leonardian (Permian) Bone Springs Formation. Location about 21 km SSE of El Capitan in the Guadalupe Mountains, West Texas, U.S.A.

Penecontemporaneous soft-sediment folds with associated axial-surface cleavage occur in several places in the Bone Springs Formation (Newell et al. 1972). The examples figured are from near the north-eastern edge of the Permian Delaware Basin. This oval sedimentary basin (220 km × 120 km) was bounded to the northwest by the reefal complexes of the Guadalupe Mts, and evidently formed by gradual down-warping on marginal flexures (Newell et al. 1972, figs 4 and 5). The basin fill is mainly fine-grained gray, limy sandstone and siltstone, with minor amounts of black, bituminous limestone deposited in comparatively deep and stagnant waters. Regional dips are subhorizontal, and the zones of soft-sediment folds are enclosed by non-deformed horizontal strata.

The folds are generally recumbent with great diversity of profile shape. Box folds, meter-scale open concentric folds, centimeter-scale harmonic similar folds, and small crenulations and kink folds all occur within a few tens of meters. Slide surfaces truncating both upper and lower limbs of folds are common. Fold-hinge orientations are variable, but in the illustrated outcrops axial surfaces are either recumbent or northwestward-dipping with an asymmetry indicating slumping toward the southeast away from the margin of the Delaware Basin.

An axial-surface cleavage, continuous at outcrop scale, is developed in some folds, but is not present in others with equally tight fold profiles. In thin section, the cleavage appears as a parallel alignment of elongate detrital fragments (some of them probably sponge spicules) and local planar concentrations of dark materials, presumably formed by incipient pressure solution.

A

B

# 225. Similarity Between Process of Formation of Axial Planar Cleavage and Fanning Cleavage

N. B. Lebedeva

Comparison of experimental results with natural cleavage gives the probable explanation of the phenomenon of refraction of cleavage. Initially, cleavage is a result of longitudinal flattening of rocks as a whole and especially of the most deformable elements. Subsequently the cleavage develops under conditions of interaction of two processes:

1. continuing longitudinal compression and
2. simultaneous bending of strata.

Longitudinal compression tends to keep the cleavage perpendicular to the axis of compression. Due to the bending of strata, cleavage tends to rotate, creating a fan of surfaces converging downwards beneath antiformal folds. The interaction of the two processes, in layers with different properties, leads to different results. The first process prevails in strata with greater deformability, the second one dominates in more rigid layers.

These samples were made up with three horizontal layers formed by grains of a mixture of resin with engine oil. The grains had the following viscosities: in the lower layer of the order of $10^5$ poises; in the middle layer about $10^7$poises; in the upper layer about $10^6$ poises. The samples were compressed horizontally. The samples **A** and **B** show different stages of deformation. (Sketches from photographs).

---

**A.** The sample was compressed by 25%. It underwent a small shortening and a small increase of thickness while the horizontal position of the strata was preserved. A few grains and bubbles of air flattened and stretched by 3 to 5 times, i.e., much more than the sample as a whole. They created initial oriented structure of "cleavage" especially in layers with low viscosity.

**B.** The sample was compressed by 35%. The layer with high viscosity was deformed mostly by bending (*middle layer*). Together with the rotation of layers, the grains, which formerly were flat, also underwent some rotation. They formed a fan-shaped structure. After flattening was followed by bending, the subsequent development of "cleavage" was interrupted and curved layers with "cleavage" remained. At the same time the lower and upper layers continued to be deformed mostly by horizontal flattening. The length of grains increased, some of them amalgamated, forming long chains oriented perpendicular to the axis of compression and parallel to the axial surfaces of folds.

These experiments show the genetic relationship between "cleavage" parallel to the axial surfaces of folds and cleavage fans; their different positions and intensities are due to the nature of a medium with layers of different mechanical properties

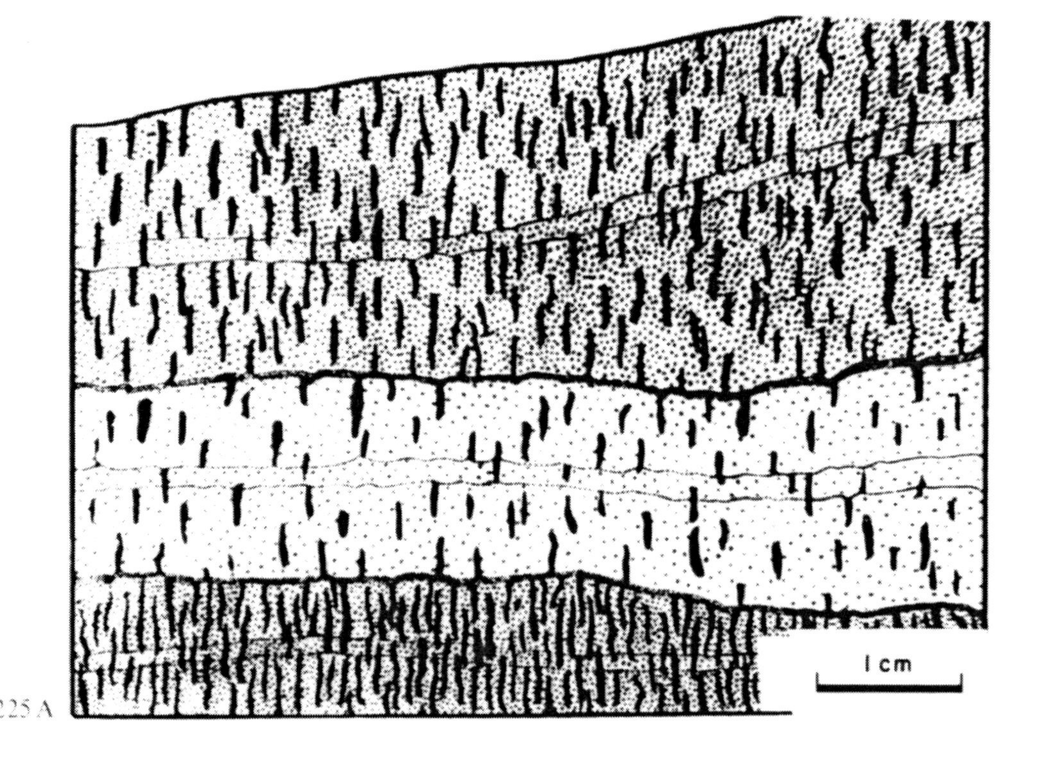


225 A

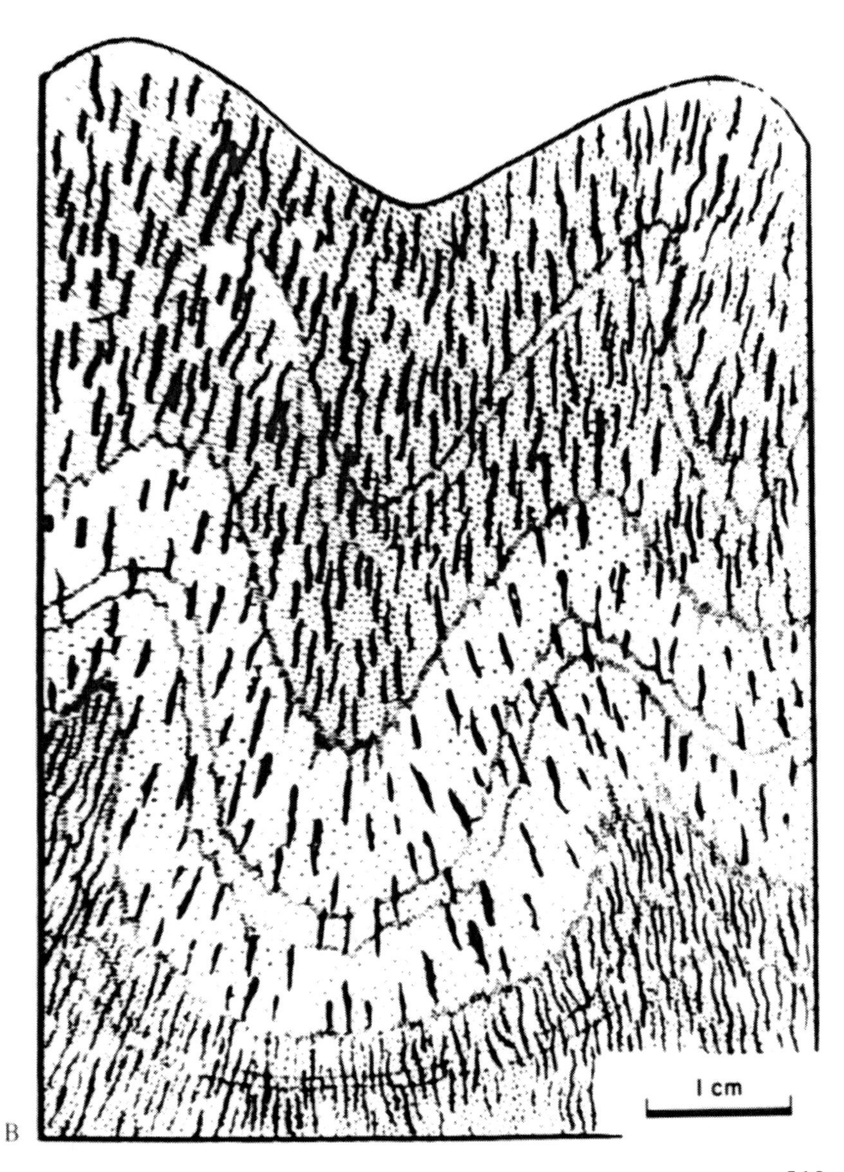


225 B

## 226. Cleavage/Bedding Relations and Structural Facing (I)

G. J. Borradaile

From the Lower Paleozoic rocks of the Malaguide Complex of the Betic Cordilleras, near Vélez Rubio, Spain (Borradaile 1976a). See also following plate.

The synform illustrated in the photograph and sketch has an axial planar cleavage *S*.

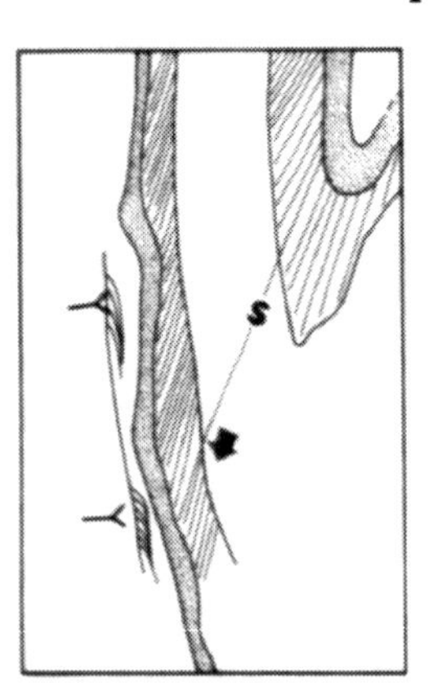


Ripple-cross lamination at two places in the siltstones youngs to the left (located just *right of the symbols* Y in the photograph). This indicates that the synform is in fact an anticline (older beds in the core of the fold) and that the fold has downwards *structural facing* (in the sense of Shackleton 1958, Borradaile 1976a). When projected onto the axial planar cleavage the younging component is downwards (*heavy arrow* in the sketch). This is an alternative demonstration of the downwards structural facing. The folds may be downward-facing for a variety of reasons. The beds may have been overturned before folding; the folds may have been overturned during or after formation. The first alternative is preferred in this case because of regional geological considerations.

226

# 227. Cleavage/Bedding Relations and Structural Facing (II)

G. J. Borradaile

From the same area as the preceding plate.
In this outcrop the folds are not visible but assuming the cleavage is axial planar, as elsewhere in the area, it is possible to determine the sense of structural facing. The bedding dips steeply to the left ($S_0$). Cleavage dips steeply to the right (S). Truncated channels in the sandstone on the left side of the photograph young to the right (shown by the *direction of the* **Y** *symbol*). The younging direction, projected onto the cleavage, has a downward component so that the structural facing is downwards.
The antiform to the right of this outcrop would be a syncline and the synform to the left of this outcrop would be an anticline.

227

## 228. Arcuate Hinge-cleavage

D. Roberts

Kongsfjord Formation, Barents Sea Group, probable Upper Riphean age; Risfjord, Varanger Peninsula, Finnmark, North Norway (Roberts 1971, Roberts and Strömgård 1971).

This arcuate development of slaty cleavage is confined to the inner arcs of mesoscopic fold hinge zones in pelite layers in a multilayered greywacke-sandstone/mudstone sequence. The cleavage is the earliest fabric recognized in the area, and is generally steeply disposed and axial planar to local, open to close, $D_1$ folds of variable wavelength. The cleavage frequently shows text-book examples of refraction (see, e.g., Roberts 1972), with divergent fanning in the pelites. Metamorphic grade is of lowest greenschist facies. In the very fine-grained pelites the cleavage is seen as a pervasive, apparently non-domainal fabric defined by the preferred orientation of phyllosilicates; occasional quartz grains are dimensionally oriented within the cleavage. With increasing grain size tightly packed, discrete, cleavage surfaces make an appearance and the fabric is visibly domainal with spacing $\leqq 1$ mm.

In outcrop, the arcuate cleavage patterns are seen on variable arc-length scales from the cm-scale up to 2 m. The distance from the pelite/psammite interface to the finite neutral point is also variable. The concentriform cleavage is rarely perfectly parallel to the folded primary bedding surfaces. Thus it produces a curvilinear intersection lineation around the fold hinge on the upper surfaces of psammite beds in anticlinal hinge zones (Roberts 1971). Where the cleavage is normal and divergent the cleavage/bedding intersection lineation is parallel or subparallel to the fold axis.

The bulk strain for the multilayered lithology has not been determined at the arcuate cleavage localities, but in a nearby area of identical lithology and structure deformed carbonate concretions in psammite layers have permitted strain calculations to be made (Roberts 1972, figs. 11, 12) from two-dimensional measurements in different planes. Shortening normal to the $D_1$ cleavage (in Z) is here in the region of 35%–40% with a principal elongation (X) within the cleavage of ca. 70%, and a decrease of ca. 5% in Y.

It has been argued that good examples of this unusual cleavage pattern are more likely to develop in a low-grade, parallel-folded, multilayered sequence with high viscosity contrasts between layers and with a minimal amount of layer-parallel shortening prior to actual buckling. These assumptions have been substantiated in experimental photoelastic strain analysis on gelatine-rubber models (Roberts and Strömgård 1972). Detailed finite element studies around fold hinge zones also show that the greater the viscosity contrast between layers the more likely the chance that this inner arc concentriform pattern will develop (Stephansson and Roberts unpublished data).

228

## 229. Differentiated Layering

P. F. WILLIAMS

Paleozoic turbidites near Bermagui, N.S.W., Australia. The grade of metamorphism is lower greenschist facies, below the biotite isograd.

Upward-facing, second-generation folds and portions of three graded beds are visible. There are two generations of differentiated layering; the most obvious is the second generation axial plane foliation which is mostly inclined to bedding by 60° to 90°. A first-generation axial plane foliation can also be seen. It is most obvious in the *top left corner* of the photograph where it is inclined to bedding by 30°. The two generations of differentiated layering are very similar in appearance both in the field and under the microscope. Both are defined by alternation of quartz-rich and layer silicate rich domains, quartz and layer silicates being the major constituents of the rock (WILLIAMS 1972). Numerous quartz veins are possible sinks for quartz removed from layer silicate rich layers during differentiation.

Scale: rule is 12 in (30 cm)

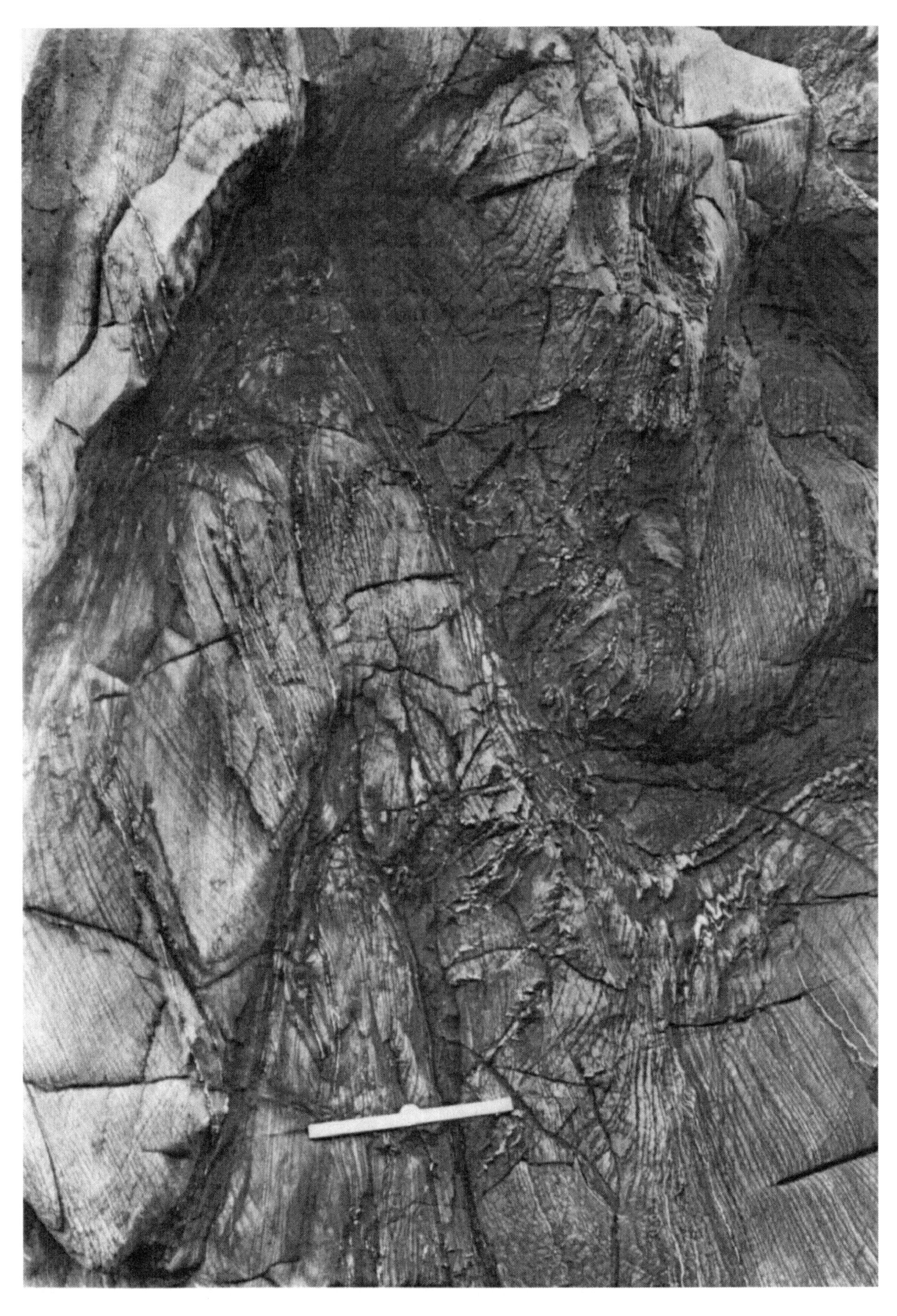

29

## 230. Transected First Folds and Transecting Cleavage

G. J. BORRADAILE

Silurian Limestones, Petit Rocher, New Brunswick, Canada.
In certain regions such as here in the northern Appalachians, folds of the first tectonic episode show the development of first cleavage which is not parallel to the axial surface of the folds (BORRADAILE 1978). Occasionally the same cleavage may vary in its degree of transection, and even become parallel to the folds' axial surfaces when traced to adjacent outcrops. The transection may be caused by lack of coupling between particulate flow and grain deformation by pressure solution in a noncoaxial strain history (BORRADAILE 1981).

---

**A.** A minor, angular, first fold is viewed in a direction nearly parallel to the fold hinge. Spaced cleavage is approximately parallel to the metre rule and transects the fold, cutting both limbs in the same sense. Scale: 1 m

**B.** Photograph of section perpendicular to spaced cleavage near **A** above. The spaced cleavage planes are stylolitic and truncate fossil fragments. The cleavage may be either synchronous with the formation of the folds or it may have been initiated somewhat later in the fold-forming episode (BORRADAILE 1978). In detail the stylolitic components of the cleavage anastomose and sometimes cut one another. This suggests that different stylolites were activated at different times. Nevertheless, distinct relative ages of stylolites have not been identified and they are considered to represent a single "episode" of cleavage. PPL

---

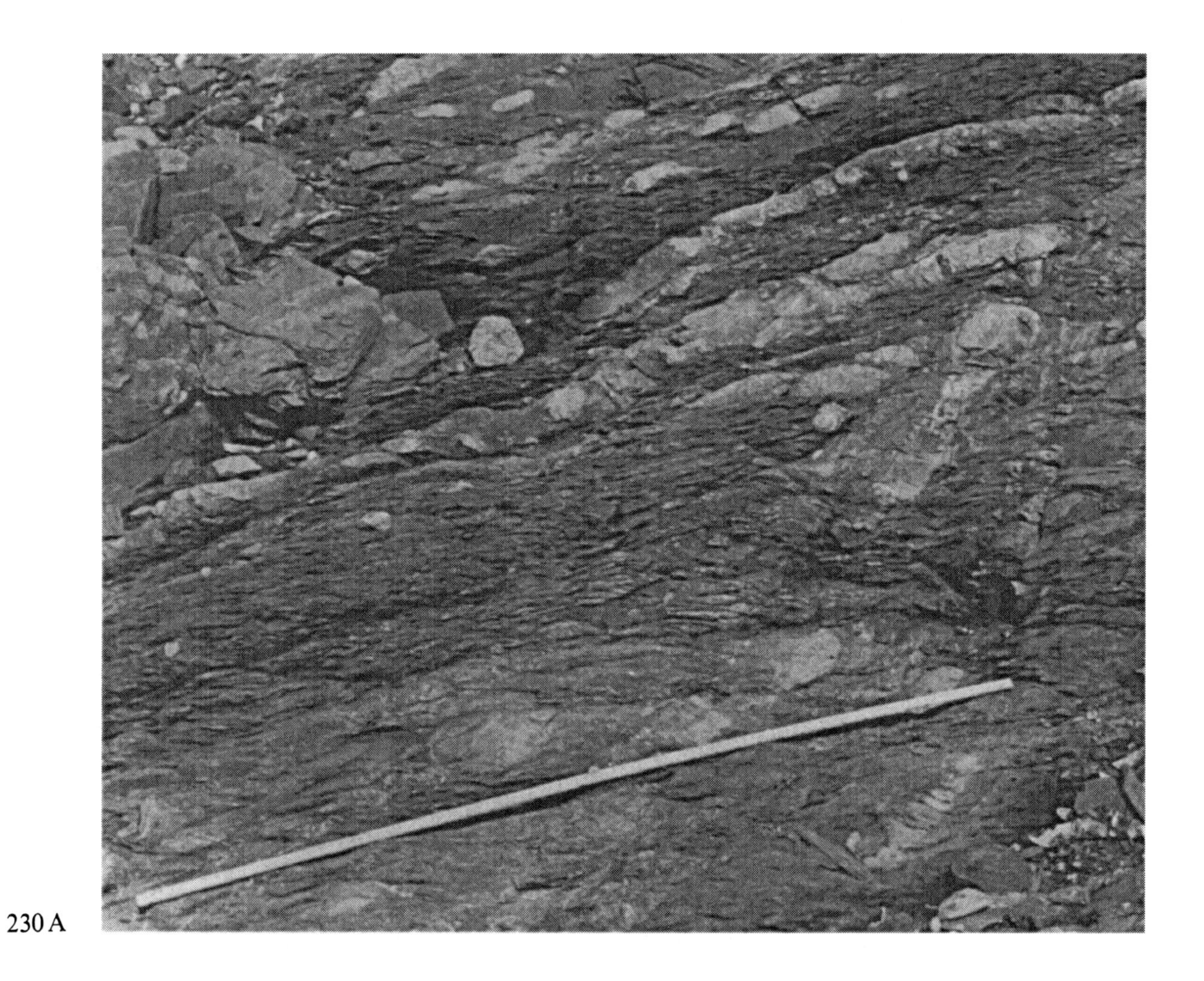

230 A

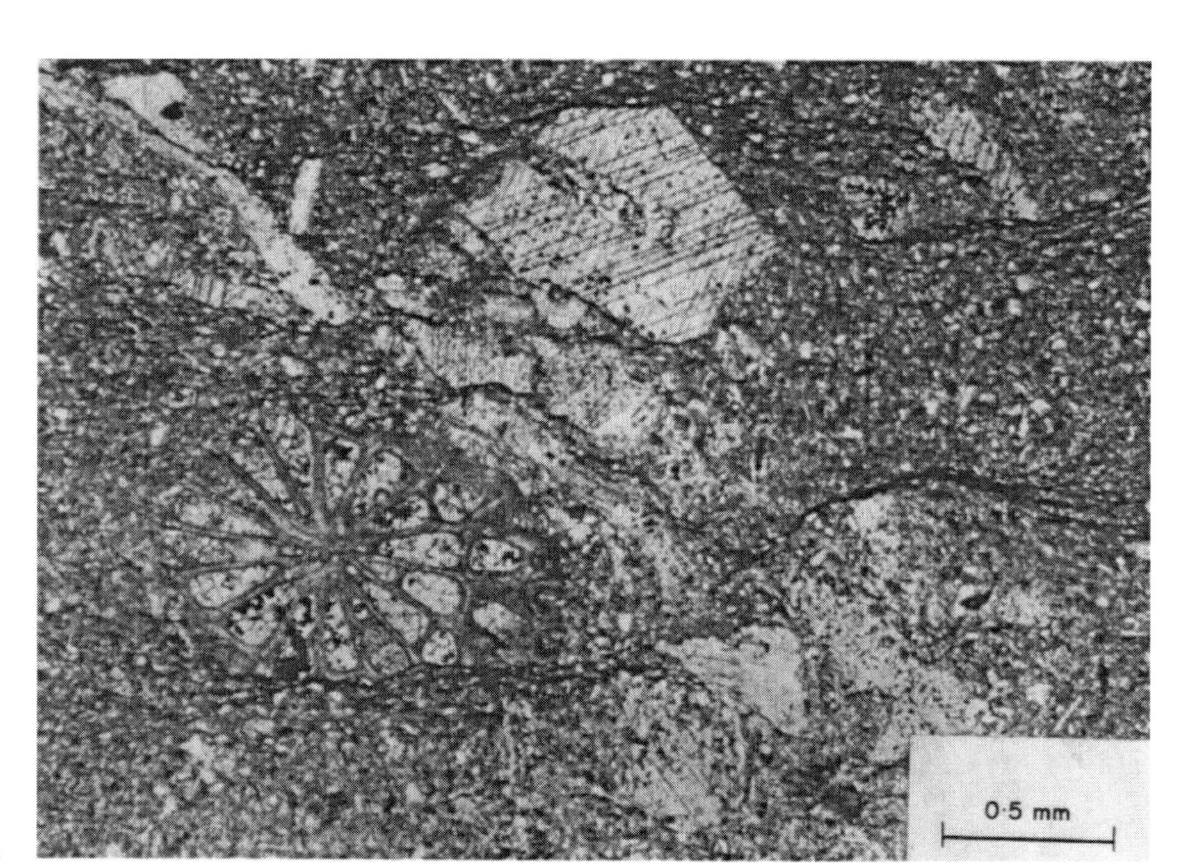


230 B

## 231. Cleavage and Transected Folds in Calcareous Mudrock (I)

D. R. GRAY and R. I. SIMON

The folds and cleavage occur in the Ordovician Moccasin Formation, at Newport, Virginia, U.S.A. A thick bedded, maroon calcareous mudrock has thin limestone layers and sandstone interbeds in the upper portion. Conodont color alteration (cf. EPSTEIN et al. 1976) suggests temperatures up to 180°–230° C for the deformation (TILLMAN 1980, pers. comm.), whereas a general lack of twinning in calcite indicates stresses below the calcite twinning yield stress for cleavage development. The folds have rounded hinges, sinusoidal form and are upright to slightly inclined with subhorizontal axes. Wavelengths are in the order of 2–5 m. Interlimb angles range from 50–90°. Fold geometry is alternating class 1B/1C and class 3. Cleavage ranges from a weak "penetrative" slaty type (**D**) to a spaced variety with smooth to stylolitic morphology (**E**). It shows noncoplanar relationships with fold axial surfaces, refraction across layers, fanning with variable cleavage-fan angles, and noncoaxial bedding-cleavage intersections. Fabric and strain data (GRAY 1981) suggest early initiation of cleavage with respect to folding during a noncoaxial deformation history.

---

**Outcrop photographs A, B, and C** are from a road outcrop, Route 601 north of Newport, Virginia. Lens cap diameter is 6 cm.

A. **Transected anticline** in sandstone (outer layers) and calcareous mudrock (inner layers).

**B. Fold (oblique view) in calcareous mudrock.** (Inner layer of fold in **A**.) Cleavage is noncoplanar with the fold axial surface.

**C. Hingeline of fold** in mudrock. (Fold in **B**) Anastomosing, irregular, spaced cleavage traces (statistically *parallel to dark-colored pen*) transecting the fold hingeline (*parallel to lighter-colored pen*).
Scale: Pen is 15 cm

**D. Primary tectonite fabric.** A domainal fabric with thin cleavage lamellae (2–10 $\mu$) alternating with thicker zones (20–30 $\mu$) where clays have almost random orientation. Clays defining the lamellae have close packing, and moderate to strong dimensional preferred orientation parallel to the lamellae traces. The traces are undulating and anastomosing but are generally subparallel. S.E.M.

**E. Spaced disjunctive cleavage.** Cleavages are thick (15–40 $\mu$), planar zones of densely packed clays. These show much tighter packing and degree of preferred orientation. S.E.M.

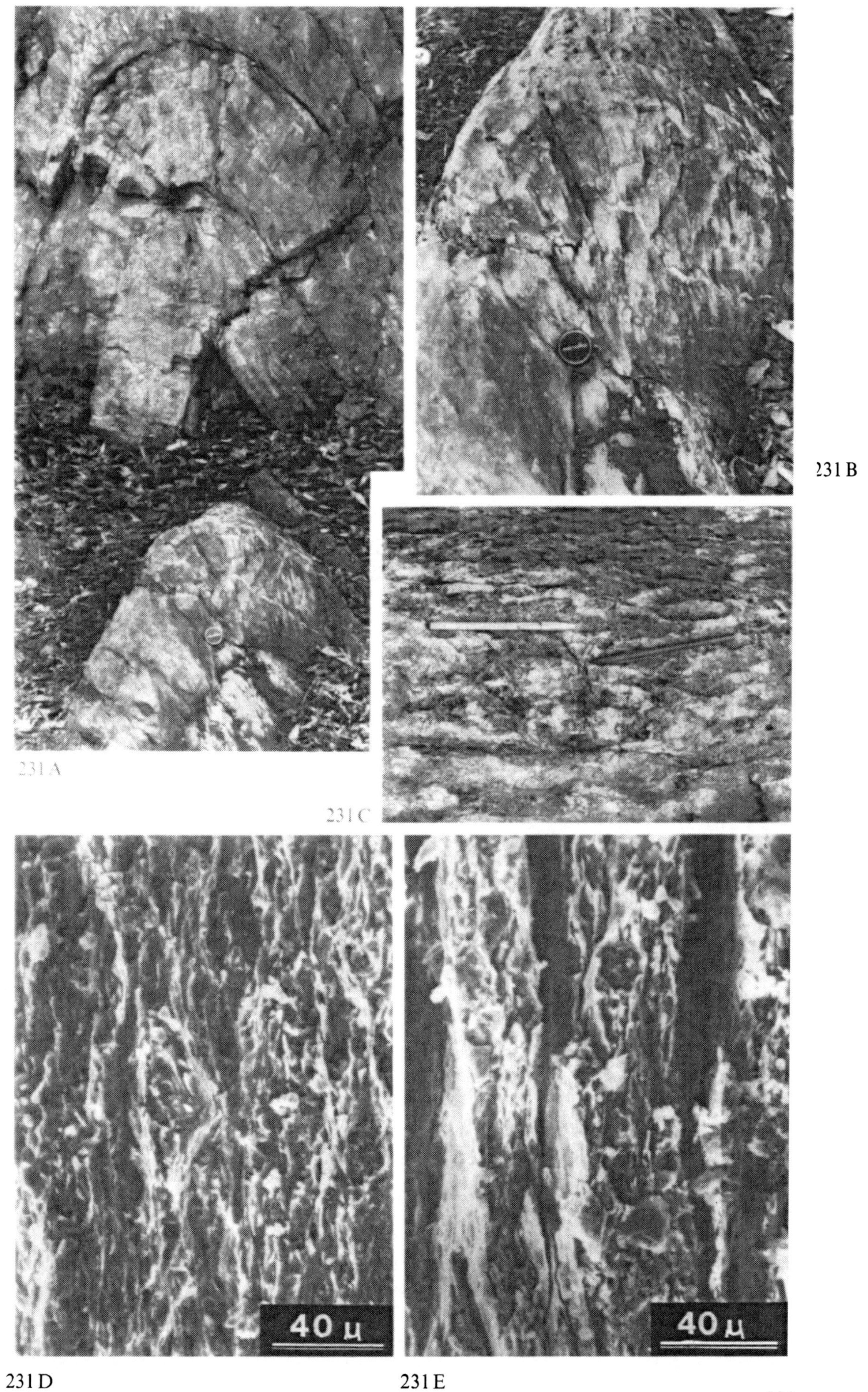
231 A
231 B
231 C
40 μ
40 μ
231 D
231 E

# 232. Cleavage and Transected Folds in Calcareous Mudrock (II)

R. I. SIMON and D. R. GRAY

**Chronology of Deformation:**

1. Layer parallel shortening within these Valley and Ridge rocks was accomodated by initiation of (i) a tectonite fabric (S≈L) with a weak fissility and down-dip lineation in the mudrock, and (ii) weakly developed spaced cleavages in the lime-mudstones and sandstones.
2. Subsequent layer-parallel slip folding: a noncoaxial history to produce transected folds.
3. Fold "flattening" and intensification of cleavage in the mudrock and hinges of folds in limestone and sandstone.

Photographs are from hillside outcrop above Sinking Creek, near Route 604, NW of Newport, Virginia, U.S.A.

---

**A. Transected anticline** (oblique view) defined by interbeds of thick (10–30 cm) maroon calcareous mudrock and lighter-colored argillaceous limestone (4–8 cm thick). A well-developed, smooth, anastomosing, spaced disjunctive cleavage transects this slightly inclined, subhorizontal fold. Cleavage fan angle is 50° and fold interlimb angle is 83°.

**B. Fold profile of fold in A.** Cleavage disrupts bedding creating apparent offsets on the macroscopic and microscopic scale. Maximum deviation (*d*) of cleavage from the axial surface is 8°. This angle decreases from the outer to the inner arc since cleavage orientation and intensity vary across the fold (see *sketch*).

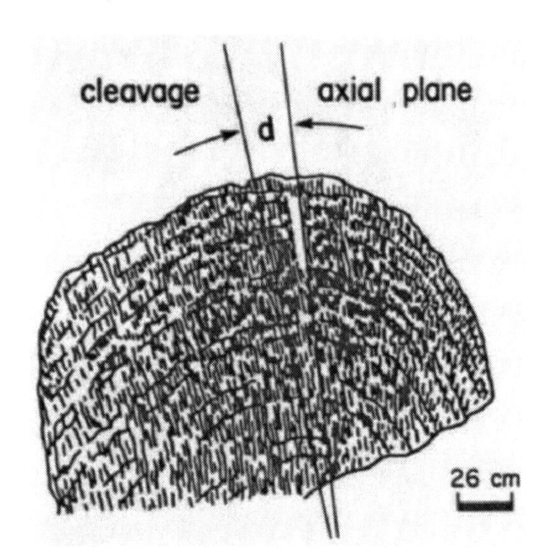


**C. Hingeline of fold in A and B.** Cleavage-bedding intersection traces on the hinge (statistically *parallel to light-colored pen*) deviate 9.5° from the fold hingeline direction (*parallel to dark-colored pen*).

**D. Cleavage refraction** (enlarged segment of right limb of fold in A and B). The spaced disjunctive cleavage has smooth morphology and shows refraction across the thin argillaceous limestone layers. Clay selvages of pressure solution origin define the cleavage

232 A

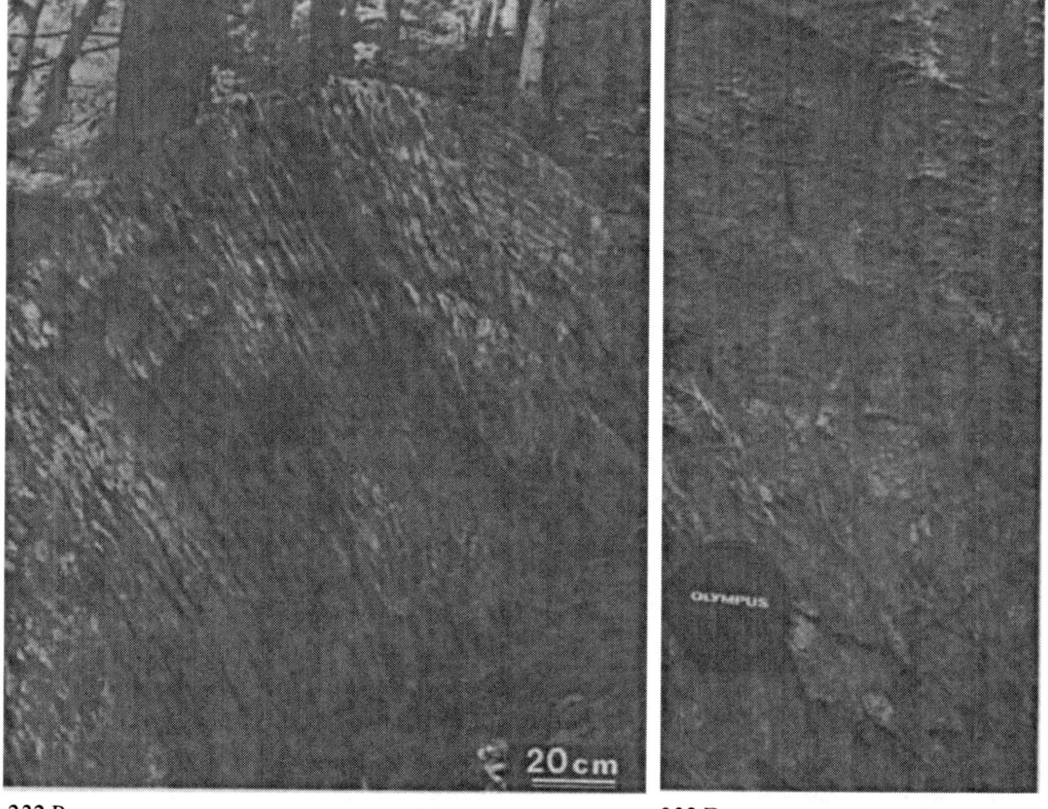


232 B

232 D

## 233. Cleavage and Transected Folds in Calcareous Mudrock (III)

D. R. GRAY and R. I. SIMON

Cleavage development must have been sequential but continuous throughout the deformation. The primary fabric, although initiated during the earliest stages of the deformation was continually modified, particularly during folding. The spaced cleavages evolved from the coalescing of one or two adjacent cleavage lamellae in the primary fabric. Collapse of the clay fabric in the inter-lamella domains due to loss of support from the rigid, but more soluble, calcite grains would give a thicker zone of oriented clays. Development of preferred orientation here is due to "pack of cards" collapse accommodated by grain boundary sliding (*gbs*). Adjacent to contraction faults the fabrics have been modified. Possible higher strain rates here may have caused *gbs* to dominate over the other operative deformation mechanisms (e.g., BORRADAILE 1978, 1981). Photographs are from a road outcrop on a property road off Route 601, north of Newport, Virginia, U.S.A. Diameter of lens cap is 6 cm.

---

**A. Transected syncline** in calcareous mudrock with thin argillaceous limestone interbeds. Cleavage fan angle is 50° and fold interlimb angle is 93°. The fold axial surface changes position and orientation across the fold but cleavage is always oblique to it.

**B. Cleavage morphology in fold profile** (enlargement of right limb segment of fold in **A**). Cleavage is a "slaty" variety defined in outcrop by a weak, approximately planar fissility. S.E.M. investigation shows this fabric is a spaced disjunctive cleavage (20–30 $\mu$ spacing) (see Plate 231 D).

**C. Anastomosing spaced disjunctive cleavages** on bedding surface of limestone layer (outer arc of fold in **A**, but not shown in the photograph). Cleavages are clay selvages of strongly oriented, tightly packed clays (see Plate 231 E)

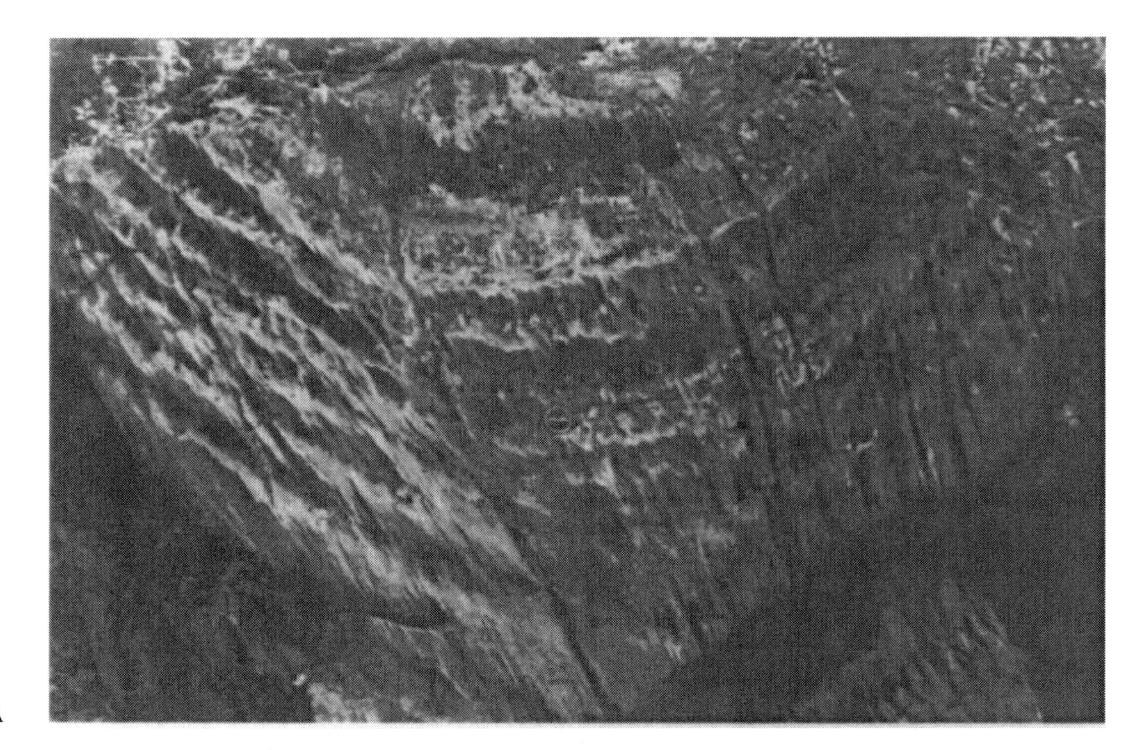

233 A

233 B

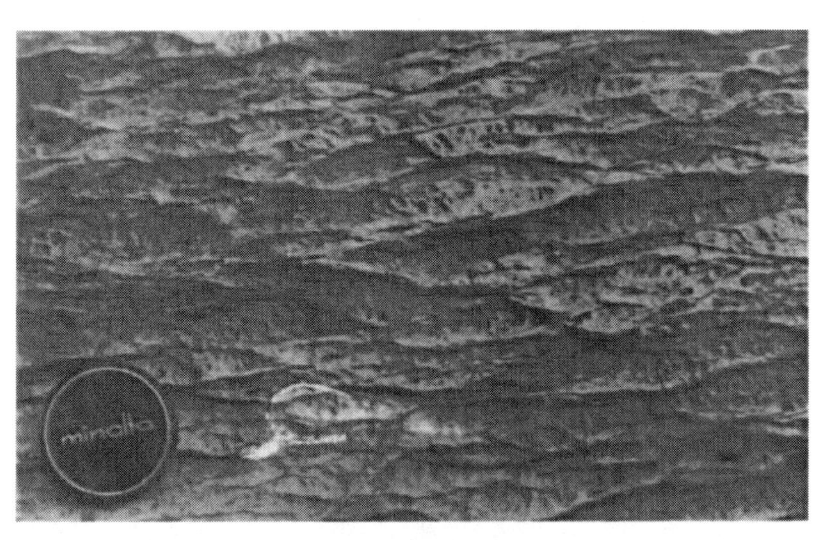


233 C

## 234. Transecting Crenulation Cleavage

G. J. Borradaile

Devonian Slate, from near Zell, Mosel Valley, W. Germany.

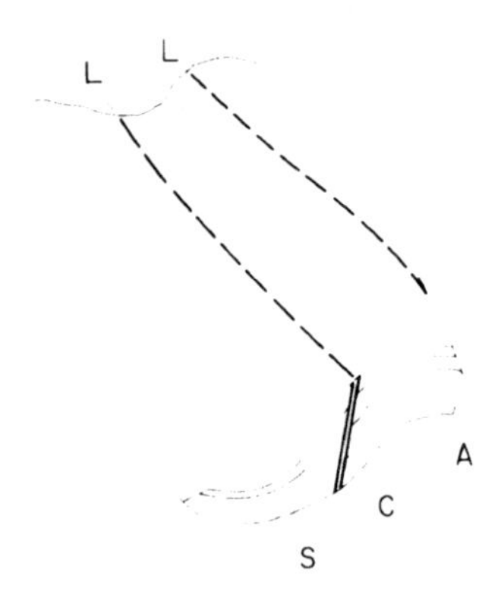


An $F_2$ synform (*S*) and antiform (*A*) of bedding also folds $S_1$ continuous cleavage. The front sawn surface of the specimen shows some folded bedding-laminae.

The $S_2$ cleavage (*C*) is not parallel to the axial surface. It intersects the folded surfaces so as to produce lineations (*L*) which plunge to the left on the left-hand flank of the antiform. *L* wraps around the hinge of the fold and plunges in the opposite direction on the other flank.

The $S_2$ cleavage is of the discrete crenulation type.

(Specimen collected by J. H. Akkermann) Scale: Specimen 20 cm long

234

# References


Aitken JD (1966) Middle Cambrian to Middle Ordovician cyclic sedimentation, southern Rocky Mountains of Alberta. Bull Can Pet Geol 14:404–441

Allen JRL, Williams BPJ (1978) The sequence of the earlier Lower Old Red Sandstone (Siluro-Devonian), north of Milford Haven, south west Dyfed (Wales). Geol J 13:113–136

Alvarez W, Engelder T, Lowrie W (1976) Formation of spaced cleavage and folds in brittle limestone by dissolution. Geology 4:698–701

Alvarez W, Engelder T, Geiser PA (1978) Classification of solution cleavage in pelagic limestones. Geology 6:263–266

Arthur MA, Fischer AG (1977) Upper Cretaceous-Paleocene magnetic stratigraphy at Gubbio, Italy: I. Lithostratigraphy and sedimentology. Geol Soc Am Bull 88:367–371

Ashby MF, Verrall RA (1973) Diffusion-accommodated flow and superplasticity. Acta Metall 21:149–163

Atkinson BK (1974) Experimental deformation of polycrystalline galena, chalcopyrite and pyrrhotite. Trans Inst Min Met 83B:B19–28

Axon HJ (1969) Pre-terrestrial deformation effects in iron meteorites. In: Millman PM (ed) Meteorite research. D. Reidel Publ Com, Dordrecht, p 796–805

Bates DEB (1975) Slaty cleavage associated with sandstone dykes in the Harlech Dome, North Wales. Geol J 10:167–175

Bathurst RGC (1975) Carbonate sediments and their diagenesis. 2nd edn. Elsevier, Amsterdam, 658 pp

Bayly MB, Borradaile GJ, Powell CMcA (eds) (1977) Atlas of rock cleavage – provisional edition: Univ Tasmania, Hobart, 110 plates +26 pp

Beach A (1974) A geochemical investigation of pressure solution and the formation of veins in a deformed greywacke. Contrib Mineral Petrol 46:61–68

Beach A (1975) The geometry of en-echelon vein arrays. Tectonophysics 28:245–263

Beach A (1977) Vein arrays, hydraulic fractures and pressure solution structures in a deformed flysch sequence, SW England. Tectonophysics 40:201–225

Beach A (1979) Pressure solution as a metamorphic process in deformed terrigenous sedimentary rocks. Lithos 12:51–58

Beach A, King M (1978) Discussion on pressure solution. J Geol Soc London 135:649–651

Bell IA, Wilson CJL (1977) Growth defects in metamorphic biotite. Phys Chem Miner 2:153–169

Bell TH (1978) The development of slaty cleavage across the Nackara Arc of the Adelaide Geosyncline. Tectonophysics 51:171–201

Bell TH, Rubenach MJ (1980) Crenulation cleavage development – evidence for progressive bulk inhomogeneous shortening from 'millipede' microstructures in the Robertson River Metamorphics. Tectonophysics 68:T9–T15

Beutner E (1978) Slaty cleavage and related strain in Martinsburg Slate, Delaware Water Gap, New Jersey. Am J Sci 278:1–23

Beutner E (1980) Slaty cleavage unrelated to tectonic dewatering: The Siamo and Michigamme slates revisited. Geol Soc Am Bull 91:171–178

Bishop DG (1972a) Progressive metamorphism from prehnite-pumpellyite to greenschist facies in the Dansey Pass area, Otago, New Zealand. Geol Soc Am Bull 83:3177–3198

Bishop DG (1972b) Transposition structures associated with cleavage formation in the Otago Schists. N Z J Geol Geophys 15:360–371

Bishop DG (1974) Stratigraphic, structural and metamorphic relationships in the Dansey Pass area, Otago, New Zealand. N Z J Geol Geophys 17:301–335

Black LP, Bell TH, Rubenach MJ, Withnal IW (1979) Geochronology of discrete structural-metamorphic events in a multiply deformed Precambrian terrain. Tectonophysics 54:103–137

de Boer RB (1975) Influence of pore solutions on rock strength. Thesis, State Univ Utrecht, 119 pp

de Boer RB (1977) Pressure solution: Theory and experiments. Tectonophysics 39:287–301

de Boer RB, Nagtegaal PJC, Duyvis EM (1977) Pressure solution experiments on quartz sand. Geochim Cosmochim Acta 41:257–264

Born A (1929) Über Druckschieferung im varistischen Gebirgskörper. Fortschr Geol Paleontol 7:329–427

Borradaile GJ (1974) Bulk finite strain estimates from the deformation of neptunian dykes. Tectonophysics 22:127–139

Borradaile GJ (1976a) "Structural Facing" (Shackleton's rule) and the Palaeozoic rocks of the Malaguide Complex near Vélez Rubio, SE Spain. Proc K Ned Akad Wet Ser B 79:330–336

Borradaile GJ (1976b) A strain study of a granite/granite-gneiss transition and accompanying schistosity formation in the Betic orogenic zone, SE Spain. J Geol Soc London 132:417–428
Borradaile GJ (1977) On cleavage and strain: results of a study in West Germany using tectonically deformed sand dykes. J Geol Soc London 133:146–164
Borradaile GJ (1978) Transected folds: a study illustrated with examples from Canada and Scotland. Geol Soc Am Bull 89:481–493
Borradaile GJ (1979a) Strain study of the Caledonides in the Islay region, SW Scotland: implications for strain histories and deformation mechanisms in greenschists. J Geol Soc London 136:77–88
Borradaile GJ (1979b) Pre-tectonic reconstruction of the Islay Anticline: implications for the depositional history of Dalradian rocks in the SW Highlands. In: Harris AL, Holland CH, and Leake BE (eds) The Caledonides of the British Isles – reviewed. Geol Soc London Spec Publ 8:229–238
Borradaile GJ (1981) Particulate flow of rock and the formation of cleavage. Tectonophysics 72:305–321
Borradaile GJ, Poulsen KH (1981) Tectonic deformation of pillow lava. Tectonophysics, 79:T17–T26
Borradaile GJ, Tarling DH (1981) The influence of deformation mechanisms on magnetic fabrics in weakly deformed rocks. Tectonophysics, 77:151–168
Bouchez JL (1977) Plastic deformation of quartzites at low temperature in an area of natural strain gradient. Tectonophysics 39:25–50
Bouchez JL (1978) Preferred orientations of quartz ⟨a⟩ axes in some tectonites: kinematic inferences. Tectonophysics 49:T25–T30
Bouchez JL, Blaise J (1976) Une structure hercynienne liée à un accident ductile: L'anticlinal de Lanvaux-les-Ponts-de-Cé aux environs d'Angers (massif Armoricain). Bull Soc Geol Fr 18:145–157
Bouchez JL, Pécher A (1976) Plasticité du quartz et sens de cisaillement dans des quartzites du Grand Chevauchement Central Himalayen. Bull Soc Geol Fr 18:1377–1385
Boullier AM, Bouchez JL (1978) Le quartz en rubans dans les mylonites. Bull Soc Geol Fr 20:253–262
Boulter CA (1979) On the production of two inclined cleavages during a single folding event; Stirling Range, S.W. Australia. J Struct Geol 1:207–219
Bradbury HJ, Harris AL, Smith RA (1979) Geometry and emplacement of nappes in the Central Scottish Highlands. In: Harris AL, Holland CH, and Leake BE (eds) The Caledonides of the British Isles – reviewed. Geol Soc London Spec Publ 8:213–220
Breaks FW, Bond WD, Stone D (1978) Preliminary geological synthesis of the English River Subprovince, Northwestern Ontario and its bearing upon mineral exploration. Ont Geol Surv Misc Pap MP72:55 pp
Brebrick RF, Scanlon WW (1957) Chemical etches and etch pit patterns on PbS crystals. J Chem Phys 27:607–608
Buchwald VF (1975) Handbook of iron meteorites, vol I. Univ California Press, Berkeley, 243 pp
Burrett CF (1978) Middle-Upper Ordovician conodonts and stratigraphy of the Gordon Limestone Sub-Group, Tasmania. Unpubl Ph D Thesis, Univ Tasmania
Caire A (1973) The Calabro-Sicilian Arc. In: de Jong KA, Scholten R (eds) Gravity and tectonics. John Wiley & Sons, New York, pp 157–173
Carter NL, Raleigh CB (1969) Principal stress directions from plastic flow in crystals. Geol Soc Am Bull 80:1231–2164
Cas RAF, Powell CMcA, Crook KAW (1980) Ordovician palaeogeography of the Lachlan Fold Belt: a modern analogue and tectonic constraints. J Geol Soc Aust 27:19–31
Champness PE, Lorimer GW (1974) A direct lattice-resolution study of precipitation (exsolution) in orthopyroxene. Philos Mag 30:357–365
Chidester AH (1962) Petrology and geochemistry of selected talc-bearing ultramafic rocks and adjacent country rocks in north central Vermont. US Geol Surv Prof Pap 345:207 pp
Clark BR, Kelly WB (1973) Sulphide deformation studies: I. Experimental deformation of pyrrhotite and sphalerite to 2000 bars and 500° C. Econ Geol 68:332–352
Clark BR, Price FR, Kelly WC (1977) Effects of annealing on deformation textures in galena. Contrib Mineral Petrol 64:149–165
Coe RS, Kirby SH (1975) The orthoenstatite to clinoenstatite transformation by shearing and reversion by annealing: mechanism and potential applications. Contrib Mineral Petrol 52:29–55
Conolly JR (1965) The stratigraphy of the Hervey Group in central New South Wales. J Proc R Soc NSW 98:37–83
Cook DG (1975) Stuctural style influenced by lithofacies, Rocky Mountain Ranges, Alberta – British Columbia. Geol Surv Canada, Bulletin 233, 73 pp
Cosgrove JW (1976) The formation of crenulation cleavage. J Geol Soc Lond 132:155–178
Coward MP (1976) Strain within ductile shear zones. Tectonophysics 34:181–197
Craig JR, Scott SD (1974) Sulphide phase equilibria in sulphide mineralogy. In: Ribbe PH (ed) Min Soc Am Short Course Notes, CS1–CS110

Crook KAW (1964) Cleavage in weakly deformed mudstones. Am J Sci 262:523–531

Damberger HH (1974) Coalification patterns of Pennsylvanian coal basins of the eastern United States. In: Dutcher RR, Hacquebard PA, Schopf JM, Simon JA (eds) Carbonaceous materials as indicators of metamorphism. Geol Soc Am Spec Pap 153:53–74

Dennis JG (1956) The geology of the Lyndonville area, Vermont. Vt Geol Surv Bull 8:98 pp

Dennis JG (1967) International Tectonic Dictionary. Am Assoc Petroleum Geologists Memoir 7, 196 pp

Dennis JG (1972) Structural geology. Ronald, New York, 532 pp

Dewey JF (1965) Nature and origin of kink bands. Tectonophysics 1:459–494

Donaldson JA, Ojakangas RW (1977) Orthoquartzite pebbles in Archaean conglomerate, North Spirit Lake, Northwestern Ontario. Can J Earth Sci 14:1980–1990

Dunnet D (1969) A technique of finite strain analysis using elliptical particles. Tectonophysics 7:117–136

Dunnington HV (1954) Stylolite development postdates rock induration: J Sediment Petrol 24:27–49

Durney DW (1972a) Deformation history of the Western Helvetic Nappes, Valais, Switzerland. Ph D Thesis, Univ London, 326 pp

Durney DW (1972b) Solution-transfer, an important geological deformation mechanism. Nature (London) 235:315–317

Durney DW (1974) Relations entre les temperatures d'homogeneisation d'inclusion fluides et les mineraux metamorphiques dans les nappes helvetiques du Valais. Bull Soc Geol Fr 7:269–272

Durney DW (1976) Pressure-solution and crystallization deformation. Philos Trans R Soc London Ser A 283:229–240

Durney DW, Ramsay JG (1973) Incremental strains measured by syntectonic crystal growth. In: de Jong KA, Scholten R (eds) Gravity and tectonics. John Wiley & Sons, New York, 67–96

Elliott D (1970) Determination of finite strain and initial shape from deformed elliptical objects. Geol Soc Am Bull 81:2221–2236

Epstein AG, Epstein JB, Harris LD (1976) Conodont color alteration – an index to organic metamorphism. US Geol Surv Prof Pap 995:27 pp

Fisher DW, Isachsen YW, Rickard LV (1970) Geological map of New York 1:250,000. New York State Museum and Science Service

Flinn D (1965) On the symmetry principle and the deformation ellipsoid. Geol Mag 102:36–45

Francis PW, Sibson RH (1973) The Outer Hebrides thrust. In: Park RG, Tarney J (eds) The early Precambrian of Scotland and related rocks of Greenland, Univ Keele, pp 95–104

Frey M (1978) Progressive low-grade metamorphism of a black shale formation, Central Swiss Alps, with special reference to pyrophyllite and margarite-bearing assemblages. J Petrol 19:95–135

Frey M, Hunziker JC, Frank W, Bocquet J, Dal Piaz GV, Jäger E, Niggli E (1974) Alpine metamorphism of the Alps, a review. Schweiz Mineral Petrogr Mitt 54:247–290

Friedman M, Higgs NG (1981) Calcite fabrics in experimental shear zones. In: Carter NL, Friedman M, Logan JM, Stearns DW (eds) Mechanical behavior of crustal rocks. Am Geophys Union Monogr 24 – the Handin Volume

Friedman M, Sowers GM (1970) Petrofabrics – a critical review. Can J Earth Sci 7:477–497

Gansser A (1964) Geology of the Himalayas. Wiley Interscience, London New York, 289 pp

Gee RD (1971) Geological atlas 1 mile series. Zone 7, sheet 22 (80165), Table Cape. Geol Surv Tasmania, Explanatory Rep

Geiser PA (1975) Slaty cleavage and the dewatering hypothesis – an examination of some critical evidence. Geology 3:717–720

Gill WD (1951) The tectonics of the Sub-Himalayan fault zone in the northern Potwar region and in the Kangra district of the Punjab. Q J Geol Soc London 107:395–421

Granath JW (1976) Petrogenesis of metamorphically layered tectonites at Cooma, New South Wales. Ph D Thesis, Monash Univ, Clayton, 226 pp

Gray DR (1976) The origin and development of crenulation cleavages in low to medium-grade metamorphic rocks from southeastern Australia. Unpubl Ph D Thesis, Macquarie Univ, Sydney, 325 pp

Gray DR (1977a) Morphologic classification of crenulation cleavages. J Geol 85:229–235

Gray DR (1977b) Some parameters which affect the morphology of crenulation cleavages. J Geol 85:763–780

Gray DR (1978) Cleavages in psammitic rocks from southeastern Australia: their nature and origin. Geol Soc Am Bull 89:577–590

Gray DR (1979) Microstructure of crenulation cleavages: an indicator of cleavage origin. Am J Sci 279:97–128

Gray DR (1981) Cleavage-fold relationships and their implications for transected folds: an example from southwest Virginia, U.S.A. J Struct Geol 3:265–277

Gray DR, Durney DW (1979) Crenulation cleavage differentiation: implications of solution-deposition processes. J Struct Geol 1:73–80

Gregg WJ (1978) The production of tabular grain shapes in metamorphic rocks. Tectonophysics 49:T19–T24

Gregg WJ (1979) The redistribution of pre-cleavage clastic dikes by folding at New Paltz, New York. J Geol 87:99–104

Griffin JR, Lindsley-Griffin N (1974) Sedimentary and slump structures of Central Maine. 1974 N Engl Inter-collegiate Geol Conf Guideb, 32–47

Groshong RH (1975) Slip-cleavage caused by pressure solution in a buckle fold. Geology 3:411–413

Guidotti CV (1974) Transition from staurolite to sillimanite zone, Rangeley quadrangle, Maine. Geol Soc Am Bull 85:475–490

Hancock PL (1973) Structural zones in Variscan Pembrokeshire. Proc Ussher Soc 2:509–520

Hancock PL, Dunne WM, Tringham ME (1981) Variscan structures in southwest Wales. Geol Mijnbouw 60:81–88

Harris AL, Bradbury HJ, McGonigal MH (1976) The evolution and transport of the Tay Nappe. Scott J Geol 12:103–113

Harris FR (1974) Geology of the Rainy Lake Area, District of Rainy River. Ont Div Mines Geol Rep 115:94 pp

Hatch NL (1975) Tectonic, metamorphic and intrusive history of part of the east side of the Berkshire Massif, Massachusetts. In: Tectonic studies of the Berkshire Massif, western Massachusetts, Connecticut and Vermont. US Geol Surv Prof Pap 888-C:33–49

Heim A (1878) Untersuchungen über den Mechanismus der Gebirgsbildung im Anschluss an die geologische Monographie der Tödi-Windgällen-Gruppe vols I & II & Atlas: Bruno Schwabe Verlagsbuchhandlung, Basel, 592 pp

Helm DG, Siddans AWB (1971) Deformation of a slaty lapillar tuff in the English Lake District: Discussion. Geol Soc Am Bull 82:523–531

Hooke RLeB (1973) Flow near the margin of the Barnes Ice Cap, and the development of ice-cored moraines. Geol Soc Am Bull 84:3929–3948

Hudleston PJ (1977) Progressive deformation and development of fabric across zones of shear in glacial ice. In: Saxena S, Bhattacharji S (eds) Energetics of geological processes. Springer, Berlin Heidelberg New York, 121–150

Hudleston PJ (1980) The progressive development of inhomogeneous shear and crystallographic fabric in glacial ice. J Struct Geol 2:189–196

James HL (1955) Zones of regional metamorphism in the Precambrian of northern Michigan. Geol Soc Am Bull 66:1455–1488

Johnson GD, Vonda CF (1972) Siwalik sediments in a portion of the Punjab re-entrant: the sequence at Haritalyangar, District Bilaspur. Himalayan Geol 2:118–144

Joplin GA (1942) Petrological studies in the Ordovician of New South Wales, I: the Cooma Complex. Proc Linn Soc N S W 67:156–196

Kehlenbeck MM (1972) Tectonic evolution of the Lac Rouvray anorthosite mass, Quebec. Can J Earth Sci 9:1640–1649

Kieffer SW, Phakey PP, Christie JM (1976) Shock processes in porous quartzite: transmission electron microscope observations and theory. Contrib Mineral Petrol 59:41–93

Kirby SH (1977) State of stress in the lithosphere: inferences from the flow laws of olivine. Pure App Geophys 115:245–258

Knipe RJ (1979) Chemical changes during slaty cleavage development. Bull Mineral 105:206–209

Knipe RJ, White SH (1977) Microstructural variation of an axial plane cleavage around a fold – H.V.E.M. study. Tectonophysics 39:355–380

Langheinrich G (1967) Die tektonische Deformation von *Schellwienella umbraculum* im Gebiet von Niedersalwey (Westfalen). Geol Mitt Aachen 7:159–172

Le Fort P (1975) Himalaya: the collided range. Present knowledge of the continental arc. Am J Sci 275A:1–44

Leitch EC (1975) Zonation of low grade regional metamorphic rocks, Nambucca Slate Belt, northeastern New South Wales. J Geol Soc Aust 22:413–422

Lipschutz ME (1968) Shock effects in iron meteorites: a review. In: French BM, Short NM (eds) Shock metamorphism of natural materials. Mono Book Comp, Baltimore, 571–583

Lisle RJ (1977) Estimation of the tectonic strain ratio from the mean shape of deformed elliptical markers. Geol Mijnbouw 56:140–144

Lisle RJ (1979) Strain analysis using deformed pebbles: the influence of initial pebble shape. Tectonophysics 60:263–277

Macculloch J (1819) A description of the Western Islands of Scotland. Hurst Robinson & Co, London

Maltman AJ (1977) Some microstructures of experimentally deformed argillaceous sediments. Tectonophysics 39:417–436

Mancktelow NS (1979) The development of slaty cleavage, Fleurieu Peninsula, South Australia. Tectonophysics 58:1–20

Marjoribanks RW (1976) The relation between microfabric and strain in a progressively deformed quartzite sequence from central Australia. Tectonophysics 32:269–293

Marlow PC, Etheridge MA (1977) Development of a layered crenulation cleavage in mica schists of the Kanmantoo Group near Macclesfield, South Australia. Geol Soc Am Bull 88:873–882

Mascle G (1974) Le système de failles de Comino, bordure nord-occidentale du haut-plateau ibléen (Sicile). CR Somm Seances Soc Geol Fr 16:134–136

Mathias BV, Clark GJ (1975) Mount Isa copper and silver-lead-zinc orebodies – Isa and Hilton Mines. In: Knight CL (ed) Economic geology of Australia and Papua-New Guinea, vol V, Metals. Aust I M M, Melbourne, 351–72

Mattauer M (1975) Sur le mécanisme de formation de la schistosité dans l'Himalaya. Earth Planet Sci Lett 28: 144–154
Maxwell JC (1962) Origin of slaty and fracture cleavage in the Delaware Water Gap area, New Jersey and Pennsylvania, In: Engel AEJ, James HL, Leonard BF (eds) Petrologic studies: a volume in honor of A. F. Buddington. Geol Soc Am, Boulder, Colorado, 281–311
McClay KR (1977) Dislocation etch pits in galena. Tectonophysics 40: T1–T8
McClay KR (1980) Sheared galena, textures and microstructures. J Struct Geol 2: 227–234
McClay KR, Atkinson BK (1977) Experimentally induced kinking and annealing of single crystals of galena. Tectonophysics 39: 175–189
Means WD (1968) Experimental folding of synthetic schistose material. Trans Am Geophys Union 49: 756
Means WD (1975) Natural and experimental microstructures in deformed micaceous sandstones. Geol Soc Am Bull 86: 1221–1229
Means WD (1977) A deformation experiment in transmitted light: Earth Plan Sci Lett 35: 169–179
Mehnert KR (1971) Migmatites and origin of granitic rocks. Elsevier, Amsterdam, 405 pp
Mercier JC, Nicolas A (1975) Textures and fabrics of upper mantle peridotites as illustrated by xenoliths from basalts. J Petrol 16: 454–487
Milnes AG, Pfiffner OA (1977) Structural development of the Infrahelvetic complex, eastern Switzerland. Eclogae Geol Helv 70: 83–95
Mimran Y (1976) Strain determination using a density distribution technique and its application to deformed Upper Cretaceous Dorset Chalks. Tectonophysics 31: 175–192
Mimran Y (1977) Chalk deformation and large-scale migration of calcium carbonate. Sedimentology 24: 333–360
Moench RH (1966) Relation of $S_2$ schistosity to metamorphosed clastic dikes, Rangeley-Phillips area, Maine. Geol Soc Am Bull 77: 1449–1462
Moench RH (1970) Premetamorphic down-to-basin faulting, folding and tectonic dewatering, Rangeley area, western Maine. Geol Soc Am Bull 81: 1463–1496
Moench RH, Zartman, RE (1976) Chronology and styles of multiple deformation, plutonism, and polymetamorphism in the Merrimack synclinorium of western Maine. Geol Soc Am Mem 146: 203–237
Morritt RFC (1979) Structural analysis of Palaeozoic rocks near Taralga, N.S.W. Unpubl BA (Hons) Thesis, Macquarie Univ, 149 pp + appendices
Müller WF (1974) One-dimensional lattice imaging of a deformation-induced lamellar intergrowth of orthoenstatite and clinoenstatite [(Mg, Fe) $SiO_3$]. Neues Jahrb Mineral Monatsh 2: 83–88
Newell ND, Rigby JK, Fischer AG, Whiteman AJ, Hickox JE, Bradley JS (1972) The Permian reef complex of the Guadalupe Mountains Region, Texas and New Mexico: a study in paleoecology. Hafner, New York, 226 pp
Nickelsen RP (1972) Attributes of rock cleavage in some mudstones and limestones of the Valley and Ridge Province, Pennsylvania. Proc Pa Acad Sci 46: 107–112
Nickelsen RP (1979) Sequence of structural stages of the Alleghany orogeny, at the Bear Valley Strip Mine, Shomokin, Pennsylvania. Am J Sci 279: 225–271
Nicolas A, Boudier F, Boullier AM (1973) Mechanisms of flow in naturally and experimentally deformed peridotites. Am J Sci 273: 853–876
Oertel G, Phakey PP (1972) The texture of a slate from Nantlle, Caernarvon, North Wales. Texture 1: 1–8
Pécher A (1978) Déformations et métamorphisme associés à une zone de cisaillement. Exemple du Grand Chevauchement Central Himalayen. Thèse, Univ Grenoble, 354 pp
Pfiffner OA (1977) Tektonische Untersuchungen im Infrahelvetikum der Ostschweiz. Mitt Geol Inst, Eidg Tech Hochsch Univ Zurich NF 217: 432 pp
Pfiffner OA (1980) Strain analysis in folds (Infrahelvetic complex, Central Alps). Tectonophysics 61: 337–362
Pfiffner OA (1981) Fold-and-thrust tectonics in the Helvetic nappes. In: McClay K, Price NJ (eds) Thrust and nappe tectonics. Blackwells Scientific Publ, Oxford, Spec. Publ. no 9. Geol Soc Lond 319–327.
Philpotts AR (1964) Origin of pseudotachylites. Am J Sci 262: 1008–1035
Plessmann W (1964) Gesteinslösung, ein Hauptfaktor beim Schieferungsprozeß, Geol Mitt Aachen 4: 69–82
Plessmann W (1965) Laterale Gesteinsverformung vor Faltungsbeginn im Unterkarbon des Edersees (Rheinisches Schiefergebirge). Geol Mitt Aachen 5: 271–285
Plessmann W (1972) Horizontal-stylolithen im französisch-schweizerischen Tafel- und Faltenjura und ihre Einpassung in den regionalen Rahmen. Geol Rundsch 61: 332–347
Potter DB, Lane MA (1969) Some major structural features of the Taconic allochthon in the Hoosick Falls area, New York-Vermont. In: Bird JM (ed) Guidebook for field trips in New York, Massachusetts and Vermont. State Univ New York at Albany 12.0–12.23
Powell CMcA (1969) Intrusive sandstone dykes in the Siamo Slate near Negaunee, Michigan. Geol Soc Am Bull 80: 2585–2594

Powell CMcA (1970) Relict diagenetic textures and structures in regional metamorphic rocks, Northern Michigan. Northwestern Univ Rep 20: NASA Res Grant NGR14-007-027, 4-35
Powell CMcA (1972 a) Tectonically dewatered slates in the Ludlovian of the Lake District, England. Geol J 8:95-110
Powell CMcA (1972 b) Tectonic dewatering and strain in the Michigamme Slate, Michigan. Geol Soc Am Bull 83:2149-2158
Powell CMcA (1973) Clastic dikes in the Bull Formation of Cambrian age, Taconic allochthon, Vermont. Geol Soc Am Bull 84:3045-3050
Powell CMcA (1979) A morphological classification of rock cleavage. Tectonophysics 58:21-34
Powell CMcA, Conaghan PJ (1973 a) Polyphase deformation in Phanerozoic rocks of the Central Himalayan Gneiss, Northwest India: J Geol 81:127-143
Powell CMcA, Conaghan PJ (1973 b) Plate tectonics and the Himalayas. Earth Planet Sci Lett 20:1-12
Powell, CMcA, Vernon RH (1979) Growth and rotation history of garnet porphyroblasts with inclusion spirals in the Karakoram schist. Tectonophysics 54:25-43
Powell CMcA, Hordern MJ, Willis IL (1976) Multiple deformation associated with the Wiagdon Fault Zone along the Turon River, near Sofala. Bull Aust Soc Explor Geophys 7:26-27
Powell CMcA, Edgecombe DR, Henry NM, Jones JG (1977) Timing of regional deformation of the Hill End Trough: a reassessment. J Geol Soc Aust 23:407-422
Powell CMcA, Crawford AR, Armstrong RL, Prakash R, Wynne-Edwards HR (1979) Reconnaissance Rb-Sr dates for the Himalayan Central Gneiss, Northwest India. Indian J Earth Sci 6:139-151
Powell CMcA, Fergusson CL, Williams AJ (1980) Structural relationships across the Lambian Unconformity in the Hervey Range – Parkes area, N.S.W. Proc Linn Soc NSW 104:195-210
Price NJ, Hancock PL (1972) Development of fracture cleavage and kindred structures. Proc 24th Int Geol Congr Sect 3:584-592
Price RA, Mountjoy EW (1970) Geologic structure of the Canadian Rocky Mountains between Bow and Athabasca Rivers – a progress report. Geol Assoc Canada Spec Pap 6:7-26
Puigdefábregas C (1975) La sedimentación molásica en la Cuenca de Jaca. Monogr Inst Estud Pirenaicas 104:1-188
Puppolo DG (1979) A study of fold behavior by means of microstructural features. Unpubl MS Thesis, Rensselaer Polytechnic Inst, Troy NY
Ramsay JG (1967) Folding and fracturing of rocks. McGraw-Hill, New York, 568 pp
Rickard MJ (1961) A note on cleavages in crenulated rocks. Geol Mag 98:324-332
Roberts D (1971) Abnormal cleavage patterns in fold hinge zones from Varanger Peninsula, northern Norway. Am J Sci 271:170-180
Roberts D (1972) Tectonic deformation in the Barents Sea Region of Varanger Peninsula, Finnmark. Nor Geol Unders Publ 282:1-39
Roberts D, Strömgård, KE (1971) A comparison of natural and experimental strain patterns around fold hinge zones. Tectonophysics 14:105-120
Roper PJ (1972) Structural significance of "Button" or "Fish Scale" texture in phyllonitic schist of the Brevard Zone, Northwestern South Carolina. Geol Soc Am Bull 83:853-860
Russell-Head DS, Budd WF (1981) Ice sheet flow properties derived from borehole shear measurements combined with ice-core studies. J Glaciol 24:117-130
Scheibner E (1973) Geology of the Taralga 1 : 100,000 sheet 8829. Geol Surv NSW, 79 pp
Schmid SM, Boland JN, Paterson MS (1977) Superplastic flow in fine-grained limestone. Tectonophysics 43:257-291
Schoneveld C (1977) A study of some typical inclusion patterns in strongly paracrystalline-rotated garnets. Tectonophysics 39:453-471
Shackleton RM (1958) Downward-facing structures of the Highland Border. Q J Geol Soc London 113:361-392
Sherwin JA, Chapple WM (1968) Wavelengths of single layer folds: a comparison between theory and observation. Am J Sci 266:167-179
Shimamoto T (1975) The finite element analysis of the deformation of a viscous spherical body embedded in a viscous medium. J Geol Soc Jpn 81:255-267
Sibson RH (1977) Fault rocks and fault mechanisms. J Geol Soc London 133:191-213
Siemens H (1977) Fabric analysis and fabric development in ores. Geol Foeren Stockholm Foerh 99:172-185
Simpson A (1968) The Caledonian history of the northeastern Irish Sea region and its relation to surrounding areas. Scott J Geol 4:135-163
Solé-Sugrañes L (1976) Formación de equistosidad por crenulación y disolución. Acta Geol Hisp 11:113-119
Spang JH, Oldershaw AE, Stout MZ (1976) Development of cleavage in the Banff Formation, Front Ranges, Canadian Rocky Mountains. Geol Assoc Can Prog Abstr 1:70
Spry, A. (1969) Metamorphic textures. Pergamon Press, Oxford, 350 pp
Stanley RS (1975) Time and space relationships of structures associated with the domes of southwestern Massachusetts and western Connecticut. In: Tectonic studies of the Berkshire Massif, western

Massachusetts, Connecticut and Vermont. US Geol Surv Prof Pap 888-F:69–96
Stephens MB, Glasson MJ, Keays RR (1979) Structural and chemical aspects of metamorphic layering development in metasediments from Clunes, Australia. Am J Sci 279:129–160
Stöffler D (1974) Deformation and transformation of rock-forming minerals by natural and experimental shock processes. II. Physical properties of shocked minerals. Fortschr Mineral 51:256–289
Talbot CJ (1979) Fold trains in a glacier of salt, S. Iran. J Struct Geol 1:5–18
Trommsdorff V, Wenk HR (1968) Terrestrial metamorphic clinoenstatite in kinks of bronzite crystals. Contrib Mineral Petrol 19:158–168
Tullis J, Christie JM, Griggs DT (1973) Microstructures and preferred orientations of experimentally deformed quartzites. Geol Soc Am Bull, 84:297–314
Turner FJ, Heard F, Griggs DT (1960) Experimental deformation of enstatite and accompanying inversion to clinoenstatite. Int Geol Congr 21 Sess, Copenhagen, Part 18: 399–408
Vernon RH (1976) Metamorphic processes. George Allen & Unwin Ltd, London, 247 pp
Vernon RH (1977) Microfabric of mica aggregates in partly recrystallized biotite. Contrib Mineral Petrol 61:175–185
Vernon RH (1978) Porphyroblast-matrix microstructural relationships in deformed metamorphic rocks. Geol Rundsch 67:288–305
Vernon RH, Ransom DM (1971) Retrograde schists of the amphibolite facies at Broken Hill, New South Wales. J Geol Soc Aust 18:267–277
Wenk HR (1978) Are pseudotachylites products of fracture or fusion? Geology 6:507–511
White SH (1976) The effects of strain on the microstructures, fabrics and deformation mechanisms in quartzite. Philos Trans R Soc London Ser A 283:69–86
White SH, Knipe RJ (1978) Microstructure and cleavage development in selected slates. Contrib Mineral Petrol 66:165–174
White SH, Wilson CJL (1978) Microstructure of some quartz pressure fringes. Neues Jahrb Mineral Abh 134:33–51
White WS (1949) Cleavage in east-central Vermont. Am Geophys Union Trans 30:587–594
Williams PF (1972) Development of metamorphic layering and cleavage in low-grade metamorphic rocks at Bermagui, Australia. Am J Sci 262:1–47
Williams PF (1976) Relationships between axial plane foliations and strain. Tectonophysics 30:181–196
Williams PF (1977) Foliation: a review and discussion. Tectonophysics 39:305–328
Williams PF, Means WD, Hobbs BE (1977) Development of axial plane slaty cleavage and schistosity in experimental and natural materials. Tectonophysics 42:139–158
Williams VA (1979) Structural evolution in Lower Palaeozoic rocks on the Abercrombie River, NSW Unpubl BA (Hons) Thesis, Macquarie Univ, 57 pp
Wilson CJL (1975) Preferred orientation in quartz ribbon mylonites. Geol Soc Am Bull 86:968–974
Wilson CJL (1977) Combined diffusion-infiltration of uranium in micaceous schists; a study using the fission track method. Contrib Mineral Petrol 65:171–181
Wilson CJL (1979) Development of schistosity in phengite schists from Zermatt, Switzerland – a discussion. Tectonophysics 56:305–316
Wilson CJL, Bell IA (1979) Deformation of biotite and muscovite: optical microstructure. Tectonophysics 58:179–200
Wilson CJL, Russell-Head DS (1979) Experimental folding in ice and the resultant *c* axis fabrics. Nature (London) 279:49–51

# Subject Index

The numbers cited are *Plate* numbers

Volume 14
A.K. Gupta, K. Yagi

## Petrology and Genesis of Leucite-Bearing Rocks

1980. 99 figures, 43 tables. XV, 252 pages
ISBN 3-540-09864-X

This volume is a much needed review-synthesis of the extensive geochemical, petrological and experimental studies on leucite-bearing mafic and ultramafic rocks.

The first five chapters of the book summarize the mineralogy, major and minor element geochemistry, strontium and oxygen isotopic studies, distribution, and conditions surrounding the formation of leucitic rocks. The next twelve chapters present a detailed account of the phase equilibria studies of synthetic and natural leucite-bearing rock systems in air and under variable pressures (in presence or absence of water), analcitization of leucite, formation of pseudoleucite, genetic relationship between kimberlites and leucitic rocks, and structure and tectonic control of volcanism associated with leucite-bearing rocks. In the last chapter, trace element geochemistry as well as field and laboratory data are used to elucidate the origin of this interesting suite of rocks.
This book will be of great value to a wide range of earth scientists and advanced students with particular interest in mineralogy, geochemistry and experimental petrology.

Volume 15
J.B. Dawson

## Kimberlites and Their Xenoliths

1980. 84 figures, 35 tables. XII, 252 pages
ISBN 3-540-10208-6

In the past two decades there has been a rapid growth of interest in kimberlite, not only as a source of diamonds, but also as a rock type that has sampled the upper-mantle more thoroughly than any other type of igneous activity. Whereas earlier books published on the subject in recent years have concentrated on specific topics such as the geology of diamond or upper-mantle mineralogy, the present volume provides on overall coverage of the geology of kimberlite from which inferences on the nature of the earth's upper-mantle can be drawn. This synthesis of up-to-date results and opinions from the recent geological literature will be highly appreciated by researchers in the field.

Volume 16
J.B. Gill

## Orogenic Andesites and Plate Tectonics

1981. 109 figures. XIV, 390 pages
ISBN 3-540-10666-9

The solution to the question of andesite genesis is a major, multidisciplinary undertaking facing geoscientists in the 1980's. *Orogenic Andesites and Plate Tectonics* was written in response to the growing need of researchers in this area for a clarification of the long-standing problem and identification of profitable areas for future investigations. This book critically summarizes the vast relevant literature on the tectonics, geophysics, volcanology, geology, geochemistry, and mineralogy of andesites and their volcanoes. The author cites over 1100 references in these specialties and includes information on the location and rock composition of more than 300 recently active volcanoes. In addition, he provides a systematic and original evaluation of genetic hypotheses. Numerous cross references enhance the integrated subject development which consistently emphasizes the implications of data for theories of magma genesis. *Orogenic Andesites and Plate Tectonics* will prove an invaluable reference source to researchers and graduate students in the geosciences seeking a careful evaluation of genetic hypotheses and key data, arguments or gaps in this promising field.

Springer-Verlag
Berlin
Heidelberg
New York

CPSIA information can be obtained at www.ICGtesting.com
Printed in the USA
LVOW09s0502131113

361108LV00003B/14/P